Hadis Morkoç

Nitride Semiconductor Devices

Related Titles

Brütting, W., Adachi, C. (eds.)
Physics of Organic Semiconductors
2012
ISBN: 978-3-527-41053-8

Mottier, P.
LED for Lighting Applications
2009
ISBN: 978-1-84821-145-2

Würfel, P.
Physics of Solar Cells
From Basic Principles to Advanced Concepts
2009
ISBN: 978-3-527-40857-3

Hofmann, P.
Solid State Physics
An Introduction
2008
ISBN: 978-3-527-40861-0

Paskova, T. (ed.)
Nitrides with Nonpolar Surfaces
Growth, Properties, and Devices
2008
ISBN: 978-3-527-40768-2

Coleman, C. C.
Modern Physics for Semiconductor Science
2008
ISBN: 978-3-527-40701-9

Neumark, G. F., Kuskovsky, I. L., Jiang, H. (eds.)
Wide Bandgap Light Emitting Materials And Devices
2007
ISBN: 978-3-527-40331-8

Piprek, J. (ed.)
Nitride Semiconductor Devices: Principles and Simulation
2007
ISBN: 978-3-527-40667-8

Adachi, S.
Properties of Group-IV, III-V and II-VI Semiconductors
2005
ISBN: 978-0-470-09032-9

Ng, K. K.
Complete Guide to Semiconductor Devices
2002
ISBN: 978-0-471-20240-0

Hadis Morkoç

Nitride Semiconductor Devices

Fundamentals and Applications

WILEY-VCH Verlag GmbH & Co. KGaA

The Author

Prof. Dr. Hadis Morkoç
Virginia Commonwealth Univ.
Dept. of Electric. Engineering
601 W. Main St. Room 338
Richmond, VA 23284-3072
USA

Cover
Background: GaN based laser
(courtesy of Prof. U. Schwarz).
Insets: SEM images of a GaN based HFET with 80 nm gate length (courtesy of Prof. T. Palacio).

All books published by **Wiley-VCH** are carefully produced. Nevertheless, authors, editors, and publisher do not warrant the information contained in these books, including this book, to be free of errors. Readers are advised to keep in mind that statements, data, illustrations, procedural details or other items may inadvertently be inaccurate.

Library of Congress Card No.: applied for

British Library Cataloguing-in-Publication Data
A catalogue record for this book is available from the British Library.

Bibliographic information published by the Deutsche Nationalbibliothek
The Deutsche Nationalbibliothek lists this publication in the Deutsche Nationalbibliografie; detailed bibliographic data are available on the Internet at <http://dnb.d-nb.de>.

© 2013 Wiley-VCH Verlag GmbH & Co. KGaA, Boschstr. 12, 69469 Weinheim, Germany

All rights reserved (including those of translation into other languages). No part of this book may be reproduced in any form – by photoprinting, microfilm, or any other means – nor transmitted or translated into a machine language without written permission from the publishers. Registered names, trademarks, etc. used in this book, even when not specifically marked as such, are not to be considered unprotected by law.

Print ISBN: 978-3-527-41101-6
ePDF ISBN: 978-3-527-64903-7
ePub ISBN: 978-3-527-64902-0
mobi ISBN: 978-3-527-64901-3
oBook ISBN: 978-3-527-64900-6

Cover Design Adam-Design, Weinheim, Germany

Typesetting Thomson Digital, Noida, India

Printing and Binding Markono Print Media Pte Ltd, Singapore

Printed on acid-free paper

To those who advance the frontiers of science and engineering

Contents

Preface *XIII*

1 **General Properties of Nitrides** *1*
1.1 Crystal Structure of Nitrides *1*
1.2 Gallium Nitride *5*
1.3 Aluminum Nitride *6*
1.4 Indium Nitride *10*
1.5 AlGaN Alloy *13*
1.6 InGaN Alloy *14*
1.7 AlInN Alloy *14*
1.8 InAlGaN Quaternary Alloy *15*
1.9 Electronic Band Structure and Polarization Effects *18*
1.9.1 Introduction *18*
1.9.2 General Strain Considerations *22*
1.9.3 k·p Theory and the Quasicubic Model *23*
1.9.4 Temperature Dependence of Wurtzite GaN Bandgap *26*
1.9.5 Sphalerite (Zincblende) GaN *26*
1.9.6 AlN *28*
1.9.6.1 Wurtzite AlN *28*
1.9.6.2 Zincblende AlN *28*
1.9.7 InN *29*
1.10 Polarization Effects *31*
1.10.1 Piezoelectric Polarization *32*
1.10.2 Spontaneous Polarization *35*
1.10.3 Nonlinearity of Polarization *35*
1.10.3.1 Nonlinearities in Piezoelectric Polarization *42*
1.10.4 Polarization in Heterostructures *46*
1.10.4.1 Ga-Polarity Single AlGaN–GaN Interface *51*
1.10.4.2 Polarization in Quantum Wells *56*
1.11 Nonpolar and Semipolar Orientations *59*
Further Reading *61*

2 Doping: Determination of Impurity and Carrier Concentrations 63
2.1 Introduction 63
2.2 Doping 63
2.3 Formation Energy of Defects 65
2.3.1 Hydrogen and Impurity Trapping at Extended Defects 67
2.4 Doping Candidates 69
2.5 Free Carriers 70
2.6 Binding Energy 70
2.7 Conductivity Type: Hot Probe and Hall Measurements 71
2.8 Measurement of Mobility 71
2.9 Semiconductor Statistics, Density of States, and Carrier Concentration 74
2.10 Charge Balance Equation and Carrier Concentration 78
2.10.1 n-Type Semiconductor 79
2.10.2 p-Type Semiconductor 84
2.11 Capacitance–Voltage Measurements 87
Appendix 2.A. Fermi Integral 94
Further Reading 95

3 Metal Contacts 97
3.1 Metal–Semiconductor Band Alignment 97
3.2 Current Flow in Metal–Semiconductor Junctions 101
3.3 Ohmic Contact Resistance 107
3.3.1 Specific Contact Resistivity 107
3.4 Semiconductor Resistance 108
3.4.1 Determination of the Contact Resistivity 109
Further Reading 113

4 Carrier Transport 115
4.1 Introduction 115
4.2 Carrier Scattering 117
4.2.1 Impurity Scattering 118
4.2.2 Acoustic Phonon Scattering 120
4.2.2.1 Deformation Potential Scattering 121
4.2.2.2 Piezoelectric Scattering 124
4.2.3 Optical Phonon Scattering 126
4.2.3.1 Nonpolar Optical Phonon Scattering 126
4.2.3.2 Polar Optical Phonon Scattering 127
4.2.4 Alloy Scattering and Dislocation Scattering 134
4.3 Calculated Mobility of GaN 143
4.4 Scattering at High Fields 147
4.4.1 Transport at High Fields: Energy and Momentum Relaxation Times 152
4.4.2 Energy-Dependent Relaxation Time and Large B 153
4.4.3 Hall Factor 155
4.5 Delineation of Multiple Conduction Layer Mobilities 156

4.6	Carrier Transport in InN	*158*
4.7	Carrier Transport in AlN	*159*
4.8	Carrier Transport in Alloys	*161*
4.9	Two-Dimensional Transport in n-Type GaN	*164*
4.9.1	Scattering in 2D Systems	*166*
4.9.1.1	Electron Mobility in AlGaN/GaN 2D System	*168*
4.9.1.2	Numerical Two-Dimensional Electron Gas Mobility Calculations	*170*
4.9.1.3	Magnetotransport and Mobility Spectrum	*173*
	Further Reading	*174*
5	**The p–n Junction**	*177*
5.1	Introduction	*177*
5.2	Band Alignment	*177*
5.3	Electrostatic Characteristics of p–n Heterojunctions	*179*
5.4	Current–Voltage Characteristics of p–n Junctions	*185*
5.4.1	Diode Current under Reverse Bias	*186*
5.4.1.1	Poole–Frenkel and Schottky Effects	*187*
5.4.1.2	Avalanching	*188*
5.4.2	Diffusion Current	*189*
5.4.2.1	Diffusion Current under Reverse Bias	*190*
5.4.2.2	Diffusion Current under Forward Bias	*190*
	Further Reading	*191*
6	**Optical Processes**	*193*
6.1	Introduction	*193*
6.2	Einstein's *A* and *B* Coefficients	*194*
6.3	Absorption and Emission	*196*
6.4	Band-to-Band Transitions and Efficiency	*198*
6.5	Optical Transitions in GaN	*200*
6.5.1	Excitonic Transitions in GaN	*200*
6.5.1.1	Strain Effects	*203*
6.5.1.2	Bound Excitons	*204*
6.6	Free-to-Bound Transitions	*205*
6.7	Donor–Acceptor Transitions	*206*
	Further Reading	*207*
7	**Light-Emitting Diodes and Lighting**	*209*
7.1	Introduction	*209*
7.2	Current Conduction Mechanism in LED-Like Structures	*211*
7.3	Optical Output Power and Efficiency	*214*
7.3.1	Efficiency and Other LED Relevant Terms	*215*
7.3.2	Optical Power and External Efficiency	*217*
7.3.3	Internal Quantum Efficiency	*218*
7.3.3.1	Auger Recombination	*219*
7.3.3.2	SRH Recombination	*220*

7.3.3.3	Radiative Recombination *222*
7.3.3.4	Continuity or Rate Equations as Pertained to Efficiency *223*
7.3.3.5	Carrier Overflow (Spillover, Flyover, Leakage) *231*
7.4	Effect of Surface Recombination *244*
7.5	Effect of Threading Dislocation on LEDs *247*
7.6	Current Crowding *247*
7.7	Perception of Color *250*
7.8	Chromaticity Coordinates and Color Temperature *251*
7.9	LED Degradation *253*
7.10	Packaging *255*
7.11	Luminescence Conversion and White Light Generation *257*
7.11.1	Color-Rendering Index *258*
7.11.2	White Light from Multichip LEDs *259*
7.11.3	Combining LEDs and Phosphor(s) *262*
	Further Reading *266*

8 Semiconductor Lasers: Light Amplification by Stimulated Emission of Radiation *267*

8.1	Introduction *267*
8.2	A Primer to the Principles of Lasers *268*
8.2.1	Waveguiding *270*
8.2.2	Analytical Solution to the Waveguide Problem *273*
8.2.2.1	TE Mode *274*
8.2.2.2	TM Mode *276*
8.2.3	Far-Field Pattern *280*
8.3	Loss, Threshold, and Cavity Modes *281*
8.4	Optical Gain *283*
8.5	A Glossary for Semiconductor Lasers *286*
8.5.1	Optical Gain in Bulk Layers: a Semiconductor Approach *289*
8.5.1.1	Relating Absorption Rate to Absorption Coefficient *290*
8.5.1.2	Relating Stimulated Emission Rate to Absorption Coefficient *290*
8.5.1.3	Relating Spontaneous Emission Rate to Absorption Coefficient *290*
8.5.2	Semiconductor Realm *291*
8.5.3	Gain in Quantum Wells *299*
8.5.3.1	Optical Gain *302*
8.5.3.2	Measurement of Gain in Nitride Lasers *304*
8.5.4	Gain Measurement via Optical Pumping *304*
8.6	Threshold Current *306*
8.7	Analysis of Injection Lasers with Simplifying Assumptions *307*
8.7.1	Recombination Lifetime *309*
8.7.2	Quantum Efficiency *311*
8.8	GaN-Based LD Design and Performance *312*
8.8.1	Gain Spectra of InGaN Injection Lasers *317*
8.8.2	Mode Hopping *321*
8.9	Thermal Resistance *322*

8.10	Nonpolar and Semipolar Orientations *323*	
8.11	Vertical Cavity Surface-Emitting Lasers (VCSELs) *325*	
8.11.1	Microcavity Fundamentals *328*	
8.11.2	Polariton Lasers *333*	
8.12	Degradation *337*	
	Appendix 8.A: Determination of the Photon Density and Photon Energy Density in a Cavity *343*	
	Further Reading *348*	

9 Field Effect Transistors *349*

9.1	Introduction *349*
9.2	Operation Principles of Heterojunction Field Effect Transistors *350*
9.2.1	Heterointerface Charge *350*
9.2.2	Analytical Description of HFETs *358*
9.3	GaN and InGaN Channel HFETs *364*
9.4	Equivalent Circuit Models: De-embedding and Cutoff Frequency *366*
9.4.1	Small-Signal Equivalent Circuit Modeling *367*
9.4.2	Cutoff Frequency *370*
9.5	HFET Amplifier Classification and Efficiency *373*
9.6	Drain Voltage and Drain Breakdown Mechanisms *378*
9.7	Field Plate for Spreading Electric Field for Increasing Breakdown Voltage *383*
9.8	Anomalies in GaN MESFETs and AlGaN/GaN HFETs *384*
9.8.1	Effect of the Traps in the Buffer Layer *386*
9.8.2	Effect of Barrier States *392*
9.8.3	Correlation between Current Collapse and Surface Charging *393*
9.9	Electronic Noise *396*
9.9.1	FET Equivalent Circuit with Noise *398*
9.9.2	High-Frequency Noise in Conjunction with GaN FETs *402*
9.10	Self-Heating and Phonon Effects *405*
9.10.1	Heat Dissipation and Junction Temperature *406*
9.10.2	Hot Phonon Effects *409*
9.10.2.1	Phonon Decay Channels and Decay Time *411*
9.10.2.2	Implications for FETs *416*
9.10.2.3	Heat Removal in View of Hot Phonons *418*
9.10.2.4	Tuning of the Hot Phonon Lifetime *421*
9.11	HFET Degradation *427*
9.11.1	Gated Structures: Reliability *434*
9.11.2	Reliability Tests *438*
9.12	HFETs for High-Power Switching *440*
	Appendix 9.A. Sheet Charge Calculation in AlGaN/GaN Structures with AlN Interface Layer (AlGaN/AlN/GaN) *444*
	Further Reading *446*

Index *449*

Preface

This book aims to describe the fundamentals of light emitters and field effect transistors based on GaN and related semiconductors with supporting material. The book is intended to provide the know-how for the reader to be well versed in the aforementioned devices with selective further reading material for additional material.

Chapter 1 deals with the structural properties of nitride-based semiconductors and their band structure and polarization with extensive tables. Chapter 2 discusses defects and doping, electron and hole concentrations along with applicable statistics as affected by temperature, and Hall and C–V measurements. Metal semiconductor junctions along with the current conduction mechanisms, contact resistivity and its determination are discussed in Chapter 3. Scattering and carrier transport at low and high electric fields are discussed in Chapter 4, which embodies ionized impurity, deformation potential, piezoelectric, optical phonon, and alloy scattering, among others, in bulk and to a lesser extent in two-dimensional systems. Hall effect/Hall factor and magneto transport along with delineation of mobility for each of the contributing layers in multichannel constructs are also included in the discussion. Chapter 5 is devoted to p–n junctions, beginning with the discussion of band lineups and leading to consideration of current conduction mechanisms, such as diffusion, generation–recombination, and Poole–Frenkel current. Avalanche multiplication, pertinent to the high-field region of FETs and avalanche photodiodes is also covered in a concise form. Chapter 6 contains a succinct discussion of optical processes in semiconductors such as absorption and emission *vis-à-vis* Einstein's A and B coefficients to pave the way for discussion of light emitters in the follow-up chapters. Chapter 7 delves into the fundamentals and practice of light-emitting diodes, perception of vision and color by human eye and methodologies, both used and proposed, for generation of white light and presents an in-depth discussion of efficiency and mechanisms responsible for its degradation at high injection levels. Chapter 8 focuses on lasers including the relevant theory and practical operation. Integral concepts such as gain and loss along with their measurement, threshold current, efficiency, polar- and nonpolar-specific processes, and microcavity-based lasers are also discussed. The final chapter, Chapter 9, treats field effect transistor fundamentals, which are applicable to any semiconductor material with points specific to GaN-based varieties. The discussion primarily focuses on 2DEG channels formed at heterointerfaces and their use for FETs including polarization effects. A succinct analytical model is provided for calculating the carrier densities at the interfaces for various scenarios and current–voltage characteristics of FETs with several examples. Hot phonon effects responsible for this shortfall and attainable carrier velocity are uniquely discussed with sufficient theory and experimental data and design approaches to mitigate the problem, along with their effect on heat dissipation and reliability.

This book would not have been possible without the support of many of our colleagues, namely, Profs. Ü. Özgür, V. Avrutin, R. Shimada, and A. Matulionis and Drs. N. Izyumskaya, J. Xie, J. Leach, and C. Kayis, who helped with material, figures, proofreading, and extensive discussions during its production.

Richmond, VA USA *Hadis Morkoç*
September 2012

1
General Properties of Nitrides

1.1
Crystal Structure of Nitrides

GaN and its binary cousins InN and AlN as well as their ternary and quaternary are considered one of the most important groups of semiconductors after Si. This follows from their ample applications in lighting and displays, consumer electronics, lasers, detectors, and high-power RF/switching devices owing to their excellent optical and electrical properties. The pertinent properties and materials parameters upon which to build the chapters on devices are succinctly discussed.

Group III nitrides can be of *wurtzite* (Wz), *zincblende* (ZB), and *rocksalt* structure. Under ambient conditions, the thermodynamically stable structure is wurtzite for bulk AlN, GaN, and InN. The space grouping for the zincblende structure is $F\bar{4}3m$ in the Hermann–Mauguin notation and T_d^2 in the Schoenflies notation and has a cubic unit cell containing four group III elements and four nitrogen elements. (The term zincblende originated in describing compounds such as ZnS that could be cubic or hexagonal. But the term has been used ubiquitously for compound semiconductors with cubic symmetry. The correct term for the cubic phase is sphalerite.). The position of the atoms within the unit cell is identical to that in the diamond crystal structure. The stacking sequence for the (111) close-packed planes in this structure is AaBbCc. Small and large letters stand for the two different kinds of constituents. The rocksalt structure (with space group $Fm\bar{3}m$ in the Hermann–Mauguin notation and O_h^5 in the Schoenflies notation) can be induced under very high pressures, but not through epitaxial growth.

The wurtzite structure has a hexagonal unit cell and thus two lattice constants c and a. It contains six atoms of each type. The space grouping for the wurtzite structure is $P6_3mc$ in the Hermann–Mauguin notation and C_{6v}^4 in the Schoenflies notation. The point group symmetry is $6mm$ in the Hermann–Mauguin notation and C_{6v} in the Schoenflies notation. The Wz structure consists of two interpenetrating *hexagonal close-packed* (hcp) sublattices, each with one type of atom, offset along the c-axis by 5/8 of the cell height ($5c/8$). The Wz structure consists of alternating biatomic close-packed (0001) planes of Ga and N pairs, thus the *stacking sequence* of the (0001) plane is AaBbAa in the (0001) direction.

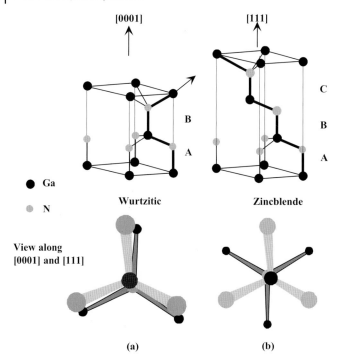

Figure 1.1 Ball-and-stick stacking model of crystals with (a) (both top and bottom) 2H wurtzitic and (b) (both top and bottom) 3C zincblende polytypes. The bonds in an a-plane ($11\bar{2}0$) are indicated with thicker lines to accentuate the stacking sequence. The figures on the top depict the three-dimensional view. The figures at the bottom indicate the projections on (0001) and (111) planes for wurtzitic and cubic phases, respectively. Note the rotation in the zincblende case along the $\langle 111 \rangle$ direction.

The Wz and zincblende structures differ only in the bond angle of the second-nearest neighbor (Figure 1.1). The stacking order of the Wz along the [0001] c-direction is AaBb, meaning a mirror image but no in-plane rotation with the bond angles. In the zincblende structure along the [111] direction, there is a 60° rotation that causes a stacking order of AaBbCc. The point with regard to rotation is illustrated in Figure 1.1b. The nomenclature for various commonly used planes of hexagonal semiconductors in two- and three-dimensional versions is presented in Figure 1.2. The Wz group III nitrides lack an inversion plane perpendicular to the c-axis; thus, nitride surfaces have either a group III element (Al, Ga, or In) polarity (referred to as the Ga-polarity) with a designation of (0001) or (0001)A plane or a N-polarity with a designation of ($000\bar{1}$) or (0001)B plane. The former notations for each are used here. The distinction between these two directions is essential in nitrides due to implications in the polarity of the polarization charge. Three surfaces and directions are of special importance in nitrides, which are (0001) c-, ($11\bar{2}0$) a-, and ($1\bar{1}00$) m-planes and the directions associated with them: $\langle 0001 \rangle$, $\langle 11\bar{2}0 \rangle$, and $\langle 1\bar{1}00 \rangle$.

1.1 Crystal Structure of Nitrides | 3

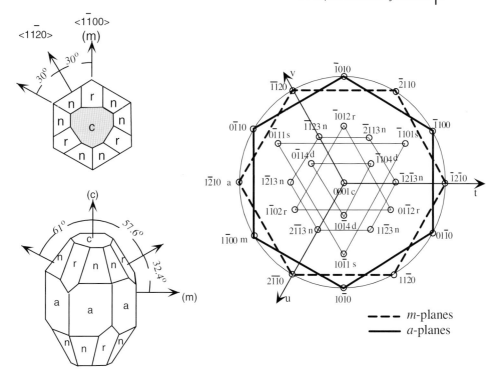

Figure 1.2 Labeling of planes in hexagonal symmetry (for sapphire), a telescopic view of labeling of planes in hexagonal symmetry in the (*tuvw*) coordinate system with *w* representing the unit vector in the *c*-direction is shown on the right. The lines are simply to show the symmetry only. If the lines connecting *m*-points among each other and *a*-points among each other were to be interpreted as the projection of those planes on the *c*-plane, the roles would be switched in that the lines connecting the *m*-points would actually represent the *a*-planes, and lines connecting the *a*-points would actually represent the *m*-planes that are normal to the plane of the page.

Delving further into the Wz structure, it can be represented by lattice parameters *a* in the basal plane and *c* in the perpendicular direction and the *internal parameter u*, as shown in Figure 1.3. The *u* parameter is defined as the anion–cation bond length (also the nearest-neighbor distance) divided by the *c* lattice parameter. The *c* parameter depicts the unit cell height. The wurtzite structure is a hexagonal

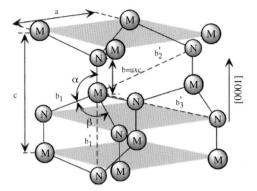

Figure 1.3 Schematic representation of a wurtzitic metal nitride structure having lattice constants a in the basal plane and c in the basal direction. u parameter is the bond length or the nearest-neighbor distance (b) divided by c (0.375 in ideal crystal), α and β (109.47° in ideal crystal) are the bond angles, and b'_1, b'_2, and b'_3 are the three types of second-nearest-neighbor distances. M denotes metal (e.g., Ga) and N denotes N.

close-packed lattice, comprising vertically oriented M–N units at the lattice sites. The *basal plane lattice parameter* (the edge length of the basal plane hexagon) is universally depicted by a and the *axial lattice parameter* perpendicular to the basal plane is universally described by c. In an ideal wurtzite structure represented by four touching hard spheres, the values of the axial ratio and the internal parameter are $c/a = \sqrt{8/3} = 1.633$ and $u = 3/8 = 0.375$, respectively. The crystallographic vectors of wurtzite are $\vec{a} = a(1/2, \sqrt{3}/2, 0)$, $\vec{b} = a(1/2, -\sqrt{3}/2, 0)$, and $\vec{c} = a(0, 0, c/a)$. In Cartesian coordinates, the basis atoms are (0, 0, 0), (0, 0, uc), $a(1/2, \sqrt{3}/6, c/2a)$, and $a(1/2, \sqrt{3}/6, [u+1/2]c/a)$.

In all Wz III nitrides, experimentally observed c/a ratios are smaller than ideal parameters and a strong correlation exists between the c/a ratio and the u parameter such that when c/a decreases, the u parameter increases in a manner that the four tetrahedral distances remain nearly constant through a distortion of tetrahedral angles. For the equal bond length to prevail, the following relation must hold:

$$u = (1/3)(a^2/c^2) + 1/4. \tag{1.1}$$

The nearest-neighbor bond length along the c-direction (expressed as b in Figure 1.3) and off c-axis (expressed as b_1 in Figure 1.3) can be calculated as

$$b = cu \quad \text{and} \quad b_1 = \sqrt{\frac{1}{3}a^2 + \left(\frac{1}{2} - u\right)^2 c^2}. \tag{1.2}$$

Most commonly used planes of nitride semiconductors, namely, the polar *c-plane* and nonpolar *a-* and *m-planes*, are graphically shown in Figure 1.4. Other planes, semipolar planes, that are gaining some attention are shown in Figure 1.5.

Table 1.1 gives the calculated as well as the experimentally observed structural parameters discussed above, including the lattice parameters, the nearest- and

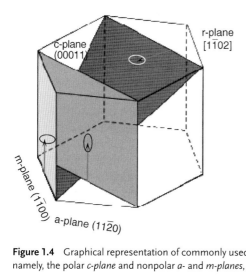

Figure 1.4 Graphical representation of commonly used planes of nitride semiconductors, namely, the polar *c-plane* and nonpolar *a-* and *m-planes*, and *r-plane*.

second-nearest-neighbor distances, and the bond angles for three end binaries: GaN, AlN, and InN. The distances are in ångströms.

1.2
Gallium Nitride

The parameters associated with electrical and optical properties of wurtzitic GaN and AlN are given in Table 1.2.

The elastic stiffness coefficients and the bulk modulus are compiled in Table 1.3.

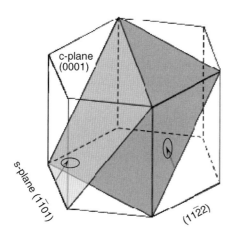

Figure 1.5 Graphical representation of semipolar $(10\bar{1}0)$ plane and $(11\bar{2}2)$ plane.

Table 1.1 Calculated (for ideal crystal) and experimentally observed structural parameters for wurtzitic GaN, AlN, and InN.

	GaN		AlN		InN	
	Ideal	Exp	Ideal	Exp	Ideal	Exp
u	0.375	0.377	0.375	0.382	0.375	0.379
a (Å)	3.199	3.199	3.110	3.110	3.585	3.585
c/a	1.633	1.634	1.633	1.606	1.633	1.618
b (Å)	1.959	1.971	1.904	1.907	2.195	2.200
b_1 (Å)	1.959	1.955	1.904	1.890	2.195	2.185
b'_1 (Å)	3.265	3.255	3.174	3.087	3.659	3.600
b'_2 (Å)	3.751	3.757	3.646	3.648	4.204	4.206
b'_3 (Å)	3.751	3.749	3.646	3.648	4.204	4.198
α	109.47	109.17	109.47	108.19	109.47	108.69
β	109.47	109.78	109.47	110.73	109.47	110.24

1.3 Aluminum Nitride

AlN has a molar mass of 40.9882 g/mol. Reported wurtzite lattice parameters range from 3.110 to 3.113 Å for the a parameter (3.1106 Å for bulk, 3.1130 Å for powder, and 3.110 Å for AlN on SiC) and from 4.978 to 4.982 Å for the c parameter. The c/a ratio thus varies between 1.600 and 1.602. The deviation from that of the ideal wurtzite crystal ($c/a = 1.633$) is plausibly due to lattice instability and ionicity. The u parameter for AlN is 0.3821, which is larger than the calculated value of 0.380. This means that the interatomic distance and angles differ by 0.01 Å and 3°, respectively, from the ideal parameters. Refer to Table 1.4 for electronic properties of Wz AlN.

The measured *bulk modulus* B and *Young's modulus* E or Y_0 are compiled in Table 1.5 along with the entire set of elastic stiffness coefficients.

The phonon energies measured by Raman scattering apply to Raman active modes (Table 1.6). Raman-active optical phonon modes belong to the A_1, E_1, and E_2 group representations.

The thermal expansion of AlN is isotropic with a room temperature value of $2.56 \times 10^{-6} \, K^{-1}$. The equilibrium N_2 vapor pressure above AlN is relatively low compared to that above GaN that makes AlN easier to synthesize. The calculated temperatures at which the equilibrium N_2 pressure reaches 1, 10, and 100 atm are 2836, 3088, and 3390 K, respectively. The *thermal conductivity* κ of AlN at room temperature has been predicted as $\kappa = 3.19$ W/(cm K) in O-free simulated material. The values of the *refractive index* n are in the range 1.99–2.25 with several groups reporting $n = 2.15 \pm 0.05$. The *dielectric constant* of AlN (ε_0) lies in the range 8.3–11.5 and most of the values fall within $\varepsilon_0 = 8.5 \pm 0.2$. Other measurements in the high-frequency range produced dielectric constants of 4.68 and $\varepsilon_\infty = 4.84$. AlN has also been examined for its potential for second-harmonic generation.

Table 1.2 Parameters related with electrical and optical properties of wurtzitic GaN.

Wurtzite polytype GaN	Parameter value/comments
Bandgap energy E_g (eV), direct	3.42 at 300 K
	3.505 at 1.6 K
Breakdown field (cm^{-1})	3–5 × 10^6 at 300 K
Electron affinity (eV)	4.1
Energy separation between Γ valley and M–L valleys (eV)	1–1.9 at 300 K
Energy separation between M–L valleys	0.6 at 300 K
degeneracy (eV)	0.6 at 300 K
Energy separation between Γ valley and A valleys (eV)	1.3–2.1 at 300 K
Index of refraction	2.3 at 300 K away from band edge
Dielectric constants, static	10.4 ($E\|c$), 9.5 ($E \perp c$) or 8.9 in c-direction ($E\|c$) at 300 K
Dielectric constants, high frequency	5.35 or 5.47 ($E \perp c$) at 300 K, 5.8 ($E\|c$) at 300 K
Optical LO phonon energy (meV)	91.2
A_1 – LO, ν_{A1}(LO) (cm^{-1})	710–744
A_1 – TO, ν_{A1}(TO$\|$) (cm^{-1})	533–534
E_1 – LO, ν_{E1}(LO$\perp$) (cm^{-1})	741–742
E_1 – TO, ν_{E1}(TO$\perp$) (cm^{-1})	556–559
E_2 (low) (cm^{-1})	143–146
E_2 (high) (cm^{-1})	560–579
Energy of spin–orbit splitting E_{so} (meV)	11(+5, −2) at 300 K calculated from the values of energy gap $E_{g,dir}$ (given in this table)
Energy of crystal field splitting E_{cr} (meV)	40 at 300 K, 22 calculated from the values of energy gap $E_{g,dir}$ (given in this table)
Effective electron mass, m_e or $m_e^{//}$	0.20m_0 at 300 K, 0.27m_0 by Faraday rotation
Effective electron mass, $m_{e\perp}$ or $m_e^{\perp}$	0.20m_0 at 300 K; 0.15–0.23m_0 fit of reflectance spectrum
Effective hole mass	0.8m_0 at 300 K
Effective hole masses (heavy) m_{hh}	$m_{hh} = 1.4m_0$ at 300 K
	$m_{hhz} = m_{hh}^{//} = 1.1m_0$, $m_{hh\perp} = m_{hh}^{\perp} = 1.6m_0$ at 300 K
	$m_{hh}^{//} = 1.1$–2.007m_0
	$m_{hh}^{\perp} = 1.61$–2.255m_0
Effective hole masses (light)	$m_{lh} = 0.3m_0$ at 300 K, $m_{lhz} = m_{lh}^{//} = 1.1m_0$ at 300 K
	$m_{lh}^{\perp} = m_{lh\perp} = 0.15m_0$ at 300 K
	$m_{lh}^{//} = 1.1$–2.007m_0, $m_{lh\perp} = 0.14$–0.261m_0
Effective hole masses (split-off band) m_s	$m_{sh} = 0.6m_0$ at 300 K
	$m_{shz} = m_{ch}^{//} = 0.15m_0$, $m_{sh\perp} = m_{ch}^{\perp} = 1.1m_0$ at 300 K
	$= m_{ch}^{//} = 0.12$–0.16m_0, $m_{ch}^{\perp} = 0.252$–1.96m_0
Effective mass of density of state m_v	1.4m_0
Effective conduction band density of states (cm^{-3})	2.3 × 10^{18} at 300 K
Effective valence band density of states (cm^{-3})	4.6 × 10^{19} at 300 K
Electron mobility [cm^2/(V s)]	∼1400 experimental at 300 K
Hole mobility [cm^2/(V s)]	<20
n-Doping range (cm^{-3})	10^{16} cm^{-3}–high 10^{19}
p-Doping range (cm^{-3})	10^{16} cm^{-3}–mid-10^{18}
Diffusion coefficient for electrons (cm^2/s)	25
Diffusion coefficient for holes (cm^2/s)	Not well defined, but 5, 26, 94 have been reported

Table 1.3 Experimental and calculated elastic coefficients (C_{ij}), bulk modulus (B_0) and its pressure derivative (B', dB/dP), and Young's modulus (E or Y_0) (in GPa) for WzGaN.

Technique	C_{11}	C_{12}	C_{13}	C_{33}	C_{44}	B_0	B'	E
X-ray	296	130	158	267	24.1	195		150
XAS						245	4	
EDX						188	3.2	
ADX						237	4.3	
Brillouin	315–390	118–145	70–114	324–398	88–109	175–210		281–362
Ultrasonic	370 377	145 160	110 114	209 390	81.4, 90	173 208		161 343
Most commonly used values	380	110			105			
PWPP (1)	367	135	103	405	95	202		363
FP-LMTO	396	144	100	392	91	207		355
PWPP (2)	C_{11}	515–C_{11}	104	414		207		373
ZB GaN	253–264	153–165			60–68	200–237	3.9–4.3	

EDX: energy dispersive X-ray; ADX: angular dispersive X-ray diffraction; XAS: X-ray absorption spectroscopy; PWPP: plane wave pseudopotential; and FP-LMTO: full-potential linear muffin-tin orbital.

1.3 Aluminum Nitride

Table 1.4 Parameters related to optical and electrical properties of wurtzitic AlN.

Wurtzite polytype AlN	Parameter
Bandgap energy (eV)	~6 at 300 K and ~6.1 at 5 K
Breakdown field (V/cm)	$1.2–1.8 \times 10^6$
dE_g/dP (eV l/bar)	3.6×10^{-3}
Conduction band energy separation between Γ valley and M–L valleys (eV)	~0.7–1
Conduction band energy separation between Γ valley and K valleys (eV)	~1.0
Valence band energy of spin–orbit splitting E_{so} (eV)	0.019: 0.036 at 300 K
Valence band energy of crystal field splitting E_{cr} (eV), Γ_7 on top of Γ_9	−0.225
Effective conduction band density of states (cm^{-3})	6.3×10^{18}
Effective valence band density of states (cm^{-3})	4.8×10^{20}
Index of refraction	$n\,(3\,eV) = 2.15 \pm 0.05$
Dielectric constant, static	7.34: 9.14 at 300 K
	9.32 for $E//c$ (modeling)
	7.76 for $E \perp c$ (experiment)
Dielectric constant, high frequency	4.6–4.84 at 300 K
	4.35 from $E//c$ (modeling)
	4.16 from $E \perp c$ (experiment)
Infrared refractive index	1.8–2.2 at 300 K
	3 from $E//c$ (modeling)
	2.8 from $E \perp c$ (experiment)
Effective electron mass m_e	$0.25–0.4 m_0$
	$m_e^{//} = 0231–0.35 m_0$
	$m_e^{\perp} = 0242–0.25 m_0$
Effective hole masses (heavy)	$m_{hh}^{//} = 3.53 m_0$ at 300 K
for k_z direction m_{hz} or $m_{hh}^{//}$	$2.02–3.13 m_0$ at 300 K
	$m_{hh}^{\perp} = 10.42 m_0$ at 300 K
for k_x direction m_{hx} or $m_{hh}^{\perp}$	$m_{hh}^{//} = 1.869–4.41 m_0$
	$m_{hh}^{\perp} = 2.18–11.14 m_0$
Effective hole masses (light)	$3.53 m_0$
	$0.24 m_0$
for k_z direction m_{lz} or $m_{lh}^{//}$	$m_{lh}^{//} = 1.869–4.41 m_0$
for k_x direction m_{lx} or $m_{lh}^{\perp}$	$m_{lh}^{\perp} = 0.24–0.350 m_0$
Effective hole masses (split-off band)	$0.25 m_0$ at 300 K
or k_z direction m_{soz} or $m_{ch}^{//}$	$3.81 m_0$ at 300 K
for k_x direction m_{sox} or $m_{ch}^{\perp}$	$m_{ch}^{//} = 0.209–0.27 m_0$
	$m_{ch}^{\perp} = 1.204–4.41 m_0$

(continued)

1 General Properties of Nitrides

Table 1.4 (Continued)

Wurtzite polytype AlN	Parameter
Effective mass of density of states m_v	$7.26m_0$ at 300 K
Optical phonon energy, meV	99.2
$v_{TO}(E_1)$ phonon wavenumber (cm^{-1})[a]	895, 614, 608
$v_{LO}(E_1)$ phonon wavenumber (cm^{-1})	671.6, 821, 888.9
$v_{TO}(A_1)$ phonon wavenumber (cm^{-1})	888, 514, 667.2
$v_{LO}(A_1)$ phonon wavenumber (cm^{-1})	659.3, 663, 909
$v(E_2)$ phonon wavenumber (cm^{-1})	303,[a] 426
$n_{TO}(E_1)$ phonon wavenumber (cm^{-1})	657–673
$n_{TO}(A_1)$ phonon wavenumber (cm^{-1})	607–614 or 659–667
$n_{LO}(E_1)$ phonon wavenumber (cm^{-1})	895–924
$n_{LO}(A_1)$ phonon wavenumber (cm^{-1})	888–910
$n^{(1)}(E_2)$ phonon wavenumber (cm^{-1})	241–252
$n^{(2)}(E_2)$ phonon wavenumber (cm^{-1})	655–660

a) Room temperature Raman, tentative.

1.4
Indium Nitride

The parameters associated with electrical and optical properties of wurtzitic InN are given in Table 1.7.

The zincblende (cubic) form has been reported to occur in films containing both polytypes. The measured Wz InN lattice parameters using powder technique are in the range $a = 3.530$–3.548 Å and $c = 5.704$–5.960 Å with a consistent c/a ratio of about 1.615 ± 0.008. The density of InN deduced from Archimedean displacement measurements is 6.89 g/cm^3 at 250 °C. Table 1.8 summarizes the measured and the calculated elastic coefficients for Wz InN.

As in the cases of Wz GaN and AlN, Wz InN has 12 phonon modes at the zone center (symmetry group: C_{6v}), 3 acoustic and 9 optical ones with the acoustic branches near 0 at $k = 0$. The infrared active modes are of the E_1 (LO), E_1 (TO), A_1 (LO), and A_1 (TO) type. Raman spectroscopy has yielded four optical phonons characteristic for InN with wavenumbers 190 (E_2), 400 (A_1), 490 (E_1), and 590 (E_2) cm^{-1} in InN layers grown by atomic layer epitaxy (ALE). Moreover, a transverse optical (TO) mode has been observed at 478 cm^{-1} (59.3 meV) by reflectance and 460 cm^{-1} (57.1 meV) by transmission measurements. From other reflectance data, the existence of a TO phonon mode at 478 cm^{-1} and an LO mode at 694 cm^{-1} was deduced.

Thermal conductivity derived from the Leibfried–Schloman scaling parameter, assuming that the thermal conductivity is limited by intrinsic phonon–phonon scattering, is about 0.80 ± 0.20 W/(cm K). The estimated effective mass of $m_e^* = 0.11m_0$ and an index of refraction of $n = 3.05 \pm 0.05$, while the long–wavelength limit of the *refractive index* has been reported to be 2.88 ± 0.15.

1.4 Indium Nitride

Table 1.5 Experimental bulk modulus and elastic coefficients (in GPa) of AlN.

Method	C_{11}	C_{12}	C_{13}	C_{33}	C_{44}	B	B'	E or Y_0
Ultrasonics	345, 419	125, 140	100, 120	390, 395	118, 120	160, 201, 209	5.2	308, 334, 354
Brillouin	410.5, 419	148.5, 177	98.9, 140	388.5, 392	110, 124.6	210.1, 237		354
PWPP	396	137	108	373	116	207		329
FP-LMTO	398	140	127	382	96	218		322
HF	464	149	116	409	128	231		365
PWPP	C_{11}FIX THIS	538	113	370		210		322

ADX: angular dispersive X-ray diffraction; EDX: energy dispersive X-ray; PWPP: plane wave pseudopotential; FP-LMTO: full-potential linear muffin-tin-orbital; HF: Hartree–Fock.

Table 1.6 Optical phonon energies and phonon deformation potentials for AlN.

Mode	Wz (Raman) unstrained		Wz (Raman range) (cm^{-1})	Unstrained (calculated) (cm^{-1})	Deformation potentials (cm^{-1})			
					Raman		Calculated	
	(meV)	(cm^{-1})			a_λ	b_λ	a_λ	b_λ
E_1-TO	82.8	667.2	667–673	677	-982 ± 83	-901 ± 145	-835	-744
A_1-TO	75.4	608.5	614–667	618	-930 ± 94	-904 ± 163	-776	-394
A_1-LO	75.4	888.9						
E_1-LO	112.8	909.6	895–921	924			-867	-808
E_2^1 (E_2 low)	30.5	246.1		246				
E_2^2 (E_2 high)	81.2	655.1		655				

Table 1.7 Parameters related to electrical and optical properties of wurtzitic InN.

Wurtzitic InN	Value
Bandgap energy, E_g (300 K)	0.6–0.7 eV
Dielectric constant (static)	15.3
Dielectric constant (static, ordinary direction)	$\varepsilon_{0,ort} = 13.1$
Dielectric constant (static, extraordinary direction)	$\varepsilon_0, \| = 14.4$
Dielectric constant (high frequency)	8.4–9.3
Infrared refractive index	Reported range: 2.80–3.05
Energy separation between Γ valley and M–L valleys (eV)	2.9 ÷ 3.9
	~4.8
Energy separation between Γ valley and A valleys (eV)	0.7–2.7
	~4.5
Energy separation between Γ valley and Γ_1 valleys (eV)	1.1–2.6
Energy separation between Γ_1 valley degeneracy (eV)	1
Effective conduction band density of states	9×10^{17} cm^{-3}
Effective valence band density of states	5.3×10^{19} cm^{-3}
Valence band crystal field splitting E_{cr}	0.017 eV
Valence band spin–orbit splitting E_{so}	0.003 eV
Index of refraction	2.5–2.9 at 300 K
Effective electron mass m_e^*	$0.11 m_0$
	$m_e^{\|} = 0.1-0/138 m_0$, $m_e^\perp = 0.1-0.141 m_0$
Effective hole masses (heavy) m_h	$1.63 m_0$ at 300 K
	$m_{hh}^{\|} = 1.350-2.493 m_0$, $m_{hh}^\perp = 1.410-2.661 m_0$
Effective hole masses (light) m_{lp}	$0.27 m_0$ at 300 K
	$m_{lh}^{\|} = 1.350-2.493 m_0$, $m_{lh}^\perp = 0.11-0.196 m_0$
Effective hole masses (split-off band) m_s	$0.65 m_0$ at 300 K
	$m_{ch}^{\|} = 0.092-0.1'4 m_0$, $m_{ch}^\perp = 0.202-3.422$
Effective mass of density of state m_v	$1.65 m_0$ at 300 K
Optical LO phonon energy (meV)	73 at 300 K

Table 1.8 Theoretical and experimental elastic coefficients and bulk modulus (in GPa) of the various forms of InN.

Method	Structure	C_{11}	C_{12}	C_{13}	C_{33}	C_{44}	B	dB/dP = B'
X-ray	Wz	190	104	121	182	10	139, 140, 165	
LMTO	Wz	271	124	94	200	46	165	
PWPP	Wz	223	115	92	224	48	138, 141, 165	3.8, 3.9
LMTO	Wz						165	

PWPP: plane wave pseudopotential; LMTO: linear muffin-tin orbital.

1.5 AlGaN Alloy

The ternary alloy of GaN with AlN forms a continuous system with a wide range of bandgap and a relatively small change in the lattice constant. An accurate knowledge of the compositional dependence of AlGaN, which is often used as barrier material in devices and to a lesser extent as active layer in, for example, UV detectors and emitters, is a prerequisite for analyzing heterostructures in general and quantum wells (QWs) and superlattices in particular. The compositional dependence of the lattice parameters follows the Vegard's law:

$$\begin{aligned} a_{Al(x)Ga(1-x)N} &= [(3.189 \pm 0.002) - (0.086 \pm 0.004)x] \text{ Å}, \\ c_{Al(x)Ga(1-x)N} &= [(5.188 \pm 0.003) - (0.208 \pm 0.005)x] \text{ Å}. \end{aligned} \quad (1.3)$$

However, the bond lengths exhibit a nonlinear behavior, deviating from the virtual crystal approximation. Essentially, the nearest-neighbor bond lengths are not as dependent on the composition as might be expected.

The compositional dependence of the principal (fundamental) bandgap of $Al_xGa_{1-x}N$ can be calculated from the following empirical expression provided that the bowing parameter b is known accurately:

$$E_g(x) = xE_g(\text{AlN}) + (1-x)E_g(\text{GaN}) - bx(1-x), \quad (1.4)$$

where E_g (GaN) = 3.4 eV, E_g (AlN) = 6.1 eV, x is the AlN molar fraction, and b is the bowing parameter. There is dispersion in the value of the bowing parameter owing to difficulties in retaining quality, particularly around the 50:50 composition, and the experimental procedures employed. This dispersion ranges from -0.8 eV (upward bowing) to $+2.6$ eV (downward bowing). More refined layers and the associated techniques seem to yield a bowing parameter of $b = 1.0$ eV for the entire range of alloy composition. It is still possible that as the quality of the films improves, the bowing parameters may have to be revised.

As for the InGaN alloy, the most celebrated one among the nitride family, difficulties/challenges associated with the growth of high-quality InN and the earlier controversy regarding its bandgap aggravated determination of the compositional dependence of its bandgap. As in the case of AlGaN, the calculated lattice parameter of this alloy follows Vegard's law:

$$\begin{aligned} A_{In(x)Ga(1-x)N} &= (3.1986 + 03862x) \text{ Å}, \\ c_{In(x)Ga(1-x)N} &= (5.2262 + 0.574x) \text{ Å}. \end{aligned} \quad (1.5)$$

The compositional dependence of InGaN bandgap is a crucial parameter in heterostructure design. As such, the topic has attracted a number of theoretical and experimental (to be discussed below) investigations and reports. Similar to the case of AlGaN, the energy bandgap of $In_xGa_{1-x}N$ over $0 \le x \le 1$ can be expressed by the empirical expression:

$$\begin{aligned} E^g_{In_xGa_{1-x}N} &= xE^g_{InN} + (1-x)E^g_{GaN} - b_{InGaN}x(1-x) \\ &= 0.7x + 3.4(1-x) - b_{InGaN}x(1-x) \text{ eV}, \end{aligned} \quad (1.6)$$

where $E^g_{GaN} = 3.40$ eV and $E^g_{InN} \approx 0.7$ eV.

The compositional dependence of the bandgap in the entire composition range can be well fit by a bowing parameter of $b = 1.43$ eV, assuming 0.7 eV for the bandgap of InN.

1.6
InGaN Alloy

InGaN alloy, together with allied nitride semiconductors, forms the backbone of emitters in wide wavelength range. With the crucial role InN plays in nitride devices comes the complexities associated with this ternary such as the great disparity between Ga and In could result in anomalies such as phase separation and instabilities. As in the case of AlGaN, the calculated lattice parameter of this alloy follows the Vegard's law:

$$A_{In(x)Ga(1-x)N} = (3.189 \pm 0.3862x) \text{ Å} \quad \text{and} \quad C_{In(x)Ga(1-x)N}(5.2262 + 0.574x) \text{ Å}. \tag{1.7}$$

By employing various tools such as high-resolution X-ray diffraction (HRXRD), the experimental data for various AlGaN support the applicability of Vegard's law in that the experimental data $a_{Al(x)Ga(1-x)N} = [(3.560 \pm 0.019)(0.449 \pm 0.019)x]$ Å and $c_{Al(x)Ga(1-x)N} = [(5.195 \pm 0.002) + (0.512 \pm 0.006)x]$ Å are within about 2% of that predicted by linear interpolation, Vegard's' law. As in the case of AlGaN but to a larger extent, the bond lengths exhibit a nonlinear behavior, deviating from the virtual crystal approximation. Essentially, the nearest-neighbor bond lengths are not as dependent on composition as might be expected from the virtual crystal approximation.

The compositional dependence of InGaN bandgap is a crucial parameter in designs of any heterostructure utilizing it. As such, the topic has attracted a number of theoretical and experimental (to be discussed below) investigations and reports. Similar to the case of AlGaN, the energy bandgap of $In_xGa_{1-x}N$ over $0 \leq x \leq 1$ can be expressed by the empirical expression:

$$\begin{aligned} E^g_{In_xGa_{1-x}N} &= xE^g_{InN} + (1-x)E^g_{GaN} - b_{InGaN}x(1-x) \\ &= 0.7x + 3.4(1-x) - b_{InGaN}x(1-x) \text{ eV}, \end{aligned} \tag{1.8}$$

where $E^g_{GaN} = 3.40$ eV and $E^g_{InN} \approx 0.7$ eV. For the nomenclature $Ga_xIn_{1-x}N$, the terms x and $1-x$ in Equation 1.6 must be interchanged. Another point of caution is that the sign in front of the bowing parameter is changed to positive in some reports. When a comparison is made, the sign of the b parameter must be changed.

1.7
AlInN Alloy

$In_{1-x}Al_xN$ can provide a lattice–matched barrier to GaN, low mole fraction of AlGaN, and InGaN and, consequently, can yield lattice-matched AlInN/AlGaN or AlInN/InGaN heterostructures, including Bragg reflectors. The lattice-matched composition naturally depends on the strain state of the GaN on which the InAlN layer is grown. As in the case of AlGaN and InGaN, the calculated lattice parameter of this alloy follows Vegard's law:

$$a_{Al(x)In(1-x)N} = (3.5848 - 0.4753x)\,\text{Å} \quad \text{and}$$

$$c_{Al(x)In(1-x)N} = (5.8002 - 0.8063x)\,\text{Å}. \tag{1.9}$$

The compositional dependence of the bandgap of AlInN can be expressed with the following empirical expression using a bowing parameter b_{AlInN}:

$$E^g_{Al_xInN} = xE^g_{AlN} + (1-x)E^g_{InN} - b_{AlInN}x(1-x)$$
$$= 6.1x + 0.7(1-x) - b_{AlInN}x(1-x)\,\text{eV}. \tag{1.10}$$

Using a bandgap of 6.0 eV for AlN, the bowing parameter value is about 2 eV.

1.8
InAlGaN Quaternary Alloy

By alloying InN together with GaN and AlN, the bandgap of the resulting alloy(s) can be increased from near 0.7 to 6 eV, which is critical for making high-efficiency visible light sources and detectors. In addition, the bandgap of this quaternary can be changed while keeping the lattice constant matched to GaN.

The relationships between composition and bandgap (or lattice constant) can be predicted by

$$Q(x,y,z) = \frac{xyT_{12}[(1-x+y)/2] + yzT_{23}[(1-y+z)/2] + zxT_{31}[(1-z+x)/2]}{xy + yz + zx},$$

$$T_{ij}(\alpha) = \alpha B_j + (1-\alpha)B_i + \alpha(1-\alpha)b_{ij}. \tag{1.11}$$

The parameters x, y, and z represent the composition of GaN, InN, and AlN in the quaternary alloy. If GaN, InN, and AlN are represented by 1, 2, and 3, then $T_{1,2}$ would represent Ga_xIn_yN. Furthermore, the term T_{12} can be expressed as $T_{12}(\alpha) = \alpha B_2 + (1-\alpha)B_1 + \alpha(1-\alpha)b_{12}$, where b_{12} is the bowing parameter for the Ga_xIn_yN alloy, $\alpha = [(1-x+y)/2]$ or $[(1-x+y)/2]$ or $[(1-z+x)/2]$ is the effective molar fraction for GaInN, InAlN, and AlGaN, respectively, B_2 is the bandgap of InN, and B_1 is the bandgap of GaN. Similar expressions can be constructed for T_{23} and T_{31} by appropriate permutations. An empirical expression similar to that used for the ternaries can also be constructed for the quaternary:

$$E^g_{Al_xIn_yGa_{1-x-y}N} = xE^g_{AlN} + yE^g_{InN} + (1-x-y)E^g_{GaN} - b_{AlGaN}x(1-x)$$
$$- b_{InGaN}y(1-y), \tag{1.12}$$

where the first three parameters on the right-hand side of the equation are contributions by the binaries to the extent of their presence in the lattice, the fourth term represents the bowing contribution related to Al, and the last term

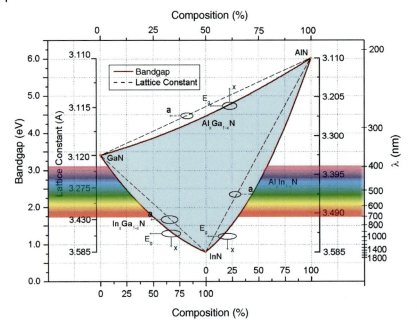

Figure 1.6 The lattice bandgap versus the lattice parameter for AlGaN, InGaN, and InAlN using bowing parameters in the same order, 1, 1.43, and 3.1 eV, and bandgap values of 6 eV for AlN, 3.4 eV for GaN, and 0.8 eV for InN. The lattice constants used for the binary AlN, GaN, and InN are 3.11, 3.199, and 3.585 Å, respectively.

depicts the bowing contribution due to In. The bowing parameters b_{AlGaN} and b_{InGaN} in Equation 1.12 are the same as those discussed in conjunction with InGaN and AlInN.

The traditional bandgap versus the composition for the GaN family of semiconductors, including the appropriate bowing is shown in Figure 1.6. The locus indicates the ternaries with end points representing the binaries. The area within the locus represents the quaternary compounds of (Ga, In, Al)N.

In nitride semiconductors, thermal, mechanical, electrical, and optical properties are interdependent. Changing one property affects one or more of the others. The Heckmann diagrams are typically used to describe the pathways between external forces, such as mechanical stress (σ), electric field (E), optical field (optical E), and thermal field, and the associated material properties, such as strain (ε) and electrical polarization (P). The trigonal diagram describing the above-mentioned interrelationships are shown in Figure 1.7. Nitride semiconductors exhibit spontaneous strain, pyroelectricity, and polarization, which can be switched hysteretically by applied stress, electric field, or heat (shown by arrows).

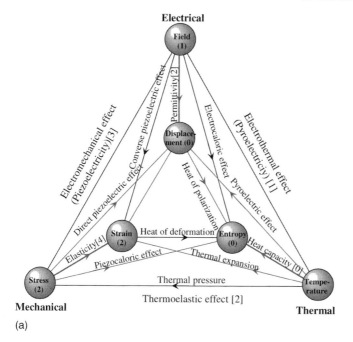

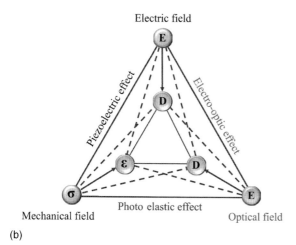

Figure 1.7 The well-known triangle diagrams, Heckman diagrams applied to the GaN family. (a) The interdependence of electrical, thermal, and mechanical properties. (b) The interdependence of the electrical, mechanical, and optical properties. Here, σ and ε represent stress and strain, respectively; D and E represent electrical polarization and electric field, respectively.

1.9
Electronic Band Structure and Polarization Effects

1.9.1
Introduction

The band structure of a given semiconductor is pivotal in the realm of devices. One group of nitride semiconductors pertains to stoichiometric systems where N represents 50% of the constituents, while the other 50% is made of metal constituents that can be in wurtzitic, which is the matter of discussion here, and zincblende forms. The other class of nitrides is the dilute compound semiconductors wherein very small amounts of N are added to the host lattice, such as GaAs, with resultant remarkably large negative bowing of the bandgap making the dilute nitride systems to compete in relatively long-wavelength applications. The latter group is not discussed here and one may review Chapter 2 of Morkoç (2008) for more details.

The structure and the first Brillouin zone of a wurtzite and zincblende crystal along with the irreducible wedges, calculated using the local density approximation (LDA) within the FP-LMTO method at the experimental lattice constant and optimized u value, are displayed in Figure 1.8a and b, respectively. In a crystal with Wz symmetry, the conduction band wavefunctions are formed of the atomic s-orbitals and transform the Γ point congruent with the Γ_7 representation of the space group C_{6v}^4. The upper valence band states are constructed out of appropriate linear combinations of products of p^3-like (p_x, p_y, and p_z-like) orbitals with spin functions.

Under the influence of the crystal field and spin–orbit interactions, the hallmark of the wurtzite structure, the sixfold degenerate Γ_{15} level associated with the cubic system, splits into Γ_9^v, upper Γ_7^v, and lower Γ_7^v levels (Figure 1.9).

The influence of the crystal field splitting, which is present only in the wurtzite structure, transforms the semiconductor from ZB to Wz, which is represented in the section on the left-hand side in Figure 1.9. The crystal field splits the Γ_{15} band of the ZB structure into Γ_5 and Γ_1 states of the wurtzite structure. These two states are further split into Γ_9^v, upper Γ_7^v, and lower Γ_7^v levels by spin–orbit interactions. Application of the spin–orbit splitting, from right to left, splits the Γ_{15} band of the ZB crystal into Γ_8 and Γ_7 states, while the crystal possesses the zincblende symmetry. Application of a crystal field further splits these states into Γ_9^v, upper Γ_7^v, and lower Γ_7^v levels, and the crystal now possesses the wurtzite symmetry.

We should mention that a carryover tradition from the zincblende nomenclature is still used for wurtzite symmetry by referring to the crystal filed split-off band with the nomenclature "SO" as if it is the spin–orbit split-off band because it happens to be the farthest from the HH band. In the zincblende symmetry, the crystal field splitting is nonexistent, making the top of the valence band degenerate, and the spin–orbit splitting is large. Portions of this book, unfortunately, participate in the misuse of this nomenclature. The solace is that the reader has been warned.

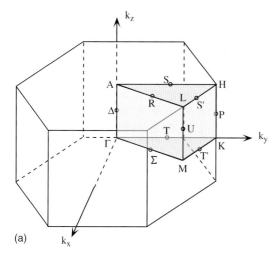

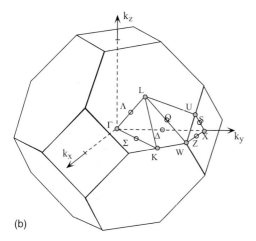

Figure 1.8 (a) Structure and the first Brillouin zone of a wurtzite crystal. Schematics of the irreducible wedges of WZ structure, indicating the high-symmetry points and lines. The U_{min} point of the WZ phase is located on the M–L line at two-thirds a distance away from the M point. (b). Structure and the first Brillouin zone of a zincblende crystal. Schematics of the irreducible wedges of ZB structure indicating the high-symmetry points and lines.

Figure 1.10 shows the dispersion of the uppermost valence band and lowermost conduction band structures in Wz GaN (a and b) and ZB GaN (c) ((a) near the Γ band for WZ, (b) including M, L, and A minima in WZ, and (c) including Γ, L, and X minima in ZB GaN).

Without the spin–orbit interaction, the valence band would consist of three doubly degenerate bands: HH, LH, and CH bands. The spin–orbit interaction removes this degeneracy and yields six bands.

20 | *1 General Properties of Nitrides*

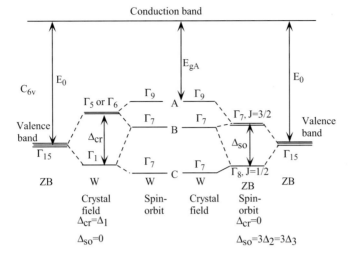

Figure 1.9 Schematic representation of the splitting of the valence band in Wz crystals due to crystal field and spin–orbit interaction. From left to right, the crystal field splitting is considered first. From right to left, the spin–orbit splitting is considered first. Regardless of which is considered first, the end result is the same in that there are three valence bands that are sufficiently close to one another for band mixing to be nonnegligible.

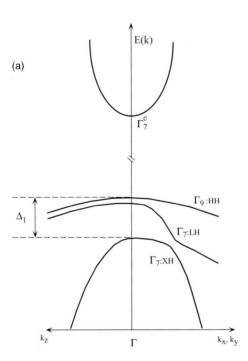

Figure 1.10 (Continued)

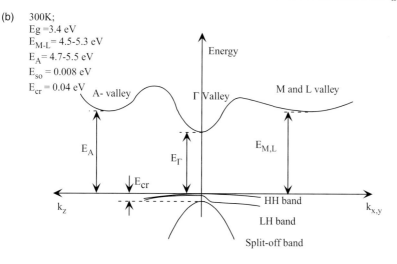

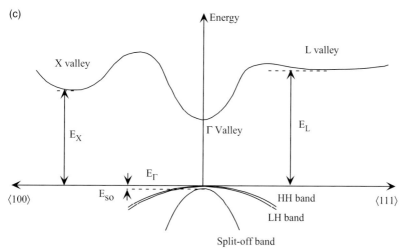

Figure 1.10 (a) Schematic representation of the Γ point valence and conduction bands in crystal with wurtzite symmetry such as GaN where the spin–orbit splitting leads to the bands labeled as HH and LH. The band caused by splitting due to crystal field is labeled as XH. Courtesy of M. Suzuki. (b). Schematic representation of the band diagram for wurtzite GaN showing the separation between the Γ, A, and M–L band symmetry points at 300 K. The values with respect to the top of the valence band are $E_\Gamma = 3.4$ eV, $E_{M-L} = 4.5–5.3$ eV, $E_A = 4.7–5.5$ eV, $E_{so} = 0.008$ eV, $E_{cr} = 0.04$ eV, $E_\Gamma = 3.4$ eV, $E_{M-L} = 4.5–5.3$ eV, $E_A = 4.7–5.5$ eV, $E_{so} = 0.008$ eV, $E_{cr} = 0.04$ eV. The values of $E_\Gamma = 6$ eV, $E_{M-L} = 7$ eV, and $E_A = 8$ eV. (c). Schematic representation of the band diagram for zincblende GaN showing the separation between the Γ, X, and L band symmetry points at 300 K. The values with respect to the top of the valence band are $E_\Gamma = 3.2$ eV, $E_L = 4.8–5.1$ eV, $E_X = 4.6$ eV, $E_{so} = 0.008$ and 0.02 eV. Note that in the ZB structure, the valence band is degenerate. The values of $E_\Gamma = 3.2$ eV, $E_L = 5.1$ eV, and $E_X = 4.3$ eV.

1.9.2
General Strain Considerations

Strain–stress relationship or Hooke's law describes the deformation of a crystal ε_{kl} due to external or internal forces or stresses σ_{ij},

$$\sigma_{ij} = \sum_{k,l} C_{ijkl}\varepsilon_{kl}, \tag{1.13}$$

where C_{ijkl} is the fourth ranked elastic tensor and represents the elastic stiffness coefficients in different directions in the crystal, which due to the C_{6v} symmetry can be reduced to a 6×6 matrix using the Voigt notation: $xx \rightarrow 1$, $yy \rightarrow 2$, $zz \rightarrow 3$, yz, $zy \rightarrow 4$, zx, $xz \rightarrow 5$, xy, $yx \rightarrow 6$. The elements of the elastic tensor can be rewritten as $C_{ijkl} = C_{mn}$ with $i, j, k, l = x, y, z$ and $m, n = 1, \ldots, 6$. With this notation, Hooke's law can be reduced to

$$\sigma_i = \sum_j C_{ij}\varepsilon_j, \tag{1.14}$$

or in the expanded form

$$\begin{vmatrix} \sigma_{xx} \\ \sigma_{yy} \\ \sigma_{zz} \\ \sigma_{xy} \\ \sigma_{yz} \\ \sigma_{zx} \end{vmatrix} = \begin{vmatrix} C_{11} & C_{12} & C_{13} & 0 & 0 & 0 \\ C_{12} & C_{22} & C_{13} & 0 & 0 & 0 \\ C_{13} & C_{13} & C_{33} & 0 & 0 & 0 \\ 0 & 0 & 0 & C_{44} & 0 & 0 \\ 0 & 0 & 0 & 0 & C_{55} & 0 \\ 0 & 0 & 0 & 0 & 0 & C_{66} \end{vmatrix} \begin{vmatrix} \varepsilon_{xx} \\ \varepsilon_{yy} \\ \varepsilon_{zz} \\ \varepsilon_{xy} \\ \varepsilon_{yz} \\ \varepsilon_{zx} \end{vmatrix} \tag{1.15}$$

with $C_{66} = (C_{11} - C_{12})/2$. If the crystal is strained in the (0001) plane and allowed to expand and constrict in the [0001] direction, the $\sigma_{zz} = \sigma_{xy} = \sigma_{yz} = \sigma_{zx} = 0$, $\sigma_{xx} \neq 0$ and $\sigma_{yy} \neq 0$, and the strain tensor has only three nonvanishing terms, namely,

$$\varepsilon_{xx} = \varepsilon_{yy} = \frac{a - a_0}{a_0}, \quad \varepsilon_{zz} = \frac{c - c_0}{c_0} = -\frac{C_{13}}{C_{33}}(\varepsilon_{xx} + \varepsilon_{yy}), \tag{1.16}$$

where a and a_0 and c and c_0 represent the in-plane and out-of-plane lattice constants of the epitaxial layer and the relaxed buffer (substrate), respectively. The above assumes that the in-plane strains in x- and y-directions are identical, namely, $\varepsilon_{xx} = \varepsilon_{yy}$. When the crystal is uniaxially strained in the (0001) c-plane and free to expand and constrict in all other directions, σ_{zz} is the only nonvanishing stress term and the strain tensor is reduced to

$$\begin{vmatrix} \varepsilon_{yy} \\ \varepsilon_{zz} \end{vmatrix} = \frac{1}{C_{13}^2 - C_{11}C_{33}} \begin{vmatrix} C_{12}C_{33} - C_{13}^2 \\ C_{11}C_{13} - C_{12}C_{13} \end{vmatrix} \varepsilon_{xx}. \tag{1.17}$$

Lack of any force in the growth direction and the fact that the crystal can relax freely in this direction lead to a biaxial strain $\varepsilon_1 = \varepsilon_2$, which in turn causes $\sigma_1 = \sigma_2$, with $\sigma_3 = 0$.

The internal strain is defined by the variation of the internal parameter under strain $(u - u_0)/u_0$. In the limit of small deviations from the equilibrium, Hooke's law gives the corresponding diagonal stress tensor σ with the elements:

$$\sigma_{xx} = \sigma_{yy} = (C_{11} + C_{12})\varepsilon_{xx} + C_{13}\varepsilon_{zz}, \quad \sigma_{zz} = 2C_{13}\varepsilon_{xx} + C_{33}\varepsilon_{zz}. \quad (1.18)$$

In Equation 1.18, four of the five independent stiffness constants C_{ij} of the wurtzite crystal are involved. The modification of Equation 1.18 by the built-in electric field due to the spontaneous and piezoelectric polarizations is neglected, as the effect is small.

In the case of uniaxial stress, for example, along the c-direction, there is an elastic relaxation of the lattice in the c-plane. The ratio of the resulting in-plane strain to deformation along the stress direction is expressed by the Poisson ratio, which in general can be anisotropic. For the wurtzite lattice subjected to a uniaxial stress σ_{zz} parallel to the c-axis, $\sigma_{xx} = \sigma_{yy} = 0$ holds. Then, Equation 1.18 gives the relation:

$$\varepsilon_{xx} = -[C_{13}/(C_{11} + C_{12})]\varepsilon_{zz} = -\nu\varepsilon_{zz}, \quad (1.19)$$

with $\nu = C_{13}/(C_{11} + C_{12})$ being the Poisson's ratio.

1.9.3
k·p Theory and the Quasicubic Model

The conduction and valence bands of nitride semiconductors are comprised of s- and p-like states, respectively. Unlike the conventional ZB III–N semiconductors and the lack of a high degree of symmetry, the crystal field present removes the degeneracy at the top of the conduction band. The spin–orbit splitting is very small and makes all three bands in the valence band closely situated in energy, making an 8×8 k·p Hamiltonian imperative. Because the bandgaps of nitrides are very large, the coupling between the conduction and valence bands can be treated as a second-order perturbation that allows the 8×8 Hamiltonian to be split into one 6×6 Hamiltonian dealing with the valence band and another 2×2 Hamiltonian dealing with the conduction band. The conduction band dispersion relation is

$$E(k) = E_{c0} + \frac{\hbar^2 k_z^2}{2m_c^{//}} + \frac{\hbar^2(k_x^2 + k_y^2)}{2m_c^{\perp}} + a_c^{\perp}(\varepsilon_{xx} + \varepsilon_{yy}) + a_c^{//}(\varepsilon_{zz}), \quad (1.20)$$

where $a_c^{\perp}$ and $a_c^{//}$ represent the in-plane and out-of-plane deformation potentials, respectively. For an isotropic parabolic conduction band, Equation 1.20 reduces to

$$E(k) = E_{c0} + \frac{\hbar^2 k^2}{2m_c} + a_c\varepsilon. \quad (1.21)$$

E_{c0} is the conduction band energy at the $k=0$ point, ε is the strain, and a_c is the deformation potential for the conduction band. The other terms have their usual meanings. It should be pointed out the system under discussion is a linear. The details of the Hamiltonian mentioned above and related manipulations can be found in Morkoç (2008) and references therein.

Results of the above-mentioned calculations are shown in Figure 1.11 with Luttinger-like parameter $A_7 = 93$ meV. The effective masses for all the three valence

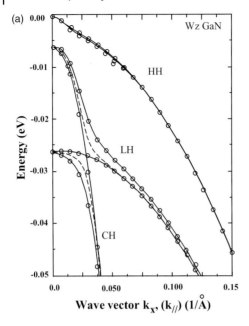

Figure 1.11 Valence band structures of wurtzite GaN with the k·p theory fitting including the spin–orbit interaction with $A_7 = 93.7$ meV Å in the solid line. The empirical pseudopotential method (EPM) calculation data. Courtesy of Y.C. Yeo, T.C. Chong, M.F. Li, G.B. Ren, Y.M. Liu, and P. Blood.

bands, both parallel and perpendicular, $m^{//}$ and $m^{\perp}$, also ensue from such calculations.

The effective masses can be expressed in terms of their dependence on the Luttinger-like parameters:

$$m_0/m_{hh}^{//} = -(A_1 + A_3),$$
$$m_0/m_{lh}^{//} = -(A_1 + A_3), \quad (1.22)$$
$$m_0/m_{so}^{//} = -A_1.$$

As indicated in the schematic of Figure 1.9, both the spin–orbit and the crystal field splitting affect the structure of the valence band in wurtzitic crystals. Typically the relevant parameters are correlated to one another as $\Delta_{so} = 3\Delta_2 = 3\Delta_3$ (in spite of the fact that a small Δ_2/Δ_3 anisotropy has sometimes been reported and $\Delta_{cr} = \Delta_1$). Experimentally, the splitting parameters are obtained from the energy differences of the A, B, and C free excitons, which have nonlinear dependencies on the various splittings. It should be pointed out that the nomenclature for the three valence bands for hexagonal system is A, B, and C for HH, LH, and SO (CH) bands when including A_7 terms because spin splitting and strain can significantly alter which band of eigenstates is "heavy" or "light" at various k-values, particularly in the c-plane. Experiments led to values of $\Delta_{cr} = 16$ meV and $\Delta_{so} = 12$ meV, and $\Delta_{cr} = 25$ meV and $\Delta_{so} = 17$ meV, the latter set from a fit to exciton energies, A and B determined by PL and C determined by reflection, but with a geometry not fully ideal in terms that some error is introduced in the value of C exciton energy. There is a sizable dispersion in the calculations reported so far particularly in the value of Δ_{cr}. This treatment is provided for its simplicity; for more accurate data, the reader should refer to full band calculations provided.

A practical approach is to modify the well-established treatment developed for the cubic system and make it applicable to the Wz system, which is referred to as quasicubic approximation. The genesis of the quasicubic approximation relies on the fact that both the Wz and ZB structures are tetrahedrally coordinated and hence are closely related. The nearest-neighbor coordination is the same for Wz and ZB structures, but the next-nearest-neighbor positions differ between the two systems. The basal plane (0001) of the Wz structure corresponds to one of the (111) planes of the ZB. When the in-plane hexagons are lined up in Wz and ZB structures, the Wz [0001], [11$\bar{2}$0], and [1$\bar{1}$00] planes are parallel to ZB [111], [10$\bar{1}$], and [1$\bar{2}$1] planes. This, in turn, leads to correlations between the symmetry direction and the k-points for the two polytypes. There are, however, twice as many atoms in the Wz unit cell as there are in the ZB one. In addition to the band structure similarities between the doubled ZB and Wz structures, one can establish a correlation between the Luttinger parameters in the ZB system and parameters of interest in the Wz system by taking the z-axis along the [111] direction and the x- and y-axes along the [11$\bar{2}$] and [$\bar{1}$10] directions, respectively. For details regarding the symmetry relations between the ZB and Wz polytypes, refer to Morkoç (2008) and references therein.

The large effective mass and the small dielectric constant of GaN, relative to more conventional group-III–V semiconductors, lead to relatively large exciton binding energies and make excitons, together with large exciton recombination rates, clearly observable even at room temperature. The bottom of the conduction band of GaN is predominantly formed from the s-levels of Ga and the upper valence band states from the p-levels of N. Even though sophisticated methods have been introduced and discussed, the method of *Hopfield* and *Thomas*, which treats the wurtzite energy levels as a perturbation to the zincblende structure, is discussed briefly as it provides a physical picture of band splitting in the valence band. Using the quasicubic model of *Hopfield*, one obtains

$$E_1 = 0, \tag{1.23}$$

$$E_2 = \frac{\delta + \Delta}{2} + \sqrt{\left[\left(\frac{\delta + \Delta}{2}\right)^2 - \frac{2}{3}\delta\Delta\right]}, \tag{1.24}$$

$$E_3 = \frac{\delta + \Delta}{2} - \sqrt{\left[\left(\frac{\delta + \Delta}{2}\right)^2 - \frac{2}{3}\delta\Delta\right]}, \tag{1.25}$$

where δ and Δ represent the contributions of uniaxial field and spin–orbit interactions, respectively, to the splittings $E_{1,2}$ and $E_{2,3}$.

1.9.4
Temperature Dependence of Wurtzite GaN Bandgap

The temperature dependence of the bandgap in semiconductors is often described by an imperial expression (assuming no localization),

$$E(T) = E(0) - \alpha T^2/(\beta + T). \tag{1.26}$$

In the case of localization, which can also be construed as bandtail effect, the temperature dependence deviates from the above equation. In the framework of the bandtail model and Gaussian-like distribution of the density of states for the conduction and valence bands, the temperature-dependent emission energy could be described by the following modified expression, which is based on a model developed for Stokes shift in GaAs/AlGaAs QWs:

$$E(T) = E(0) - [\alpha T^2/(\beta + T)] - [\sigma^2/(kT)], \tag{1.27}$$

where the last term represents the localization component with σ indicating the extent of localization or bandtailing, which is nearly imperative for In-containing alloys. The parameters α is in units of energy over temperature and β is in units of temperature.

1.9.5
Sphalerite (Zincblende) GaN

The valence band of zincblende GaN has been the topic of various theoretical efforts. Although the hole effective masses in zincblende GaN have apparently not been measured, a number of theoretical predictions of Luttinger parameters are available in the literature (Table 1.9) (refer to Morkoç (2008) and references therein). Once the Luttinger parameters are known, the full picture in terms of the hole effective masses can be determined. First, it should be pointed out that in polar semiconductors such as the III–V compounds in general and GaN in particular, it is the nonresonant polaron mass that is actually measured. The polaron mass exceeds the bare electron mass by about 1–2%, the exact value of which depends on the strength

Table 1.9 Luttinger parameters γ_1, γ_2, and γ_3 for zincblende GaN obtained from a fit along the [110] direction.

Parameter	Value
γ_1	2.67–3.07
γ_2	0.75–0.90
γ_3	1.07–1.26

1.9 Electronic Band Structure and Polarization Effects

of the electron–phonon interaction. Because the band structure is governed by the *bare* electron mass, this is the quantity that is typically reported whenever available.

At the valence band edge, the heavy hole (hh) effective masses in the different crystallographic directions are related to the free mass through the Luttinger parameters in the following manner:

$$(m_0/m^*_{hh/lh})^{[100]} = \gamma_1 \pm 2\gamma_2,$$
$$(m_0/m^*_{hh/lh})^{[111]} = \gamma_1 \pm 2\gamma_3, \qquad (1.28)$$
$$(m_0/m^*_{hh/lh})^{[1100]} = \frac{2\gamma_1 \pm \gamma_2 \pm \gamma_3}{2}.$$

Here, the z-direction is perpendicular to the growth plane of (001). These expressions described by Equation 1.28 show the relationship of the Luttinger parameters to the hh effective masses that can typically be measured in a more direct manner. The light hole (lh) and so hole effective masses are given by

$$m^z_{lh} = \frac{m_0}{\gamma_1 + 2\gamma_2}, \quad m^{[110]}_{lh} = \frac{2m_0}{2\gamma_1 + \gamma_2 + 3\gamma_3}, \quad m^{[111]}_{lh} = \frac{m_0}{\gamma_1 + 2\gamma_3}, \qquad (1.29)$$

$$\frac{m_0}{m_{so}} = \gamma_1 - \frac{E_P \Delta_{so}}{3E_g(E_g + \Delta_{so})}. \qquad (1.30)$$

To restate, although the hole effective masses in zincblende GaN have apparently not been measured, a number of theoretical predictions of Luttinger parameters are available in the literature (Table 1.10). The values are based on averages of the heavy hole and light hole masses along [001], as well as the degree of anisotropy in $\gamma_3 - \gamma_2$. The parameter set used is $\gamma_1 = 2.70$, $\gamma_2 = 0.76$, and $\gamma_3 = 1.11$. Similarly, averaging all the reported split-off masses leads to $m^*_{so} = 0.29 m_0$. In its simplest form, the Luttinger parameters can be used to determine quickly the effective masses in various valence bands both in equilibrium and also under biaxial strain. In fact, with biaxial strain, the valence band degeneracy can be removed and, most strikingly, the heavy hole in-plane mass can be made smaller with compressive strain, a notion that has been exploited in the InGaAs/GaAs system very successfully, particularly for low-threshold lasers.

Table 1.10 Effective masses for electrons (e), and heavy holes (hh), light holes (lh) and spin–orbit split-off holes (so) in units of the free electron mass m_0 along the [100], [111], and [110] directions for zincblende GaN.

m_e	$m^{[100]}_{hh}$	$m^{[100]}_{lh}$	$m^{[111]}_{hh}$	$m^{[111]}_{lh}$	$m^{[110]}_{hh}$	$m^{[110]}_{lh}$	m_{so}
0.14	0.84	0.22	2.07	0.19	1.52	0.20	0.35
0.14	0.86	0.21	2.09	0.19	1.65	0.19	0.30
0.13	0.76	0.21	1.93	0.18	1.51	0.19	0.32
0.17	0.85	0.24	1.79	0.21	1.40	0.21	0.37
0.15	0.85	0.24	2.13	0.21	1.55	0.21	0.29
0.12	1.34	0.70	1.06	0.63	1.44	0.58	0.20

1.9.6
AlN

AlN forms the larger bandgap binary used in conjunction with GaN for increasing the bandgap for heterostructures. As in the case of GaN, AlN also has wurtzitic and zincblende polytypes, the latter being very unstable and hard to synthesize. Owing to increasing interest in solar blind devices, UV emitters and detectors, and the expectation that AlGaN with large mole fractions of AlN would have relatively large breakdown properties, this material has been steadily gaining attention. It should also be mentioned that the N overpressure on Al is the smallest among those over Ga and In, paving the way for equilibrium growth of AlN bulk crystals, albeit not without O contamination.

1.9.6.1 Wurtzite AlN

Wurtzite AlN is a direct bandgap semiconductor with a bandgap near 6.1 eV and still considered to be semiconductor because it is dopable. The zincblende polytype is not stable with a predicted indirect bandgap. The AlN derive its technological importance from its providing the large bandgap binary component of the AlGaN alloy, which is commonly employed in both optoelectronic and electronic devices based on the GaN semiconductor system.

Absorption measurements carried out early on indicated a large energy gap of 6.1 eV for wurtzite AlN at 5 K and about 6 eV at room temperature. Averaging all the available theoretical crystal field splittings, one obtains a value of $\Delta_{cr} = -169$ meV. Using optical reflectance data performed on a- and c-plane bulk AlN and a quasicubic model developed for the wurtzite crystal structure, the crystal field splitting was determined to be $\Delta = -225$ meV. *Note that the negative sign for the crystal field splitting has important implications, namely, that the Γ_7 valence band is on top of Γ_9 valence band, which is opposite of that in GaN.* As for the spin–orbit splitting, the literature values range from 11 to 36 meV, the latter having been determined by optical reflectance spectra in high-quality bulk AlN. The bare mass values of $m_e^\perp = 0.30 m_0$ and $m_e^{//} = 0.32 m_0$ obtained by averaging the available theoretical masses may represent a good set of default values at this stage.

The effective masses for both the conduction band and various valence bands are compiled in Table 1.11.

1.9.6.2 Zincblende AlN

Due to lack of sufficient experimental data, the treatment here primarily relies on theoretical projections. The only quantitative experimental study of the bandgap indicated a Γ valley indirect gap of 5.34 eV at room temperature. Assuming that the Varshni parameters for the wurtzitic AlN hold for the zincblende polytype, the aforementioned room-temperature bandgap translates to a low-temperature gap of 5.4 eV. The values for the X and L valley gaps are 4.9 and 9.3 eV. The spin–orbit splitting is expected to be nearly the same as in wurtzite AlN at 19 meV. Averaging the available theoretical results one arrives at a Γ valley effective mass of $0.25 m_0$. The longitudinal and transverse masses for the X valley have been predicted to be $0.53 m_0$

Table 1.11 Effective masses and band parameters for wurtzitic AlN.

Parameter	aniso	iso	a)	b)	c)	d)	e)	f)
$m_e^{\parallel}$	0.231	0.232	0.33	0.33	0.24	0.24	0.35	0.33
$m_e^{\perp}$	0.242	0.242	0.25	0.25	0.25	0.25		
$m_{hh}^{\parallel}$	2.370	2.382	3.68	3.53	1.949	1.869	3.53	4.41
$m_{lh}^{\parallel}$	2.370	2.382	3.68	3.53	1.949	1.869	3.53	4.41
$m_{ch}^{\parallel}$	0.209	0.209	0.25	0.25	0.229	0.212	0.26	0.27
$m_{hh}^{\perp}$	3.058	3.040	6.33	10.42	2.584	2.421	11.14	2.18
$m_{lh}^{\perp}$	0.285	0.287	0.25	0.24	0.350	0.252	0.33	0.29
$m_{ch}^{\perp}$	1.204	1.157	3.68	3.81	0.709	1.484	4.05	4.41
A_1	−4.789	−4.794	−3.95	−4.06	−4.367	−4.711	−3.86	−3.74
A_2	−0.550	−0.571	−0.27	−0.26	−0.518	−0.476	−0.25	−0.23
A_3	4.368	4.374	3.68	3.78	3.854	4.176	3.58	3.51
A_4	−1.511	−1.484	−1.84	−1.86	−1.549	−1.816	−1.32	−1.76
A_5	1.734	1.726	1.95	2.02	1.680	1.879	1.47	1.52
A_6	−1.816	−1.788	−2.91		−2.103	−2.355	1.64	1.83
A_7	0.134	0.153	0	0	0.204	0.096		0
Δ_1	−0.128	−0.160	−0.059	−0.059	−0.093	−0.093	−0.215	

Effective masses in units of free electron mass m_0, Luttinger-like parameters A_i ($i = 1, \ldots, 6$) in units of $\hbar^2/2m_0$, and A_7 in units of eV Å The crystal field splitting energy Δ_1 is given in units of meV. The term *aniso* represents the values derived using a band structure calculation with anisotropically screened model potentials, whereas the term *iso* describes a comparative band structure calculation on the basis of isotropically screened model potentials using an averaged ε_0 value by taking the spur of the dielectric tensor. Courtesy of D. Fritsch and M. Grundmann.
a) Full-potential linearized augmented plane wave (FPLAPW) band structure calculations.
b) Another FPLAPW band structure calculations.
c) A_i obtained through a Monte Carlo fitting procedure to the band structure and effective masses.
d) Direct k·p calculations for A_i and effective masses obtained from A_i.
e) Direct fit of A_i to first-principles band structures.
f) A_i and effective masses obtained in the quasicubic model from zincblende parameters.

and $0.31m_0$, respectively. If the method used previously for the GaN is applied to zincblende AlN, one arrives at recommended Luttinger parameters of $\gamma_1 = 1.92$, $\gamma_2 = 0.47$, and $\gamma_3 = 0.85$, and $m_{so} = 0.47m_0$. These as well as the other literature values of the Luttinger parameters are listed in Table 1.12.

The calculated effective masses for conduction and valence bands, the latter involving the light and heavy holes, as well as the spin–orbit split-off mass, and taking the anisotropy into account, are listed in Table 1.13 for zincblende AlN.

1.9.7
InN

As in the case of AlN, the main impetus in InN is due to the InGaN alloy that is used in lasers and LEDs operative in the visible and violet regions of the optical spectrum. The properties, particularly the fundamental parameters of InGaN for a given composition, depend very much on the InN parameters, particularly on its bandgap.

Table 1.12 Luttinger parameters γ_1, γ_2, and γ_3 for zincblende AlN obtained from a fit along the [110] direction in addition to those available in the literature.

Parameter	a)	b)	c)	d)	e)
γ_1	1.85	1.54	1.50	1.91	1.92
γ_2	0.43	0.42	0.39	0.48	0.47
γ_3	0.74	0.64	0.62	0.74	0.85

Courtesy of D. Fritsch, and M. Grundmann.
a) Empirical pseudopotential calculations.
b) Self-consistent FPLAPW method within LDA.
c) First-principles band structure calculations.
d) Empirical pseudopotential calculations.
e) Recommended values.

The initially accepted bandgap value of 1.98 eV has given way to about 0.7 eV for Wz InN, which motivated some to consider this material system for photovoltaic cells. Estimates of the crystal field splitting in wurtzite InN range from 17 to 301 meV, but a value of 40 meV can be adopted. Based on the calculations, spin–orbit splittings vary from 1 to 13 meV, but a value of $\Delta_{so} = 5$ meV can be assumed.

Turning our attention to other electronic properties affected by the band structure, measurements of the electron effective mass in wurtzitic InN produced values of 0.11–0.14m_0. Accounting for the substantial nonparabolicity that can cause an overestimate of the mass because of high doping leads to a recommended band edge effective mass of 0.07m_0. The dispersion in the effective mass for both the conduction band and various valence bands as obtained by various computational methods and the parameters used in the description of the bandgap for wurtzitic InN, particularly in the context of empirical pseudopotential method, are compiled in Table 1.14.

Table 1.13 Effective masses for electrons (e), heavy holes (hh), light holes (lh), and spin–orbit split-off holes (so) in units of the free electron mass m_0 along the [100], [111], and [110] directions for zincblende AlN.

Methodology	m_e	$m_{hh}^{[100]}$	$m_{lh}^{[100]}$	$m_{hh}^{[111]}$	$m_{lh}^{[111]}$	$m_{hh}^{[110]}$	$m_{lh}^{[110]}$	m_{so}
a)	0.23	1.02	0.37	2.64	0.30	1.89	0.32	0.54
b)	0.28	1.44	0.42	4.24	0.36	3.03	0.37	0.63
c)	0.21	1.05	0.35	2.73	0.30	2.16	0.31	0.51
d)	0.30	1.39	0.44	3.85	0.36	2.67	0.38	0.67
e)	0.25	1.02	0.35	4.55	0.28	2.44	0.29	0.47

Courtesy of D. Fritsch, and M. Grundmann.
a) Pseudopotential calculations.
b) Self-consistent FPLAPW method within LDA.
c) Empirical pseudopotential calculations.
d) Calculated from Luttinger parameters.
e) Calculated from recommended Luttinger parameters.

Table 1.14 Effective masses and band parameters for wurtzitic InN.

Parameters	Anisotropic	Isotropic	a)	b)	c)	d)
$m_e^{//}$	0.138	0.137	0.11	0.11	0.10	0.10
$m_e^{\perp}$	0.141	0.140	0.10	0.10	0.10	0.10
$m_{hh}^{//}$	2.438	2.493	1.56	1.67	1.431	1.350
$m_{lh}^{//}$	2.438	2.493	1.56	1.67	1.431	1.350
$m_{ch}^{//}$	0.140	0.137	0.10	0.10	0.106	0.092
$m_{hh}^{\perp}$	2.661	2.599	1.68	1.61	1.410	1.449
$m_{lh}^{\perp}$	0.148	0.157	0.11	0.11	0.196	0.165
$m_{ch}^{\perp}$	3.422	1.446	1.39	1.67	0.209	0.202
A_1	−7.156	−7.298	−9.62	−9.28	−9.470	−10.841
A_2	−0.244	−0.441	−0.72	−0.60	−0.641	−0.651
A_3	6.746	6.896	8.97	8.68	8.771	10.100
A_4	3.340	3.064	4.22	4.34	4.332	−4.864
A_5	−3.208	−3.120	−4.35	−4.32	−4.264	−4.825
A_6	−4.303	−3.948		−6.08	−5.546	−6.556
A_7	0.072	0.103	0	0	0.278	0.283
Δ_1	0.214	0.084			0.0375	0.0375

Effective masses in units of free electron mass m_0, Luttinger-like parameters A_i ($i = 1, \ldots, 6$) in units of $\hbar^2/2m_0$, and A_7 in units of eV Å The crystal field splitting energy Δ_1 is given in units of meV. The term *aniso* represents the values derived using a band structure calculation with anisotropically screened model potentials, whereas the term *iso* describes a comparative band structure calculation on the basis of isotropically screened model potentials using an averaged ε_0 value by taking the spur of the dielectric tensor. Courtesy of D. Fritsch, and M. Grundmann. Aniostropically screened (*aniso*) and isotropically screened (*iso*) values.
a) Effective masses and A_i obtained through a line fit to the band structure.
b) Direct k·p calculation in a 3D fit.
c) A_i obtained through a Monte Carlo fitting procedure to the band structure and effective masses.
d) Direct k·p calculations for A_i.

1.10
Polarization Effects

Group III–V nitride semiconductors exhibit highly pronounced polarization effects, the genesis of which has to do with large electronegativity in the case of single layers and with differential electronegativity in the case of heterostructures. Semiconductor nitrides lack center of inversion symmetry and exhibit large piezoelectric effects when strained along $\langle 0001 \rangle$. The strain-induced piezoelectric charge and spontaneous polarization, the latter caused by compositional gradient, have profound effects on device structures. Spontaneous polarization was understood fully only recently.

Polarization is dependent on the polarity of the crystal, namely, whether the bonds along the c-direction are from cation sites to anion sites, or vice versa. The convention is that when the single bonds are from cation (Ga) to anion (N) atoms, the [0001] axis points from the face of the N plane to the Ga plane and marks the positive z-direction, or c^+- or $+z$-direction. When the single bonds are along the c-direction or from anion (N) to cation (Ga) atoms, the polarity is said to be the

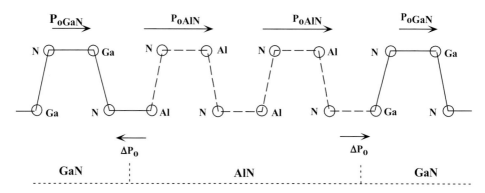

Figure 1.12 Schematic depicting the convention used for determining the polarity and crystalline direction in wurtzitic nitride films. The diagram shows the case for a Ga-polarity film with its characteristic bonds parallel to the c-axis (horizontal in the figure) going from the cation (Ga or Al) to the anion (N). The spontaneous polarization components P_{0Ga} and P_{0AlN} for a periodic GaN/AlN structure are also indicated with that for AlN having a larger magnitude. The spontaneous polarization is negative and thus points in the [000$\bar{1}$] direction. Caution must be exercised here as there is no long-range polarization field, just that it is limited to the interface. The polarization in AlN is larger in magnitude than that in GaN. There exists a difference in polarization at the interface, ΔP_0 pointing in the [000$\bar{1}$] direction for both GaN–AlN interfaces. The axial Born factor is defined as $Z^* = (\sqrt{3}a_0^2/4q)(\partial P_3/\partial u) = Z_3^T$. Courtesy of V. Fiorentini.

N-polarity, and the direction is said to be the c^-- or $-z$-direction. A schematic representation of the spontaneous polarization in a model GaN/AlN/GaN wurtzitic crystal is shown in Figure 1.12.

The magnitude of the polarization charge, converted to number of electrons, can be in the mid 10^{13} cm^{-2} level for AlN–GaN heterointerfaces, which is huge by any standard, some 10 times larger than the doping-induced electron density in the GaAs/AlGaAs system. The magnitude of the polarization charge is compiled in Table 1.15 along with elastic coefficients taken from the literature.

The data in bold letters are recommended.

1.10.1
Piezoelectric Polarization

In a polarizable medium, the displacement vector can be expressed in terms of two components due to both the dielectric and polarizability nature of the medium:

$$\vec{D} = \varepsilon \vec{E} + 4\pi \vec{P} \text{ in cgs} \quad \text{and} \quad \vec{D} = \varepsilon \vec{E} + \vec{P} \text{ in mks units,} \tag{1.31}$$

where $\vec{E}$ and $\vec{P}$ represent the electric field and polarization vectors, respectively. Considering only the piezoelectric component, the piezoelectric polarization vector

Table 1.15 Elastic constants and spontaneous polarization charge in nitride semiconductors.

	AlN	GaN	InN
$e_{33}*$ (C/m^2)	1.46	0.73	0.97
LDA	1.8	0.86	1.09
GGA	**1.5**	**0.67**	**0.81**
e_{31} (C/m^2)	−0.60	−0.49	−0.57
LDA	−0.64	−0.44	−0.52
GGA	**− 0.53**	**−0.34**	**−0.41**
e_{31}^p LDA	−0.74	−0.47	−0.56
GGA	**−0.62**	**−0.37**	**−0.45**
C_{33} (Gpa) GGA	377	354	205
C_{31} (Gpa) GGA	94	68	70
P_0 (C/m^2)	−0.081	−0.029	−0.032
LDA	−0.10	−0.032	−0.041
GGA	**−0.090**	**−0.034**	**−0.042**
P_0 (C/m^2), ideal wurtzite structure	−0.032	−0.018	−0.017
$R = -C_{31}/C_{33}$ LDA	−0.578	−0.40	−0.755
GGA	**−0.499**	**−0.384**	**−0.783**
$[e_{31} - (C_{13}/C_{33})e_{33}]$ in (C/m^2)	−0.86	−0.68	−0.90

The data in bold letters are associated with density functional theory (DFT) in the generalized gradient approximation (GGA) that are more accurate than others reported earlier. Moreover, the resultant predictions are in relatively better agreement with experimental data as well as the bowing parameters observed in polarization charge in alloys. Courtesy of F. Bernardini and V. Fiorentini. e_{31} and e_{33} are piezoelectric constants. C_{31} and C_{33} are elastic stiffness coefficients or elastic constants. e_{31}^p is the proper piezoelectric constant.

is given by

$$\vec{P}_{PE} = \overleftrightarrow{e}\,\vec{\varepsilon}, \tag{1.32}$$

where $\overleftrightarrow{e}$ and $\vec{\varepsilon}$ are the piezoelectric and stress tensors, respectively. In hexagonal symmetry, electric polarization is related to strain through electric piezoelectric tensor:

$$\begin{pmatrix} P_x \\ P_y \\ P_z \end{pmatrix} = \begin{pmatrix} 0 & 0 & 0 & 0 & e_{15} & 0 \\ 0 & 0 & 0 & e_{24} & 0 & 0 \\ e_{31} & e_{31} & e_{33} & 0 & 0 & 0 \end{pmatrix} \begin{pmatrix} \varepsilon_{xx} \\ \varepsilon_{yy} \\ \varepsilon_{zz} \\ \varepsilon_{yx} \\ \varepsilon_{zx} \\ \varepsilon_{xz} \end{pmatrix}, \tag{1.33}$$

where P_i, e_{ij}, and ε_{ij} represent the electric polarization, electric piezoelectric coefficient, and strain, respectively. Note that for hexagonal symmetry, $e_{24} = e_{15}$. Without shear, $\varepsilon_{yz} = \varepsilon_{zx} = \varepsilon_{xy} = 0$. For biaxial strain only,

$$\varepsilon_{xx}(\text{or } \varepsilon_{11}) = \varepsilon_{yy}(\text{or } \varepsilon_{22}) = \frac{a - a_0}{a_0}. \tag{1.34}$$

Here, $a_0 \equiv a_{\text{buffer}}$ and $a \equiv a_{\text{epi}}$ represent the relaxed (equilibrium) in-plane lattice constants of the buffer layer or the substrate, depending on layers and their thicknesses, and of the epitaxial layer of interest, the strained epitaxial layer, respectively. The expression for the out-of-plane strain is

$$\varepsilon_{33} \equiv \varepsilon_{zz} = \frac{c - c_0}{c_0}. \tag{1.35}$$

Similarly, c_0 and c represent the relaxed and the out-of-plane lattice parameters that would correspond to the buffer layer and epitaxial layer, respectively. In case the in-plane strain is anisotropic, $\varepsilon_{11} \neq \varepsilon_{22}$.

The components of the piezoelectric polarization tensor given by Equation 1.33 can be expressed in terms of a summation, using P_i^{pz} instead of P_{PE}, as

$$P_i^{pz} = \sum_j e_{ij} \varepsilon_j, \quad \text{with} \quad i = 1, 2, 3 \quad \text{and} \quad j = 1, \ldots, 6, \tag{1.36}$$

where P_i^{pz} is the ith component of the piezoelectric polarization.

The wurtzite symmetry reduces the number of independent components of the elastic tensor e to three, namely, e_{15}, e_{31}, and e_{33}, and the negligence of the shear strain makes $e_{15} = 0$. In this case the electric polarization can be expressed as

$$P_z = e_{31}\varepsilon_{xx} + e_{31}\varepsilon_{xx} + e_{33}\varepsilon_{zz} = 2e_{31}\varepsilon_{xx} + e_{33}\varepsilon_{zz}. \tag{1.37}$$

For isotropic basal plane strain, the strain component $\varepsilon_{xx} \equiv \varepsilon_\perp/2$, and thus Equation 1.37 can be written as

$$P_z \equiv P_3^{pe} = e_{31}\varepsilon_\perp + e_{33}\varepsilon_{zz}. \tag{1.38}$$

In hexagonal symmetry, strain in the z-direction can be expressed in terms of the basal plane strain $\varepsilon_\perp$ by using Poisson's ratio, which is expressed in terms of the elastic coefficients C_{ij} as $\varepsilon_{zz} = -2(C_{13}/C_{33})\varepsilon_{xx} = -(C_{13}/C_{33})\varepsilon_\perp$. In the case of externally applied pressure in addition to mismatch strain, the out-of-plane strain can be related to the in-plane strain through $\varepsilon_{zz} = -[(p + 2C_{13}\varepsilon_{xx})]/C_{33}$, where p is the magnitude of compressive pressure (in the same unit as the elastic coefficients). In terms of the nomenclature again, it should also be noted that $\varepsilon_1 \equiv \varepsilon_{11} \equiv \varepsilon_{xx}$ and $\varepsilon_3 \equiv \varepsilon_{33} \equiv \varepsilon_{zz}$ in the other notation used in the literature and also in this text:

$$P_z = \left(e_{31} - e_{33}\frac{C_{13}}{C_{33}}\right)\varepsilon_\perp. \tag{1.39}$$

Knowing the piezoelectric parameters of the end binary points is generally sufficient, to a first order, to discern parameters for more complex alloys. For example, in the case of Al$_x$Ga$_{1-x}$N, the piezoelectric polarization vector expression, using linear interpolation within the framework of Vegard's law, can be described:

$$\mathbf{P}^{pe} = [x\overleftrightarrow{e}_{AlN} + (1-x)\overleftrightarrow{e}_{GaN}]\vec{\varepsilon}(x). \tag{1.40}$$

The same argument can be extended to piezoelectric polarization in quaternary alloys such as Al$_x$In$_y$Ga$_{1-x-y}$N in a similar fashion:

$$\vec{P}^{pe} = [x\overleftrightarrow{e}_{AlN} + (1-x)\overleftrightarrow{e}_{InN} + (1-x-y)\overleftrightarrow{e}_{GaN}]\vec{\varepsilon}(x). \tag{1.41}$$

The linear interpolation is very convenient and accurate. However, as will be discussed in the following, while the Vegard's law applies to the alloys, the polarization charge itself is not a linear function of composition.

1.10.2
Spontaneous Polarization

Spontaneous polarizations calculated for the binary nitride semiconductors are compiled in Table 1.15. For ternary and quaternary alloys, the simplest approach is to use a linear combination of the binary end points taking into account that the mole fraction can be used under the auspices of the Vegard's law. However, this linear interpolation falls short of agreeing with the experimental variation of spontaneous polarization with respect to the mole fraction motivating the development of nonlinear models (to be discussed next). Within the framework of the linear interpolation, the spontaneous polarization in quaternary alloys such as $Al_xIn_yGa_{1-x-y}N$ can be expressed as

$$\bar{P}^{sp}(x, y) = x\bar{P}^{sp}_{AlN} + y\bar{P}^{sp}_{InN} + (1 - x - y)\bar{P}^{sp}_{GaN}. \tag{1.42}$$

The ternary cases can be obtained by simply setting either x or y to zero. This is again predicated on the assumption that polarization charge obeys Vegard's law, as shown below. Later in this section, the nonlinearity involved is discussed. Using the GGA calculation results for the spontaneous polarization and linear interpolation, as in Equation 1.42, for ternaries, one gets

$$\begin{aligned}
\mathbf{P}^{sp}_{Al_xGa_{1-x}N} &= -0.09x - 0.034(1 - x), \\
\mathbf{P}^{sp}_{In_xGa_{1-x}N} &= -0.042x - 0.034(1 - x), \\
\mathbf{P}^{sp}_{Al_xIn_{1-x}N} &= -0.09x - 0.042(1 - x).
\end{aligned} \tag{1.43}$$

1.10.3
Nonlinearity of Polarization

Linear interpolation from the binary end points serves well to a first degree. However, more accuracy requires consideration of nonlinearity, particularly in AlInN and InGaN alloys for which the binary constituents are very largely lattice mismatched. The polarization in an alloy can be described in a generic measurable quantity to a first approximation by a parabolic model involving a *bowing* parameter, similar to that used for the bandgap of alloys. In this vein, Bernardini and Fiorentini considered ordered structures to obtain at the spontaneous polarization. The chalcopyrite (CH)-like structure, used for the 0.5 composition, is formed by each anion site being surrounded by two cations of one species and two of the other, with the overall condition of conforming to the periodical ($2 \times 2 \times 2$) wurtzite supercell. Among the possible ordered structures, CH is in a sense the most homogeneous for a given composition. A further useful ordered structure considered is a luzonite (LZ)-like structure, used for the 0.25 and 0.75 points resembling zincblende-based alloys. In this structure, each nitrogen atom is surrounded by three cations of one species and one of the other: in a sense, this is analogous to the CH structure for molar fractions $x = 0.25$ and 0.75. The comparison between CH-like and random structures, the latter used for the 0.5 composition, provides insight into the effect of randomness versus ordering without the biases due to specific superlattice ordering as in the CuPt (CP) structure, used for the 0.5 composition point. LZ-like structure can provide values of the polarization in the intermediate molar fractions.

For freestanding random alloys, the values of the lattice constants within the realm of Vegard's law are given by

$$a_{Al_xGa_{1-x}N} = xa_{AlN} + (1-x)a_{GaN},$$
$$c_{Al_xGa_{1-x}N} = xc_{AlN} + (1-x)c_{GaN}.$$
(1.44)

Figure 1.13 shows a comparison between calculated equilibrium basal and axial lattice parameters a and c for three binaries and their alloys and those determined in the realm of Vegard's law (dashed lines). The agreement between the calculated equilibrium a and c lattice parameters and those determined using Vegard's law is

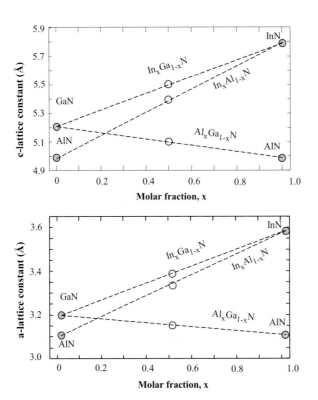

Figure 1.13 The basal plane lattice parameter a and the axial lattice parameter c of wurtzite nitride binaries and alloys directly calculated versus those determined by Vegard's law. The open circles denote the random alloy, used for 0.5 molar fraction. The dashed lines are Vegard's law. Courtesy of F. Bernaridini and V. Fiorentini.

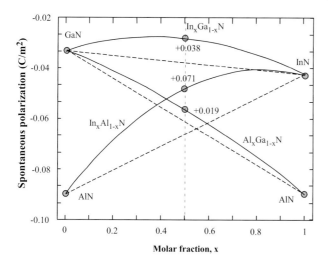

Figure 1.14 Spontaneous polarization versus molar fraction in freestanding (strain free) random nitride alloys (solid circles). The solid lines represent the results of the bowing model described (Equation 1.49). The dashed lines are determined by Vegard's law. The numbers shown in the figure are the bowing parameters expressed in units of C/m². Courtesy of F. Bernaridini and V. Fiorentini.

quite good. The dependence of the polarization on composition is then the same as that on the lattice parameter(s), with modulo a multiplicative factor.

Figure 1.14 also shows the spontaneous polarization values for the aforementioned ternaries in strain-free form. The solid lines represent interpolations utilizing Equation 1.51 for AlInN, the binary points of which were determined using 32-super cell calculations. The dashed lines represent the simple Vegard's law-based interpolations and the numbers indicate the bowing parameters in terms of C/m².

Figure 1.15 shows the spontaneous polarization values versus the molar fraction through which to the a lattice parameter for the three main alloys in nitrides, namely, AlGaN, AlInN, and InGaN. The solid circles, squares, and triangles represent the values for the random, CH/LZ-like, and CuPt-like alloys, respectively. The dashed lines represent the values calculated using Vegard's law. The numbers shown in the figure represent the bowing parameters for the CH/LZ- and CP-like ordered alloys.

The CH and LZ calculations can be used to verify applicability of both the interpolation model and the bowing parameter. The ordered LZ structure is analogous to CH for molar fractions of 0.25 and 0.75. The extent to which the LZ values deviate from those calculated with the aid of Equation 1.51 and the CH value ($x = 0.5$) indicates whether or not nonparabolicity occurs in the $P(x)$ relation for CH-like order. Because the polarization of the CH structure behaves qualitatively as that of the random structure (Figure 1.15), the conclusions drawn for CH are applicable to the random phase.

The values of the polarization calculated for the LZ structures at molar fractions 0.25 and 0.75, shown in Figure 1.15, are very close to the parabolic curve for InGaN

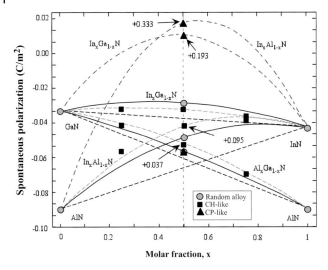

Figure 1.15 Spontaneous polarization versus the molar fraction in all three ternary nitride alloys. Solid circles, squares, and triangles represent random, CH/LZ, and CuPt (CP)-like structures, respectively. In other words, the dashed and dotted lines with solid triangles are for the CP-like alloys, the dashed lines with solid squares are for CH-like alloys, and solid lines with filled circles are for random alloys. The dashed lines represent Vegard's law. Numbers indicated in the figure are for CP- and CH/LZ-like ordered alloy bowing parameters in terms of C/m^2. Courtesy of F. Bernaridini and V. Fiorentini.

and AlGaN calculated using the quadratic expression in Equation 1.51, paving the way for the use of the analytical (quadratic) expression for polarization calculations. However, for AlInN, the calculated values are somewhat above the quadratic relation for $x = 0.25$ and below it for $x = 0.75$, indicating some nonparabolicity in polarization. Specifically, the bowing is higher for low In concentration in AlInN. This nonparabolicity is relatively modest, of order 10%, compared to the quadratic nonlinearity (Equation 1.51) for AlInN. One can then conclude that the analytical (quadratic) expression would predict polarization in AlGaN and InGaN fairly accurately, and also for AlInN but with about 10% accuracy.

The physical origin of the bowing in spontaneous polarization may be due to the internal structural and bond alternation, volume deformation, and disorder. The bond length and angle alternation will affect the polarization bowing only if it changes the projection of the bond length along the [0001] axis. The volume deformation can be due to compression or dilation of the bulk binaries. The disorder is due to the random distribution of the chemical elements on the cation sites. In ordered alloys, the structural contribution has been shown to be dominant. The volume deformation accounts for one-third of the bowing found in random alloys, and that the effect of disorder appears insignificant in terms of its effect on the bowing of spontaneous polarization.

As for the volume deformation, it is assumed that Vegard's law holds for the lattice constant a, establishing a linear relationship between the composition and the lattice

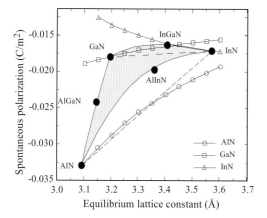

Figure 1.16 Spontaneous polarization versus the lattice constant in *ideal wurtzite* structures. Solid circles depict the values of binary compounds and random ternary alloys. Open circles, squares, and triangles represent the polarization calculated as a function of the lattice constant in bulk AlN, GaN, and InN. Solid lines correspond to Vegard interpolations based on the ideal binaries under hydrostatic pressure. The dashed lines represent the Vegard interpolation of the polarization using the values for the binaries at equilibrium. The data show that the polarization in the alloys is a direct result of the hydrostatic pressure and thus volume deformation. Courtesy of F. Bernaridini and V. Fiorentini.

constant. This segues into the calculation of the polarization in each of the *binary* nitrides in their ideal structure as a function of the lattice constant $a(x)$. Finally, they could express the alloy polarization as a composition-weighted Vegard-like average of the polarizations of the binary end points (through a reduction of Equation 1.42):

$$P_{sp}(Al_xGa_{1-x}N) = xP_{sp}^{a(x)}(AlN) + (1-x)P_{sp}^{a(x)}(GaN). \tag{1.45}$$

In this approach, any nonlinearity must have its origin in the different response of polarization to perturbations in $a(x)$, and hence to hydrostatic compression. To illustrate this point, Figure 1.16 shows calculated polarizations in the *ideal* wurtzite structure for the three ternary alloys, polarization of the binaries in the ideal structure, and the polarization by the Vegard interpolation of Equation 1.45. The calculated polarization and interpolation by Vegard's prediction agree well. Because the Vegard interpolation intrinsically accounts for the volume deformation, the origin of the volume deformation component of the nonlinearity and its large values in In-containing alloys become clear. Essentially, this has its genesis in the fact that polarization decreases with hydrostatic pressure in AlN and GaN, while it increases in InN. Note that the polarizations in the ideal structure are between 35% and 50% of their values in freestanding (relaxed) alloys, and despite the absence of bond alternation, which is designed for the purpose of separating the components in effect, the bowing is still very large.

Expression of the bowing parameter b_{model} for an *ideal* wurtzite structure alloy as function of the polarization response to hydrostatic pressure (for the model case of

AlGaN) can now be developed as

$$b_{\text{model}}^{\text{AlGaN}} = (a_{\text{GaN}} - a_{\text{AlN}})\left(\frac{\partial P_{\text{GaN}}}{\partial a} - \frac{\partial P_{\text{AlN}}}{\partial a}\right)\bigg|_{a=a(1/2)}$$
$$+ \frac{1}{4}(a_{\text{GaN}} - a_{\text{AlN}})^2 \left(\frac{\partial^2 P_{\text{GaN}}}{\partial a^2} - \frac{\partial^2 P_{\text{AlN}}}{\partial a^2}\right)\bigg|_{a=a(1/2)}. \quad (1.46)$$

Any nonlinearity in the spontaneous polarization can be treated by using a bowing parameter as commonly employed in interpolating the bandgap of an alloy from the binary point. In this vein, the spontaneous polarization for a ternary P_{sp} (A_xB_{1-x}N) with A and B representing the metal components and N representing nitrogen has been provided by Bernardini and Fiorentini:

$$P_{A_xB_{1-x}N}^{\text{sp}} = xP_{AN}^{\text{sp}} + (1-x)P_{BN}^{\text{sp}} - b_{AB}x(1-x). \quad (1.47)$$

P_{AN}^{sp} and P_{BN}^{sp} are the spontaneous polarization terms for the end binaries forming the alloy. The bowing parameter is by definition

$$b_{AB} = 2P_{AN} + 2P_{BN} - 4P_{A_{0.5}B_{0.5}N}, \quad (1.48)$$

which requires only the knowledge of the polarization of the ternary alloy at the midpoint, that is, molar fraction $x = 0.5$. Knowledge of the bowing parameter from Equation 1.48 would lead to determination of the spontaneous polarization at any composition. For $Al_xGa_{1-x}N$, Equations 1.47 and 1.48 take the form

$$P_{Al_xGa_{1-x}N}^{\text{sp}} = xP_{AlN}^{\text{sp}} + (1-x)P_{GaN}^{\text{sp}} - b_{Al_xGa_{1-x}N}x(1-x), \quad (1.49)$$

with $\quad b_{Al_xGa_{1-x}N} = 2P_{AlN} + 2P_{GaN} - 4P_{Al_{0.5}Ga_{0.5}N}. \quad (1.50)$

The first two terms on the right-hand side in Equation 1.49 are the usual linear interpolation terms between the binary constituents. The third term, quadratic, represents the nonlinearity. Higher-order terms are neglected because their contribution is estimated to be less than 10%. Using the numerical GGA values in Table 1.15 for the spontaneous polarization in AlN and GaN and the bowing parameter for random alloy AlGaN leads to

$$P_{Al_xGa_{1-x}N}^{\text{sp}} = -0.09x - 0.034(1-x) + 0.0191x(1-x)$$

and

$$P_{In_xGa_{1-x}N}^{\text{sp}} = xP_{InN}^{\text{sp}} + (1-x)P_{GaN}^{\text{sp}} - b_{In_xGa_{1-x}N}x(1-x),$$
with $\quad b_{In_xGa_{1-x}N} = 2P_{InN} + 2P_{GaN} - 4P_{In_{0.5}Ga_{0.5}N}.$

Again, using the numerical GGA values in Table 1.15 and the bowing parameter for random alloy $In_xGa_{1-x}N$ leads to

$$P_{In_xGa_{1-x}N}^{\text{sp}} = -0.042x - 0.034(1-x) + 0.0378x(1-x)$$

and

$$P_{Al_xIn_{1-x}N}^{\text{sp}} = xP_{AlN}^{\text{sp}} + (1-x)P_{InN}^{\text{sp}} - b_{Al_xIn_{1-x}N}x(1-x),$$
with $\quad b_{Al_xIn_{1-x}N} = 2P_{AlN} + 2P_{InN} - 4P_{Al_{0.5}In_{0.5}N}.$

1.10 Polarization Effects

Using the numerical GGA values in Table 1.15 and the bowing parameter for random alloy $In_xAl_{1-x}N$ leads to

$$P^{sp}_{Al_xIn_{1-x}N} = -0.090x - 0.042(1-x) + 0.0709x(1-x). \quad (1.51)$$

The lattice parameter (a), spontaneous polarization (P_{sp}), and bowing parameter (b_{AB}) for the three ternary nitride alloys in the form of random alloy and CH-, LZ-, and CuPt-like ordered alloys are tabulated in Tables 1.16–1.22 for convenience in addition to the data shown in Figures 1.13–1.15.

Table 1.16 The lattice parameter (a) and spontaneous polarization (P_{sp}) for AlN, GaN, and InN determined by 32-atom supercell calculations by F. Bernardini and V. Fiorentini.

	AlN	GaN	InN
a	3.1058 Å	3.1956 Å	3.5802 Å
P_{sp}	−0.0897 C/m²	−0.0336 C/m²	−0.0434 C/m²

Table 1.17 The lattice parameter (a), spontaneous polarization (P_{sp}), and the bowing parameter (b_{AB}) for random alloy ternaries with a molar fraction of $x = 0.5$ determined by 32-atom supercell calculations by F. Bernardini and V. Fiorentini.

R 50%	$Al_{0.5}Ga_{0.5}N$	$In_{0.5}Ga_{0.5}N$	$Al_{0.5}In_{0.5}N$
a	3.1500 Å	3.3872 Å	3.3352 Å
P_{sp}	−0.0569 C/m²	−0.0290 C/m²	−0.0488 C/m²
b_{AB}	+0.0191 C/m²	+0.0378 C/m²	+0.0709 C/m²

Table 1.18 The lattice parameter (a), spontaneous polarization (P_{sp}), and the bowing parameter (b_{AB}) for CuPt (CP-like) ordered alloy with a molar fraction of $x = 0.5$ determined by 32-atom supercell calculations by F. Bernardini and V. Fiorentini.

CP 50%	$Al_{0.5}Ga_{0.5}N$	$In_{0.5}Ga_{0.5}N$	$Al_{0.5}In_{0.5}N$
a	3.1489 Å	3.3884 Å	3.3222 Å
P_{sp}	−0.0573 C/m²	+0.0098 C/m²	+0.0168 C/m²
b_{AB}	+0.0176 C/m²	+0.1934 C/m²	+0.3336 C/m²

Table 1.19 The lattice parameter (a) and spontaneous polarization (P_{sp}) for the LZ-and CH-like alloys with a molar fraction of $x = 0.25$ determined by 32-atom supercell calculations by F. Bernardini and V. Fiorentini.

LZ 25%	$Al_{0.25}Ga_{0.75}N$	$In_{0.25}Ga_{0.75}N$	$Al_{0.25}In_{0.75}N$
a	3.1724 Å	3.2920 Å	3.4510 Å
P_{sp}	−0.0413 C/m²	−0.0323 C/m²	−0.0385 C/m²

Table 1.20 The lattice parameter (a) and spontaneous polarization (P_{sp}) for the LZ-and CH-like alloys with a molar fraction of $x=0.5$ determined by 32-atom supercell calculations by F. Bernardini and V. Fiorentini.

CH 50%	Al$_{0.5}$Ga$_{0.5}$N	In$_{0.5}$Ga$_{0.5}$N	Al$_{0.5}$In$_{0.5}$N
a	3.1474 Å	3.3949 Å	3.3369 Å
P_{sp}	-0.0523 C/m^2	-0.0328 C/m^2	-0.0427 C/m^2
b_{AB}	$+0.0374$ C/m^2	$+0.0226$ C/m^2	$+0.0952$ C/m^2

Table 1.21 The lattice parameter (a) and spontaneous polarization (P_{sp}) for the LZ- and CH-like alloys with a molar fraction of $x=0.75$ determined by 32-atom supercell calculations by F. Bernardini and V. Fiorentini.

LZ 75%	Al$_{0.75}$Ga$_{0.25}$N	In$_{0.75}$GaN$_{0.25}$	Al$_{0.75}$In$_{0.25}$N
a	3.1276 Å	3.4828 Å	3.2146 Å
P_{sp}	-0.0690 C/m^2	-0.0366 C/m^2	-0.0564 C/m^2

Table 1.22 The bowing parameter (a) for the random alloy and ordered alloy (LZ-like, CH-like, and CuPt (CP-like) ordered alloys) calculated by F. Bernardini and V. Fiorentini.

	AlGaN	InGaN	AlInN
Random	$+0.019$	$+0.038$	$+0.071$
CH and LZ	$+0.037$	$+0.023$	$+0.095$
CP	$+0.018$	$+0.193$	$+0.333$

1.10.3.1 Nonlinearities in Piezoelectric Polarization

Let us consider a coherently strained alloy on a relaxed binary buffer layer (bulk for this purpose) in which the *in-plane* lattice parameter $a_{alloy} = a_{GaN}$. The piezoelectric component is the difference between the total polarization to be obtained and the spontaneous polarization discussed above. Figure 1.17 shows the piezoelectric polarization as a function of the alloy composition. Symbol designations are similar to those for spontaneous polarization and represent the calculated polarizations for AlGaN, InGaN, and InAlN alloys as a function of compositions. It is clear that contrary to the spontaneous component of polarization, the piezoelectric polarization component weakly depends on the microscopic structure of the alloy. One then might ask whether the piezoelectric polarization of the alloy can be reproduced by a Vegard's-like model interpolated from the binaries in the form

$$P^{pe}_{AlGaN}(x) = x P^{pe}_{AlN}[\varepsilon(x)] + (1-x) P^{pe}_{GaN}[\varepsilon(x)], \tag{1.52}$$

where $P^{pe}_{AlN}[\varepsilon(x)]$ and $P^{pe}_{GaN}[\varepsilon(x)]$ represent the strain-dependent bulk piezoelectric polarization of the binary end points. With obvious permutations, this expression can also be constructed for InGaN and InAlN ternaries. To a first approximation, one

1.10 Polarization Effects

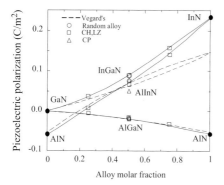

Figure 1.17 Piezoelectric component of the macroscopic polarization in ternary nitride alloys epitaxially strained on a relaxed GaN layer (template). Open symbols represent the directly calculated values for random alloy (circles), CH- and LZ-like (squares), and CP-like (triangles) structures. Dashed lines represent the prediction of linear piezoelectricity, while the solid line is the prediction of Equation 1.52. Courtesy of F. Bernaridini and V. Fiorentini.

may calculate the piezoelectric polarization of the binary compounds for symmetry conserving in-plane and axial strains:

$$P^{pe}_{AlN} = e_{33}\varepsilon_3 + 2e_{31}\varepsilon_1. \tag{1.53}$$

The piezoelectric constants e can be calculated for the equilibrium state of the binary, AN, and as such they do not depend on strain. The dashed lines in Figure 1.17 represent the piezoelectric term computed from using the piezoelectric constants for the binaries. The Vegard's law of Equation 1.52, when combined with Equation 1.53, clearly fails to reproduce the calculated polarization and misses the strong nonlinearity of the piezoelectric term evident in Figure 1.17. This is due to a nonlinearity of the bulk piezoelectricity of the binary constituents, which is of nonstructural origin. Bowing due to the microscopic structure of the alloys is negligible. Figure 1.18 clearly indicates that the piezoelectric polarization of the

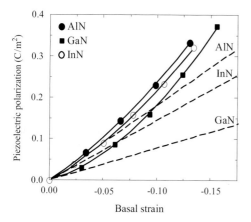

Figure 1.18 Piezoelectric polarization in binary nitrides as a function of basal strain (symbols and solid lines) compared to linear piezoelectricity prediction (dashed lines). The c and u lattice parameters are optimized for each strain. Courtesy of F. Bernaridini and V. Fiorentini.

binaries is an appreciably nonlinear function of the lattice parameter a, which is related basal strain. Because all lattice parameters closely follow Vegard's law, the nonlinearity cannot be related to deviations from linearity in the structure. Substitution of the *nonlinear* piezoelectric polarization computed for the binaries into the Vegard interpolation, Equation 1.52, led to excellent agreement with the polarization calculated directly for the alloys, as shown with solid lines in Figure 1.17. Therefore, the piezoelectric polarization of any nitride alloy at any strain can be found by noting the value for x (the composition), followed by calculating the basal plane strain, $\varepsilon(x)$ from Vegard's law and P_{pe} from Equation 1.52 using the nonlinear piezoelectric polarization of the binaries (Figure 1.18).

The nonlinear piezoelectricity of the binaries can be described by the following relations (in C/m^2):

$$P^{pz}_{AlN} = -1.808\varepsilon + 5.624\varepsilon^2, \quad \text{for} \quad \varepsilon < 0, \quad P^{pz}_{AlN} = -1.808\varepsilon - 7.888\varepsilon^2, \quad \text{for} \quad \varepsilon > 0,$$
$$P^{pz}_{GaN} = -0.918\varepsilon + 9.541\varepsilon^2,$$
$$P^{pz}_{InN} = -1.373\varepsilon + 7.559\varepsilon^2. \tag{1.54}$$

The calculation of the piezoelectric polarization of an $A_xB_{1-x}N$ alloy for any level of strain would proceed by calculating the strain $\varepsilon = \varepsilon(x)$ for a given molar fraction x, using the Vegard's law, and the piezoelectric polarization by

$$P^{pz}_{A_xB_{1-x}N} = xP^{pz}_{AN}(\varepsilon) + (1-x)P^{pz}_{BN}(\varepsilon), \tag{1.55}$$

where $P^{pz}_{AN}(\varepsilon)$ and $P^{pz}_{BN}(\varepsilon)$ are the end binary strain-dependent piezoelectric polarizations that can be calculated for a given strain $\varepsilon(x)$, using Equation 1.54. Application of this process to each of the three ternaries for all the possible cases of ternaries is as follows: Using Equation 1.39 and linear interpolation for the elastic constants strain determined from Equation 1.34, the piezoelectric polarization between a given ternary and binary can be calculated, as demonstrated in the following equations (in C/m^2):

$$P^{pz}_{Al_xGa_{1-x}N/GaN} = -0.0525x + 0.0282x(1-x),$$
$$P^{pz}_{Al_xGa_{1-x}N/AlN} = -0.026x + 0.0282(1-x), \tag{1.56}$$
$$P^{pz}_{Al_xGa_{1-x}N/InN} = -0.28x - 0.113(1-x) + 0.042x(1-x).$$

$$P^{pz}_{In_xGa_{1-x}N/GaN} = -0.148x + 0.0424x(1-x),$$
$$P^{pz}_{In_xGa_{1-x}N/AlN} = -0.182x + 0.026(1-x) - 0.0456x(1-x), \tag{1.57}$$
$$P^{pz}_{In_xGa_{1-x}N/InN} = 0.113(1-x) + 0.0276x(1-x).$$

$$P^{pz}_{Al_xIn_{1-x}N/GaN} = -0.0525x + 0.148(1-x) + 0.0938x(1-x),$$
$$P^{pz}_{Al_xIn_{1-x}N/AlN} = -0.182(1-x) + 0.092x(1-x), \tag{1.58}$$
$$P^{pz}_{Al_xIn_{1-x}N/InN} = -0.028x + 0.104x(1-x).$$

The calculated values as a function of molar fraction for the ternaries for the epitaxial layer and template combinations are represented in Figure 1.19 assuming that the template is fully relaxed and the epitaxial layer is fully and coherently strained. For partial relaxation, unless the degree to which the relaxation that occurs

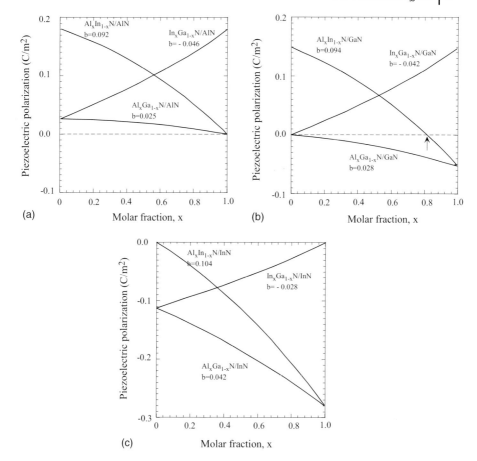

Figure 1.19 (a) Piezoelectric polarization of fully and coherently strained ternary alloys on fully relaxed AlN templates in which case the ensuing positive piezoelectric polarization and the negative spontaneous polarization are antiparallel. The bowing parameters describing the nonlinearity in the compositional dependence of P^{pz} are also indicated. (b) The piezoelectric polarization of coherently strained ternary alloys on fully relaxed GaN template. Note that $Al_{0.82}In_{0.18}N$/GaN heterojunction (indicated with an arrow) is lattice matched, thus strain and piezoelectric polarization vanishes Also note that other experiments indicate the lattice-matching composition for AlInN on GaN where the piezoelectric polarization charge vanishes to be different. (c) The piezoelectric polarization of coherently strained ternary alloys on fully relaxed InN template. For ternary alloys grown on InN, the negative piezoelectric polarization and the spontaneous polarization are parallel and oriented along the c-axis. Courtesy of O. Ambacher.

is known, the calculations cannot be made. The extent of relaxation depends on whether strain-relieving defects propagate from the template to the epitaxial layers. In addition, the effect of cool down induced thermal mismatch strain due to cooling from the growth temperature down to the operating temperature of the structure must also be considered. As stated, the results shown are for a perfect system with fully relaxed template and fully strained epitaxial layer on top of it. We should point out that in a particular pair with a particular composition, $Al_{0.82}In_{0.18}N/GaN$ heterostructure, there is a perfect lattice match, thus the misfit-induced piezoelectric polarization is equal to zero. It should also be pointed out that other experiments indicate the lattice-matching composition for AlInN on GaN where the piezoelectric polarization charge vanishes to be different.

1.10.4
Polarization in Heterostructures

Heterostructures such as QW and single heterointerfaces are ideal platforms to test the knowledge base of spontaneous and piezoelectric polarizations. Depending on the polarity of the sample, that is, Ga or N, on the order of growth, meaning GaN on AlGaN or the other way round, and on the buffer layer that determines whether the barrier, the well, or both are in strain, the piezoelectric and spontaneous polarization charge add or subtract. Regardless of these, the resultant band bending leads to a redshift in energy due to the presence of electric field, known as quantum confined Stark effect (QCSE), and the ensuing deformation of the wavefunctions that are pushed to opposite ends of the interface. This leads to reduction of radiative efficiency, increase in lifetime, and also blueshift with injection that is screening. In parallel, if the well size is small enough, carrier confinement induces a blueshift. Consequently, the redshift induced by polarization and the blueshift induced by localization compete each other in determining the transition energy. Here, the manifestation of polarization charge in heterostructures is treated.

The electric field resulting from the polarization in GaN would cause a large Stark shift in optical measurements and reduce the effective bandgap in, for example, QWs that is of paramount importance in optoelectronic devices. The Stark shift can be screened on a length scale on the order of the Debye length, $\sqrt{(\varepsilon kT)/(q^2 n)}$, by injecting free carriers in the GaN layer(s) as is the case in LEDs, lasers, and PL experiments with large excitation intensities. In such a case, the strain-induced field causes carrier separation, which, in turn, creates an internal field that opposes the strain field. Strain-induced polarization in such a heterostructure can also lead to a net electric field, which can be measured as a voltage drop across the sample.

For an alternating sequence of wells (W) and barriers (B), the total electric field in the well can be calculated by recognizing that the normal component of the displacement vector **D** of Equation 1.31 is continuous (provided that there is no interface charge).

$$\varepsilon_W E_W + 4\pi P_W^{total} = D_W = D_B = \varepsilon_B E_B + 4\pi P_B^{total}, \tag{1.59}$$

in cgs units (for MKS units, remove the 4π terms).

Utilizing Equation 1.59 together with the periodicity-imposed equality of voltage drops but with opposite sign,

$$L_W E_W = -L_B E_B, \tag{1.60}$$

one arrives at

$$E_W = -4\pi L_B \left(P_W^{total} - P_B^{total}\right)/(L_W \varepsilon_B + L_B \varepsilon_W)$$
$$\text{or } E_W = -4\pi \left(P_W^{total} - P_B^{total}\right)/(L_W/L_B \varepsilon_B + \varepsilon_W),$$
$$E_B = -4\pi L_W \left(P_B^{total} - P_W^{total}\right)/(L_B \varepsilon_W + L_W \varepsilon_B) \qquad (1.61)$$
$$\text{or } E_B = -4\pi \left(P_B^{total} - P_W^{total}\right)/\left(\frac{L_B}{L_W \varepsilon_W} + \varepsilon_B\right),$$

where ε_W and ε_B are the dielectric constants of the well and barrier layers, respectively. Likewise, L_W and L_B represent the well and barrier thicknesses. The same notation is used for polarization also. Both W and B superscripts and subscripts relate to wells and barriers, respectively. The total polarization term can be changed to piezoelectric term and spontaneous polarization term for cases when only the former or the latter, respectively, is in effect.

Explicit in Equation 1.61 is that whenever $P_W^{total} \neq P_B^{total}$, there will be an electric field. Due to strain or screening, the electric field will be present in both the well and the barrier. If we limit ourselves to piezoelectric polarization and somehow achieve relaxed heterostructures, the piezoelectric polarization charge and therefore the field will be zero. The electric field in wells and barriers in a QW has two components, one from spontaneous polarization and the other from piezoelectric polarization. If the thickness of both the wells and the barrier are same, the field in wells and barriers are related to each other:

$$E_W^{sp} + E_W^{pe} = E_W = -E_B \equiv -\left(E_B^{sp} + E_B^{pe}\right), \qquad (1.62)$$

where E_W and E_B represent the electric field in the well and barrier material, respectively. The superscripts sp and pe indicate the field due to spontaneous polarization and piezoelectric polarization, respectively. Additional comments that can be made are that if lattice-matched AlInGaN alloy is used, the piezoelectric component in that material is eliminated. However, the spontaneous component would still be present. Another point to be recognized is that the piezoelectric field induced in InGaN and AlGaN layers of the same composition of In in the former and Al in latter, if grown on relaxed GaN, is larger in the former because of the larger lattice mismatch between GaN and InN compared to that between GaN and AlN.

The spontaneous polarization in alloys can be found using Equations 1.42 and 1.43 together with values listed for the binary end points in Table 1.15. The spontaneous polarization so formulated together with piezoelectric polarization of Equation 1.40 would allow the computation of total polarization charge. This must be done for both the barrier and well material. The electric field can then be calculated using Equation 1.61.

Any free carrier present in the well and barrier regions as well as those injected by optical and/or electrical means tends to screen the polarization-induced field. A complete picture can be obtained by solving self-consistently a set of effective mass theory or tight-binding theory and simultaneously the Schrödinger–Poisson equation. Combining the *ab initio* calculations and Schrödinger–Poisson solver allowed calculation of the charge distribution and field profiles in QWs in the presence of free carriers. Poisson's equation is solved using the boundary condition that the electric field is zero at the ends of the simulated regions. This corresponds to L_W reaching the infinity limit in Equation 1.61. The potential thus obtained across the structure is plugged into the tight-binding Schrödinger equation, which is solved to obtain energies and wavefunctions. The new quasi-Fermi levels are then calculated, which then lead to carrier concentrations and the procedure is iterated until self-consistency is obtained.

Of interest for heterojunction field effect transistors (HFETs) and parts of QWs is one that features a ternary on the surface where polarization charge would exist. The same is true for the interface between $A_xB_{1-x}N$ and GaN heterostructures. The characteristic of this charge is that it changes abruptly leading to a fixed two-dimensional polarization charge density σ, on the surface and at the interface, which is given by

$$\sigma_{A_xB_{1-x}N} = P^{sp}_{A_xB_{1-x}N} + P^{pz}_{A_xB_{1-x}N}, \quad \text{on the surface,}$$

$$\sigma_{A_xB_{1-x}N/GaN} = (P^{sp}_{GaN} + P^{pz}_{GaN}) - (P^{sp}_{A_xB_{1-x}N} + P^{pz}_{A_xB_{1-x}N}) \text{ at the interface.}$$

(1.63)

Figure 1.20 shows the polarization-induced surface and interface sheet density σ/q for relaxed and coherently strained binary nitrides and ternary–binary $A_xB_{1-x}N$–GaN interfaces, such as $Al_xGa_{1-x}N/GaN$.

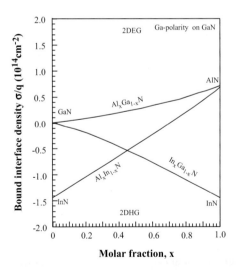

Figure 1.20 Bound interface charge density for coherent $Al_xGa_{1-x}N$, $In_xGa_{1-x}N$, and $Al_xIn_{1-x}N$ grown on relaxed Ga-polarity GaN. Polarization-induced interface charge (positive in n-type samples and negative in p-type samples), which is bound, is screened by free carriers such as electrons in n-type and holes in p-type samples leading to two-dimensional gas. Note that the interface charge is converted to carrier concentration by dividing it with the electronic charge of $q = 1.602 \times 10^{-19}$ C. It should be pointed out that there is dispersion in the literature composition of $Al_xIn_{1-x}N$ that lattice matches to GaN. Courtesy of O. Ambacher.

With the above in hand, the sheet charge distribution and band edge at an AlGaN ($A_xB_{1-x}N$)–GaN HFET heterointerface can be calculated. The criterion is that the bandgap of $A_xB_{1-x}N$ ternary must be larger than that of GaN. A point of interest, the $Al_xIn_{1-x}N$ ternary lattice matches GaN. For completeness, the expressions for the compositional dependence of the alloy bandgap are repeated below. It should be pointed out that the bandgap bowing parameter $b_{A_xB_{1-x}N}$ for AlGaN is converging onto a value of nearly 1 eV and that for InGaN to 2.53 eV. When all the available data for InAlN are considered in aggregate, a bowing parameter of about 3 eV appears to be a very good value:

$$E^g_{Al_xGa_{1-x}N} = xE^g_{AlN} + (1-x)E^g_{GaN} - b_{AlGaN}x(1-x) = 6.1x + 3.42(1-x) - x(1-x)\,\text{eV},$$
$$E^g_{In_xGa_{1-x}N} = xE^g_{InN} + (1-x)E^g_{GaN} - b_{InGaN}x(1-x) = 6.1x + 0.7(1-x) - 1.43x(1-x)\,\text{eV},$$
$$E^g_{Al_xInN} = xE^g_{AlN} + (1-x)E^g_{InN} - b_{AlInN}x(1-x) = 6.1x + 0.7(1-x) - b_{AlInN}x(1-x)\,\text{eV}.$$
(1.64)

The presence of displacement or polarization gradient leads to a volume charge density ρ_v given by

$$\nabla \cdot \vec{D} = \nabla \cdot (\varepsilon \vec{E} + \vec{P}^{\text{total}})$$
$$= -\rho_v \text{ in MKS units. For cgs units, add } 4\pi \text{ before } \vec{P}. \quad (1.65)$$

In a one-dimensional system, which is considered here along the c-axis and utilizing $E = -dV/dz$, Equation 1.65 can be rewritten as the Poisson equation:

$$dD/dz = d/dz(-\varepsilon(z)dV/dz + P^{\text{total}}(z)) = -q\rho_z$$
$$= -q[N_D^+(z) + p(z) - n(z) - N_A^-(z)]. \quad (1.66)$$

The term on the left-hand side of Equation 1.66 is the volume density of net charge, and the position-dependent quantities D, ε, and V are the displacement field, dielectric constant, and potential, respectively. The term P^{total} is the position-dependent total transverse polarization (along the c-axis and perpendicular to the interfaces). $N_D^+(z)$ and $N_A^-(z)$ represent the ionized donor and acceptor concentrations, and $p(z)$ and $n(z)$ represent the hole and electron concentrations, respectively. The effects of composition, polarization, and free carrier screening are thus fully included.

With the knowledge of the band profile for both the valence band and the conduction band, one can determine the electronic states in the heterostructure by solving the Schrödinger equation, as a function of the spatial coordinate z. In the effective mass approximation, one needs to solve the following eigenvalues problem:

$$-\frac{\hbar^2}{2m_0}\frac{d}{dz}\left(\frac{1}{m_z^*}\frac{d\Psi_i(z)}{dz}\right) + V_c(z)\Psi_i(z) = E_i\Psi_i(z), \quad (1.67)$$

for which the appropriate choice of the effective masses in the conduction band and in the valence band is done by preliminary calculation using the tight-binding approximation. The solution of Equation 1.67 determines the eigenstates E_i in the conduction band and in the valence band, and the corresponding eigenfunctions Ψ_i.

It is imperative to note that free carriers cannot eradicate the polarization charge, which is bound and invariable (unless structural changes are made), but one can screen it to a degree determined by carrier concentration. Likewise, polarization charge is bound charge and cannot by itself be the source of free carriers, but it can cause a redistribution of free carriers that would tend to screen the polarization charge.

For calculations, one must know the dielectric constants, doping levels, effective masses, and band discontinuities as well as the compositions and the layer thicknesses used. The bandgap of the ternaries is found from Equation 1.64 with the appropriate bowing parameter and a certain fraction of the total bandgap difference is assigned to the conduction band and the rest to the valence band. Typically, 60–70% of the total band discontinuity is assigned to the conduction band. Relative dielectric constants of 10.28, 10.31, and 14.61 in GaN, AlN, and InN, respectively, have been calculated. The dielectric constant for ternaries can be expressed:

$$\varepsilon_{Al_xGa_{1-x}N}/\varepsilon_0 = 10.4 - 0.03x,$$
$$\varepsilon_{In_xGa_{1-x}N}/\varepsilon_0 = 10.4 + 4.2x, \quad (1.68)$$
$$\varepsilon_{Al_xIn_{1-x}N}/\varepsilon_0 = 14.6 - 4.3x.$$

The Schottky barrier height for the ternaries can also be interpolated from the binary end points. It should also be noted that the barrier height depends on the particular metal used, the details of which is discussed in Chapter 3. Because the Fermi level on the surface of GaN is not fully pinned in that heavier metals with large work functions lead to larger barrier heights, it is plausible and natural to assume that the barrier height increases with increasing AlN content and decreases with InN content. However, for InN, the Fermi level is already in the conduction band as in InSb and InAs. Comparing the data deduced for Ni using IV, C–V, and photoemission methods, barrier heights of 0.95 (0.84 eV if not corrected for the nonideal ideality factor), 0.96, and 0.91 eV have been obtained for GaN (this figure increases to about 1.2 eV for Pt). For $Al_{0.15}Ga_{0.85}N$, the same figures are 1.25 (1.03 eV if not corrected for the nonideal ideality factor), 1.26, and 1.28 in order and using the same methods. These figures represent an increase of about 0.3 eV in barrier height for $Al_{0.15}Ga_{0.85}N$ over GaN. Assuming that Ni is used for Schottky barriers, the barrier height for GaN is 0.95 eV, and more boldly the reported figures for one mole fraction do actually represent a figure consistent with a linear interpolation, one can express the molar fraction dependence of the barrier height on $Al_xGa_{1-x}N$ as

$$\phi_{Al_xGa_{1-x}N} = 0.95 + 2xV. \quad (1.69)$$

The linear interpolations for the other two ternaries, modified for barrier height on GaN used here, are

$$\phi_{In_xGa_{1-x}N} = 0.95 - 0.36xV,$$
$$\phi_{Al_xIn_{1-x}N} = 0.59 + 1.36xV. \quad (1.70)$$

These expressions lead to barrier heights ($q\phi$ or $e\phi$) of 0.95, 2.95, and 0.59 eV on GaN, AlN, and InN, respectively. With the exception of the value for GaN, the rest is really speculative at this point.

1.10.4.1 Ga-Polarity Single AlGaN–GaN Interface

Returning to single interface structures, such as those used in HFETs, the total interface charge, for example, at a gated $Al_xGa_{1-x}N$ and GaN interface, ρ_s, for an n-type case would be sum of total polarization charge ρ_p and free carrier charge n_s.

$$\rho_s = \rho_p - n_s = \rho_p - \frac{\varepsilon_r \varepsilon_0}{qd_{AlGaN}}\left[V_G - \left(\varphi_B - V_{p^2} + \frac{E_{Fi} - \Delta E_C}{q}\right)\right], \quad (1.71)$$

where V_G is the applied gate bias in terms of V, ϕ_B is the Schottky barrier height in terms of V on $Al_xGa_{1-x}N$, E_{Fi} is the Fermi level in GaN at the interface with respect to the edge of conduction at the interface, ΔE_C is the conduction band discontinuity between the ternary $Al_xGa_{1-x}N$ and GaN, and V_{p^2} is the voltage drop across the doped AlGaN, which in turn is given by $V_{p^2} = qN_d d_d^2/2\varepsilon_r\varepsilon_0$, where N_d is the donor concentration in $Al_xGa_{1-x}N$, all assumed to be ionized, d_d is the thickness of the doped AlGaN, and $\varepsilon_r\varepsilon_0$ is the dielectric constant of $Al_xGa_{1-x}N$. V_{p2} is negative for depleting voltage in which case it would add to the Schottky barrier height. We should point out at this stage that the form of Equation 1.71 is good for determining the sheet carrier concentration at the interface, but when used for field effect transistors (FETs), both the polarization and the charge induced by any doping must be lumped into the threshold or off voltage; for details, see Chapter 9. The effect of any undoped $Al_xGa_{1-x}N$ layer designated as having a thickness of d_i is small because its thickness is several nanometers and has been neglected in Equation 1.71. This set back layer was originally employed by the author in the GaAs system to further reduce remote Coulomb scattering for increased mobility. For an undoped AlGaN/GaN heterostructure, the V_{p2} term can be set to zero. In this case, the boundary conditions for potential or the Fermi level is made consistent with the Schottky barrier height on the $Al_xGa_{1-x}N$ surface. In the bulk of the structure, due to the special nature of the quasi-triangular barrier at the interface, shown in Figure 1.21, the Fermi level is generally taken to be near the midgap of the smaller bandgap material, which in this case is GaN.

One-dimensional Schrödinger–Poisson solver can be iteratively used to determine the band and carrier profile (simultaneous solutions of Equations 1.66 and 1.67 either in the tight-binding realm or the effective mass approximation, which is the case here). The bound charge can be represented by a thin and heavily doped interfacial layer, the thickness of which is about 1 nm or less, keeping in mind the total charge associated with such a fictitious layer must be equal to the bound charge. In order to solve this pair of equations, boundary conditions at the interface and surface as well as the structural parameters must be known. In a typical undoped AlGaN–GaN HFET, the GaN buffer layer is unintentionally doped with a level of $N_D \approx 10^{16}$ cm^{-3} and the thickness of this buffer layer spans 1–3 μm. The barrier

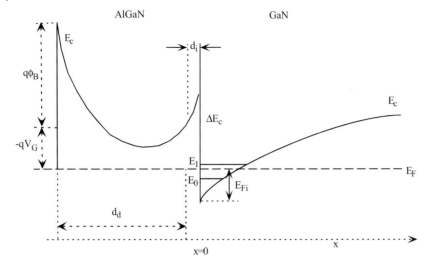

Figure 1.21 Conduction band edge of what is generically referred to as a modulation-doped structure based on the AlGaN/GaN system with an intentionally doped AlGaN barrier (for undoped barrier, see Chapter 9). The origin of the interface charge is in aggregate due to polarization and free carriers. In Ga-polarity samples, AlGaN grown on GaN produces polarization charge (due to spontaneous polarization and piezoelectric polarization because AlGaN is under tensile strain) causing accumulation of electrons at the interface in addition to any electron donated by any donor impurities, intentional or unintentional, in AlGaN. The diagram is shown for a doped AlGaN, doped portion of which is indicated with d_d and the undoped part is indicated by d_i.

thickness spans 10–20 nm. Because the structure is not intentionally doped, it is assumed that electrons are provided by surface states and other impurities/defects, such as O, Si, and V_N.

The sheet charge at the interface of such an undoped structure is dominated by the polarization-induced charge. On Ga-polarity surfaces, this charge increases with increasing AlN molar fraction in the barrier because both the piezo and spontaneous components of the polarization charge increase, assuming of course that the barrier is coherently strained. The conduction band and the electron concentration profiles for an undoped Ga-face $Al_{0.3}Ga_{0.7}N/GaN$ (30 nm/2 µm) heterostructure with a Ni Schottky contact on top are shown in Figure 1.22a. Electric field strength of about 0.4 MV/cm in the barrier and a sheet electron concentration of 1.2×10^{13} cm^{-2} are induced by polarization. The bound sheet density and 2DEG sheet carrier density induced by polarization in heterostructures (identical to that shown in Figure 1.22a with the exception that the alloy compositions in the barrier have been changed) are shown in Figure 1.22b and c for both the linear and nonlinear polarization cases with solid lines. The underlying assumptions are that the GaN buffer layer is relaxed, the barriers are coherently strained, and the physical parameters of importance (C_{ij}, e_{ij}, and P^{SP}) linearly scale from binaries to ternaries.

The error estimated, depicted by the gray area in Figure 1.22c, is primarily due to uncertainties in the barrier thickness and the conduction band offsets. The sheet carrier concentrations of 2DEGs confined in $Al_xGa_{1-x}N/GaN$ heterostructures for $x = 0.5$ have been measured by C–V profiling using Ti/Al ohmic and Ni Schottky contacts. The highest measured and calculated sheet carrier concentration for $Al_xGa_{1-x}N/GaN$ heterostructures is 2×10^{13} cm^{-2} for $x = 0.37$, as higher AlN molar fractions in the alloy for 30 nm barrier causes a partial relaxation lowering the piezo contribution. The extrapolation, indicated by dashed lines in Figure 1.22c, represents the case when the mole fraction is sufficiently large to cause some degree of

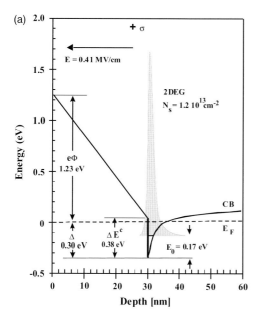

Figure 1.22 (a) Self-consistent calculation of the Schrödinger–Poisson equations for the conduction band edge and the electron density profile for an undoped Ga-polarity $Al_{0.3}Ga_{0.7}N$ (30 nm)/GaN (2 µm) single-interface heterostructure. Also shown are the Schottky metal on the surface and the polarization-induced surface and interface charges. (b) The polarization-induced bound interface charge density as a function of the alloy composition in the barrier for Ga-polarity $Al_xGa_{1-x}N/GaN$ heterostructures for the case of relaxed buffer layer and coherently strained barrier layer. The upper solid line for sheet charge corresponds to the case of linear interpolation of physical parameters (C_{ij}, e_{ij}, and P^{SP}) from the binary compounds. The lower solid line for the sheet charge, on the other hand, corresponds to the case of nonlinear extension of polarization from binary end points. The dashed lines depict the lagging sheet density with increasing molar fraction, x, due to partial relaxation that is accounted for by the measured degree of barrier relaxation. (c) 2DEG density as a function of alloy composition for the structure of (b). Again, the upper solid line corresponds to linear interpolation while the lower solid line corresponds to the non linear case. The shaded area depicts the uncertainty in estimates of the sheet density, n_s. The measured 2DEG density by C–V-profiling is shown as open symbols. Finally, the dashed lines represent the dependence of sheet charge on the mole fraction, x, in AlGaN, again by taking the measured degree of barrier relaxation into account. The dashed lines account for the experimentally observed strain relaxation of the barrier for $x >$ about 0.4. Courtesy of O. Ambacher.

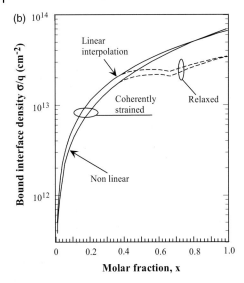

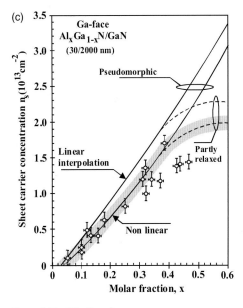

Figure 1.22 (Continued)

relaxation. As expected, the sheet density decreases nonlinearly with reducing mole fraction for the 30 nm barrier modeled. Among the two curves that bend over, the upper one is the total calculated polarization charge and the lower one is the calculated screening charge using the one-dimensional Schrödinger–Poisson equations. When full or partial relaxation occurs, the piezo component of the polarization charge is reduced, while the spontaneous component remains unchanged. The experimental data along with the error bars are shown in open circles. Overall, there is then a reduction of the charge. Likewise, the screening sheet carrier concentration is also reduced. The obvious conclusion that can be made is that a much better agreement is attained between calculations and experiments when the nonlinear polarization is used. The case of nonlinear polarization is stronger in the 2DEG case than it is in the QW case due to the direct nature of the measurement.

An empirical expression relating to the sheet carrier concentration for a nominally undoped AlGaN/GaN single-interface heterostructure having barrier thicknesses of greater than 15 nm and AlN molar fractions over 6% has been provided:

$$n_s(x) = [-0.169 + 2.61x + 4.50x^2] \times 10^{13} \text{ cm}^{-2}. \tag{1.72}$$

Naturally, the sheet charge becomes affected by parameters other than the AlN molar fraction when the barriers are made much thinner, as can be deduced from Equation 1.71. Using an expression similar to Equation 1.71, the dependence of the sheet density on barrier thickness and barrier molar fraction can be calculated. A priori, it is clear that beyond a certain thickness of the barrier, the density should saturate even if coherent strain prevails, as shown in Figure 1.23. The sheet density for a set of samples with $x = 0.3$ by C–V profiling for barrier thicknesses spanning the range 1–50 nm, are also presented.

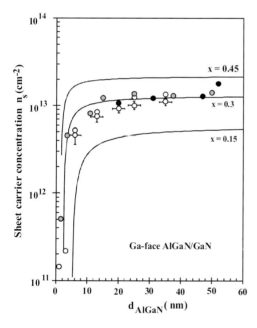

Figure 1.23 Barrier thickness d, dependence of sheet density in nominally undoped and coherently strained $Al_xGa_{1-x}N$–GaN heterointerfaces for $x = 0.15$, 0.30, and 0.45 (solid lines). Experimental data available for $x = 0.3$ measured by C–V profiling for barrier thicknesses spanning 1 and 50 nm, representing an aggregate from several reports. Courtesy of O. Ambacher.

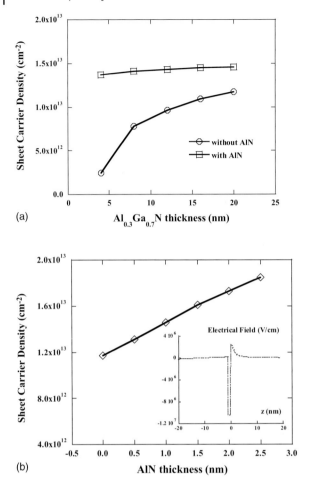

Figure 1.24 (a) Sheet carrier density dependence on AlGaN barrier thickness, with and without 1 nm AlN spacer. (b) Sheet carrier density dependence on AlN thickness in AlGaN/AlN/GaN HFETs structure; the electrical field distribution in AlGaN/AlN/GaN HFETs structure with 1 nm AlN spacer is also shown. Courtesy of Qian Fan.

With the help of Schrödinger–Poisson solvers, one can also include the AlN interface barrier, typically 1 nm, which is commonly used to enhance both the electron mobility in the channel and the electron confinement. Doing so leads to the dependence of the sheet density on the barrier thickness as well as on the AlN interface layer thickness, as shown in Figure 1.24.

1.10.4.2 Polarization in Quantum Wells

For multiple interface heterostructures, the sheet carrier density, barrier thickness, and width of QWs are of interest because the total potential drop across the structure is directly proportional to the product of polarization field and well width within the

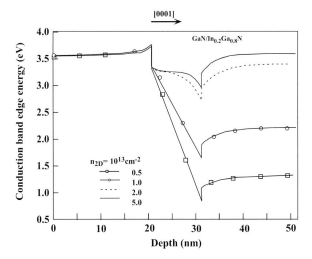

Figure 1.25 Conduction band profile of a 10 nm GaN/In$_{0.2}$Ga$_{0.8}$N QW for various levels of free carriers ranging from 5×10^{12} cm^{-2} to 5×10^{13} cm^{-2} present by doping of injection. The effect of polarization is nearly on all, but wiped out for the largest sheet electron concentration. Courtesy of Aldo Di Carlo.

constant field approximation if free carrier screening is neglected. At high electron levels, the polarization-induced field is screened, which is illustrated by the band profile of a 10 nm GaN/In$_{0.2}$Ga$_{0.8}$N QW shown in Figure 1.25 for several sheet densities. Even at a substantial sheet density of $n_{2D} = 5 \times 10^{12}$ cm^{-2}, a nearly uniform electrostatic field of strength 2.5 MV/cm is still present in the well. One needs to increase n_{2D} to 5×10^{13} cm^{-2} before recovering the quasi-field-free shape of the QW. This is achieved much earlier in thinner QWs.

1.10.4.2.1 Nonlinear Polarization in Quantum Wells

Assuming that the ternary nitride alloys have random microscopic structure, the spontaneous polarization of random ternary nitride alloys, in unit C/m^2, is already expressed in Equation 1.51. The first two terms in all three equations contained in Equation 1.51 are the usual linear interpolation between the binary compounds represented by Equation 1.43 with the third term, representing the bowing component. Higher order terms are neglected, as their effect is smaller than 10% in the worst case being the AlInN alloy.

For a model Al$_x$Ga$_{1-x}$N alloy, the piezoelectric polarization can be related to the binary end points using the Vegard's law, but recognizing that the terms for the bulk binaries must contain the nonlinear terms, as indicated in Equation 1.52. Such polarizations can be expressed accurately and compactly (in units of C/m^2) as

$$\begin{aligned}
P^{pe}_{AlN} &= -1.808\varepsilon + 5.624\varepsilon^2, \quad \text{for } \varepsilon < 0, \\
P^{pe}_{AlN} &= -1.808\varepsilon - 7.888\varepsilon^2, \quad \text{for } \varepsilon > 0, \\
P^{pe}_{GaN} &= -0.918\varepsilon + 9.541\varepsilon^2, \\
P^{pe}_{InN} &= -1.373\varepsilon + 7.559\varepsilon^2,
\end{aligned} \quad (1.73)$$

as a function of the basal strain of the alloy layer in question, with $a(x)$ and a_{subs} or a_0 the lattice constants of the unstrained alloy at composition x, and of the relaxed buffer layer or the substrate.

$$\varepsilon(x) = [a_{subst} - a(x)]/a(x). \quad (1.74)$$

In the case of pseudomorphic layers on GaN buffer layers, basal strain ε can be calculated directly from the lattice constants, which follow Vegard's law as a function of composition (depending on the lattice constants used):

$$\begin{aligned} a_{Al_xGa_{1-x}N}(x) &= a_{GaN} - x(a_{GaN} - a_{AlN}) = 0.31986 - 0.00891x \text{ nm} \\ a_{In_xGa_{1-x}N}(x) &= a_{GaN} + x(a_{InN} - a_{GaN}) = 0.31986 + 0.03862x \text{ nm} \\ a_{Al_xIn_{1-x}N}(x) &= a_{InN} - x(a_{InN} - a_{AlN}) = 0.35848 - 0.04753x \text{ nm} \end{aligned} \quad (1.75)$$

The combination of Equations 1.73–1.75 provides a convenient way of determining the polarization dependence on basal strain. The coefficients in Equation 1.73 are related (not equal) to piezoelectric constants and come about from the *ab initio* calculations.

The polarization charge values calculated using Equations 1.73–1.75 for heterostructures can be used together with a self-consistent Schrödinger–Poisson solver based, for example, on effective mass theory (not as accurate but efficient) or tight binding (more accurate, but computation-intensive) to determine field and charge distribution in the entire heterostructure as well as the effect of free carriers.

1.10.4.2.2 InGaN/GaN Quantum Wells

Cn-type GaN/In$_{0.13}$Ga$_{0.87}$N/GaN structures with Ga-polarity, where the width of the QW $d_{In_xGa_{1-x}N}$ varied between 0.9 and 54 nm, have been considered. The spontaneous polarization of the InGaN layer $P^{sp}_{In_{0.13}Ga_{0.87}N}$ is -0.031 C/m^2 and points in the $[000\bar{1}]$ direction, while the piezoelectric polarization is 0.016 C/m^2, which is antiparallel to the spontaneous polarization because the In$_x$Ga$_{1-x}$N QW is under compressive strain. The bound charge at the GaN–InGaN interface near the surface is positive and that at the lower interface is negative for Ga-polarity sample due to the compressive strain In$_x$Ga$_{1-x}$N. This implies that the electron accumulation caused by screening would occur at the interface near the surface in n-type samples. If p-type samples were considered, a hole accumulation would occur at the other interface.

The total polarization-induced interface sheet density is then given by

$$(P^{sp}_{GaN} + P^{pz}_{GaN}) - (P^{sp}_{In_xGa_{1-x}N} + P^{pz}_{In_xGa_{1-x}N}). \quad (1.76)$$

Noting relaxed GaN leads to $P^{pz}_{GaN} = 0$ and that substituting the numerical values, one gets for the total polarization

$$\begin{aligned} &(-0.034 + 0) - (-0.031 + 0.016) \text{ C/m}^2 \\ &= q1.18 \times 10^{13} \text{ C/cm}^2 \quad \text{or} \quad 1.18 \times 10^{13} \text{ electrons/cm}^2. \end{aligned}$$

The GaN/In$_{0.13}$Ga$_{0.87}$N/GaN heterostructures with GaN top layer and In$_{0.13}$Ga$_{0.87}$N QW with thicknesses of 130 and 20 nm, respectively, have been examined particularly in terms of their electron profiles, as determined by C–V measurements. The 2DEG sheet carrier concentration with well width, which follows a 2.5 power of well width, is much faster than the well width (the volume), which may be attributed to the system not being in equilibrium in terms of screening for thinner wells. The discrepancy between predictions and experiments raises an interesting question if the polarization-induced charge is fully screened by electrons and ionized donors. Assuming that the polarization-induced charge at the free surface and the GaN–substrate interface are fully screened, an electric field in the constant field approximation forms:

$$E_{well} = \frac{P^{total}}{\varepsilon_0(\varepsilon_r^{InGaN} - 1)} = \frac{P_{GaN}^{total} - P_{InGaN}^{total}}{\varepsilon_0(\varepsilon_r^{InGaN} - 1)}. \tag{1.77}$$

The terms here have their usual meanings.

1.11
Nonpolar and Semipolar Orientations

Motivation for semipolar and nonpolar orientation of GaN and related heterostructure lies in the fact that particularly at low carrier densities, wavefunction separation due to polarization induced by compositional gradient or strain causes a decrease in recombination probability in QW with the popular nomenclature of quantum confined Stark effect (QCSE) (see Figure 1.26). The planes of interest, nonpolar and semipolar, are shown in the wurtzitic system in Figures 1.4 and 1.5. The QCSE effect vanishes in nonpolar orientations and reduces in semipolar orientations, a measure of which is shown in Figure 1.27 in terms of the piezoelectric field. QCSE limits the thickness of the QWs to be used in light emitters, although the situation is assuaged somewhat at high injection levels. However, using thin QWs introduces a limit for the maximum total density of states available in a single QW and reduces

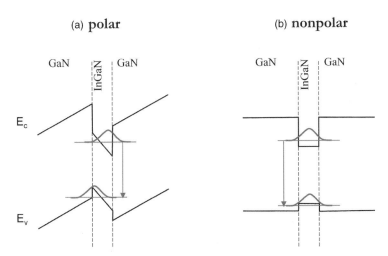

Figure 1.26 Schematic band diagrams of (a) polar c-plane and (b) nonpolar GaN/InGaN/GaN QWs. Spontaneous and piezoelectric polarization fields in the polar-oriented structure result in spatial separation electrons and hole wavefunctions and decreased transition energy.

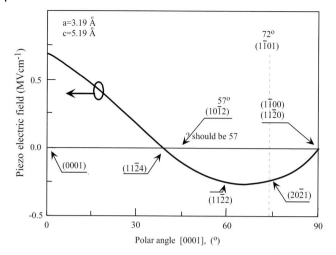

Figure 1.27 Piezoelectric field strength versus orientation as measured by deviation from the c-axis in GaN including highly polar [0001], semipolar, nonpolar directions. Adapted from Takeuchi et al.

the light output. This sets in motion the need to increase the number of QWs. Increased number of QWs is problematic because all the wells may not be populated equally under electrical injection, particularly by holes due to their heavy mass and relatively lower concentration compared to electrons. In addition, composition and thickness fluctuations result in inhomogeneous broadening of the QW absorption, the degree of which is amplified by the QCSE in wide QWs. Therefore, nonpolar orientations, a- and m-planes, are of particular interest, as they would allow the use of thicker QWs due to the absence of polarization fields. The oscillator strengths would be higher and the inhomogeneous broadening lower in thick nonpolar QWs, which bode well for stronger light–matter coupling and large Rabi splittings in microcavities. Semipolar orientations also eliminate the polarization field effects to varying extents.

The a-plane suffers from very low quality as it is generally produced on r-plane sapphire with highly mismatched interfaces; this point is somewhat muted if a-orientation GaN or SiC is used as the substrate. Although both a- and m-plane GaN suffer from relatively high stacking fault density, the m-plane variety is relatively of better quality. If corroborated widely by experiments, the calculated relatively low valence band density of states for m-plane is of paramount interest. The associated reduced effective mass would lead to increased hole concentration, which is problematic in GaN to say the least. In fact, increased hole concentrations have been observed in Mg-doped m-plane GaN layers grown by MBE on m-plane SiC ($7 \times 10^{18}\,\mathrm{cm}^{-3}$) and freestanding m-plane GaN templates ($8.7 \times 10^{18}\,\mathrm{cm}^{-3}$).

Although it will be discussed in Chapter 7 in detail, the polarization field is detrimental to carrier flyover in LEDs that cause efficiency degradation at high injection levels. The associated efficiency degradation becomes more severe as the InN mole fraction is increased to attain green emission. This, combined with green emission being extremely difficult to achieve with conventional compound semiconductors, forms the basis for the nomenclature "green gap" wherein as efficient LEDs are not yet possible.

Further Reading

Ambacher, O., Majewski, J., Miskys, C., Link, A., Hermann, M., Eickhoff, M., Stutzmann, M., Bernardini, F., Fiorentini, V., Tilak, V., Schaff, B., and Eastman, LF. (2002) *J. Phys.: Condens. Matter*, **14**, 3399–3434.

Ashcroft, N.W. and Mermin, N.D. (1976) *Solid State Physics*, Saunders College.

Bernardini, F. and Fiorentini, V. (2001) *Phys. Rev. B*, **64**, 085207.

Bernardini, F., Fiorentini, V., and Vanderbilt, D. (1997) *Phys. Rev. B*, **56**, R10024.

Bir, G.L. and Pikus, G.E. (1974) *Symmetry and Strain-Induced Effects in Semiconductors*, John Wiley & Sons, Inc., New York.

Blakemore, J.S. (1985) *Solid State Physics*, 2nd edn, Cambridge University Press.

Davydov, V.Yu., Klochikhin, A.A., Seisyan, R.P., Emtsev, V.V., Ivanov, S.V., Bechstedt, F., Furthmüller, J., Harima, H., Mudryi, A.V., Aderhold, J., Semchinova, O., and Graul, J. (2002) *Phys. Status Solidi B*, **229** (3), R1–R3.

Dingle, R., Sell, D.D., Stokowski, S.E., and Ilegems, M. (1971) *Phys. Rev. B*, **4** (4), 1211–1218.

Fritsch, D., Schmidt, H., and Grundmann, M. (2003) *Phys. Rev. B*, **67**, 235205.

Hopfield, J.J. (1960) Fine structure in the optical absorption edge of anisotropic crystals, *J. Phys. Chem. Solids*, **15**, 97.

Hopfield, J.J. and Thomas, D.G. (1963) Theoretical and experimental effects of spatial dispersion on optical properties of crystals, *Phys. Rev.*, **132**, 563.

Ludwig, W. and Falter, C. (1996) *Symmetries in Physics: Group Theory Applied to Physical Problems*, 2nd edn, Springer Series in Solid-State Sciences, vol. **64**, Springer, Berlin.

McKelvey, J.P. (1986) *Solid State Physics*, 4th edn, Krieger Publishing Company.

Morkoç, H. (2008) *Handbook on Nitride Semiconductors and Devices*, vol. 1, Wiley-VCH Verlag GmbH, Weinheim.

Nye, J.F. (1985) *Physical Properties of Crystals: Their Representation by Tensors and Matrices*, Clarendon Press, Oxford.

Pikus, G.E. (1962) New method of calculating energy spectrum of current carriers in semiconductors—2 Soviet Physics—*JETP*, **14** (5), 1075–1085.

Ren, G.B., Liu, Y.M., and Blood, P. (1999) *Appl. Phys. Lett.*, **74** (8), 117–119.

Suzuki, M., Uenoyama, T., and Yanase, A. (1995) *Phys. Rev. B*, **52** (11), 8132–8139.

Takeuchi, T., Amano, H., and Akasaki, I. (2000) Theoretical study of orientation dependence of piezoelectric effects in wurtzite strained GaInN/GaN heterostructures and quantum wells, *Jpn. J. Appl. Phys.*, **39**, 413–416.

Vurgaftman, I., Meyer, J.R., and Ram-Mohan, L.-R. (2001) *J. Appl. Phys.*, **89** (11), 5815–5875.

Yeo, Y.C., Chong, T.C., and Li, M.F. (1998) *J. Appl. Phys.*, **83** (3), 1429–1436.

2
Doping: Determination of Impurity and Carrier Concentrations

2.1
Introduction

The property that distinguishes semiconductors from others and defines them is their dopability with shallow donors and acceptors. For the instant case, control of the electrical properties of GaN and related materials is critical for device development. Doping is further complicated by the influence of defects present or generated in the process of doping. Controlled p-type doping in a wide range of concentrations in wide-bandgap semiconductors in general and GaN in particular is very difficult. All efforts, until the early 1990s, to obtain p-type doping have resulted in heavily compensated and highly resistive films. With resilience and clever engineering, p-type doping with room-temperature hole concentration near or above $10^{18}\,\text{cm}^{-3}$ is possible. Steady progress even for AlGaN and InGaN, for which the task is even more challenging, has been made. In this chapter, various impurities/dopants in terms of their formation energies, which shed light on their likelihood of incorporation, and experimental particulars are discussed. Only a brief discussion will be given here leaving the curious reader to refer to Morkoç (2008).

2.2
Doping

Generally speaking, elements such as C, Si, and Ge on the Ga sites and O, S, and Se on the N sites can potentially form shallow donors in GaN. Elements such as Be, Mg, Ca, Zn, and Cd on the Ga sites and C, Si, and Ge on the N sites have the potential of forming acceptors in GaN. In practice, controlled n-type conductivity in the nitride semiconductor family is generally achieved by Si doping in both vacuum deposition and metalorganic chemical vapor deposition techniques. Evidence seems to indicate that at high fluences Si might affect defect generation/propagation whereas Ge does not, a notion that is somewhat controversial, at least for the time being. Focusing on the workhorse Si, it substitutes a Ga atom in the lattice and provides a loosely *bound electron*. Measurements on high-quality GaN layers grown on freestanding GaN wafers indicate the binding energies of Si and O (residual impurity) in GaN to be

30.18 and 33.20 meV, respectively. The determination is based on magneto-optical studies as well as on detailed analysis of the two-electron transitions in photoluminescence experiments: $(D_1^0, X_A^{n=1})_{2e}$ and $(D_2^0, X_A^{n=1})_{2e}$ in which a donor-bound exciton transition is accompanied by excitation of an electron to the $n=2$ state of another donor-bound exciton. The difference between the ground-state and first excited-state transition energies of donor-bound excitons is then equal to 3/4 of the donor binding energy in the framework of the hydrogenic model, which applies within reason. The solubility of Si in GaN is high, on the order of $10^{20}\,\mathrm{cm}^{-3}$, making it suitable for group III n-doping and is most frequently used, but high concentrations lead to compensation/defect generation.

The relative difficulty in achieving one type of conductivity as opposed to the other, doping asymmetry, arises from the fact that wide-bandgap semiconductors have either a low valence band maximum or a high conduction band minimum. Some semiconductors such as ZnTe, CdTe, and diamond in which the valence band is relatively close to the vacuum level have preferable p-type conductivity. Consequently, achieving n-type conductivity in these semiconductors is very difficult, diamond being a case in point. In contrast, GaN and related nitrides, ZnO, ZnSe, ZnS, and CdS, with their valence bands relatively far from the vacuum level, have preferable n-type conductivity.

Many potential p-type dopants have been attempted for incorporation into GaN. Some impurities have been observed to effectively compensate electrons in GaN, leading to highly resistive mate. In addition, much theoretical work has been undertaken in an effort to determine which elements are more likely to lead to p-type doping. Only after post-growth *low-energy electron beam irradiation* (LEEBI), reported in 1991, or thermal annealing has p-type GaN become possible. No other dopant has been as successful as Mg. The maximum hole concentration in equilibrium is generally in the low $10^{18}\,\mathrm{cm}^{-3}$ range. Annealing is said to release H, liberating Mg acceptors, details of which can be found in Blakemore (1985). Codoping with, for example, O has been reported to increase the hole concentration, but this technique is not believed to be applied in practice.

The large activation energy of Mg in GaN and $Al_xGa_{1-x}N$ (160–200 meV in GaN and increases with increasing Al fraction) results in low hole concentration and therefore low conductivity, degrading the performance of light-emitting diodes, lasers, and heterojunction bipolar transistors. The large scatter in the measured activation energy of Mg in GaN is still not well understood, although concepts such as screening are suggested. In addition, presence of donor-like impurities such as O and Si can cause complex formation that reduces the activation energy (ionization energy), dubbed as *codoping*, and paves the way for attaining relatively high hole concentrations.

Carrier concentration and doping in semiconductors are inextricably connected and as such a description of point defects is provided to segue into the discussion on doping. Point defects, also known as native defects or intrinsic defects, manifest themselves as background doping or autodoping, compensated dopants, unless they are neutral, and complicate if not aggravate attempts to dope the semiconductor of interest in order to control its conductivity. In addition, depending on their energy position and optical activity, they can also act as nonradiative and in some cases radiative recombination centers, again complicating attempt in light emitters as well. Point defects can also cause charged centers, which scatter carriers, degrade the electron mobility, and affect FET performance and characteristics. Associated trapping causes degradation of diffusion lengths and thus adversely affects devices relying on minority carrier transport.

There are three basic types of native point defects: vacancies, self-interstitials, and antisites. *Vacancies* may be interpreted as the lattice sites missing their atoms. *Self-interstitials* are the additional atoms in between the lattice sites. *Antisites*, which are unique to compound semiconductors, are the cations sitting on anion sites, or vice versa. *Native defects* result when bonds in a semiconductor are either broken or distorted; they often give rise to deep levels within the forbidden gap. The *Fermi level* determines the charge state of a particular point defect. Point defects can be donor-type, acceptor-type, or amphoteric.

Wide-bandgap semiconductors exhibit self-compensation caused by defects. For example, when Si donors are introduced into GaN, the lattice may attempt to create Ga vacancies, V_{Ga}, which are acceptors, in order to reduce the total energy. This is one path through which point defects, that is, vacancies, interstitials, and/or antisites, could be formed. Another avenue is through defective or incomplete kinetic processes on the growing surface of an epitaxial layer. For example, insufficient N flux at the growing surface could result in N vacancies, V_N. Another driving force for the creation of point defects may be the polarization field that is always present in nitrides. Even in the absence of strain (which leads to piezoelectric fields), spontaneous polarization is present, which can give rise to significant fields. Ultimately, these fields are screened in macroscopic samples because of free carriers, but one may wonder whether on microscopic length scales the fields can provide a driving force for defect creation. Not all imaginable point defects are energetically favorable in a given semiconductor. To gain some insight, formation energies of defects are calculated and their likelihood therefore is determined.

Point defects in GaN have been tentatively identified, including V_N, V_{Ga}, and Ga_i. A wide range of analysis tools have been brought to bear to investigate the point defects and their impact on physical properties of GaN. Among the techniques employed are *photoluminescence* (PL), *deep-level transient spectroscopy* (DLTS), minority carrier lifetime measurements, *positron annihilation* (PA), *electron paramagnetic resonance* (EPR), and *optically detected magnetic resonance* (ODMR). A discussion of these techniques and data produced by them is beyond the scope of this book. However, a curious reader can refer to Morkoç (2008) and references cited therein.

2.3
Formation Energy of Defects

Theoretical investigations in the form of formation energies of point defects versus the Fermi level have been carried out. The first-principles calculations of defects and impurities using the local density approximation (LDA) and generalized gradient

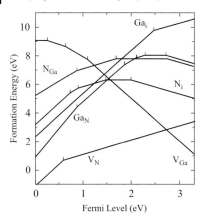

Figure 2.1 Calculated formation energies as a function of Fermi level for native point defects in GaN. Ga-rich conditions are assumed. The zero of Fermi level corresponds to the top of the valence band. Only segments corresponding to the lowest energy charge states are shown. Courtesy of C. Van de Walle.

approximation (GGA) to the density functional theory (DFT) have been used and also been the subject of intense debate. Hybrid functional calculations can substantially improve the accuracy of the calculated defect formation energies at an affordable cost for illuminating the defect physics. In terms of the ability to reproduce the experimental bandgaps, hybrid functional calculations are inferior only to the quasiparticle GW method, which is very computationally expensive. For point of reference, larger formation energy implies reduced likelihood of forming that particular defect.

Figure 2.1 shows the formation energies for all native point defects in GaN as a function of the Fermi level. The slope of each line represents the charge of the defect. For each charge state, only the segment giving the overall lowest energy is shown. Thus, the change in slope of the lines represents a change in the charge state of the defect, and the corresponding Fermi energy represents the energy level of the defect that can be measured experimentally. It is clear from Figure 2.1 that *self-interstitial* and *antisite* defects have very high formation energies and are thus unlikely to occur in GaN during growth. However, electron irradiation or ion implantation can create such defects in large numbers, and identification of a complex involving Ga_i has been reported. A *divacancy* ($V_{Ga}V_N$) has relatively high formation energy in GaN and as such is also unlikely to form in large concentrations. Nitrogen *antisite* defect (N_{Ga}) is expected to be metastable, with large spontaneous Jahn–Teller displacement in the [111] direction, both in cubic and in wurtzite GaN. In this treatment, the formation energy of nitrogen vacancies is high in n-type GaN, a topic that will be revisited later on, and low in p-type GaN. Therefore, nitrogen vacancies are unlikely to be formed in n-type material, and can be abundant in p-type GaN.

Mg acceptor has low formation energy and can easily incorporate into GaN. It prefers Ga sites and its shallow acceptor level provides good p-type conductivity. At high Mg flux, Mg_2N_3 complex can be formed as the solubility-limiting phase. V_N is

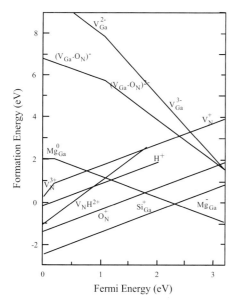

Figure 2.2 Formation energy as a function of the Fermi level for O_N, Si_{Ga}, Mg_{Ga}, H_i, V_N, V_{Ga}, and complexes of V_N and V_{Ga} with H and O, respectively. Courtesy of C. Van de Walle.

expected to autocompensate Mg_{Ga}, but if hydrogen is present in the system, compensation by hydrogen will dominate (Figure 2.2), and formation of V_N will be suppressed. In p-type GaN, the formation energy of V_N is significantly lowered, making it a likely compensating center during acceptor doping. The +/3+ energy level of V_{Ga} is estimated at about 0.5 eV above the valence band. The +/3+ transition of V_{Ga} is characterized by a large lattice relaxation. The 2+ charge state is always higher in energy than either the + or 3+ state, and thus thermodynamically unstable (the so-called negative-U defect). For more details, the reader is referred to Morkoç (2008).

2.3.1
Hydrogen and Impurity Trapping at Extended Defects

Monatomic interstitial hydrogen may exist in two charge states in GaN: H^+ and H^-, whereas the H^0 state is unstable (Figure 2.3), which exhibits a very large negative-U effect. H^+ prefers the nitrogen *antibonding* site, whereas H^- prefers the Ga *antibonding* site as it is most energetically stable. In a similar vein, H^+ is expected to be mobile even at room temperature due to a small migration barrier (~0.7 eV), while H^- has a very limited mobility in GaN owing to a very large migration barrier (~3.4 eV). It follows from Figure 2.2 that the solubility of H is considerably higher in p-type conditions (where it exists as H^+) than in n-type conditions (H^-). Hydrogen can form complexes with other defects in GaN, and often the formation energies of the hydrogenated defects are lower. In p-type GaN, the formation of the $(V_NH)^{2+}$ complex becomes very favorable. These complexes may incorporate only during

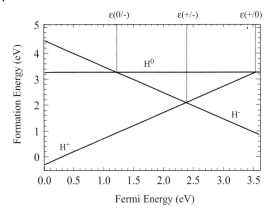

Figure 2.3 Calculated formation energy of interstitial hydrogen in wurtzite GaN as a function of Fermi level. $E_F = 0$ corresponds to the valence band maximum, and formation energies are referenced to the energy of an H_2 molecule. Courtesy of C. Van de Walle.

growth: after growth, both H and V_N are donors and would repel each other. The energy level of the V_NH donor is very close to the conduction band, either at resonance or slightly below the conduction band minimum.

Hydrogen readily forms complexes with defects in GaN, and often the formation energies of the hydrogenated defects are lower. In n-type GaN, up to four hydrogen atoms can be bound to Ga vacancy to form the complexes $(V_{Ga}H)^{2-}$, $(V_{Ga}H_2)^{-}$, $(V_{Ga}H_3)^0$, and $(V_{Ga}H_4)^+$. The first three complexes are acceptors, whereas the latter one is a single donor. The calculated energy levels of the $(V_{Ga}H)^{2-}$ and $(V_{Ga}H_2)^{-}$ complexes are about 1.0 eV above the valence band, which are close to that for an isolated V_{Ga}, while the level of $(V_{Ga}H_3)^0$ is near the valence band maximum. Further, the formation of complexes with several H atoms, such as $(V_{Ga}H_3)^0$ and $(V_{Ga}H_4)^+$, is unlikely because isolated hydrogen exists as H^- in n-type GaN and it would be repelled from the negatively charged Ga vacancy. Dissociation of the $V_{Ga}H_n$ complexes is unlikely due to large values of the binding energies. Consequently, once formed during growth in the presence of hydrogen, these complexes cannot dissociate in any post-growth thermal annealing.

In p-type GaN, hydrogen is known to passivate the dominant acceptor (Mg_{Ga}), as well as the dominant compensating donor (V_N). Naturally, H is liberated during the dopant activation process to obtain p-type GaN. In Mg-doped GaN grown by OMVPE, the electrically neutral Mg–H complex has a binding energy of 0.7 eV, with the H atom located in an *antibonding* site behind the N neighbor of the acceptor. During the post-growth annealing, the Mg–H complex dissociates, and H diffuses either to the surface or to extended defects. Similarly, in Be-doped GaN grown by OMVPE, the Be–H complex may form with a binding energy of 1.81 eV and dissociation energy of 2.51 eV. A post-growth annealing would also be required to remove hydrogen from the Be acceptors for dopant activation. This is a much studied topic with further details being available in Morkoç (2008).

2.4
Doping Candidates

Unintentionally doped GaN has, in all cases, been observed to be n-type with the best samples still showing a compensated electron concentration, which during the early stages of development approached 10^{20} cm^{-3}, but fortunately got reduced to about 2×10^{16} cm^{-3} in freestanding samples. With the exception of oxygen and silicon, no impurity has been found to be present in a sufficient quantity to account for the carriers, so researchers have initially, by nature, attributed the background to impurities and/or native defects. p-type doping above 10^{18} cm^{-3} for GaN remains a primary challenge and point of focus for researchers. Steady progress even for AlGaN for which the task is even more challenging and also for InGaN has been made.

Generally speaking, elements such as C, Si, and Ge on the Ga sites and O, S, and Se on the N sites can potentially form shallow donors in GaN. Elements such as Be, Mg, Ca, Zn, and Cd on the Ga sites and C, Si, and Ge on the N sites have the potential of forming relatively shallow acceptors in GaN. A brief review of the formation energies and energy levels calculated from the first principles and by using the effective mass method is provided herewith. While the former can predict which defect can be easier to form, the latter is much more accurate in terms of determining the ionization energy.

For p-type doping in GaN, Zn, Cd, Be, Mg, C, Ca, Hg, and Li on the Ga site and C, Si, and Ge on the N site could potentially give rise to relatively shallow acceptors. Ionization energies of main substitutional acceptors in wurtzite and zincblende GaN have been calculated in the effective mass approximation. Based on the calculations, main candidates for shallow acceptors in GaN are Be_{Ga}, Mg_{Ga}, C_N, and Si_N. From an analysis of the electronegativity differences between the acceptor atoms and the host atoms, it was deduced that the ionization energy of Be_{Ga} should be only slightly larger than the effective mass value, estimated to be 85 meV. In increasing order of ionization energies, the other Ga-substitutional acceptors are Mg, Zn, Cd, and Hg, according to increasing electronegativity difference between these impurities and a Ga atom. Consideration of the atomic and electronic structure of substitutional Be, Mg, and C acceptors in GaN through first-principles calculations suggests that these impurities should result in effective mass, not AX states in GaN (which represent some unknown acceptor-like defect states, the term for which is coined after the DX centers that represent some unknown deep defect of donor character). The calculated formation energies of some of the substitutional acceptors are shown in Figure 2.2. The formation energies of Mg_{Ga} and Be_{Ga} (the latter is not shown) and their ionization energies are the lowest. However, Be atom is very small, which paves the way for its efficient incorporation into the interstitial site where it acts as a double donor. Therefore, among group II impurities Mg and Be are the most promising p-type dopants in GaN, and Be appears the best candidate provided that formation of Be_i double donor can be suppressed. Formation energies of acceptors from group IV impurities, such as Si_N and Ge_N, are relatively high, so that under equilibrium conditions formation of these acceptors is unlikely in GaN.

While the theory suggests Be to be the best available p-type doping, the experimental data are heavily in favor of Mg.

2.5
Free Carriers

Let us now turn our attention to the analysis of donors and acceptors as well as the free carrier behavior. When impurities such as donors and acceptors are introduced into a semiconductor, they produce levels within the energy gap. The energy of a level with respect to the edge of the conduction band in the case of donors, and the valence band in the case of acceptors, is called the ionization energy. The simplest calculation of an impurity energy level is based on the hydrogenic model.

2.6
Binding Energy

The ionization energy of a hydrogen atom is given by

$$E_H = \frac{m_0 q^4}{8\varepsilon_0^2 h^2} = 13.6 \text{ eV}. \tag{2.1}$$

The ionization energy for a donor or an acceptor atom can be simply found by replacing the mass m_0 with the *conduction band conductivity effective mass* and the *valence band conductivity effective mass*, respectively, and substituting the *free space* dielectric constant ε_0 with the *dielectric constant of the semiconductor*. This operation leads to donor and acceptor binding energies as

$$E_D = 13.6(\varepsilon_0/\varepsilon_s)^2 (m_n^*/m_0) \text{ eV} \quad \text{and} \quad E_A = 13.6(\varepsilon_0/\varepsilon_s)^2 (m_p^*/m_0) \text{ eV}. \tag{2.2}$$

Choosing the relative dielectric constant of 8.9 and an effective conduction band mass of 0.22, the *hydrogenic donor binding energy* in GaN would be 38 meV, which compares with 29 meV determined from electrical measurements and 30 meV deduced from optical two-electron transitions. If we adopt an effective hole mass of 2, the acceptor binding energy would be about 200 meV, which is close to the calculated and deduced values for somewhat dilute Mg in GaN. When the effect of *screening* on the binding energy, which is generally determined from the dependence of the electron/hole concentration on temperature, is taken into consideration, the effective binding energy is lowered to some extent. The value thus determined is the screened value and related to that in a *dilute semiconductor* by $E_D = E_D^0 - \alpha N_D^{1/3}$, where E_D^0 is the donor binding energy for extremely small donor concentrations (dilute semiconductor), N_D is the total donor concentration, and α is a constant, which depends on the semiconductor. It should be mentioned that the binding energy calculated in the context of the hydrogenic model is for dilute semiconductors.

2.7
Conductivity Type: Hot Probe and Hall Measurements

Hot probe and Hall measurement is perhaps the easiest method to determine the conductivity polarity of compound semiconductors, the premise of which depends on unimpeded motion of carriers and as such requires lack of boundaries except that set by sample or pattern geometry. The hot probe measurement takes advantage of the Seebeck effect, named after the Estonian–German physicist Thomas Johann Seebeck, which in general is known as one component of the thermoelectric effect. This effect is also the basis for the electronic component, next to the vibrational or phonon component, of the thermal conductivity. The Seebeck effect states that the voltage difference between the two points is proportional to the temperature difference between the same two points through a coefficient called the Seebeck coefficient, S.

As can be inferred, the Seebeck effect is caused by electron diffusion from the hot zone to the cold zone. This effect is balanced with the electric field that sets up and opposes further carrier diffusion. The field would point from the hot region to the cold region, as the hot region would locally be in net positive charge. In a p-type GaN, all the donors are already ionized as the Fermi level is near the bottom of the forbidden band. When local heating is applied, additional acceptors would be ionized and generate excess holes under the hot tip. Excess holes set up a diffusion process with holes moving away from the hot tip. Consequently, a positive electrostatic voltage with respect to the hot probe (at negative) is formed and can be measured by a high-impedance electrometer. If, on the other hand, the sample is n-type, the voltage at the hot probe with respect to the cold one is positive. The polarity of the electric field is then used to determine whether the semiconductor is n-type or p-type.

2.8
Measurement of Mobility

To determine the hole and the electron concentrations and Hall mobility, Hall measurements are performed. Here a conducting bar, with an arbitrary pattern, is placed in a magnetic field that is normal to the *epitaxial layer*, and a *Hall voltage* develops normal to the current flow. The Lorentz force, which causes the Hall voltage to develop, causes electrons as well as holes to accumulate on the same side of the structure. In a semiconductor with mixed conduction (sizeable electron and hole contribution), the polarity of the Hall voltage may not correlate with the polarity of the semiconductor.

If the magnetic field is in the z-direction and the current flow is in the x-direction, for the magnetic field B, the electric field E, and the velocity v, we can write

$$\vec{B} = B\vec{z}, \qquad \vec{E} = E_x\vec{x} + E_y\vec{y}, \qquad \vec{v} = v_x\vec{x} + v_y\vec{y}. \qquad (2.3)$$

The Lorentz force affecting the carriers is

$$\vec{F} = q\vec{E} + q(\vec{v} \times \vec{B}) \quad \text{(in the SI system)}$$
$$= q\vec{E} + \frac{q}{c}(\vec{v} \times \vec{B}) \quad \text{(in the cgs system).} \quad (2.4)$$

The nature of the geometry is such that the z component of the source vanishes. Constructing the expressions for the x and y components of the current as

$$J_x = qE_x(n\mu_n + p\mu_p) \quad \text{and} \quad J_y = q\mu_n n(E_y - v_{xn}B) + q\mu_p p(E_y - v_{xp}B) \quad (2.5)$$

and recognizing that the y component of the current vanishes, one gets an expression for the field developed in the direction perpendicular to the current direction, which is given by

$$E_y = \frac{E_x B(\mu_p^2 p - \mu_n^2 n)}{(\mu_n n + \mu_p p)}.$$

Utilizing Equation 2.5,

$$E_y = \frac{J_x B(p - b^2 n)}{q(p + bn)^2} = R_H J_x B \quad \text{with} \quad b \equiv \frac{\mu_n}{\mu_p} \quad (2.6)$$

and

$$R_H = \frac{r}{q} \frac{(p - b^2 n)}{(p + bn)} \quad \text{and} \quad r_H \equiv \frac{\langle \tau^2 \rangle}{\langle \tau \rangle^2}.$$

If the mean free time between collisions is not energy dependent, the Hall factor $r_H = 1$. Moreover, the energy dependence depends on the scattering mechanism in place, and thus the Hall factor. See Chapter 3 of Morkoç (2008) for details regarding the Hall factor.

For $n \gg p$,

$$R_H = \frac{-r_H}{qn}. \quad (2.7)$$

For $p \gg n$,

$$R_H = \frac{r_H}{qp}. \quad (2.8)$$

Multiplying Equation 2.6 with the width of the sample and converting the current density to current through $J_x = I_x/Wt$, we obtain for the Hall voltage

$$V_y = \frac{I_x B(p - b^2 n)}{qt(p + bn)^2}, \quad (2.9)$$

where I_x is the current in the x-direction (the product of the current density and the vertical area) and t is the thickness of the epitaxial layer.

The situation is complex for a semiconductor, which is near intrinsic. In that case for the semiconductor to be p-type, the Hall voltage V_y must be larger than zero, which means that

$$(\mu_p^2 p - \mu_n^2 n) \geq 0 \quad \text{or} \quad p \geq n\frac{\mu_n^2}{\mu_p^2} \tag{2.10}$$

in order for the Hall measurement to accurately indicate that the polarity is p-type.
Let us summarize the practical expressions related to Hall measurements:

$$\rho = \frac{E_x}{j_x} = \frac{V_l/l}{I_x/tw} = \frac{V_l tw}{I_x l} = \sigma^{-1} = \frac{V_l[V] t[\text{cm}] w[\text{cm}]}{I_x[A] l[\text{cm}]} [\Omega\,\text{cm}],$$

$$\mu_H = \left|\frac{R_H}{\rho}\right| \doteq \frac{E_y}{BE_x} = \frac{V_H l}{B V_l w} = \frac{V_H[V] l[\text{cm}]}{B[V\,\text{s/cm}^2 \equiv T] V_l[V] w[\text{cm}]} [\text{cm}^2/(V\,s)],$$

$$n_H = \left|\frac{1}{eR_H}\right| = \frac{j_x B}{eE_y} = \frac{IB}{eV_l t} = \frac{I[A] B[V\,\text{s/m}^2 \equiv T]}{e[C] V_l[V] t[m]} [m^{-3}]$$

$$= \frac{I[A] B[G]}{e[C] V_l[V] t[\text{cm}]} [\text{cm}^{-3}], \tag{2.11}$$

where l, w, and t represent the length along the x-direction, the width along the y-direction, and the thickness of the sample, respectively.

In the laboratory, Hall measurements are carried out using either the Hall bar geometry, shown in Figure 2.4, or van der Pauw geometry, shown in Figure 2.5. The allure of the van der Pauw method is in its applicability to arbitrary shape of the contacts and the sample. However, contacts should be as close as possible to the edges of the sample to not allow current conduction around the outside skirt of the contacts. Measurements are performed with magnetic field in the z-direction, normal to sample surface, as well as in the $-z$-direction. In addition, the measurements are conducted with and without magnetic field with the role of the contacts being rotated and the resultant data are properly averaged. The Hall coefficient is

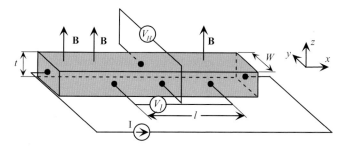

Figure 2.4 A schematic of the Hall bar geometry used for Hall measurement of mobility and carrier concentration. Current is passed along the length of the bar (x-direction) and the voltage drop caused by carrier motion and the magnetic field between the ends of the sample in the transverse direction is measured. While the schematic is for a bar, the technique can be used for layers on high-resistivity templates by forming the bar pattern through lithographic means in the conducting layer.

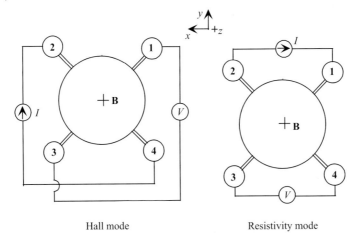

Figure 2.5 van der Pauw pattern, arbitrary (a) and patterned (b), along with current–voltage measurement details.

given by

$$R_H = \frac{[V_{12}(B) - V_{12}(0)]t}{I_{34}B} = \frac{[V_{12}(B) - V_{12}(-B)]t}{2I_{34}B}, \quad (2.12)$$

where t represents the thickness of the epitaxial layer.
The resistivity is obtained by

$$R_{34,12} = \frac{|V_{34}|}{I_{12}} \quad \text{and} \quad R_{23,41} = \frac{|V_{23}|}{I_{41}}, \quad (2.13)$$

$$\rho = \frac{\pi t(R_{34,12} + R_{23,41})f}{2\ln 2}, \quad (2.14)$$

where f is a factor that ideally would be equal to 1 if the sample and contacts are homogeneous. If, for example, the ratio $R_{34,12}/R_{23,41} = 10$, the factor f would be 0.7. More details regarding the f factor can be found in Figure 2.6 and the text related to it. As in the Hall bar case, the Hall mobility $\mu_H = R_H\sigma$, where σ is the conductivity of the sample that is a measurable quantity and comes out of Hall measurements, or $\mu = R_H/\rho$. The electron or hole concentration can be calculated using Equation 2.7 for n-type and Equation 2.8 for p-type samples. The Hall factor r_H is generally assumed to be unity, which presumes a dearth of distribution for the mean free time. This factor, however, depends on the scattering mechanism, which will be discussed below. The drift mobility is given by $\mu_d = \mu_H/r_H$.

2.9
Semiconductor Statistics, Density of States, and Carrier Concentration

The carrier concentration in a semiconductor is related to the semiconductor parameters through statistics. The topic has been the discussion of many

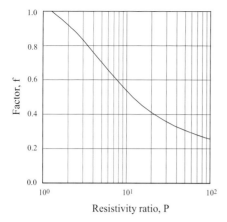

Figure 2.6 The dependence of the f factor on the resistance function ratio $P = R_{34,12}/R_{23,41}$ for the range from 1 to 100. It should be pointed out that for many films the resistivity ratio does not deviate from unity sufficiently to impact the calculations if the factor f is taken unity.

publications, including many reference and textbooks; a case in point for the latter is that by Blakemore.

Typically, the energy distribution function is given by the well-known *Fermi–Dirac distribution function*:

$$f(E) = \frac{1}{1 + g\exp[(E - E_F)/kT]}, \qquad (2.15)$$

where $f(E)$ is the distribution function and E_F is the Fermi energy. Here the term g represents the degeneracy factor, which practically takes the values of 1 for conduction electrons, 1/2 for electrons at a donor center, and 4 for electrons at an acceptor site. For localized levels such as impurities and defects, the distribution function is not well known.

The electron concentration can be found from the integral of the product of density of states and the Fermi–Dirac distribution function over the entire conduction band:

$$n = \int_{E_C}^{\infty} \frac{\sqrt{2}(E - E_C)^{1/2}(m_n^*)^{3/2}}{\pi^2 \hbar^3} \frac{1}{1 + \exp[(E - E_F)/kT]} dE, \qquad (2.16)$$

where m_n^* is the *conduction band density of states effective mass*. Note the conduction band electron degeneracy factor is taken to be 1. For a spherical energy surface, the band density of states effective mass is the same as the effective mass. Equation 2.16 reduces to

$$n = N_C \frac{2}{\sqrt{\pi}} F_{1/2}\left(-\frac{E_C - E_F}{kT}\right), \qquad (2.17)$$

where N_C is the effective density of states in the conduction band and is given by

$$N_C = 2(2\pi m_n^* kT/h^2)^{3/2} = 2.5 \times 10^{19} (m_n^*/m_0)^{3/2} (T/300)^{3/2} \text{ cm}^{-3} \quad (T \text{ in K}), \tag{2.18}$$

and

$$F_{1/2}(x_0) = \int_0^\infty \frac{x^{1/2} \, dx}{1 + \exp(x - x_0)}, \tag{2.19}$$

which for $E_C - E_F \geq 3kT$ reduces to

$$F_{1/2}(x_0) = \sqrt{\pi} \exp\left(\frac{x_0}{2}\right). \tag{2.20}$$

Then,

$$n = N_C \exp\left(-\frac{E_C - E_F}{kT}\right). \tag{2.21}$$

Note that the term $E_C - E_F$ is taken as a positive quantity when the Fermi level is below the conduction band.

If the n-type semiconductor is degenerate (Fermi level is in the conduction band), $\exp[(E_C - E_F)/kT] \ll 1$ and the expression for electron concentration takes on the form

$$n \approx \frac{4}{3\sqrt{\pi}} N_C \left(\frac{E_F - E_C}{kT}\right)^{3/2}. \tag{2.22}$$

Note that for the degenerate case, the quantity $E_F - E_C > 0$ and the electron concentration is larger than the total density of states N_C. It is interesting to solve Equation 2.22 for $E_F - E_C$, which together with the use of the total density of states expression $N_C = 2(m_n kT/2\pi\hbar^2)^{3/2}$ leads to

$$E_F - E_C = \left(\frac{\hbar^2}{2m_n^*}\right) (3\pi^2 n)^{2/3}. \tag{2.23}$$

The significance of Equation 2.23, as compared to the corresponding expression for the electron concentration for a nondegenerate semiconductor given by Equation 2.21, is that the Fermi level is not a function of temperature, but rather a function of electron concentration only.

A similar treatment to that performed in conjunction with Equation 2.17 but for holes leads to

$$p = N_V \frac{2}{\sqrt{\pi}} F_{1/2}\left(\frac{E_V - E_F}{kT}\right). \tag{2.24}$$

Again when the Fermi level is above the valence band by several kT values, the Fermi–Dirac distribution can be replaced by the Boltzmann distribution leading to

$$p = N_V \exp\left(\frac{E_V - E_F}{kT}\right), \qquad (2.25)$$

where N_V is the total density of states in the valence band and is given by

$$N_V = 2(2\pi m_p^* kT/h^2)^{3/2} = 2.5 \times 10^{19} (m_p^*/m_0)^{3/2} (T/300)^{3/2} \text{ cm}^{-3} \quad (T \text{ in K}). \qquad (2.26)$$

Note that the quantity $E_V - E_F < 0$ when the Fermi level is within the bandgap.

As in the case of n-type semiconductor, if the p-type semiconductor is degenerate, which is not yet technologically feasible under equilibrium conditions but could be under high-injection nonequilibrium cases (Fermi level is in the valence band), $\exp[(E_V - E_F)/kT] \ll 1$ (here $E_V - E_F$ is taken as a positive quantity) and the expression for electron concentration takes on the form

$$p \approx \frac{4}{3\sqrt{\pi}} N_V \left(\frac{E_V - E_F}{kT}\right)^{3/2}. \qquad (2.27)$$

Note that $E_V - E_F > 1$ and the hole concentration is larger than the total density of states N_V. It is interesting to solve Equation 2.27 for $E_C - E_F$, which together with the total valence band density of states expression $N_V = 2(m_p kT/2\pi\hbar^2)^{3/2}$ leads to

$$E_V - E_F = \left(\frac{\hbar^2}{2m_p^*}\right)(3\pi^2 p)^{2/3}. \qquad (2.28)$$

As in the case of n-type degenerate semiconductor, the significance of Equation 2.28 is that the Fermi level is not a function of temperature, but rather a function of hole concentration only.

Under equilibrium, the intrinsic electron concentration is related to the conduction and valence band densities of states by

$$n_i^2 = np = N_C N_V \exp\left(-\frac{E_g}{kT}\right), \qquad (2.29)$$

which can be expressed in a more convenient fashion as

$$n_i = 2.5 \times 10^{19} (m_n^* m_p^*/m_0^2)^{3/4} (T/300)^{3/2} \exp(-E_g/2kT) \text{ cm}^{-3} \quad (T \text{ in K}). \qquad (2.30)$$

When the semiconductor contains both donors and acceptors (they can be donor-like and acceptor-like defects as well), the charge balance (charge neutrality) requires that the negative charge concentration be equal to the positive charge concentration as discussed below.

If the *compensated semiconductor* is n-type, all the acceptors would be ionized making $N_A^- = N_A$ and thus the equilibrium electron concentration from the charge balance and the electron–hole product being equal to the square of the intrinsic carrier concentration (Equation 2.29) can be expressed as

$$n_0 = \frac{1}{2}\left[(N_D^+ - N_A) + \sqrt{(N_D^+ - N_A)^2 + 4n_i^2}\right]. \tag{2.31}$$

Because GaN has a large bandgap and the intrinsic carrier concentration n_i is very small, the electron concentration can be written as

$$n_0 = N_D^+ - N_A. \tag{2.32}$$

On the other hand, if the semiconductor is p-type, the equilibrium hole concentration is given by

$$p_0 = \frac{1}{2}\left[(N_A^- - N_D) + \sqrt{(N_A^- - N_D)^2 + 4n_i^2}\right], \tag{2.33}$$

with an analogous interpretation of the donors in a p-type semiconductor, that is, all the donors are ionized. Again, because GaN has a large bandgap and the intrinsic carrier concentration n_i is very small, the hole concentration can be written as

$$p_0 = N_A^- - N_D. \tag{2.34}$$

The relationship between the ionized acceptor and total acceptor concentrations must be determined using the statistics applicable to occupancy. The statistics depend on whether band states or donor or acceptor states are involved. The energy distribution of electrons in an energy band is well represented by the Fermi–Dirac distribution function. As mentioned, the distribution function for localized levels such as those for impurities and defects is not well known, but Equation 2.15 can be used to the first extent.

It is more common and useful to relate the donor energies to the conduction band in which case $E_{1CB} = E_1 - E_g \equiv -E_D$, where E_D is the donor binding energy. Moreover, one can let $E_{FCB} = E_F - E_g \equiv -E_F$ (taking the conduction band edge as reference), then Equation 2.15 becomes

$$N_D^+ = \frac{N}{1 + 2\exp[(E_D - E_F)/kT]}, \tag{2.35}$$

where the degeneracy factor became 2.

The case for shallow acceptors, however, is more complicated due to contributions from s-like and p-like states at the Γ-point. The p_z states are often lowered by a few kT from the p_+ and p_- states in which case the p_+ and p_- states can hold up to four electrons. In this case, s and p_z states would have two electrons each. If the acceptor in question is in its ground state, it can supply five electrons, two of which are taken by the p_z state, and only three electrons out of a possible four can be present in the p_+ and p_- states. In this case, the ground-state degeneracy factor is 4.

2.10
Charge Balance Equation and Carrier Concentration

In equilibrium charge balance prevails, meaning the sum of all negative and positive charges is zero. In *wide-bandgap semiconductors* such as GaN, the intrinsic carrier concentration n_i is extremely small (10^{-7} cm^{-3}) at room temperature. When the semiconductor is even moderately n- or p-doped, the opposite free carrier concentration is negligible. This means that in n-type GaN the hole concentration, p, can be neglected and in p-type the electron concentration, n, can be neglected. The Fermi energy, which is determined by solving the charge balance or charge neutrality equation, together with occupation probabilities is needed to determine free carrier concentration.

The *charge neutrality equation* calls for the positive and negative charges to be equal:

$$n + N_A^- = p + N_D^+, \tag{2.36}$$

where the right-hand side represents the positive charges with p and N_D^+ depicting the hole and ionized donor concentrations, respectively, and the left-hand side represents the negative charges with n and N_A^- depicting electron and the ionized acceptor concentrations, respectively. In n-type semiconductors, the Fermi level is above the midgap and thus all the acceptors whose states are well below the Fermi level would be ionized under equilibrium. Likewise, in a p-type sample the Fermi level is in the lower half of the bandgap and therefore all the donors whose states are well above the Fermi level are ionized.

In case of multiple donors and acceptors, the expression for charge balance takes the form

$$p + \sum_i N_{Di}^+ = n + \sum_j N_{Aj}^-, \tag{2.37}$$

where $\sum_i N_{Di}^+$ and $\sum_j N_{Aj}^-$ represent the total number of ionized donors and acceptors, respectively.

2.10.1
n-Type Semiconductor

When $n \gg p$, p can be neglected, easily satisfied in GaN as the intrinsic concentration is very small, 10^{-7} cm^{-3} at room temperature. Let us assume that the Fermi level is above all the donor levels by a few kT except the *dominant donor*. In this scenario, the Fermi distribution function associated with all the acceptors and donors with the exception of the dominant donor is temperature independent. The ionized acceptor concentration associated with a particular acceptor A_j in this case would be equal to the acceptor concentration if the Fermi level is above the acceptor level and zero if the acceptor level is above the Fermi level.

Employing the ionized donor to total donor concentration relationship, while skipping the details, which can be found in Chapter 2 of Morkoç (2008), the electron concentration can be expressed as

$$n = \frac{1}{2}(n_1 + N_A^{\text{net}}) \left\{ \left[1 + \frac{4n_1(N_D - N_A^{\text{net}})}{(n_1 + N_A^{\text{net}})^2} \right]^{1/2} - 1 \right\} \text{ with}$$

$$n_1 \sim N_C' T^{3/2} \left(\exp \frac{\alpha_D}{k} \right) \left(\exp -\frac{E_{D0}}{kT} \right), \tag{2.38}$$

where $N_C = N'_C T^{3/2} = 2(2\pi m^* k)^{3/2}/h^3 T^{3/2}$ and $E_D = E_{D0} - \alpha_D T$, the latter assuming a linear temperature dependence of effective donor binding energy.

At *low temperatures* and for $n_1 \ll N_A^{\text{net}}$ and $n_1 \ll (N_A^{\text{net}})^2/(N_D - N_A^{\text{net}})$, the plot of $\ln(n/T^{3/2})$ versus $1/T$ would be a straight line with a slope of $-E_{D0}/k$.

At *low temperatures* but with large donor concentration, particularly in relation to the acceptor concentration, $n_1 \gg N_A^{\text{net}}$ and $n_1 \gg (N_A^{\text{net}})^2/(N_D - N_A^{\text{net}})$, and thus the plot of $\ln(n/T^{3/2})$ versus $1/T$ would be a straight line with a slope of $-E_{D0}/2k$ as opposed to $-E_{D0}/k$ for the previous consideration.

At *high temperatures* with $n_1 \gg N_A^{\text{net}}$ and $n_1 \gg N_D - N_A^{\text{net}}$, the second term in the square root term in Equation 2.38 can be expanded and n becomes

$$n \approx \frac{1}{2}(n_1 + N_A^{\text{net}})\frac{2n_1(N_D - N_A^{\text{net}})}{(n_1 + N_A^{\text{net}})^2} = N_D - N_A^{\text{net}}. \qquad (2.39)$$

This shows that at high temperatures the electron concentration approaches a constant $N_D - N_A^{\text{net}}$.

Let us calculate the temperature dependence of electron concentration in GaN for a given donor concentration with varying degrees of compensating acceptors. Let us also further assume for simplicity that the donor binding energy is temperature independent. Using $N_A^{\text{net}} = \sum_j N_{Aj}$ and $g = g_{D10}/g_{D00} = 2$, the electron concentration can be expressed as

$$n = \sqrt{\frac{1}{4}\left(n_1 + \sum N_{Aj}\right)^2 + n_1\left(N_D - \sum N_{Aj}\right)} - \frac{1}{2}\left(n_1 + \sum N_{Aj}\right), \qquad (2.40)$$

where $n_1 = (N_C/g)e^{-E_D/kT}$.

Figure 2.7 shows the solution of Equation 2.40 for a donor concentration of $2 \times 10^{16}\,\text{cm}^{-3}$, donor binding energy of 30 meV, and effective electron mass of $0.22 m_0$, as a function of temperature for a range of acceptor concentrations from

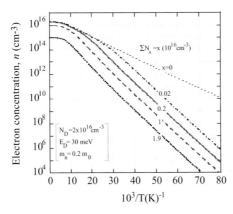

Figure 2.7 Temperature dependence of electron concentration for a donor concentration of $2 \times 10^{16}\,\text{cm}^{-3}$, donor binding energy of 30 meV, and effective electron mass of $0.22 m_0$, as a function of temperature for a range of acceptor concentrations from 0 to $1.9 \times 10^{16}\,\text{cm}^{-3}$ in GaN. Courtesy of M. Reshchikov.

0 to 1.9×10^{16} cm^{-3} in GaN. If the acceptor concentration could be taken as zero or negligibly small, the donor concentration as well as the donor binding energy can be determined from Figure 2.7 alone without the need for temperature-dependent mobility and scattering analysis.

Experimentally, the temperature dependence of the electron concentration can be determined from Hall measurements. Because there are acceptor-like native defects, determination of the donor concentration requires knowledge of the acceptor concentration. This can be determined by knowledge of the temperature dependence of the Hall mobility, electron concentration, and fitting experiments using the scattering mechanisms discussed in Chapter 4. In addition to shallow donor in GaN, there may be deeper donor-like states. If so, determination of the donor concentrations from the temperature dependence of electron and mobility requires simulations as well as use of Equation 2.40. The best-fit parameters are then taken as the effective parameters. An exercise of Equation 2.40 for two cases, one with two donors at 10 and 100 meV with concentrations of 2×10^{16} and 3×10^{16} cm^{-3}, respectively, and the other with two donors at 30 and 100 meV with concentrations of 2×10^{16} and 3×10^{16} cm^{-3}, respectively, is shown in Figure 2.8. The electron effective mass is taken as $0.2m_0$ and a donor degeneracy of factor of 2.

Consider experimentally measured temperature-dependent electron concentration in a sample presumed to contain a single donor. Using expressions described in this section, one can determine the donor binding energy additionally assuming that there are no acceptors. If there are acceptors present, then one must make a determination as which of the two cases, that is, $n_1 \ll N_A^{net}$ and $n_1 \ll (N_A^{net})^2/(N_D - N_A^{net})$ or $n_1 \gg N_A^{net}$ and $n_1 \ll (N_A^{net})^2/(N_D - N_A^{net})$, is applicable at low temperatures. If the

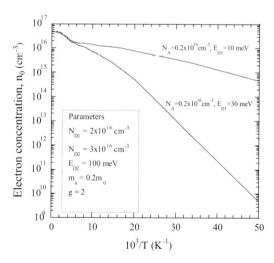

Figure 2.8 Temperature dependence of electron concentration in GaN for two cases: one with two donors at 10 and 100 meV with concentrations of 2×10^{16} and 3×10^{16} cm^{-3}, respectively, and the other with two donors at 30 and 100 meV with concentrations of 2×10^{16} and 3×10^{16} cm^{-3}, respectively. Courtesy of M. Reshchikov and S.S. Chevchenko.

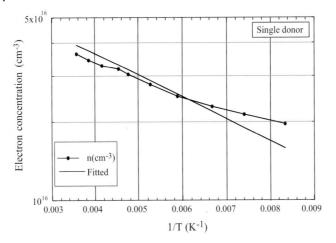

Figure 2.9 The measured and calculated electron concentrations as a function of temperature with the donor and acceptor concentrations and the donor binding energy as fitting parameters.

former applies, then the measured activation energy from the plot of $\ln(n/T^{3/2})$ versus $1/T$ is equal to the donor binding energy for that particular donor concentration. If the latter is the case, then the measured activation energy determined from a plot of $\ln(n/T^{3/2})$ versus $1/T$ would give only half the donor binding energy. At very high temperatures, the temperature dependence vanishes as all the donors ionize as illustrated in Figures 2.7 and 2.8.

The results of such a fitting exercise are depicted in Figure 2.9. It is important to keep in mind that the donor binding energy thus determined is the screened binding energy that is related to the binding energy E_D^0 in a dilute semiconductor through $E_D = E_D^0 - \alpha N_D^{1/3}$. Because there are many fitting parameters with possible nonunique solutions, the accuracy of the fitting can be improved by utilizing the temperature dependence of the Hall mobility together with an appropriate theory (Chapter 4). The donor activation energy and N_D for various samples are tabulated in Table 2.1. Figure 2.10 shows a plot of donor binding energy as a function of $N_D^{1/3}$ obtained from a series of samples with donor concentration as the variable. Clearly, the screened donor binding energy follows the empirical expression for a dilute semiconductor with $E_D^0 = 29.7$ meV and $\alpha = 2.59 \times 10^{-5}$ meV cm. The binding energy figure is in excellent agreement with 28.8 and 32.6 meV determined with what are believed to be Si and O from detailed optical measurements (see Chapter 5 of Morkoç (2008) and references cited therein).

For some samples, the single-donor model does not fit the temperature dependence of the measured electron concentration. Invoking the notion of a shallow donor and a deeper donor allows for a better fit to the data. It should be pointed out that a two-donor model is not needed in all samples for a satisfactory fit. Figure 2.11 displays a case where a single-donor model is adequate for fitting the data. On the other hand, Figure 2.12 shows the temperature-dependent electron concentration wherein a two-donor model must be invoked for a good fit. Note the shallow and deeper donors and their concentrations.

2.10 Charge Balance Equation and Carrier Concentration

Table 2.1 Compilation of measured electron concentration (donor concentration) N_D, $N_D^{1/3}$, and donor binding energy E_D in various GaN layers and templates.

N_D (cm^{-3})	$N_D^{1/3}$ (cm^{-1})	E_D (meV)	Comment
1.76×10^{16}	2.60×10^5	25.20	
3.01×10^{16}	3.11×10^5	23.50	
1.1×10^{17}	4.79×10^5	15.00	
1.25×10^{17}	5.00×10^5	17.00	
2.05×10^{17}	5.90×10^5	16.00	
2.1×10^{17}	5.94×10^5	10.00	
2.3×10^{17}	6.13×10^5	14.00	
2.33×10^{17}	6.15×10^5	10.65	
3.1×10^{17}	6.77×10^5	17.00	
4×10^{17}	7.37×10^5	7.00	
4.13×10^{17}	7.45×10^5	7.90	
4.84×10^{17}	7.85×10^5	10.60	
7.41×10^{17}	9.05×10^5	12.00	
8.31×10^{17}	9.40×10^5	5.05	
9.1E + 17	9.69×10^5	2.00	
1.37×10^{17}	1.11×10^6	1.00	
1.4×10^{18}	1.12×10^6	2.00	
Least-squares fitting	$E_D^0 = 29.7$ meV	$\alpha = 2.59 \times 10^{-5}$	$E_D = E_D^0 - \alpha N_D^{1/3}$

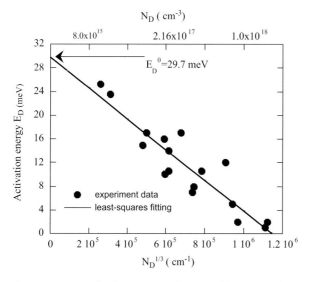

Figure 2.10 Donor binding energy as determined for a series of samples with varying donor concentrations. Measured, screened, and donor binding energies with a dilute value of 29.7 meV and α of 2.14×10^{-5} meV cm.

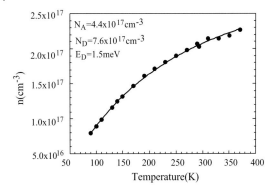

Figure 2.11 Temperature dependence of the electron concentration, for which the measured data can be fitted with a single-donor model.

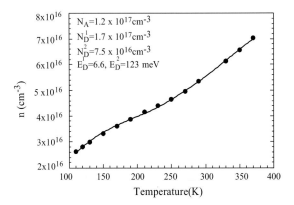

Figure 2.12 Temperature dependence of the electron concentration for which the measured data have been fitted with a model invoking the presence of a shallow donor and a deeper donor.

2.10.2
p-Type Semiconductor

As done for an n-type semiconductor, charge neutrality for a p-type sample with $p \gg n$ (*acceptor dominant*) is

$$N_A^- = N_D^{net} + p, \tag{2.41}$$

where N_D^{net} represents the temperature-independent parts of the ionized donors. Following a treatment similar to that for an n-type semiconductor, charge neutrality for a p-type semiconductor with negligible electron concentration can be written as

$$N_A^- = N_D^{net} + p = \frac{N_A}{1 + (g_{A00}/g_{A10})\exp[(E_{A1} - E_F)/kT]} = \frac{N_A}{1 + p/p_1}, \tag{2.42}$$

where as in the case of Equation 2.38

$$p_1 = \frac{g_{A10}}{g_{A00}} N'_V T^{3/2} \exp\left(\frac{-E_A}{kT}\right) = \frac{g_{A10}}{g_{A00}} N'_V T^{3/2} \left(\exp\frac{\alpha_A}{k}\right)\left(\exp-\frac{E_{A0}}{kT}\right), \quad (2.43)$$

with $N'_V = 2(2\pi m_p^* k)^{3/2}/h^3$ and $E_A = E_{A0} - \alpha_A T$, the latter assuming a linear temperature dependence of effective donor binding energy. Equation 2.42 together with Equation 2.43 is too complex to allow for the simple observation of the temperature dependence. However, for low and high temperatures and for the conditions shown for an n-type sample, the temperature dependences can be discerned just as in n-type samples.

Let us calculate the temperature dependence of electron concentration in GaN for a given donor concentration with varying degrees of compensating acceptors. Let us assume for simplicity that the acceptor binding energy is temperature independent. Using $N_D^{net} = \sum_i N_{Di}$ and $g = g_{A00}/g_{A10} = 4$, the solution of Equation 2.42 yields

$$p = \sqrt{\frac{1}{4}\left(p_1 + \sum N_{Di}\right)^2 + p_1\left(N_A - \sum N_{Di}\right)} - \frac{1}{2}\left(p_1 + \sum N_{Di}\right), \quad (2.44)$$

where $p_1 = (N_V/g)e^{-E_A/kT}$

Figure 2.13 shows the solution of Equation 2.44 for an acceptor concentration of 2×10^{18} cm^{-3}, acceptor binding energy of 180 meV, and effective electron mass of $2m_0$, as a function of temperature for a range of donor concentrations from 0 to 1.9×10^{18} cm^{-3} in GaN. If the donor concentration were zero or negligibly small (uncompensated material), the acceptor concentration as well as the acceptor binding energy can be determined from Figure 2.13 alone. Otherwise, a combination of temperature-dependent hole concentration and temperature-dependent hole mobility with hole scattering theory must be used in principle to attempt to determine the acceptor binding energy and concentration. This is difficult to do because of the complexity of the valence band and large effective mass.

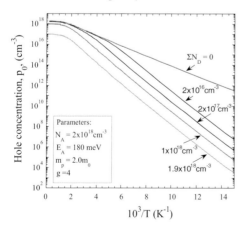

Figure 2.13 Temperature dependence of hole concentration for an acceptor concentration of 2×10^{18} cm^{-3}, donor binding energy of 180 meV, and effective electron mass of $2m_0$, as a function of temperature for a range of donor concentrations from 0 to 1.9×10^{18} cm^{-3} in GaN. Courtesy of M. Reshchikov and S.S. Chevchenko.

The temperature dependence of the hole concentration can be used to determine the acceptor concentration and its binding energy in case of zero or negligible compensation. In case of compensated semiconductor, a combination of Hall mobility and hole concentration versus temperature can, in principle, be applied to discern the binding energy and acceptor concentration. But this is much harder than that in n-type material as the effective mass is large and temperature dependence of Hall mobility is not as rich in information as in n-type semiconductor. However, marching forward, in wide-bandgap materials such as GaN, the minority free carrier concentration can be neglected because the intrinsic carrier concentration is small and this further simplifies the analysis.

Hole concentration, measured as a function of temperature and calculated with Equation 2.44, can be used together with donor concentration, with the acceptor binding energy being fitting parameter, to determine the acceptor concentration and its binding energy. Note that there is no unique solution to this fitting (in compensated semiconductor), which introduces some level of uncertainty in the parameters extracted. In GaN, additional information cannot reliably be garnered from the Hall mobility and its dependence on temperature because the Hall mobility is so low (high mass) and its temperature dependence is nondiscernible. In addition, the severe nonparabolicity of the valence band and the coupling of the various valence bands necessitate numerical approaches for mobility calculations, which are not yet developed sufficiently to be useful. In spite of this, the results of such a fitting for a degeneracy of 4 are shown in Figure 2.14, where an activation energy of 170 meV, acceptor and donor concentrations of 1.6×10^{20} and $8 \times 10^{18}\,\mathrm{cm}^{-3}$,

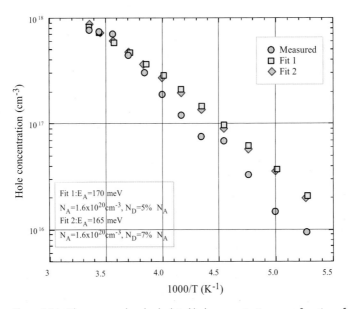

Figure 2.14 The measured and calculated hole concentrations as a function of inverse temperature with the donor and acceptor concentrations and the acceptor binding energy as fitting parameters for two sets of parameters.

respectively, and 5% of the acceptor concentration have been used. Again, the reader is cautioned that the p-type GaN has not yet been well established. It is therefore more difficult to analyze the problem by the lack of good ohmic contacts even at room temperature, let alone at low temperatures. Measurements must be done at temperatures higher than room temperature.

A brute-force approach to determine E_A is one where the semiconductor is assumed to be uncompensated, and the hole concentration is small compared to the acceptor concentration. This transforms Equation 2.44 into the approximate expression

$$p = \left(\frac{N_A N_V}{g_a}\right)^{1/2} \exp\left(\frac{E_V - E_A}{2kT}\right). \tag{2.45}$$

The slope of a plot of hole concentration versus temperature in logarithmic scale is half the acceptor activation energy neglecting the temperature dependence of the density of states. This approach is used very extensively.

2.11
Capacitance–Voltage Measurements

An ideal Schottky diode and a p–n junction with no series resistance and no leakage current can be represented by an ideal capacitor whose value is simply equal to that of the depletion capacitance, C. However, even in the reverse-bias directions these junctions do conduct some current, which means that a conductance, G, must be added to the capacitance in the parallel configuration to account for the leakage current (representing the loss term). In addition, due to semiconductor series resistance and the ohmic contact and wire resistances, there is also a series resistance, r_s, that must be added as shown in Figure 2.15a. A series or a parallel equivalent circuit model composed of an ideal capacitor and a resistor can be developed for the diode as shown in Figure 2.15b with element values of C_s and R_s ($=1/G_s$) and Figure 2.15c with element values of C_p and G_p, respectively. Typical capacitance–voltage measurements performed casually measure the parallel equivalent capacitance, C_p, not necessarily the junction depletion capacitance C. If used in determination of doping level and its profile, meaning using C_p in place of C as if it were C, it would lead to errors. The series model is not used as often, but could be very beneficial in evaluation of the semiconductor.

The equivalent circuit element values, C_s and R_s, of the series circuit or C_p and R_p of the parallel circuit can be related to the elemental values of r_s, G, and C. Other losses in the system that can be resistive in nature and also those that can present additional phase shift to the AC voltage applied to the diode during measurements must also be treated. One such source of loss is that by interface states and traps. The newly added complexity can be represented by another parallel combination of a capacitor, C_i, and a resistor, R_i, as is commonly done in the MOS realm (see, for example, Nicollian and Brews (1982)), as shown in Figure 2.15d. A capacitance–voltage behavior that is not

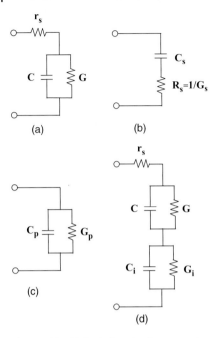

Figure 2.15 (a) Equivalent circuit representing a Schottky or p–n diode with junction capacitance C, internal conductance G, and a series resistance r_s, without an interfacial layer. (b) Series equivalent circuit of the diode in (a). (c) Parallel equivalent circuit of a diode in (a). (d) Equivalent circuit representing a Schottky or p–n diode with an interfacial layer.

congruent with space charge capacitance and its dependence on voltage could also be contributed by minority carrier injection.

In a simple circuit with a purely capacitive element, the current is ahead in terms of its phase by 90°, or the voltage lags by 90°. In a purely inductive circuit, the situation is reversed in that the voltage is 90° ahead in phase in relation to the current. If admittance for an ideal capacitor is ωC and if this is taken as positive, the admittance for an ideal inductor is $-1/\omega L$ (is negative).

In a lossy circuit such as that shown in Figure 2.15d with interface traps, at very low frequencies the traps would respond to the input while at very high frequencies they would not. What is low and high depends on the trap time constants (emission and capture rates) that are temperature dependent in that the higher the temperature the shorter the time constants and the higher the frequency the traps would respond. With manipulation of the equivalent circuit for a junction with traps, there could be a range of frequencies and biases wherein the voltage may be ahead of current in terms of phase, which is typical of an inductive circuit, not capacitive. When that occurs, the system would act as if the capacitance is negative, and thus the term "negative capacitance."

Heretofore, the discussion has been about majority carriers and majority traps. If minority carriers were injected into the system (the region of measurement), a capacitance in excess of the depletion capacitance or the space charge capacitance would be measured. The term "excess capacitance" has been coined to describe the effect of minority carriers on the junction capacitance. This is in some ways similar to the diffusion capacitance, which increases with increasing minority carrier injection in a forward-biased p–n junction. Depending on the dynamics of minority carriers, the measured capacitance could even increase with increased reverse bias, bucking the usual trend. As alluded to above, C_p and, if a deliberate attempt is made, R_p are obtained from standard C–V measurements in the parallel mode of Figure 2.15b, although the series mode could also be used to some advantage. In the series mode shown in Figure 2.15b, R_s is the measured apparent resistance and C_s is the measured apparent capacitance.

In an abrupt p–n junction and in depletion approximation, the space charge capacitance is basically given by the product of the low-frequency dielectric constant and the junction area divided by the total depletion depth that is the sum of depletion depths in n- and p-regions. If we consider the *capacitance per unit area*, C, the relationship reduces to

$$C = \varepsilon/W, \qquad (2.46)$$

where W is the total depletion depth with contributions from the n- (x_n) and the p-side (x_p). Assuming uniform doping and total ionization on both sides, x_n and x_p are given by

$$x_n^2 = \frac{2N_A \varepsilon_n \varepsilon_p (V_{bi} - V)}{qN_D(\varepsilon_n N_D + \varepsilon_p N_A)} \qquad (2.47)$$

and

$$x_p^2 = \frac{2\varepsilon_n \varepsilon_p N_D (V_{bi} - V)}{qN_A(\varepsilon_n N_D + \varepsilon_p N_A)}. \qquad (2.48)$$

If total ionization were not the case, the acceptor and donor concentrations would be replaced with ionized acceptor and donor concentrations (N_A^-, N_D^+). The total depletion depth for a homojunction ($W = x_n + x_p$) is given by

$$W^2 = \frac{2\varepsilon(N_D + N_A)(V_{bi} - V)}{q(N_D N_A)} = \frac{2\varepsilon[(1/N_D) + (1/N_A)](V_{bi} - V)}{q}. \qquad (2.49)$$

Assuming that the junction is one sided, meaning that the p-side is very heavily doped compared to the n-side ($N_A \gg N_D$) or we are dealing with a Schottky barrier on an n-type material, the depletion depth reduces to

$$W = \left[\frac{2\varepsilon(V_{bi} - V)}{qN_D}\right]^{1/2}. \qquad (2.50)$$

Substituting Equation 2.50 into Equation 2.46 leads to

$$C = \left[\frac{q\varepsilon_n N_D}{2(V_{bi} - V)}\right]^{1/2}. \qquad (2.51)$$

This means that the junction capacitance is inversely proportional to $(V_{bi} - V)^{1/2}$ for uniform doping. For an arbitrary doping concentration, the depletion capacitance can be found using

$$C = dQ_d/dV, \tag{2.52}$$

where Q_d is the depletion charge. Basically, the voltage rate of change of the depletion charge is equal to the capacitance.

With the aid of Equation 2.51, the donor concentration can be calculated by

$$N_D = \frac{2(V_{bi} - V)}{q\varepsilon} C^2. \tag{2.53}$$

Due to band bending caused by the Schottky barrier and the applied bias, the Fermi level in the depletion region, away from the transition region between the depletion region and the neutral region, is below the donor level and all the donors can be assumed ionized. Because the C–V measurements are sensitive to charge, they measure the ionized or nearly the total donor concentration. For comparison, Hall measurements measure the net electron or hole concentration.

If the doping level is nonuniform, Equation 2.51 can be reconfigured as

$$\frac{-d(1/C^2)}{dV} = \frac{2}{q\varepsilon N_D} \tag{2.54}$$

and the donor concentration (the ionized donor concentration in reality) is given by

$$N_D = \frac{2}{q\varepsilon} \left[\frac{-1}{d(1/C^2)/dV} \right]. \tag{2.55}$$

By differentiating the inverse capacitance square versus voltage, one can obtain the ionized donor profile versus voltage from Equation 2.55, which can be converted to a depth profile numerically.

By plotting C^{-2} versus the applied voltage V, one can determine the doping level N_D (ionized donor concentration is what is determined otherwise an assumption is made that all the donors are ionized) from the slope. However, if one wishes to obtain the doping profile, which is almost always the case, one must perform an iterative method to calculate the doping level from the voltage dependence of the capacitance and simultaneously the depletion depth. We have implicitly assumed that the depletion approximation holds.

Let us obtain the values of internal resistance of the diode by employing the I–V expression, which is applicable to both Schottky barriers and p–n junctions:

$$I = I_s \exp[(qV_j/nkT) - 1], \tag{2.56}$$

where I_s represents the saturation current, q is the electronic charge, n is the ideality factor, k is Boltzmann's constant, T is temperature, and V_j is the intrinsic junction voltage across the junction and expressed by $V_j = V_a - r_s I$, with V_a representing the applied voltage to the diode. The saturation current expression for a Schottky barrier is different from that for a p–n junction. At a

2.11 Capacitance–Voltage Measurements

forward-bias voltage when $V_j \gg nkT/q$, the junction conductance G consists mainly of differential conductance and is given by

$$G = \frac{dI}{dV_j} = \frac{q}{nkT}\left(1 - \frac{V_j}{n}\frac{dn}{dV_j}\right). \tag{2.57}$$

The total impedance of the circuit of Figure 2.15a is

$$Z = r_s + \frac{1}{G_p + j\omega C_p}, \tag{2.58}$$

where ω is the angular frequency of the applied alternating signal.

The impedance in series and parallel configurations of Figure 2.15b and c is given by

$$Z = R_s + \frac{1}{j\omega C_s} \quad \text{and} \quad Z = \frac{1}{G_p + j\omega C_s}, \quad \text{respectively.} \tag{2.59}$$

Recognizing that the total impedance for the circuit in Figure 2.15a is equal to the total impedance for the series circuit in Figure 2.15c and by equating real and imaginary components, one obtains

$$G_p = \frac{G(1 + r_s G) + r_s(\omega C)^2}{(1 + r_s G)^2 + (\omega r_s C)^2} \quad \text{and} \quad C_p = \frac{C}{(1 + r_s G)^2 + (\omega r_s C)^2}. \tag{2.60}$$

To obtain accurate capacitance data without detailed analysis characterized by Equation 2.60, the following conditions must be satisfied: $r_s G \ll 1$ and $(\omega r_s C)^2 \ll 1$, in which case $G = G_p$ and $C = C_p$. If the measurements are conducted at 1 MHz, which is typical, the above conditions lead to $r_s < 1.6 \times 10^{-7}/C$ (pF) to satisfy the latter condition. For $C = 50$ pF, the diode series resistance r_s should be much less than 3200 Ω for accurate measurements, the extent of which depends on how far the equation for C_p deviates from C in Equation 2.60.

Similarly, recognizing that the total impedance for the circuit in Figure 2.15a is equal to the total impedance for the series circuit in Figure 2.15b and by equating real and imaginary components, one obtains

$$R_s = r_s + \frac{G}{G^2 + (\omega C)^2} \quad \text{and} \quad C_s = \frac{G^2 + (\omega C)^2}{\omega^2 C}. \tag{2.61}$$

According to the first part of Equation 2.59, the series equivalent resistance R_s would peak when $G = \omega C$ and in turn G is determined by the junction voltage (refer to Equation 2.57). Note that higher voltages produce more current and thus larger conductance. Lower frequencies produce larger peaks at smaller bias voltages. When $G \gg \omega C$, Equation 2.59 (for large values of G due to increased current through the junction and/or smaller frequencies used) would take the form

$$R_s = r_s + \frac{1}{G} \quad \text{and} \quad C_s = \frac{G^2}{\omega^2 C}. \tag{2.62}$$

The first part of Equation 2.62 indicates that the equivalent resistance is independent of frequency and $R_s \approx r_s$ if $1/G \ll r_s$, that is, the equivalent resistance approaches the series resistance of the diode for large forward-bias voltages. By the combined use of Equations 2.56, 2.57, and 2.62, the dependence of r_s, C, n, and V_j on the forward-bias voltages can be obtained.

For extending the treatment to include interface states, refer to the equivalent circuit of Figure 2.15d, where R_i and C_i are the contributions to the resistance and capacitance, respectively, by interface states that can be considered as an interfacial layer. By defining $R_r = r_s - R_i$, $G_i = 1/R_i$, and calculating the series impedance of the equivalent circuit impedance, the R_s and C_s equivalent circuit elements in the serial mode can be determined as

$$R_s = R_r + \frac{G}{G^2 + (\omega C)^2} + \frac{R_i}{1 + (\omega C_1 R_i)^2},$$

$$C_s = \left[\frac{\omega^2 C}{G^2 + (\omega C)^2} + \frac{\omega C_1 R_i^2}{1 + (\omega C_1 R_i)^2} \right]^{-1}. \qquad (2.63)$$

It can be seen that R_s is frequency dependent, which means that a dispersion measurement in conjunction with this analysis is needed. Unlike the case with no interfacial layer or interface states, R_s may peak twice with respect to frequency, once for $G = \omega C$ (as before) and the other is due to interfacial resistance R_i. In the low-frequency limit, if $\omega C \ll G$ and $\omega C_i R_i \ll 1$, Equation 2.63 becomes

$$R_s = r_s + \frac{1}{G} \quad \text{(without interface states)} \quad \text{and} \quad C_s = \frac{1}{\omega^2} \left[\frac{C}{G^2 + C_i R_i^2} \right]^{-1}. \qquad (2.64)$$

The first parts of Equations 2.62 and 2.64 being equal can be understood by recognizing that an "open circuit" at lower frequencies would represent C_i.

In the high-frequency limit, $\omega C_i R_i \gg 1$ and $\omega C_i R_r \ll 1$, Equation 2.63 reduces to

$$R_s = R_r + \frac{G}{G^2 + (\omega C)^2} \quad \text{and} \quad C_s = \left[\frac{\omega^2 C}{G^2 + (\omega C)^2} + \frac{1}{C_i} \right]^{-1}. \qquad (2.65)$$

The first part of Equation 2.65 indicates that R_s approaches R_r for large forward-bias voltages. This simply means that the resistance due to the interface states or the interfacial layer is shunted by the capacitance due to the same.

GaN-based Schottky barriers have been measured in terms of C–V characteristics using a structure of n$^+$-GaN ($\approx 5 \times 10^{18}$ cm^{-3})/n-GaN ($\approx 6 \times 10^{16}$ cm^{-3})/sapphire. The dependence of equivalent series resistance R_s, extracted using the analysis discussed above, on forward bias at frequencies of 1 kHz–1 MHz for a device annealed at 700 °C for 40 s following the deposition of the Ti–Al electrode is shown in Figure 2.16. For small bias voltages where $G = \omega C$ holds, there is a peak in the R_s–V curve below about 0.5 V, indicated by a series of four curves at the bottom-left side of the figure. At large voltages, the value of R_s is almost independent of frequency and decreases slightly with increasing voltage (the upper curve). The data do not point to any obvious interfacial layer in this particular Schottky device.

Figure 2.17 shows frequency-dependent C–V data for a Schottky barrier on GaN in the frequency range 10 kHz–10 MHz. Note the excess capacitance above a bias

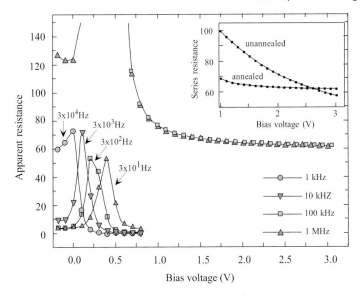

Figure 2.16 Dependences of equivalent series resistance, R_s, on forward-bias voltages with different frequencies for Ni/n-GaN Schottky diodes with an n-layer, for the annealed Schottky diode. The inset shows the measured series resistance r_s for annealed and unannealed devices versus forward voltage. Courtesy of C.D. Wang.

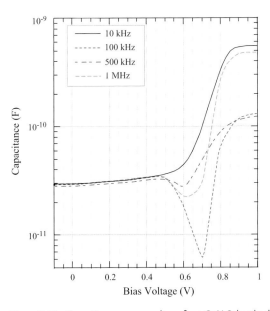

Figure 2.17 Capacitance versus voltage for a GaN Schottky device in the frequency range 10 kHz–10 MHz.

voltage of about 0.6 V, which is indicative of minority carrier injection. In addition, the data for 500 kHz, 1 MHz, and 10 MHz show negative capacitance, which is indicative of inductive modulation in the device.

As for the diffusion capacitance, it comes about from minority carrier injection and associated time delay before the minority carriers are diffused away, swept away, or recombined. Because minority carrier dynamics are not instantaneous, the associated charging effect with its delay in relation to the input frequency gives rise to excess capacitance, which early on got termed as diffusion capacitance. Even though it is traditionally applied to forward-biased p–n junctions, it is observed even in Schottky barriers wherein the minority carriers may play a role.

Appendix 2.A. Fermi Integral

The Fermi integrals $F_n(\eta)$ employed in the calculation of the Fermi–Dirac distribution function are briefly summarized and appropriate approximations are outlined. The general Fermi integral can be expressed as

$$F_n(\eta) = \frac{2}{\sqrt{\pi}} \int_0^\infty \frac{x^n \, dx}{1 + \exp(x - \eta)}. \tag{2.A.1}$$

The Fermi integral of the 1/2 kind can be written as

$$F_{1/2}(\eta) = \frac{2}{\sqrt{\pi}} \int_0^\infty \frac{x^{1/2} \, dx}{1 + \exp(x - \eta)}. \tag{2.A.2}$$

To simplify calculations, this expression can be approximated by

$$F_{1/2}(\eta) \approx \exp(\eta) \quad \text{for} \quad \eta \ll -1 \tag{2.A.3}$$

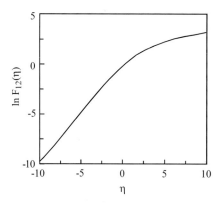

Figure 2.A.1 The Fermi integral.

and

$$F_{1/2}(\eta) = \frac{4\eta^{3/2}}{3\sqrt{\pi}} \quad \text{for} \quad \eta \gg 1. \tag{2.A.4}$$

For $-10 < \eta < 10$, the Fermi integral of the $F_{1/2}(\eta)$ kind can be approximated by the following:

$$F_{1/2}(\eta) = \exp(-0.3288 + 0.7404\eta - 0.0454\eta^2 - 8.797 \times 10^{-4}\eta^3 + 1.512 \times 10^{-4}\eta^4). \tag{2.A.5}$$

For other values of η, $F_{1/2}(\eta)$ can be obtained from Figure 2.A.1.

Further Reading

Blakemore, J.S. (1985) *Solid State Physics*, 2nd edn, Cambridge University Press.

Look, D.C. (1989) *Electrical Characterization of GaAs Materials and Devices*, John Wiley & Sons, Ltd, Chichester.

McKelvey, J.P. (1986) *Solid State Physics*, 4th edn, Krieger Publishing Company.

Morkoç, H. (2008) *Handbook on Nitride Semiconductors and Devices*, vol. 2, Wiley-VCH Verlag GmbH, Weinheim.

Neugebauer, J. and Van de Walle, C.G. (1996) *Festkörperprobleme (Advances in Solid State Physics)*, vol. 35, Vieweg, Braunschweig, p. 25.

Nicollian, E.H. and Brews, R.J. (1982) *MOS Physics and Technology*, John Wiley & Sons, Inc., New York.

Shockley, W. (1950) *Electrons and Holes in Semiconductors*, Van Nostrand, Princeton, NJ.

Sze, S.M. and Ng, K. (2007) *Physics of Semiconductor Devices*, 3rd edn, Wiley-Interscience.

Van de Walle, C.G. (1997) *Phys. Rev. B*, 56, R10020–R10023.

Van de Walle, C.G. (2003) *Phys. Status Solidi B*, 235, 89.

Van de Walle, C.G. and Neugebauer, J. (2004) *J. Appl. Phys.*, 95, 3851.

Van de Walle, C. G., et al., in *Properties, Processing and Applications of Gallium Nitride and Related Semiconductors*, edited by J. Edgar, S. Strite, I. Akasaki, H. Amano, and C. Wetzel, EMIS Datareview Series No. 23 (INSPEC, IEE, 1999), pp. 275–280; 281–283; 313–316; 317–321.

3
Metal Contacts

3.1
Metal–Semiconductor Band Alignment

Any semiconductor device must be connected to the outside world with no adverse change to its current–voltage characteristics and no additional voltage drop. This can be accomplished only through low-resistance ohmic contacts to the semiconductor. An ideal contact is one where, when combined with the semiconductor, there are no barriers to the carrier flow in either the positive or the negative direction. When a metal and a semiconductor with no surface states are brought in contact and the equilibrium is maintained, their Fermi levels will align. If the Fermi levels of the metal and semiconductor were the same before contact, then there would be no change in the band structure after contact. Because the Fermi level in the semiconductor, and thus the work function, is carrier concentration dependent, matching the work functions is nearly impossible and one must, therefore, settle for a compromise.

Let us consider the case of an n-type semiconductor and a metal with a work function that is larger than that of the semiconductor. The alignment of the Fermi levels after contact, brought about by the charge motion from the higher toward the lower energy side, creates a depletion region in the semiconductor and a barrier at the interface. The barrier height ϕ_B (before the image force lowering discussed below) is simply the difference between the metal work function ϕ_m and the electron affinity in the n-type semiconductor ($\phi_m - \chi$), as shown in Figure 3.1. In this ideal picture, the band bending in the semiconductor is simply the difference between the metal work function ϕ_m and semiconductor work function ϕ_s.

The image force lowering taken into consideration in the figure comes about from a negative charge at a distance x from the surface of, say, a metal for convenience, inducing a positive charge of equal value at a distance $-x$ from the surface. The confining barrier, which results, can be lowered by application of an electric field normal to the surface. The image force lowering, $\Delta\phi$, is given as

$$\Delta\phi = \sqrt{\frac{qE}{4\pi\varepsilon_s}}, \qquad (3.1)$$

with E representing the magnitude of applied field, which is normal to the interface.

Nitride Semiconductor Devices: Fundamentals and Applications, First Edition. Hadis Morkoç.
© 2013 Wiley-VCH Verlag GmbH & Co. KGaA. Published 2013 by Wiley-VCH Verlag GmbH & Co. KGaA.

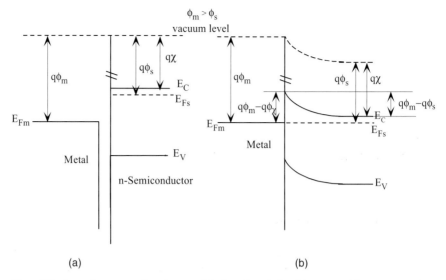

Figure 3.1 A metal n-type semiconductor pair before (a) and after (b) contact with no surface/interface states. The metal work function is greater than that for the semiconductor ($\phi_m > \phi_s$).

Clearly, the image force lowering increases through a square root dependence with the electric field. This means that the effective barrier for electron escaping the metal is given by the difference between the metal work function and the image force lowering, expressed as

$$\phi_B = \phi_m - \Delta\phi. \tag{3.2}$$

A forward bias (negative voltage applied to the n-type semiconductor with respect to the metal) lowers the overall barrier, as shown in Figure 3.2a, and a reverse bias (positive voltage applied to the n-type semiconductor with respect to the metal) increases the barrier, as shown in Figure 3.2b. The barrier to electron flow from the metal to the semiconductor remains almost unchanged except through a change in $\Delta\phi$, which increases with electric field. The band diagram for a rectifying metal p-type semiconductor system before (a) and after (b) contact (in equilibrium) is shown in Figure 3.3. The same with forward bias (positive voltage applied to the semiconductor with respect to the metal) is shown in Figure 3.4a and with reverse bias (negative voltage applied to the semiconductor with respect to the metal) is shown in Figure 3.4b.

The case where the fortuitous matching of the metal–semiconductor pair occurs, that is, metal work function is equal to or slightly smaller than that of the semiconductor metal work function, is depicted in Figure 3.5 with an automatic ohmic contact behavior. Unlike the case where the metal work function is smaller than that of the semiconductor, charge accumulation rather than

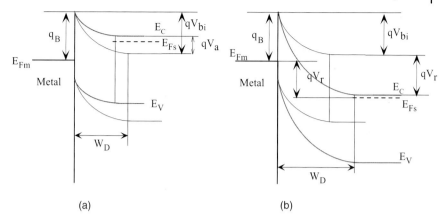

Figure 3.2 A metal n-type semiconductor system where the metal work function is greater than the semiconductor work function in forward-bias (a) and reverse-bias (b) cases. The terms V_a and V_r represent the forward- and reverse-bias voltages, respectively. The extension of depletion regions (W_D) is also shown, so is the equilibrium band diagram of the semiconductor in lighter shade. The image force lowering is not shown for simplicity.

depletion occurs here with a negligible voltage drop. The same, but impractical, is displayed in Figure 3.5 for a p-type semiconductor where the metal work function needs to be equal to or larger than that for the semiconductor. For p-type GaN, this would mean a metal with a work function of about 8 eV, which does not exist.

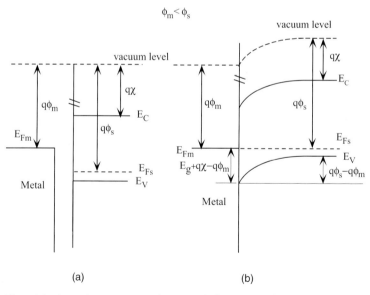

Figure 3.3 A metal p-type semiconductor pair before (a) and after (b) contact with no surface/interface states. The metal work function is smaller than that for the semiconductor ($\phi_m < \phi_s$).

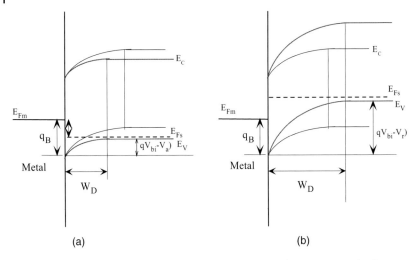

Figure 3.4 A metal p-type semiconductor system where the metal work function is greater than the semiconductor work function in forward-bias (a) and reverse-bias (b) cases. The terms V_a and V_r represent the forward- and reverse-bias voltages, respectively. The extension of depletion regions (W_D) is also shown, so is the equilibrium band diagram of the semiconductor in lighter shade. The image force lowering is not shown for simplicity.

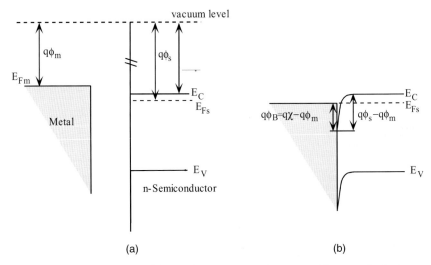

Figure 3.5 Fictitious matching of the metal p-type semiconductor pair with automatic ohmic contact behavior brought about by the assumed metal work function being equal to or greater than that for the semiconductor (a: before contact; b: after contact).

3.2
Current Flow in Metal–Semiconductor Junctions

When a metal is brought in contact with a semiconductor, there arises a potential barrier unless the work functions of the metal and the semiconductor match, which is highly unlikely. When a bias is applied, current flow takes place if the carriers in the metal or in the semiconductor gain sufficient energy. Carriers can overcome the barrier thermally or by the aid of an electric field.

However, when the barrier is sufficiently thin, they can also go through the barrier either by direct tunneling if the barrier thickness is comparable to the tunneling distance or by gaining sufficient energy with respect to the Fermi level combined with tunneling at some point in the barrier. The current conduction process over or through a barrier created by a metal–semiconductor contact is schematically shown in Figure 3.6. These processes can be circumvented by defects. In cases where

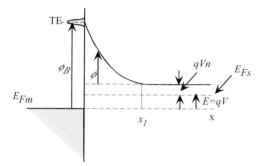

(a)

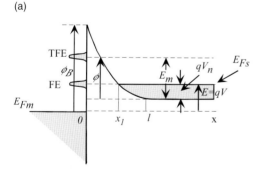

(b)

Figure 3.6 Potential energy diagram and current flow mechanisms for a forward-biased Schottky barrier, (a) for the TE process that is more likely when the doping level in the semiconductor is relatively low and the Fermi level is below the conduction band, and (b) for thermionic FE and direct tunneling, which is referred to as FE, that is more likely when the doping level in or on the semiconductor surface is sufficiently high to the extent that the Fermi level may even be in the conduction band as depicted.

defects are not involved, there are three mechanisms that govern the current flow in a metal–semiconductor system, which are discussed below.

1) *Thermionic emission* (TE). For lightly or moderately doped semiconductors, $N_D < \approx 10^{17}$ cm^{-3}, the depletion region is relatively wide. It is, therefore, nearly impossible for electrons to tunnel through the barrier unless aided by defects, which are considered not to exist in this ideal picture. In a forward-biased junction, however, the electrons can surmount the top of the barrier, which is lowered with respect to the Fermi level in the semiconductor by an amount equal to the applied bias. This is called the thermionic emission as shown in Figure 3.6a and has been treated in many papers and early texts. In reverse bias, the barrier for electrons from the semiconductor to the metal is made even larger and the electron flow from the semiconductor to the metal in this ideal picture is cut off. On the metal side, if the electrons in the metal gain sufficient energy by the applied bias, they too can overcome the barrier, which is the dominant mechanism for the reverse-bias current in an ideal picture. Naturally, an ohmic behavior is not observed. The electron flow from the metal to the semiconductor and from the semiconductor to the metal must balance for zero bias under steady-state conditions, which ensures zero net current. The thermionic process requires electrons to gain sufficient energy for current flow from which it gets its name.

The traditional current–voltage expression representing thermionic emission is given by

$$J_{te} = J_{te0}\left[\exp\left(\frac{qV}{kT}\right) - 1\right], \tag{3.3}$$

with

$$J_{te0} = A^* T^2 \exp\left[\frac{-q(\phi_B - \Delta\phi)}{kT}\right], \tag{3.4}$$

where J_{te0} is the saturation value of the current density J_{te}, A^* is the *effective Richardson constant*, ϕ_B is the barrier height, and $\Delta\phi$ is the image force *barrier lowering*. Equation 3.7 is based on the condition that the series resistance of the circuit is negligibly small. It should be pointed out that the saturation current density is typically designated by J_S in general. We do the same throughout most of this chapter. However, in this subsection, to make the point that we are discussing thermionic emission, a more descriptive nomenclature, J_{te0}, is used.

As the kT term in the exponent indicates, the slope of J_{te0}/T^2 would vary with temperature with a slope of kT in a semilogarithmic plot. The Richardson constant for free space is given by

$$A^*_{free} = \frac{4\pi q k^2 m_0}{h^3}, \tag{3.5}$$

which is equal to 120 A/(cm^2 K^2). The effective Richardson constant is $A^* = A^*_{free}(m^*_e/m_0)$ for n-type and $A^* = A^*_{free}(m^*_{hh}/m_0)$ for p-type semiconductors under the assumption of single-valley conduction bands such as n-type GaN and single and spherical valence band conduction. When both heavy and light hole bands are occupied, the effective Richardson constant is given by $A^* = A^*_{\text{free}}[(m^*_{\text{hh}} + m^*_{\text{lh}})/m_0]$. In cubic compound semiconductors, the valence band is degenerate and thus the last expression for the Richardson constant should be used.

Equation 3.3 is a representation of the carrier flux from the semiconductor to the metal for which the barrier is voltage dependent (field dependent). However, for flux from the metal to the semiconductor, the barrier is fixed at ϕ_B. Because of parasitic resistance, such as semiconductor resistance, the thermionic emission current expression is modified as

$$J_{te} = J_{te0}[e^{q(V-IR_s)/kT} - 1]. \tag{3.6}$$

Here the current I is determined by the product of the current density J and the area of the structure.

Because both A^* and $\Delta\phi$ are voltage dependent, it is customary to represent the current–voltage characteristics simply as $J \sim \exp(qV/nkT)$ for applied voltages $> 3kT/q$, with n being the ideality factor.

In the reverse direction, the barrier lowering becomes more important. In a such a case (using J_S instead of J_{te0} for saturation current),

$$J_R \approx J_S = A^* T^2 \exp\left(\frac{-q\phi_B}{kT}\right) \exp\left(\frac{q\Delta\phi}{kT}\right), \tag{3.7}$$

where the image force barrier lowering

$$\Delta\phi = \sqrt{qE/4\pi\varepsilon_s} \tag{3.8}$$

and the electric field at the metal–semiconductor interface is given by

$$E = \sqrt{\frac{2qN_D}{\varepsilon_s}\left(-V + V_{bi} - \frac{kT}{q}\right)}. \tag{3.9}$$

Neglecting the image force lowering, the barrier height and the effective Richardson constant can be experimentally determined by plotting $\ln(J_R/T^2)$ versus $1000/T$ (Richardson plot). Actually, the result would be that of the effective barrier height including the image force lowering. If the interface electric field can be determined, the image force lowering can be calculated. Performing the measurements for a range of reverse-bias conditions that would allow the determination of saturation current for that range of biases would allow the determination of the image force lowering component. The assumption here is that components of the current other than the thermionic emission are non-existent or can be separated out. Confidence can be gained if thus determined image force lowering is linearly dependent on the square root of the interface electric field. From an experimental point of view, generation–recombination current would also increase, unless negligible, which would exacerbate the determination of barrier lowering by image force.

Somewhat related, the generation–recombination current, which is discussed in more detail in Chapter 5, is given by

$$J_{gr} = \frac{qn_i W}{\tau} \exp\left(\frac{qV}{2kT}\right), \quad (3.10)$$

where n_i is the intrinsic concentration, τ is the effective carrier lifetime, and W is the depletion depth given by

$$W = \sqrt{\frac{2\varepsilon_s(V_{bi} - V)}{qN_D}}. \quad (3.11)$$

Because the intrinsic carrier concentration is nearly nil at room temperature for GaN, the generation–recombination current component should be negligibly small.

2) *Thermionic field emission* (TFE). For intermediately doped semiconductors, $\approx 10^{17}$ cm$^{-3} < N_D < \approx 10^{18}$ cm^{-3}, the depletion region is not sufficiently thin to allow direct tunneling of carriers that are more or less in equilibrium. This process requires some energy gain from the bias sufficient to raise the electron energy to a value E_m, where the barrier is sufficiently thin for tunneling, as shown in Figure 3.6b. This process is the one that incorporates elements of thermionic emission in the sense that electrons must be moderately hot or warm and tunneling, which requires penetration through a sufficiently thin barrier.

The forward current density due to TFE can be expressed as

$$J_{tfeF} = J_{SF} \exp\left(\frac{qV}{E_0}\right), \quad (3.12)$$

where $E_0 = E_{00} \coth(E_{00}/kT)$ and $E_{00} = (q\hbar/2)\sqrt{N_D/\varepsilon_s m^*}$ and J_{SF} is the saturation value of the current J_{tfeF} and expressed by

$$J_{SF} = \frac{A^* T^2 \sqrt{\pi q E_{00}(\phi_B - V + V_n)}}{kT \cosh(E_{00}/kT)} \exp\left(\frac{qV_n}{kT} - \frac{q(\phi_B + V_n)}{E_0}\right). \quad (3.13)$$

Considering the electron emission from the metal to the semiconductor at energy E_m, the total current in the forward direction (neglecting the error function term) can be expressed as

$$J_F = J_{SF}[\exp(qV/n_F kT) - 1] \quad \text{with} \quad n_F = \frac{E_{00}}{kT} \coth\left(\frac{E_{00}}{kT}\right) = \frac{E_0}{kT}. \quad (3.14)$$

In the reverse bias, the metal potential is raised, as shown in Figure 3.7. If the doping level in the semiconductor is low and the barrier width is large (keep in mind that the barrier width becomes smaller for energies above the Fermi level as compared to the forward-bias case), the current flow is through thermionic emission. This process is schematically shown in Figure 3.7a. However, in cases when the doping level is moderate or high, the dominant current mechanism in the reverse-bias direction also would be TFE and field emission (FE) currents for the forward-bias case as shown in Figure 3.7b. For the case of forward bias, we

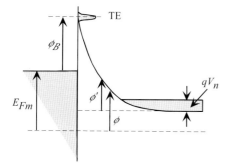

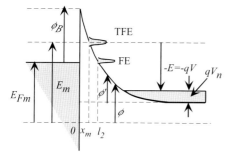

Figure 3.7 Potential energy diagram and current flow mechanisms for a reverse-biased Schottky barrier, (a) for the TE process that is more likely when the doping level in the semiconductor is relatively low and the Fermi level is below the conduction band, and (b) for thermionic FE and direct tunneling, which is referred to as FE. The latter is more likely when the doping level in or on the semiconductor surface is sufficiently high to the extent that the Fermi level may even be in the conduction band as depicted.

can think of the TFE current being dominant in an intermediate temperature range and the FE current being dominant in the low-temperature range.

Performing similarly but considering the electron emission from the semiconductor to the metal, the reverse current–voltage characteristics for the thermionic FE region can be expressed in terms of more familiar parameters as

$$J_R = J_{SR}[\exp(qV/n_R kT) - 1] \quad \text{with} \quad n_R = \frac{E_{00}}{kT}\left[\frac{E_{00}}{kT} - \tanh\left(\frac{E_{00}}{kT}\right)\right]^{-1}. \quad (3.15)$$

The term J_{SR} represents the saturation current as in the case of TE but with very different functional dependence. These relations provide the smooth transition from the TFE regime to just FE regime as the temperature is lowered, which hampers the thermionic emission. A unique property here is that the sum of the inverse of the forward and reverse ideality factors is 1:

$$n_F^{-1} + n_R^{-1} = 1. \quad (3.16)$$

3) *Field emission.* In heavily doped semiconductors, $N_D > \approx 10^{18}$ cm^{-3}, the depletion region is narrow even for cold and cool electrons at the bottom of the conduction band or at the Fermi level, the latter is for degenerate semiconductors, and direct electron tunneling from the semiconductor to the metal is allowed as shown in Figure 3.6b. In the absence of a good match between the metal and the semiconductor work functions, which is generally the case, this is the best approach to pursue ohmic contacts provided, of course, that very large doping concentrations can be attained. Direct tunneling process, that is, FE, is dominant at low temperatures, if allowed by features such as high doping concentrations. The flux from the semiconductor to the metal is proportional to the product of the transmission coefficient, the occupation probability in the semiconductor, f_s, and the unoccupation probability in the metal, $1 - f_m$:[1)]

$$J_{s \to m} = \frac{A^* T}{k} \int_0^{q\phi_B} f_s T(\xi)(1 - f_m) d\xi, \tag{3.17}$$

where $T(\xi)$ is the transmission coefficient and is given by, for low temperatures and/or high doping levels, $T(\xi) \approx \exp(-q\phi_B/E_{00})$. Similarly, the density of current flowing from the metal to the semiconductor is proportional to the product of the transmission coefficient, the unoccupation probability in the semiconductor, and the occupation probability in the metal:

$$J_{m \to s} = -\frac{A^* T}{k} \int_0^{q\phi_B} f_m T(\xi)(1 - f_s) d\xi. \tag{3.18}$$

The total density of current, which is simply the sum of the density of current flowing in both directions, can be approximated by

$$J_t \approx \exp\left(\frac{-q\phi_B}{E_{00}}\right). \tag{3.19}$$

Padovani and Stratton presented an analytical expression for the forward current for direct tunneling as

$$J_{FE} = J_{SFE} \exp\left(\frac{qV}{E_{00}}\right), \tag{3.20}$$

where

$$J_{SFE} = \frac{2\pi A^* T^2 E_{00}}{kT[\log\{2(\phi_B - V)/V_n\}]\sin[(\pi kT/2E_{00})\log\{2(\phi_B - V)/V_n\}]} \exp\left(\frac{-q\phi_B}{kT}\right). \tag{3.21}$$

---

1) The occupation probability depicts the likelihood that a state *is* occupied by an electron, and one minus the occupation probability exhibits that to be free of electrons.

For the reverse-bias case,

$$J_{FER} = \frac{\pi A E_{00} T^2}{kT\left[\sqrt{\phi_B/(\phi_B - V)}\right]\sin\{(\pi kT/E_{00})\sqrt{\phi_B/(\phi_B - V)}\}} \quad (3.22)$$

$$\exp\left\{[-2(q\phi_B)^{3/2}] \,/\, \left[3E_{00}\sqrt{q\phi_B - qV}\right]\right\},$$

where A is the Richardson constant of the metal. In the limit of zero temperature, Equation 3.22 further reduces to

$$J_{FER} = AT^2 \left(\frac{E_{00}}{kT}\right)^2 \frac{\phi_B - V}{\phi_B} \exp\left\{\left[-2(q\phi_B)^{3/2}\right] \,/\, \left[3E_{00}\sqrt{q\phi_B - qV}\right]\right\}. \quad (3.23)$$

Equation 3.23 shows that a plot of $\ln[J_{FER}/q(\phi_B - V)]$ as a function of $\sqrt{q\phi_B - qV}$ would yield a straight line with a slope of $2(\phi_B)^{3/2}/3E_{00}$.

A plot of the natural logarithm of the current versus voltage given in Equation 3.20 would yield a slope of q/E_{00} regardless of the temperature, which is a characteristic of direct tunneling current. The coefficient in front of Equation 3.20 (the saturation current), which is expanded in Equation 3.21, clearly indicates that the lower the barrier, ϕ_B, and the higher the doping level, which increases E_{00}, the higher the saturation current and thus the higher the current. This explicitly implies that the resistance is low.

3.3
Ohmic Contact Resistance

An ohmic contact is a metal–semiconductor contact that has a very small contact resistance compared to the bulk or spreading resistance of the semiconductor. It is said that the contact is ohmic when the ratio of the potential drop V across the contact versus the current I flowing through the contact is linear with a constant R_c. Ideal ohmic contacts should not contribute to the voltage drop across the device and should not alter the current–voltage relationship. In addition, the contact must remain intact and robust regardless of the environment, and the contact characteristics must not change with storage and dynamic operations. Naturally, not all of these requirements can be met simultaneously but gallant strides should be made to satisfy as many as possible and to the extent possible.

3.3.1
Specific Contact Resistivity

Although the current–voltage (I–V) expression is sufficient, it is customary to deduce the specific resistance near zero bias. Caution should be exercised, as the I–V characteristic may not be linear, thus causing a voltage-dependent resistance term. Nevertheless, the specific resistance creates impediments to current flow. It is in this context that we define the specific contact resistivity, in terms of Ω cm^2. The product of R_c and the area A of the contact is called the *specific contact resistance* ρ_c expressed as

$$\rho_c = \left[\frac{\partial J}{\partial V}\right]^{-1}_{V=0} \quad (\Omega\text{cm}^2). \tag{3.24}$$

For $kT/E_{00} \gg 1$ (*moderate doping concentrations*), the TE mechanism dominates the current conduction and the specific contact resistance near $V=0$, with the aid of Equations 3.7 and 3.8, becomes

$$\rho_c = \frac{k}{qA^*T}\exp\left(\frac{q\phi_B}{kT}\right). \tag{3.25}$$

It is clearly dependent on temperature, and at higher temperatures, there is more thermionic emission current, which results in a smaller ρ_c.

For $kT/E_{00} \approx 1$ (*intermediate doping concentrations*), a mixture of thermionic, thermionic FE, and tunneling mechanisms is in place, which would lead to

$$\rho_c \propto \exp\left[\frac{q\phi_B}{E_{00}\coth(E_{00}/kT)}\right]. \tag{3.26}$$

For $kT/E_{00} \ll 1$, which is associated with high doping levels, the tunneling current dominates and we have

$$\rho_c \propto \exp\left(\frac{q\phi_B}{E_{00}}\right). \tag{3.27}$$

In this case, ρ_c depends strongly on the doping concentration and the barrier height. As the doping concentration is increased further, the depletion width of the Schottky junction decreases. This results in an increase in the tunneling transmission coefficient and a decrease in the resistance.

If a large number of surface states exist on the semiconductor surface, the barrier height is pinned at the semiconductor surface within its energy gap, and is independent of the metal work function. This is the Bardeen limit, which contrasts the Schottky limit where the metal–semiconductor contact is assumed ideal and the surface states are ignored. In practice, the Fermi levels of most III–V compounds are pinned, and the resultant barriers must be considered. The barrier height depends on the bandgap and the surface state density of the semiconductor.

3.4
Semiconductor Resistance

In addition to the metal–semiconductor resistance, the semiconductor resistance too must be added to the total resistance. The semiconductor resistance, mainly due to the neutral region, may be defined as

$$R_s = \frac{1}{A_j}\int_{x_1}^{x_2} \rho(x)dx, \tag{3.28}$$

where x_1 represents the depletion edge, x_2 denotes the boundary of the epitaxial layer, $\rho(x)$ is the resistivity at x, and A_j is the area of the metal–semiconductor junction. The parameter x_1 depends on the depletion width W, which is a function of temperature,

$$W = \left[\frac{2\varepsilon_s}{qN_{\text{Deff}}}\left(V_{\text{bi}} - V - \frac{kT}{q}\right)\right]^{1/2}, \tag{3.29}$$

where V_{bi} is the built-in potential given by

$$V_{\text{bi}} = \phi_B - \eta = \phi_B - \frac{kT}{q}\ln\left(\frac{N_c}{N_D}\right). \tag{3.30}$$

In GaN-based p–n junctions, such as those in LEDs and lasers, p-type contact resistance dominates because of a large metal–semiconductor barrier and a large effective mass. In addition, the semiconductor resistance is also large due to a combination of the low hole concentration and the low hole mobility. The non-ohmic behavior caused by a combination of a high metal to p-semiconductor barrier and a low hole concentration, unless chemical interaction between the metal layers and the semiconductor causes the direct-tunneling-like current to dominate, will give rise to a voltage drop as well as an increased resistance exacerbating the Joule heating and the resultant rise in junction temperature. In addition, sapphire substrates, must they be used, are semi-insulating, necessitating the use of surface contacts for both n- and p-type regions. Doing so requires current conduction laterally from the n-type contact to the junction area. Due to the considerable distance involved between the metal contact and the junction area, the semiconductor resistance is considerable. It could be lowered if the n-type semiconductor is sufficiently thick. It is exasperating that highly Si-doped n-type GaN cracks if its thickness is increased beyond about 3 μm due to the residual thermal strain.

3.4.1
Determination of the Contact Resistivity

The most widely used method for determining the *specific contact resistance* is the method of *transfer length* first introduced by Shockley, which now goes with the nomenclature of *transmission line model* (TLM). In this particular approach, a linear array of contacts is fabricated with various spacings between them. The pattern used and the resistance versus the gap spacing l (l_{12}, l_{23}, l_{34}, ...) are depicted in Figure 3.8. The total resistance is given by

$$R_T = 2R_c + \frac{lR_{\text{sshr}}}{Z}, \tag{3.31}$$

where the first term represents twice the contact resistance R_c because the resistance is measured between two identical contacts, while the second term is due to the semiconductor resistance that is dependent on the contact separation or the gap between contacts. The term R_{sshr} denotes the sheet resistance of the semiconductor layer. Care must be taken to account for the resistance between the ohmic contacts and measurement setup. If an overlay metallization is used and

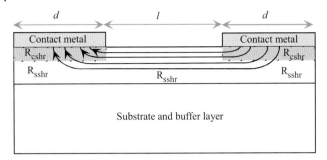

Figure 3.8 Schematic representation of alloyed ohmic contacts to a semiconductor where the filled region below the metallization indicates the altered semiconductor due to interaction between the contact material and the semiconductor. Here R_{sshr} and R_{cshr}, respectively, indicate the sheet resistance of the unaltered semiconductor and the sheet resistance of the altered semiconductor below the contact metal after annealing.

the probe to the overlay metallization contact is very good, the resistance can be negligible. The contact resistivity measurement methods that have been devised deduce not only the resistivity but also the semiconductor sheet resistance.

Contacts formed on heavily doped semiconductor, which take advantage of tunneling, do not alter the semiconductor properties under the metal. However, in contacts relying on interaction of the deposited metal and the underlying semiconductor through annealing, the semiconductor properties and thus its resistance under the contact metal are altered. Consequently, the semiconductor resistance under the contact differs from that outside the contact region as shown in Figure 3.8, where R_{sshr} and R_{cshr} represent the semiconductor sheet resistance and sheet resistance of the alloyed region under the contact metal, respectively. In addition, the current flow paths are also shown. Because the current flows through the least path of resistance, the current density is higher near the edge of the contact.

Referring to Figure 3.8, the current distribution in the alloyed region can be expressed as

$$dI(x) = -ZJ(x)dx, \tag{3.32}$$

where Z is the width of the ohmic contact, x represents the distance parallel to the surface of the semiconductor, and J is the current density that can be related to the specific contact resistance through

$$J(x) = V(x)/r_c, \tag{3.33}$$

where r_c is the specific contact resistivity and $V(x)$ is the channel potential with respect to the potential of the contact metal. The voltage distribution is expressed as

$$dV(x) = -\frac{I(x)R_{cshr}}{Z}dx, \tag{3.34}$$

where R_{cshr} is the sheet resistance per square of the region under the ohmic contact.

3.4 Semiconductor Resistance

Equations 3.32 and 3.34 may be reduced to a form that describes a transmission line model in electromagnetics as

$$\frac{d^2V}{dx^2} = \frac{V}{L_T^2}, \tag{3.35}$$

where L_T represents the transfer length that is defined as

$$L_T^2 = r_c/R_{cshr}. \tag{3.36}$$

Equation 3.35 holds if the epitaxial layer thickness is much smaller than L_T. For solving Equation 3.35, the boundary conditions are

$$\left.\frac{dV}{dx}\right|_{x=0} = I_0 \frac{R_{cshr}}{Z} \quad \text{and} \quad \left.\frac{dV}{dx}\right|_{x=d} = 0, \tag{3.37}$$

where d is the contact length. The second boundary condition states that the electric field in the alloyed region at the end of the contact opposite to the channel region is zero. The second-order differential equation represented by Equation 3.35 has a general solution of the form

$$V(x) = C_1 \exp(x/L_T) + C_2 \exp(-x/L_T). \tag{3.38}$$

The application of the boundary conditions of Equation 3.37 results in the determination of constants C_1 and C_2 as

$$C_1 = \frac{I_0 R_{cshr} L_T \exp(d/L_T)}{Z[\exp(d/L_T) - \exp(-d/L_T)]},$$

$$C_2 = \frac{I_0 R_{cshr} L_T \exp(-d/L_T)}{Z[\exp(d/L_T) - \exp(-d/L_T)]}. \tag{3.39}$$

Solving for V in Equation 3.38 at $x=0$ results in

$$V(0) = I_0 R_{cshr}(d/Z)(L_T/d)\coth(d/L_T), \tag{3.40}$$

which can be conveniently expressed as

$$V(0) = I_0 R_c, \tag{3.41}$$

where R_c is the contact resistance and is related to transfer length and sheet resistance in the contact region through $R_c = R_{cshr}(d/Z)F_{tlm}$ and $F_{tlm} = (L_T/d)\coth(d/L_T)$. For $d/L_T \ll 1$, $\coth(d/L_T) \approx L_T/d$, which leads to $F_{tlm} \approx (L_T/d)^2$ and

$$R_c = R_{cshr}(L_T^2/dZ). \tag{3.42}$$

For $d/L_T \gg 1$, which is the case for many of the patterns employed, $F_{tlm} \approx L_T/d$ and

$$R_c = R_{cshr}(L_T/Z). \tag{3.43}$$

Solving for V in Equation 3.38 at $x=d$ results in

$$V(d) = \frac{I_0 R_{cshr} L_T}{Z \sinh(d/L_T)}. \tag{3.44}$$

The end resistance R_{end} can be defined as

$$R_{end} = V(d)/I_0 \qquad (3.45)$$

and calculated using Equation 3.44 as

$$R_{end} = \frac{R_{cshr} L_T}{Z \sinh(d/L_T)} \qquad (3.46)$$

using $R_c = R_{cshr}(L_T/Z)\coth(d/L_T)$, and

$$R_c/R_{end} = \cosh(d/L_T). \qquad (3.47)$$

The resistivity under the contact region can be determined. Equations 3.45 and 3.47 are used in the measurements and calculation of ohmic contact resistivity.

In the TLM method, the intercontact spacing (gap) must be much smaller than the contact width in order to avoid edge effects. Edge effects can be eliminated altogether with the circular transmission line method (CTLM) where the contacts are circular, as opposed to rectangular, and concentric. In another method, the four-terminal (Kelvin) resistor method, the test pattern consists of four metal pads on an insulator. Two are connected to a semiconductor bar by means of large-area contacts. The other two touch the semiconductor at the contact opening. One pair is used to pass current while the other two are used to measure the voltage drop. In contrast to the TLM case, the resistance of the line outside the contact area does not contribute to the contact resistance in this test structure; this allows a more precise measurement. For very small contact resistances, this method is more accurate than the TLM method.

The value of R_c can be determined from the intercept of the resistance, R, versus l measurements. The semiconductor sheet resistance, R_{sshr}, can be found from the slope of R versus l line. When the sheet resistance of the active layer below the annealed region is different from that of the semiconductor in between the contacts, additional measurement is needed to determine sheet resistance in the alloy region and r_c. Even though this is not likely the case for n-type contacts in nitride semiconductors as they are surface oriented, the same may not be true for p-type contacts considering the very defective nature of p-type GaN and expected penetration of the ohmic contact metal. The end resistance can be determined by measuring the current–voltage characteristics of the sample structure shown in Figure 3.9 using

$$R_{end} = V_{2,3}/I_{1,2}, \qquad (3.48)$$

where $V_{2,3}$ is the voltage between contacts 2 and 3 and $I_{1,2}$ is the current flowing between contacts 1 and 2. Contact and end resistances, R_c and R_{end}, can be found using Equation 3.47. From Equation 3.37 one can express the transfer length in terms of these two resistances as

$$L_T = \frac{d}{\cosh^{-1}(R_c/R_{end})}. \qquad (3.49)$$

R_{cshr} and r_c can be found using Equation 3.36 together with Equation 3.47.

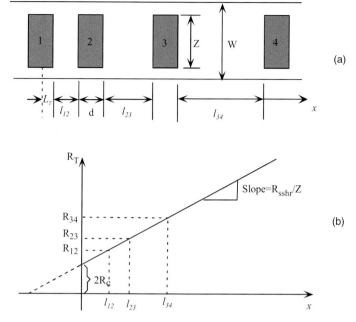

Figure 3.9 (a) Top view of a transmission line pattern commonly used to deduce the specific contact resistivity in planar contacts. (b) The variation of the resistance with respect to the gap distance.

In terms of the relevant technology to GaN and related materials, Ti/Al-based ohmic contacts for n-type materials are very popular. Electrical contact resistivity is positively affected by formation of TiN. As expected, contacts on p-type material are much more problematic with contact resistivities of 10^{-3} Ω cm^2, which is some three orders of magnitude larger than that for n-type contacts. To mitigate the problem to some extent and also increase light extraction in LEDs, transparent conductive oxides are used to coat the p-GaN in which case the contact formation relies on mainly tunneling afforded by extremely high doping levels in the conductive oxide.

Further Reading

Berger, H.H. (1972) *J. Electrochem. Soc.*, **119**, 509.

Padovani, F.A. and Stratton, R. (1966) *Solid-State Electron.*, **9**, 695.

Reeves, G.K. and Harbison, H.B. (1982) *IEEE Electron Device Lett.*, **EDL-3**, 111–113.

Shockley, W. (1964) Research and investigation of inverse epitaxial. UHF power transistor, *Final Tech. Rept.* No. AL-TDR-64-207. Air Force Avionics Lab., Air Force Systems Command, Wright. Patterson Air Force Base, OH.

4
Carrier Transport

4.1
Introduction

As the carriers travel through a semiconductor, they undergo a variety of interactions with the host material, all of which are reflected in the carrier mobility.

Transport depends on many intrinsic (processes related to innate properties of the semiconductor) and extrinsic (those imposed by the noninnate properties such as impurities and defects in the form of charged centers or in the form of boundaries) parameters, and as such is a lively field in terms of both experiments and predictions. A convenient table tabulating the various scattering mechanisms is given in Figure 4.1. Transport in semiconductors in general and nitride semiconductors and their heterostructures in particular has been treated extensively.

Transport properties of carriers in semiconductors can be described with drift–diffusion in the linear form where the diffusion constant and mobility are assumed field independent, or with nonlinear methods where the dependence on electric field is considered. The drift–diffusion model is, however, a classical model, and more importantly it cannot describe transport in very small (submicron) devices where nonlocal effects can be important because the electric field inside the device can change rapidly, thereby causing strong nonlocal effects such as velocity overshoot. To model nonlocal effects accurately, the semiclassical Boltzmann transport equation (BTE) is required, and if transport is phase coherent, meaning that the electron's quantum mechanical phase is not randomized by inelastic collisions, then quantum interference effects may dominate the transport properties. In that case, a quantum mechanical model (such as those based on the Liouville equation, the Wigner distribution function, or the nonequilibrium Green's function method) can be applied. The effective (ghost) diffusion constant and velocity can be determined by methods that take reduced dimensional effects into consideration and can be used in the standard drift–diffusion model for better accuracy.

In the drift–diffusion model, the current conduction equation, the continuity equation, and the Poisson equation are used to self-consistently determine the current–voltage characteristics in epitaxial layers, as well as both majority and

Nitride Semiconductor Devices: Fundamentals and Applications, First Edition. Hadis Morkoç.
© 2013 Wiley-VCH Verlag GmbH & Co. KGaA. Published 2013 by Wiley-VCH Verlag GmbH & Co. KGaA.

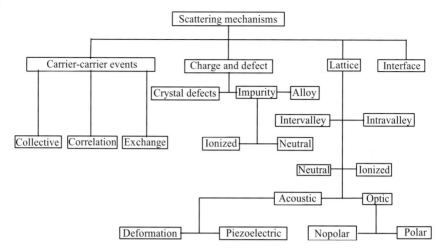

Figure 4.1 A tabulation of various scattering mechanisms.

minority devices relying on lateral or vertical transport. The continuity equation for electrons only is given by

$$\frac{\partial n}{\partial t} = \frac{1}{q}\vec{\nabla}\cdot\vec{J}_n + G_n - R_n, \qquad (4.1)$$

where the term on the left-hand side represents the time rate of change of the electron concentration, the first term on the right-hand side represents the change in electron concentration due to current flow, the second term is the generation rate, and the last term is the recombination rate. The drift–diffusion current for electrons is given by

$$\vec{J}_n = q\mu_n n \vec{E} + q D_n \vec{\nabla} n. \qquad (4.2)$$

Inclusion of the field dependence of mobility and diffusion constants renders the drift–diffusion transport equations nonlinear. However, the transport equation can be handled, for example, with the following empirical expression:

$$\mu_n(E) = \frac{\mu_n^0}{[1 + (E/E_{cr})^\beta]^{1/\beta}}, \qquad (4.3)$$

where μ_n^0 represents the zero-field mobility, β is a fitting parameter, and E_{cr} is the critical field where the velocity saturates. Equation 4.3 is well suited for semiconductors such as Si without multivalley scattering. Even so, it is still applied to semiconductors such as GaAs and GaN with multivalley scattering with negative differential resistance behavior above a critical field. In addition, when device dimensions are reduced to the nanometer scale, velocity overshoot occurs even in semiconductors such as Si; therefore, the mobility cannot be fit with Equation 4.3.

The field dependence of the diffusion constant can be treated in a manner similar to the case of mobility as

$$D_n(E) = \frac{D_n^0}{[1+(E/E_{cr})^\beta]^{(\beta-1)}}. \tag{4.4}$$

4.2 Carrier Scattering

Electrons (holes) in a periodic crystal are distributed over states in the energy bands defined by wave vector k and in real space defined by r, and band number i. The conduction band energy dispersion relation in k-space, $\xi_c(k)$, defines the average electron velocity as $\langle v_n \rangle = (1/\hbar)\nabla_k \xi_c(k)$. Under the influence of an external electric field E, the electron k-vector varies as $k = k_0 + (q/\hbar)Et$. In a spherical band and near $k = 0$, that is, the bottom of the conduction band, the electron motion is in three-dimensional real and k-space (six dimensional), with each (r, k) space having the average occupancy of the phase space. With time dependence, the problem becomes a seven-dimensional one. The electron distribution over the conduction band states can be described by $f(r, k, t)$, where t represents time. The time evolution of the distribution function can be written as

$$\frac{\partial f}{\partial t} + v \cdot \vec{\nabla}_r f + \frac{q}{\hbar} E \cdot \vec{\nabla}_k f - \left(\frac{\partial f}{\partial t}\right)_{coll} = 0, \tag{4.5}$$

which is known as the classical BTE. The collision term, the last one, represents the changes in f due to electron scattering into another phase space. The usual approach for solving the BTE is to assume that the presence of an external force, such as that induced by electric and magnetic fields, shifts the equilibrium distribution function, f_0, by a small amount $f - f_0$. Then, the last term in the BTE is related to the transition probabilities of specific scattering mechanisms. An expression can be developed for $f - f_0$ for the steady-state case. If the scattering events are elastic, meaning no energy gain or loss due to the particular collision, the last term in the BTE can be reduced to

$$\left(\frac{\partial f}{\partial t}\right)_{coll} = \frac{f - f_0}{\tau(\xi)}, \tag{4.6}$$

where $\tau(\xi)$ is the energy-dependent relaxation time. As can be inferred, such a simplification cannot be done for inelastic scattering events, such as optical phonon scattering, necessitating the solution of the BTE numerically. It will be shown later that a quasi-analytical method can be developed for optical phonon scattering, which requires lookup tables produced by numerical simulations.

If the relaxation time approximation is applicable for each of the scattering event types and each of the scattering events is independent of one another, meaning one does not change the nature of the other, the overall mobility can be calculated using Matthiessen's rule given as

$$\langle \tau \rangle = \left\langle \frac{1}{\tau_1^{-1} + \tau_2^{-1} + \cdots} \right\rangle. \tag{4.7}$$

Furthermore, if the relaxation time is energy independent as in the case of degenerate semiconductors, then

$$\tau^{-1} = \tau_1^{-1} + \tau_2^{-1} + \cdots \tag{4.8}$$

and

$$\mu_{\text{total}} = \left[\sum_i (1/\mu_i) \right]^{-1}, \tag{4.9}$$

Matthiessen's rule is commonly applied to semiconductors as a first approximation. The accuracy of Matthiessen's rule is within about 10%. It is nevertheless used very commonly for its convenience. However, when the dominant scattering mechanisms are inelastic and anisotropic, such as optical phonon scattering, the relaxation time approximation is not appropriate despite its great appeal. Even though the accuracy is compromised, the relaxation time approximation has, nevertheless, been applied to every type of scattering. It should be pointed out that the relaxation time approximation does not really apply to optical phonon scattering, particularly in GaN because of its large LO phonon energy of 91–92 meV. More accurate calculations require numerical methods, one of which is the numerical solution of the full BTE method including all the scattering events.

4.2.1
Impurity Scattering

Carrier scattering due to neutral impurities is negligible in comparison to that by charge impurities and as such we will delve deep into ionized impurity scattering. Scattering by ionized impurities was treated by Conwell and Weiskopf using a classical approach in which an electron is assumed to scatter classically via Coulomb interaction with an ion. The associated scattering cross section is calculated, which leads the way to determining the scattering rate. Because scattering of this kind relaxes the momenta of the carriers, a relaxation time and thus the mobility can be calculated. The standard textbook approach for ionized impurity scattering is the *Brooks–Herring* (BH) technique. The BH method is based on two inherent and important approximations, namely, the Born approximation and the single-ion screening approximation. Despite the enormous impact of the BH method, the predictions do not really agree well with the experimental mobility, particularly at high donor concentrations. In the case of high compensation, the single-ion screening formalism becomes less of an issue and it is more difficult for a given number of electrons to screen all the ionized donors separately while maintaining the charge neutrality condition ($n + N_A^- = N_D^+ + p$). Meyer and Bartoli improved on the BH method by invoking a phase-shift analysis of electron–impurity scattering to overcome the Born approximation. By recognizing that carriers are scattered through the screening Coulomb potential of ionized impurities, the relevant relaxation time can be determined, the details of which can be found in Morkoç (2008).

The relaxation time for ionized impurity scattering is given by (recall that ionized impurity scattering can be described within the relaxation approximation)

$$\tau_{ii}^{-1} = \frac{N_I Z^2 q^4 [\ln(1+y') - y'/(1+y')]}{16\sqrt{2}\pi\varepsilon^2 (m^*)^{1/2} \xi^{3/2}} \quad \text{or} \quad \tau_{ii} = \frac{16\sqrt{2}\pi\varepsilon^2 (m^*)^{1/2} \xi^{3/2}}{N_I Z^2 q^4 [\ln(1+y') - y'/(1+y')]}$$

with $f(y') = \ln(1+y') - y'/(1+y')$,

(4.10)

where Zq is the ionic charge (for singly charged scattering centers $Z = 1$), N_I is the ionized impurity concentration, and $y' = 8\varepsilon m^*(kT)\xi/\hbar^2 q^2 n$.

For $y' \gg 1$, the above expression becomes identical to the well-known BH expression for ionized impurity scattering and is valid for $y' \gg 1$, meaning at sufficiently high temperatures and low n. Meyer and Bartoli showed that a better criterion for applicability of the BH formula is $ka_B/2 \gg 1$, where a_B is the Bohr radius. As mentioned previously, the BH scattering theory is based on the Born approximation. If the Born approximation is not valid, a partial-wave phase-shift analysis must be used as outlined by Meyer and Bartoli.

In the case of a nondegenerate semiconductor, the $\partial f_0/\partial \xi$ term can be approximated by a Boltzmann distribution, in which case $\partial f_0/\partial \xi \propto \exp(-\xi/kT)$ leading to

$$\mu_I \cong \frac{128\sqrt{2}\pi\varepsilon^2 (kT)^{3/2}}{N_I Z^2 q^3 (m^*)^{1/2} [\ln(1+y) - y/(1+y)]}, \quad (4.11)$$

where $y = 24\varepsilon m^*(kT)^2/\hbar^2 q^2 n$ and the expression can be utilized when $y \gg 1$. The value of y is obtained by recognizing that the term in brackets is a slowly varying function of ξ and can be substituted for by the maximum of the integral, meaning $3kT$, which is the factor between y and y'.

The Brooks–Herring expression, though very widely used, breaks down for highly doped semiconductors. For GaN, it would lead to erroneous results for electron concentrations in excess of 10^{18} cm^{-3}. The situation is exacerbated at low temperatures where the intrinsic carrier concentration and the density of states drop precipitously, and degeneracy can occur readily even with moderate doping levels. A closed-form expression for ionized impurity scattering limited mobility in the degenerate case has been treated in the literature as discussed below.

For degenerate statistics, $\partial f_0/\partial \xi \approx \delta(\xi - \xi_F)$, and the mobility is given by $\mu_I \approx q\tau(\xi_F)/m^*$, where the scattering time is that at the Fermi level.

The ionized impurity scattering for a degenerate semiconductor thus becomes

$$\mu_I = \frac{24\pi^3 \varepsilon^2 \hbar^3 n}{N_I Z^2 q^3 (m^*)^2 [\ln(1+y_F) - y_F/(1+y_F)]}, \quad (4.12)$$

where $y_F = 4\sqrt[3]{3}\varepsilon\pi^{8/3}\hbar^2 n^{1/3}/q^2 m^*$. We should keep in mind that $Z = 1$ for singly charged scatterers. In contrast to the nondegenerate case, the mobility is inversely proportional to the square of the effective mass as opposed to square root. Also, there is no explicit temperature dependence.

We should keep in mind that for heavily doped semiconductors the Fermi level extends into the conduction band in which case the effective mass that is commensurate with the nonparabolic band structure should be used. In this context, the dispersion relation could be accounted for by adding a quadratic term:

$$\frac{\hbar^2 k^2}{2m_0^*} = E + CE^2. \tag{4.13}$$

Here C is the nonparabolicity parameter with m_0^* being the effective mass at the bottom of the conduction band. For dispersion described by Equation 4.13, the effective mass for electron concentration N_e can be written as

$$m^* = m_0^* \left[1 + 2C \frac{\hbar^2}{m_0^*} (3\pi^2 N_e)^{2/3}\right]^{1/2}. \tag{4.14}$$

4.2.2
Acoustic Phonon Scattering

To reiterate, acoustic phonons cause local perturbations in the lattice spacing and thus the bandgap. Beginning at about 5 K, acoustic phonon scattering becomes the main mechanism limiting the mobility through both deformation potential and piezoelectric scattering. Unless explicitly specified, acoustic phonon scattering is generally assumed to be only due to deformation potential, as the nomenclature was developed for semiconductors with no or negligible piezoelectric behavior. Because the nitrides are highly piezoelectric, both components must be taken into consideration. In this text, deformation potential-related acoustic phonon scattering is referred to as the acoustic phonon scattering. Scattering by piezoelectricity-induced processes is referred to as piezoelectric scattering.

For temperatures below about 200 K, the deformation potential-induced acoustic phonon scattering dominates over the piezoelectric components. For high temperatures, the piezoelectric component is relatively stronger. For a two-dimensional electron gas (2DEG), and at temperatures above 170 K, the mobility is limited by polar optical phonon (POP) scattering. It has also been reported that the low-temperature (below 50 K) mobility in a 2DEG is determined by deformation potential scattering, which depends on both temperature and electron density.

Scattering between an electron and an acoustic phonon is quasi-elastic with the main consequence being the relaxation of electron momentum. In order to gain additional insight into LA phonon scattering, such as its energy and temperature dependence, the probability of an electron of emitting one LA phonon can be calculated with closed-form solutions being available for nondegenerate and parabolic band conditions in the form of $T(\xi_k)^{1/2}$, which describes the temperature dependence of deformation potential-induced acoustic phonon scattering. We should mention that for piezoelectric scattering, the power of the temperature dependence expression takes the form $T(\xi_k)^{-1/2}$.

The approximations made above allow the problem to be described relatively easily despite the reduction in accuracy. However, when the accuracy is a prime consideration, numerical solutions of the Boltzmann equation and Monte Carlo calculations are performed.

4.2.2.1 Deformation Potential Scattering

Deformation potential scattering due to lattice vibrations is important, particularly in nonpolar semiconductors, and will be discussed briefly. The scattering of electrons by lattice vibrations can be construed as a wave phenomenon. One of the early treatments relies on wave transmission and reflection through and off strain-induced band undulations (compressive strain followed by tensile strain as the acoustic wave propagates), as shown in Figure 4.2. The variation of the band edges with varying lattice constant, which is caused by lattice vibrations, is also shown in Figure 4.2. The vibration of lattice atoms about their equilibrium positions causes a local perturbation in the lattice spacing and thereby the local bandgap. Because the lattice is deformed locally, the perturbation potential is called the deformation potential, which simply is the change in the bandgap due to unit change (100%) in the lattice spacing. This is termed E_{dp} or D_{ac}, has the unit of energy, and is given as 9.2 eV for the hydrostatic component determined from optical measurements, but a global consensus is lacking. The fitted value for the unscreened case of Look is 13.5 eV. However, literature values are in the range 7–13.5 eV, and the large spread may have its genesis in the way screening is accounted for or the lack thereof. Other fitting exercises of the experimental data in high-mobility 2DEG samples have led to a value of $D_{ac} = 9.1 \pm 0.7$ eV.

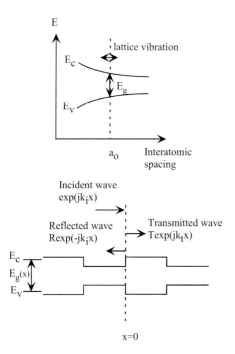

Figure 4.2 A schematic representation of the effect of lattice vibrations on the effective bandgap of a semiconductor and propagation of the wave associated with electrons.

Recognizing that the mean free path of an electron depends on the number of deflections it suffers within a given characteristic distance, the exact 3D quantum mechanical treatment leads to an expression of the relaxation time (assuming the Boltzmann distribution):

$$\tau_{dp} = \frac{\pi \hbar^4 \rho s^2}{2^{1/2} (m_n^*)^{3/2} D_{ac}^2 (k_B T)} \xi^{-1/2}. \tag{4.15}$$

It should be pointed out that this relaxation time is also referred to as τ_{ac} for acoustic phonon scattering and D_{ac}, the deformation potential for acoustic phonons, is also depicted with the symbols of E_{dp} and a_c in other texts. In terms of the other parameters, ρ and s represent the density of and sound velocity in the material, respectively.

The energy gain (phonon absorption) and energy loss (phonon emission) by phonons in a generic semiconductor E–k diagram take place while obeying momentum and energy conservation rules, as shown in Figure 4.3 for phonon emission and absorption for acoustic phonons and emission for LO phonons.

The average value of the relaxation time can be found with the aid of the distribution function for nondegeneracy and spherical band, as

$$\langle \tau_{dp} \rangle = \frac{2(2\pi)^{1/2} c_L \hbar^4}{3 D_{ac}^2 (m^*)^{3/2}} (kT)^{-3/2}, \tag{4.16}$$

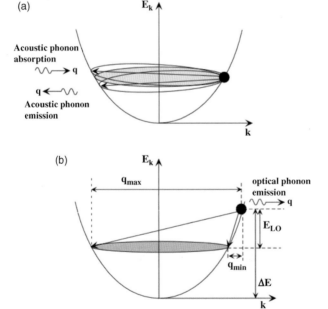

Figure 4.3 Schematic diagram of scattering events of an electron in a parabolic band by phonon emission and absorption. (a) The case for acoustic phonons. (b) The case for optical phonons (LO), but only the emission is shown. The range of phonon wavelengths, $q_{min} (k_{min} = k_e - k_1)$ and $q_{max} (k_{max} = k_e + k_1)$, is also shown. LO phonon absorption can be patterned after the acoustic phonon adsorption case.

where $c_L = \rho s^2$, the product of the material density, ρ, and the second power of the sound velocity, s^2. Now, the mobility limited by deformation potential or acoustic phonon scattering limited mobility can be calculated as

$$\mu_{dp} = \frac{q\langle\tau\rangle}{m^*} = \frac{2(2\pi)^{1/2} c_L \hbar^4 q}{3 D_{ac}^2 (m^*)^{5/2}} (kT)^{-3/2}. \tag{4.17}$$

Note that the index for temperature dependence of acoustic phonon scattering is $T^{-3/2}$. Deformation potential scattering limited mobility becomes very important in semiconductors with extremely low ionized impurity scattering and may be dominant low-temperature mobility limiter in nonpiezoelectric semiconductors such as Si and Ge. In order to gain an understanding of the extent of deformation potential limited scattering, the mobility in all three binaries of nitrides, namely, InN, GaN, and AlN, and for comparison purposes in GaAs versus temperature calculated using Equation 4.17 is displayed in Figure 4.4. The calculated mobility for GaAs and InN is almost the same and not distinguishable on the plot. The parameters used for those calculations are listed in Table 4.1.

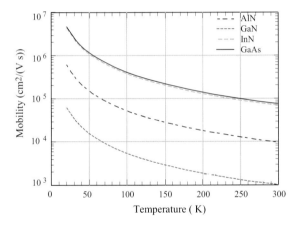

Figure 4.4 Deformation potential (acoustic phonon) scattering limited mobility in InN, GaN, and AlN and for comparison in GaAs.

Table 4.1 Parameters used for calculation of the mobility limited by deformation potential scattering.

Parameter	GaN	AlN	InN	GaAs
Mass density, ρ (kg/m^3)	6150	3230	6810	5317
Deformation potential, E (eV)	8.3	9.5	7.1	9.3
Electron effective mass, m^*	0.22	0.48	0.115	0.067
Sound velocity, s (m/s)	6.59×10^3	9.06×10^3	6.24×10^3	4.73×10^3

4.2.2.2 Piezoelectric Scattering

In noncentrosymmetric crystals, a polarization field is induced when stress is applied. Carriers then interact with the electric field induced by strain. In a sense, the carriers can be scattered by a TA phonon in addition to LA phonons through the piezoelectric coupling, as is the case in deformation potential acoustic phonon scattering. The situation is different from deformation potential scattering in that the direction of the q wave vector is also important. Nevertheless, first-order comparisons can be derived from deformation potential-related events, and the conclusion that piezoelectric scattering is more important for phonons with small wave vectors can be drawn. Because phonons with small wave vectors are less effective in relaxing carrier momentum in semiconductors where covalent bonding is predominant, one can argue that deformation potential scattering is dominant. However, in highly ionic semiconductors and semiconductors in which the piezoelectric coefficients are large, such as GaN, this statement does not hold true. On the contrary, the piezoelectric scattering would dominate over the deformation potential scattering, as is the case in GaN. This implies that at low temperatures and in GaN with minimal ionized centers, the mobility will be limited by piezoelectric scattering. The piezoelectric scattering would also dominate the low-temperature mobility in GaN-based 2DEG systems where the ionized center scattering is eliminated and/or screened. If, for the sake of argument, we are to assume that the direction of the wave vector is not important, the scattering rate for piezoelectric acoustic phonon scattering is proportional to $T(\xi_k)^{-1/2}$. This is not valid for very low energy electrons because for small ξ_k, the scattering vector $\vec{q}$ is also small.

The relaxation time for piezoelectric scattering is given by

$$\tau_{pe}(\xi) = \frac{2^{3/2}\pi\hbar^2 \rho s^2 \varepsilon^2}{q^2 h_{pz}^2 (m^*)^{1/2}(kT)} \xi^{1/2} \equiv \frac{2^{3/2}\pi\hbar^2 \varepsilon}{q^2 P_\perp^2 (m^*)^{1/2}(kT)} \xi^{1/2}, \quad (4.18)$$

where $P_\perp^2 \equiv (h_{pz}^2)/\varepsilon \rho s^2$, with h_{pz} being the piezoelectric constant. The term $P_\perp$ is the perpendicular component of the piezoelectric coefficient, and has a value of 0.113 for GaN,

$$\begin{aligned}
P_\perp^2 &= 4\varepsilon_0(21h_{15}^2 + 6h_{15}h_x + h_x^2)/105 C_t + \varepsilon_0(21h_{33}^2 - 24h_{33}h_x + 8h_x^2)/105 C_l, \\
P_\parallel^2 &= 2\varepsilon_0(21h_{15}^2 + 18h_{15}h_x + 5h_x^2)/105 C_t + \varepsilon_0(63h_{33}^2 - 36h_{33}h_x + 8h_x^2)/105 C_l, \\
h_x &= h_{33} - h_{31} - 2h_{15},
\end{aligned} \quad (4.19)$$

where h_{ij} ($\equiv e_{ij}$ in Chapter 1) is the ij component of the piezoelectric stress tensor. The spherically averaged elastic constants, C_{ij}, are related to four out of six independent elastic constants because not all the acoustic modes are piezoelectrically active:

$$\begin{aligned}
C_l &= (8C_{11} + 4C_{13} + 3C_{33} + 8C_{44})/15, \\
C_t &= (2C_{11} - 4C_{13} + 2C_{33} + 7C_{44})/15.
\end{aligned} \quad (4.20)$$

The average relaxation time, using the distribution function of $\int_0^\infty \xi^{3/2} \exp(-\xi/kT) d\xi$ (which is applicable for nondegeneracy and parabolic band), is

$$\langle \tau_{pe} \rangle = \frac{16\sqrt{2\pi}\varepsilon\hbar^2}{3q^2 P_\perp^2 (m^*)^{1/2}(kT)^{1/2}}. \quad (4.21)$$

The closed form of the mobility expression, limited by piezoelectric scattering, is given by

$$\mu_{pe} = \frac{q\langle\tau_{pe}\rangle}{m^*} = \frac{16\sqrt{2\pi}\varepsilon\hbar^2}{3qP_\perp^2(m^*)^{3/2}(kT)^{1/2}} = \frac{16(2\pi)^{1/2}\rho s^2 \hbar^2 q}{3(qh_{pz}/\varepsilon_s)^2(m^*)^{3/2}}(kT)^{-1/2}. \quad (4.22)$$

In order to gain an understanding of the extent of piezoelectric potential limited scattering, the mobility in all three binaries of nitrides, namely, InN, GaN, and AlN, and for comparison purposes in GaAs versus temperature calculated using Equation 4.22 is displayed in Figure 4.5. The parameters used for those calculations are listed in Table 4.2.

In a highly degenerate system such as a 2DEG, the scattering rates due to piezoelectric and deformation potential can be added to get the total scattering

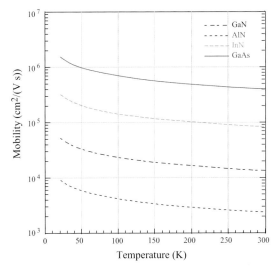

Figure 4.5 Piezoelectric phonon scattering limited mobility in InN, GaN, and AlN and for comparison in GaAs.

Table 4.2 Parameters used for calculation of the mobility limited by piezoelectric scattering.

Parameter	GaN	AlN	InN	GaAs
Piezoelectric scattering strength (piezoelectric coefficient), h_{pz} (C/m^2)	0.5	0.56	0.5	0.1913
Low-frequency dielectric constant, ϵ_s (static)	8.9	8.5	15.3	12.85
Electron effective mass, m^*	0.22	0.48	0.115	0.067
Mass density, ρ (kg/m^3)	6150	3230	6810	5317
Sound velocity, s (m/s)	6.59×10^3	9.06×10^3	6.24×10^3	4.73×10^3

rate due to these two events. This is possible if the scattering events are assumed independent. In addition, the relaxation time can be taken as independent of energy for this scenario because of degeneracy, in which case the distribution function is peaked sharply at the Fermi level, and modeled with a delta function, which results in an energy-independent relaxation time. In terms of the mobility and relaxation time, because they are both inversely proportional to the scattering rate, the inverse relaxation times or mobilities limited by each of these two scattering events can be summed to get the overall relaxation time, or the mean free time between scattering events, or the mobility limited by the total acoustic phonon scattering as

$$\tau_{ac}^{-1} = \tau_{dp}^{-1} + \tau_{pe}^{-1}, \tag{4.23}$$

where τ represents the mean free time, the inverse of which, τ^{-1}, is the scattering rate. We should mention that the above is valid when all the mean free times have the same energy distribution.

4.2.3
Optical Phonon Scattering

Optical phonons in semiconductors have energies in the tens of meV with the value of about 92 meV (91.1–91.7 meV) for LO phonons. This implies that at low temperatures, such as <100 K, most electrons do not have sufficient thermal energy to emit optical phonons. Moreover, the thermal occupation number for phonons, N_q, is very small and therefore the probability of an electron absorbing an optical phonon is also very low. This basically indicates that optical phonon scattering at low temperatures is negligible. However, at higher temperatures, for example, room temperature, the picture is very different in that electrons have sufficient energy to emit optical phonons and overtake LA acoustic phonon scattering. This statement is particularly applicable to polar semiconductors where the Fröhlich electron–phonon coupling is very strong (see Chapter 9 for more details). The distribution of final states after an electron emits an LO phonon is shown in Figure 4.3, where it is clear that electrons must have sufficient energy to emit LO phonons. For low-field mobility, the sole source of energy is thermal energy. Scattering by LA phonons relaxes primarily the electron momentum with negligible energy change; however, scattering by optical phonons relaxes both electron momentum and energy significantly. Consequently, the relaxation approximation to the Boltzmann equation $(\partial f/\partial t)_{coll} = -(f - f_0)/\tau(\xi)$ does not apply because $\tau(\xi)$ becomes dependent on the perturbation strength in addition to the energy of the electrons.

4.2.3.1 Nonpolar Optical Phonon Scattering
This is similar to the case of acoustic phonon scattering, but in this case optical mode lattice vibrations produce local perturbations in the bandgap that scatter electrons. The associated deformation potential is termed D_{op} as opposed to D_{ac}, which is the energy per unit strain by definition and is typically expressed in units of eV.

The relaxation time for nonpolar optical phonon scattering is

$$\tau_{op}(\xi) = \frac{\sqrt{2\pi}\rho\hbar^2(\hbar\omega_{LO})}{D_{op}^2(m^*)^{3/2}n_0[(\xi+\hbar\omega_{LO})^{1/2} + u(\xi-\hbar\omega_{LO})(n_0+1)n_0^{-1}(\xi-\hbar\omega_{LO})^{1/2}]}, \quad (4.24)$$

where ω_{LO} is the nonpolar optical phonon frequency, $u(x)$ is the Heaviside step function, and $n_0 = 1/[\exp(\hbar\omega_{LO}/kT) - 1]$ is the phonon occupation number. In general, the term ω_{LO} would represent the frequency of the optical phonon to which electrons couple predominantly. In this case, it is the LO phonons that are assumed to couple to electrons most strongly. In cases where the electron transport is off one of the major axes, mixed modes may be present in which case the issue here would be more complex.

It is not possible to solve for the corresponding mobility in the closed form. However, for limiting cases, closed-form approximations are available as

$$\mu_{npo} = \frac{2^{3/2}\pi^{1/2}q\rho\hbar^2(\hbar\omega_{LO})^2}{3D_{op}^2(m^*)^{5/2}(kT)^{3/2}} \quad \text{for} \quad kT \gg \hbar\omega_{LO} \quad (4.25)$$

and

$$\mu_{npo} = \frac{2^{1/2}\pi q\rho\hbar^4(\hbar\omega_{LO})^{1/2}}{D_{op}^2(m^*)^{5/2}n_0} \quad \text{for} \quad kT \ll \hbar\omega_{LO}. \quad (4.26)$$

The strength of this form of scattering is not really known in GaN as a value for D_{op} is not available. However, because the polar optical phonon scattering is so dominant at room temperature, the nonpolar optical mode scattering can be neglected.

Other versions of nonpolar optical phonon scattering limited expressions for the mobility are also available, among which is that contained in a book by Hamaguchi, which is given below for completeness:

$$\mu_{npo} = \frac{4\sqrt{2\pi}qh^2\rho\sqrt{\hbar\omega_{LO}}}{3(m^*)^{5/2}D_{op}^2} f(z_0), \quad (4.27)$$

where $f(z_0) = z_0^{5/2}(e^{z_0}-1)\int_0^\infty ze^{-z}[(1+z_0/z)^{1/2} + e^{z_0}(1-z_0/z)^{1/2}]^{-1}dz$ with $z_0 = \hbar\omega_{LO}/k_BT$ and $z = \xi/k_BT$. As would be appreciated from the above expression, the temperature dependence is quite complicated in this case, but for $z \gg 1$, the temperature dependence is known from a body of work as $T^{-3/2}$, as indicated in Equation 4.25.

4.2.3.2 Polar Optical Phonon Scattering

Just so that there is no doubt, when a scattering event is inelastic such as the case of polar optical phonon scattering, analytical expressions for the mobility limited by that event are not appropriate even though approximations can be derived in order for the relaxation time approximation to be used. In spite of this, the irresistible attraction of closed-form solutions paves the way for the development of analytical expressions as will be done here. The LO phonon energy in GaN is much larger than the electron energy at room temperature, which precludes the use of analytical descriptions and necessitates numerical solutions of the Boltzmann equation, which can also describe the variation of the distribution function in the presence of an external perturbation, such as a field. The large polar optical phonon energy in GaN is 91.1–91.7 meV (with 91.1 meV corresponding to a wavenumber of 735 cm^{-1} and a temperature of 1057 K). For transport in the basal plane, the electrons couple to E_1 LO phonons, and for transport in the c-direction, they couple to A_1 LO phonons. For transport in other planes, such as pyramidal ones, mixed phonon modes would have to be considered.

Howarth and Sondheimer's expression for polar optical phonon scattering limited mobility, after multiplying with 4π to bring it into the MKS units as done by Look, is given as

$$\mu_{pop} = \frac{2^{9/2}\pi^{1/2}\hbar^2(kT)^{1/2}(e^{T_{LO}/T}-1)\chi(T_{LO}/T)}{3q(m^*)^{3/2}(kT_{LO})(\varepsilon_\infty^{-1}-\varepsilon_s^{-1})}, \tag{4.28}$$

where ε_∞ and ε_s represent the high- and low-frequency dielectric constants of the semiconductor, respectively, and $\chi(T_{LO}/T)$ is a slowly varying function of T. In order to calculate the polar optical phonon scattering limited mobility, one must know this function. With the constraint that $0 \leq T_{LO}/T \leq 5$ (or $T \geq 220$ K for practical purposes), meaning for the upper temperature end of the scale, the approximate form of this expression is available as

$$\chi(T_{LO}/T) = 1 - 0.5841(T_{LO}/T) + 0.292(T_{LO}/T)^2 - 0.037164(T_{LO}/T)^3 + 0.0012(T_{LO}/T)^4. \tag{4.29}$$

With the constraint that $5 \leq T_{LO}/T \leq \infty$ (or $0 \leq T \leq 220$ K for GaN), meaning for the lower temperature end of the scale:

$$\chi(T_{LO}/T) = \frac{3\sqrt{\pi}}{8}\left(\frac{T_{LO}}{T}\right)^{1/2}. \tag{4.30}$$

The approximate forms of function $\chi(T_{LO}/T)$ can be plotted as a function of T_{LO}/T using approximations 1 and 2 depending on the temperature range to facilitate the calculation of the polar optical phonon scattering, as shown in Figure 4.6. Doing so for the low-temperature range ($T \leq 220$ K for GaN) leads to

$$\mu_{pop} = \frac{2^{3/2}\pi\hbar^2(e^{T_{LO}/T}-1)}{q(m^*)^{3/2}(kT_{LO})^{1/2}(\varepsilon_\infty^{-1}-\varepsilon_0^{-1})}. \tag{4.31}$$

Polar optical phonon limited mobility in GaN using Equation 4.31 is shown in Figure 4.7. In the calculations, the $\chi(T_{LO}/T)$ expression for $0 \leq T_{LO}/T \leq 5$ as $\chi(T_{LO}/T) = 1 - 0.5841(T_{LO}/T) + 0.292(T_{LO}/T)^2 - 0.037164(T_{LO}/T)^3 + 0.0012016(T_{LO}/T)^4$ and that for $5 \leq T_{LO}/T \leq \infty$ as $\chi(T_{LO}/T) = (3\pi^{1/2}/8)[T_{LO}/T]^{1/2}$ have been used.

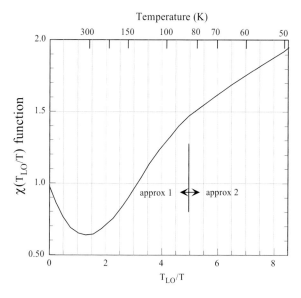

Figure 4.6 The function $\chi(T_{LO}/T)$ plotted against T_{LO}/T and also temperature T representing Equation 4.29 for the upper temperature range ($T > 220$ K) and Equation 4.30 for the lower temperature range ($T < 220$ K) in the polar optical phonon scattering mobility described by Equation 4.28.

It is always useful to define a relaxation time. For the higher temperature range, Look suggests

$$\tau_{pop}(\xi) = \frac{2^{3/2}\pi\hbar^2(e^{T_{LO}/T} - 1)\chi(T_{LO}/T)}{q^2(m^*)^{1/2}(kT_{LO})^{1/2}(\varepsilon_\infty^{-1} - \varepsilon_s^{-1})}\xi^{1/2} \qquad (4.32)$$

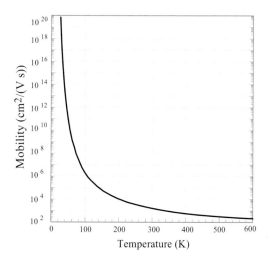

Figure 4.7 The polar optical phonon limited mobility in GaN calculated using Howarth and Sondheimer treatment described in Equation 4.31.

and for the lower temperature range

$$\tau_{pop}(\xi) = \frac{2^{3/2}\pi\hbar^2(e^{T_{po}/T}-1)}{q^2(m^*)^{1/2}(kT_{po})^{1/2}(\varepsilon_\infty^{-1}-\varepsilon_s^{-1})}. \tag{4.33}$$

At low temperatures, that is, $T_{LO}/T \gg 1$, the product $T^{1/2}\chi(T_{LO}/T)$ is temperature independent making τ_{po} energy independent, as depicted in Equation 4.33. At high temperatures, however, $T_{LO}/T < 5$, the function $\chi(T_{LO}/T)$ deviates from unity by about 0.4 while $T^{1/2}$ changes by several times more and it is reasonable to let the $T^{1/2}$ be represented by $\xi^{1/2}$ as indicated in Equation 4.32.

A pseudo-analytical expression has been developed by Ehrenreich through the use of a variational method, which includes a numerically calculated function to preserve the accuracy that would otherwise be lost in an all-analytical method. In fact, this function, $G(z)$, can be modified to account for the screening effect as well, and the polar optical phonon limited mobility becomes

$$\mu_{pop} = 0.199 \left(\frac{T}{300}\right)^{1/2} \left(\frac{q}{\varepsilon_c^*}\right)^2 \left(\frac{m}{m^*}\right)^{3/2} (10^{22}M) \cdot (10^{23}v_a) \cdot (10^{-13}\omega_{LO})$$
$$\cdot (e^{z_0}-1) \cdot G(z_0).$$

Repeating the same expression with units for all the parameters indicated in brackets,

$$\mu_{pop}[\text{cm}^2/(\text{V s})] = 0.199 \times (T/300)^{1/2} \times \left(\frac{e}{e^*}\right)^2 \times (m_0/m^*)^{3/2}$$
$$\times (10^{22}M\,[\text{g}]) \times (10^{23}v_a\,[\text{cm}^3]) \tag{4.34}$$
$$\times (10^{-13}\omega[\text{s}^{-1}]) \times (e^{\hbar\omega/k_BT}-1) \times G\left(\frac{\hbar\omega}{k_BT}\right),$$

where

$$\varepsilon_c^* = \sqrt{M\omega_{LO}^2 v_a\left(\frac{1}{\varepsilon_\infty}-\frac{1}{\varepsilon_s}\right)} \tag{4.35}$$

is the longitudinal or Callen's effective ionic charge. Here $\varepsilon_\infty = 5.47\varepsilon_0$ is the optical dielectric constant and $\varepsilon_s = 10.4\varepsilon_0$ is the low-frequency dielectric constant of GaN for an E field parallel to the c-axis. The low-frequency dielectric constant of the GaN semiconductor becomes $\varepsilon_s = 9.5\varepsilon_0$ for the E field perpendicular to the c-axis. The term ε_0 denotes the dielectric constant of free space. The low-frequency dielectric constant can be related to the optical dielectric constant by

$$\varepsilon_s = \varepsilon_\infty(\omega_{LO}/\omega_{TO})^2, \tag{4.36}$$

which is known as the Lyddane–Sachs–Teller relation. The longitudinal ω_{LO} and transverse ω_{TO} phonon frequencies are 744 and 533 cm^{-1}, respectively.

M denotes the reduced mass of the nearest-neighbor atoms in grams and is given by

$$M^{-1} = M_1^{-1} + M_2^{-1}, \tag{4.37}$$

where M_1 and M_2 designate the masses of the nearest neighbors. The value of M for wurtzite GaN is 1.936×10^{-23} g. The term v_a is the volume of the unit cell that can be calculated from the a and c lattice constants:

$$v_a = \left(\sqrt{3}a^2 c\right)/4, \tag{4.38}$$

yielding the value of 2.28×10^{-23} cm^3. Alternatively,

$$v_a = \frac{\text{molar weight}}{\rho N_a} = \frac{69.72_{\text{Ga}} + 14.0067_{\text{N}}}{6.1 \times 6.023 \times 10^{23}} = 2.28 \times 10^{-23} \text{ cm}^3, \tag{4.39}$$

where ρ is the mass density.

The LO phonon angular zone-center frequency for GaN is related to the LO phonon temperature by

$$\omega_{\text{LO}} = 1.309 \times 10^{11} \theta_{\text{LO}} [\text{s}^{-1}], \quad \text{the unit for } \theta_{\text{LO}}(= T_{\text{LO}}) \text{ is K,}$$

and has a value of 1.367×10^{14} s^{-1} for wurtzite GaN. The term θ_{LO} is the equivalent LO phonon temperature and has a value of 1057 K for wurtzite GaN if we use 91.1 meV for the LO phonon energy. The term θ_{LO} commonly goes by T_{po} as well.

In Equation 4.34, the term $G(z_0)$ is a slowly varying function of z (reduced LO phonon energy), which in turn is given by

$$z_0 = (\hbar\omega_{\text{LO}}/kT) = \theta_{\text{LO}}/T. \tag{4.40}$$

Neglecting screening effects, Hammar and Magnusson took advantage of an accurate iterative method for the solution of the BTE to determine the numerical values for this function (Table 4.3). Ehrenreich, on the other hand,

Table 4.3 Ehrenreich's relation $G(z_0) = \hbar\omega_{\text{LO}}/kT = \theta_{\text{LO}}/T$ as a function of temperature for various electron concentrations.

z_0	$G(z_0)$	z_0	$G(z_0)$
0.0	1.0	2.5	1.041
0.2	0.8957	3.0	1.194
0.4	0.8102	3.5	1.353
0.6	0.7524	4.0	1.495
0.7	0.7340	4.5	1.621
0.8	0.7219	5.0	1.733
1.0	0.7146	6.0	1.919
1.2	0.7263	7.0	2.065
1.4	0.7528	8.0	2.188
1.6	0.7909	9.0	2.296
1.8	0.8378	10.0	2.394
2.0	0.8911	11.0	2.487
2.2	0.9490		

The term θ_{LO} is commonly referred to as T_{po} and both terminologies are used throughout this book.

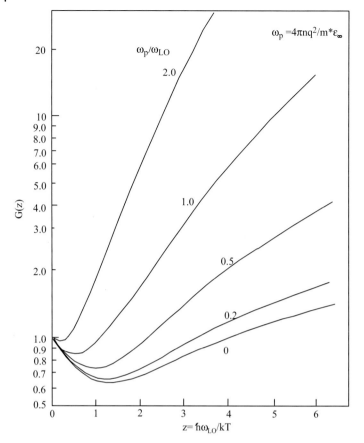

Figure 4.8 Ehrenreich's function $G(z_0) = \hbar\omega_{LO}/kT = \theta_{LO}/T$ as a function of temperature for various electron concentrations.

included the screening effects and determined $G(z_0)$ through a variational method (Figure 4.8).

There are many parameters involved in the polar optic phonon expressions above and it is hard to keep track of all of them. For convenience, the parameters and units to be used for calculating the Callen's effective ionized charge for GaN, in terms of the MKS units followed by cgs units, are summarized below:

- ε_0 permittivity of free space (8.85×10^{-12} F/m).
- M reduced ion mass in GaN (1.936×10^{-26} kg).
- ν_a volume of a Ga and N ion pair (2.283×10^{-29} m^3).
- ω_{LO} polar phonon frequency (1.367×10^{14} s^{-1}).

The parameters to be used for calculating μ_{pop}, in terms of the cgs units, are

- T temperature (in kelvin).
- q electron charge (1.60×10^{-19} C).

M	reduced ion mass in GaN (1.936×10^{-23} g).
V_a	volume of a Ga and N ion pair (2.283×10^{-23} cm³).
ω_{LO}	polar phonon frequency (1.367×10^{14} s⁻¹).

For completeness, the results of other treatments of POP scattering are also provided here. In the low-temperature limit, POP scattering is most likely not the dominant one. According to Hamaguchi, the mobility limited by polar optical phonon scattering is proportional to

$$\mu_{pop} \propto \frac{\exp(\theta_{LO}/T) - 1}{(m^*)^{1/2}\theta_{LO}(\varepsilon_\infty^{-1} - \varepsilon_s^{-1})} \quad \text{for} \quad T < \theta_{LO},$$

which is 1057 K for GaN (or $kT \ll \hbar\omega_{LO}$). (4.41)

On the other hand for $T > \theta_{LO}$ or $\xi \gg \hbar\omega_{LO}$ (high-temperature regime), the electron mobility limited by polar optical phonon scattering is given by

$$\mu_{pop} = \frac{8\sqrt{2kT}}{3\sqrt{\pi m^*}\xi_0} \frac{e^{z_0} - 1}{e^{z_0} + 1}, \quad \text{where} \quad z_0 = \frac{\hbar\omega_{LO}}{kT} \quad \text{and} \quad \xi_0$$

$$= \frac{m^* q \hbar \omega_{LO}}{4\pi\hbar^2}\left(\frac{1}{\varepsilon_\infty} - \frac{1}{\varepsilon_s}\right). \quad (4.42)$$

For very high temperatures compared to the LO phonon temperature, an expression with temperature power index of $-1/2$ is available. For completeness, the expression given by Wang is

$$\mu_{pop} \propto \frac{1}{(m^*)^{3/2}(\varepsilon_\infty^{-1} - \varepsilon_s^{-1})}\left(\frac{T}{\theta_{LO}}\right)^{-1/2} \quad \text{for} \quad T > \theta_{LO}. \quad (4.43)$$

In Equation 4.43, it should be noted, however, that the power of the mass term is inconsistent with all the other treatments. Equation 4.43 implies that polar optical phonon scattering, neglecting the temperature dependence of the $G(z_0)$ term, has a temperature dependence of $T^{-1/2}$ at temperatures above the LO phonon temperature that is nearly 1074 K, well above any practical temperature range of measurements. The expression is provided here to illustrate a sense of what would become of polar optical phonon scattering if extrapolated to high temperatures in terms of the index of temperature dependence that cannot be defined at lower temperatures. This is also implicit in the Ehrenreich treatment, which also indicates that the temperature dependence of the polar optical phonon scattering is not simple, except at very high temperatures that are well above the measurement temperatures commonly employed.

As we have done for other scattering mechanisms, polar optical phonon scattering calculated for GaN, AlN, and InN and for comparison purposes for GaAs, using Equation 4.37 with the aid of Equation 4.38, is plotted in Figure 4.9 and the parameters used for the calculations are tabulated in Table 4.4.

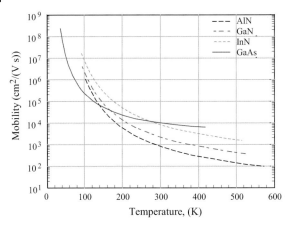

Figure 4.9 Polar optical phonon scattering limited mobility versus temperature for GaN, AlN, and InN and for comparison for GaAs using Equation 4.34 with the aid of Equation 4.35.

Table 4.4 Parameters used for calculation of polar optical phonon scattering limited mobility in GaN, AlN, and InN and for comparison in GaAs.

Parameter	GaN	AlN	InN	GaAs
High-frequency dielectric constant, ϵ_∞	5.5	4.77	8.4	10.88
Low-frequency dielectric constant, ϵ_s (static)	8.9	8.5	15.3	12.85
Lattice constant, a (Å)	3.16	3.11	3.545	5.65
Lattice constant, c (Å)	5.125	4.98	5.703	—
Volume of unit cell, $v_a = \sqrt{3}a^2c/4$ (cm^3)	2.216×10^{-23}	2.0857×10^{-23}	3.1034×10^{-23}	4.5168×10^{-23}
Electron effective mass, m^* (in m_0)	0.22	0.48	0.115	0.067
Phonon energy, $\hbar\omega_{LO}$ (meV)	91.2	99.2	89.0	36.13
Polar phonon frequency, ω_{LO} ($\times 10^{14}$ s^{-1})	1.3837	1.505	1.3503	0.5482
Reduced ion mass ($\times 10^{-23}$ g)	1.9367	1.531	2.0729	5.9969

4.2.4
Alloy Scattering and Dislocation Scattering

There are a number of additional scattering mechanisms such as short-range potential-induced scattering as in the case of alloy scattering, potential barrier scattering, potential well scattering, space charge scattering, dipole scattering, carrier–carrier scattering, plasmon scattering, boundary scattering, and dislocation scattering, as well as the impact of inhomogeneities on mobility. Because of their particular importance, alloy scattering and dislocation scattering are discussed. The details and a discussion of the other scattering mechanisms mentioned above can be found in Morkoç (2008).

Alloy scattering can be grouped under short-range interaction, as can scattering due to localized defects and space charge. The short-range potentials have constant strength over a small volume beyond which the potentials vanish rapidly to what can be considered zero. The simplest potential is that of a delta function that can represent localized potentials well (by Look):

$$\Delta V = V_\delta \xi_\delta \delta(\vec{r} - \vec{r}_0). \tag{4.44}$$

Here ξ_δ is an energy parameter associated with the potential and V_δ is the volume affected by the potential. For comparison, the scattering potential due to ionized impurities is an exponentially decaying function.

Random alloys in a crystal introduce short-range potentials, whose results can be understood by recognizing that the constituents in alloys cause a fluctuation in the band edge potential, which scatters the carriers. Whether the strength of the band edge discontinuity is determined by the difference in electron affinity of the two binaries forming the alloy or by the conduction band edge discontinuity is debated although the latter seems to be very popular. The similar covalent radii of Al and Ga encourage consideration of differences in polar strength bonds, which plays down the alloy scattering. The problem actually reduces to finding out the term $NV_\delta^2 \xi_\delta^2$, which is needed to calculate the mobility limited by this process.

The mean potential in a lattice composed of two binary compounds, if V_A and V_B represent the potential at each binary site (e.g., GaN for A and InN for B in InGaN), is given by

$$V = (1-x)V_A + xV_B, \tag{4.45}$$

where x represents the mole fraction of InN in the lattice. The potential discontinuities experienced by electrons at the A and B sites are $V - V_A$ and $V - V_B$, respectively. The scattering potentials at an A unit and a B unit are

$$V - V_A = x(V_B - V_A) \equiv \xi_A \quad \text{and} \quad V - V_B = (1-x)(V_B - V_A) \equiv \xi_B. \tag{4.46}$$

If N_c represents the number of primitive cells in a unit volume, the number of A and B elements in a primitive cell is given by $(1-x)N_c$ and xN_c, respectively. The $NV_\delta^2 \xi_\delta^2$ term can then be written as

$$\begin{aligned} NV_\delta^2 \xi_\delta^2 &= V_c^2[(1-x)N_c \xi_A^2 + xN_c \xi_B^2] \\ &= V_c^2[(1-x)x^2 N_c(V_B - V_A)^2 + x(1-x)^2 N_c(V_A - V_B)^2] \\ &= V_c^2 N_c(1-x)x[V_B - V_A]^2 \equiv V_c^2 N_c(1-x)x[V_A - V_B]^2 = x(1-x)V_c^2 N_c \xi_{AB}^2, \end{aligned} \tag{4.47}$$

where $\xi_{AB} \equiv |V_A - V_B|$, and the volume of a primitive unit cell is $V_c = \Omega = v_a \approx 2/N_A^*$. Using $N_c V_c = 1$, the alloy scattering limited mobility can be expressed as

$$\mu_{al} = \frac{2^{3/2} q \pi^{1/2} \hbar^4}{3 V_c x(1-x) \xi_{AB}^2 (m^*)^{5/2}} (kT)^{-1/2}. \qquad (4.48)$$

For the atomic concentration in GaN, N_A^*, equal to 8.76×10^{22} cm^{-3}, it yields $V_c = 2.283 \times 10^{-23}$ cm^3.

Note that the $x(1-x)$ term is maximized for $x = 0.5$, which represents the composition for the smallest mobility, all else being equal. In addition, the alloy scattering limited mobility is inversely quadratically proportional to the alloy potential $(V_A - V_B)$, and is thus affected significantly by the choice of this potential.

For a nondegenerate semiconductor with parabolic band, the average of the relaxation time can be written as

$$\langle \tau_{al} \rangle = \frac{\pi \hbar^4}{2^{1/2} V_c x(1-x) \xi_{AB}^2 (m^*)^{3/2}} \frac{\int_0^\infty \xi^{-1/2} \xi^{3/2} e^{-\xi/kT} d\xi}{\int_0^\infty \xi^{3/2} e^{-\xi/kT} d\xi} \qquad (4.49)$$

and the mobility is then given by

$$\mu_{al} = \frac{q \langle \tau \rangle}{m^*} = \frac{q \pi \hbar^4}{2^{1/2} V_c x(1-x) \xi_{AB}^2 (m^*)^{5/2}} (kT)^{-1/2}. \qquad (4.50)$$

The main issue is of course determination of alloy scattering potential for both ternary and quaternary alloys. Again, options are to use the bandgap discontinuity between the binaries forming the ternaries and quaternaries, the differences in conduction band discontinuities, the differences between the electron affinities, and as just discussed above, a determination based on the chemical trends. Until further clarifying research, the question will remain. For visualizing alloy scattering limited mobility, the mobility limited by this process using Ridley's model for $Al_xGa_{1-x}N$ and $In_xGa_{1-x}N$ is presented.

The data calculated using other models are presented in sections dealing with mobility in $Al_xGa_{1-x}N$ and $In_xGa_{1-x}N$. The electron mobility for $Al_xGa_{1-x}N$ versus temperature for compositions in the range 0.1–0.7 has been calculated and is shown in Figure 4.10. Note that the conduction band AlN effective mass is taken as 0.35 and the parameters used in Ridley's model are

$$b = 1.5, \quad m^* = (1-x)0.22 m_0 + x 0.35 m_0, \quad r_A = 1.225 \times 10^{-10} \text{ m},$$
$$r_B = 1.23 \times 10^{-10} \text{ m}, \quad Z \text{ (valency)} = 3, \quad \Omega = 2.283 \times 10^{-29} \text{ m}^3.$$

Because there is considerable debate in what alloy scattering potential to use, we present the room-temperature mobility for $Al_xGa_{1-x}N$ versus the mole fraction with scattering potential ξ_{AB} being a parameter and varied in the range 0.1–2.1 eV, as shown in Figure 4.11.

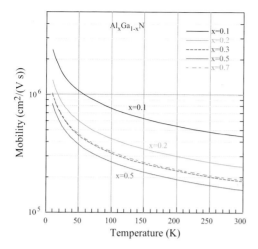

Figure 4.10 The alloy scattering limited mobility in $Al_xGa_{1-x}N$ versus temperature for AlN mole fractions in the range 0.1–0.7 using Ridley's model.

The electron mobility for $In_xGa_{1-x}N$ versus temperature for compositions in the range 0.1–0.7 has been calculated and is shown in Figure 4.12. Note that the conduction band InN effective mass m^* is taken as 0.35 and the parameters used in Ridley's model are

$$b = 1.5, \quad m^* = 0.22m_0 - 0.128m_0 + 0.047x^2 m_0, \quad r_A = 1.225 \times 10^{-10} \text{ m},$$
$$r_B = 1.41 \times 10^{-10} \text{ m}, \quad Z \text{ (valency)} = 3, \quad \Omega = 2.283 \times 10^{-29} \text{ m}^3.$$

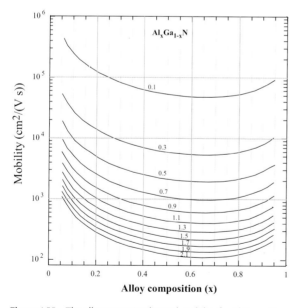

Figure 4.11 The alloy scattering limited mobility for $Al_xGa_{1-x}N$ versus the mole fraction with scattering potential ξ_{AB} varied in the range 0.1–2.1 eV using Ridley's model.

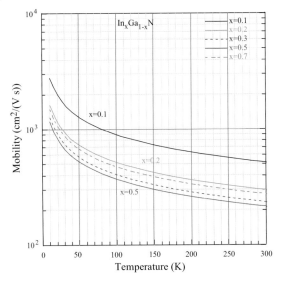

Figure 4.12 The alloy scattering limited mobility in $In_xGa_{1-x}N$ versus temperature for InN mole fractions in the range 0.1–0.7 using Ridley's model with an electron effective mass of $0.047m_0$ for InN.

The compositional dependence of the conduction band effective mass is obtained by interpolating the value for GaN ($0.22m_0$) with InN effective mass being $0.047m_0$.

The room-temperature mobility for $In_xGa_{1-x}N$ versus the mole fraction with scattering potential ξ_{AB} being a parameter and varied in the range 0.1–2.1 eV is shown in Figure 4.13.

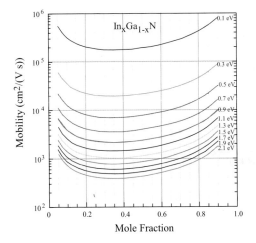

Figure 4.13 The alloy scattering limited mobility for $In_xGa_{1-x}N$ versus the mole fraction with scattering potential ξ_{AB} varied in the range 0.1–2.1 eV using Ridley's model.

Often times it is very helpful to be able to determine temperature dependence, effective mass dependence, energy dependence, or so on of the major scattering processes taking place in semiconductors. It is in this vein that Tables 4.5 and 4.6 are provided wherein the dependences of relaxation time and mobility, respectively, on

Table 4.5 Energy and mass dependence indices of relaxation time for various scattering events in nondegenerate semiconductors with parabolic band structure.

Scattering mechanism		Expression for relaxation time	Index of energy dependence	Index of mass dependence
Acoustic phonon	Deformation potential	$\dfrac{\pi \hbar^4 \rho s^2}{\sqrt{2}(m_n^*)^{3/2} D_{ac}^2 (k_B T)} \xi^{-1/2}$	$-1/2$	$-3/2$
	Piezoelectric	$\dfrac{2^{3/2} \pi \hbar^2 \rho s^2 \epsilon^2 \rho s^2}{q^2 h_{pz}^2 (m^*)^{1/2} (kT)} \xi^{1/2}$	$1/2$	$-1/2$
Impurity	Ionized	$\dfrac{16\sqrt{2}\pi\epsilon^2 (m^*)^{1/2} \xi^{3/2}}{N_I Z^2 q^4 [\ln(1+y') - y'/(1+y')]}$	$3/2$	$1/2$
	Neutral	$\mu_{nI} = \dfrac{m^* q^3}{80\pi \epsilon \hbar^3 N_{nI}}$	0	1
Nonpolar optical phonon		$\dfrac{2^{1/2} \pi \rho \hbar^2 (\hbar \omega_{LO}) F_{00}(\xi)}{D_{op}^2 (m^*)^{3/2} n_0}$; see the text for $F_{00}(E)$	$-1/2$	$-3/2$
Polar optical phonon		$\dfrac{2^{3/2} \pi \hbar^2 (e^{T_{LO}/T} - 1) \chi(T_{LO}/T)}{q^2 (m^*)^{1/2} (kT_{LO})^{1/2} (\epsilon_\infty^{-1} - \epsilon_0)} \xi^{1/2}$ for high temperatures $T > 220$ K $\dfrac{2^{3/2} \pi \hbar^2 (e^{T_{po}/T} - 1)}{q^2 (m^*)^{1/2} (kT_{po})^{1/2} (\epsilon_\infty^{-1} - \epsilon_0^{-1})} \xi^{1/2}$ for low temperatures $T \le 220$ K	$1/2$	$-1/2$
Alloy		$\dfrac{\pi \hbar^4}{2^{1/2} V_c x(1-x) E_{AB}^2 (m^*)^{3/2}} (\xi)^{-1/2}$	$-1/2$	$-3/2$
Potential barrier		$\left[\dfrac{\sqrt{2} N V_\delta^2 \xi_\delta^2 (m^*)^{3/2} \xi^{1/2}}{\pi \hbar^4}\right]^{-1}$	$-1/2$	$-3/2$
Potential well		$\dfrac{3(m^*)^{3/2} \xi_B}{2^{5/2} \pi^2 \hbar^2 N (kT)^{1/2}} \xi^{-1/2}$	$-1/2$	$-3/2$
Space charge		$\dfrac{(m^*)^{1/2}}{\sqrt{2} N \sigma_{sc}} \xi^{-1/2}$	$-1/2$	$1/2$
Dipole		$\dfrac{2^{3/2} 3\pi \hbar^2 \epsilon^2}{(m^*)^{1/2} q^2 N q_d^2} \xi^{1/2}$	$1/2$	$-3/2$

With $\xi \equiv kT$.

Table 4.6 Energy and mass dependence indices of mobility for various scattering events in nondegenerate semiconductors with parabolic band structure.

Scattering mechanism		Expression for mobility	Index of mass dependence	Index of temperature dependence
Acoustic phonon	Deformation potential	$\dfrac{2(2\pi)^{1/2} c_L \hbar^4 q}{3E_{dp}^2 (m^*)^{5/2}} (kT)^{-3/2}$	$-5/2$	$-3/2$
	Piezoelectric	$\dfrac{16\sqrt{2\pi}\epsilon\hbar^2}{3qP_\perp^2 (m^*)^{3/2} (kT)^{1/2}}$	$-3/2$	$-1/2$
Impurity	Ionized	$\dfrac{128\sqrt{2\pi}\epsilon^2 (kT)^{3/2}}{N_I Z^2 q^3 (m^*)^{1/2} [\ln(1+\gamma) - \gamma/(1+\gamma)]}$, $\gamma = 24\epsilon m^*(kT)^2/\hbar^2 q^2 n$	$-1/2$	$3/2$
	Neutral	$\dfrac{q}{20 a_B \hbar N_{nI}}$	0	0
Nonpolar optical phonon		$\dfrac{2^{3/2}\pi^{1/2} q\rho\hbar^2 (\hbar\omega_{LO})^2}{3D_{op}^2 (m^*)^{5/2} (kT)^{3/2}}$ for $kT \gg \hbar\omega_{LO}$	$-5/2$	$-3/2$
		$\dfrac{2^{1/2}\pi^{1/2} q\rho\hbar^4 (\hbar\omega_{LO})^{1/2}}{D_{op}^2 (m^*)^{5/2} n_0}$ for $kT \ll \hbar\omega_{LO}$		
Polar optical phonon		$\mu_{pop} = \dfrac{2^{9/2}\pi^{1/2}\hbar^2 (kT)^{1/2} (e^{T_{LO}/T} - 1)\chi(T_{LO}/T)}{3q(m^*)^{3/2}(kT_{LO})(\epsilon_\infty^{-1} - \epsilon_s^{-1})}$	$-3/2$	$-3/2$

$$\mu_{\text{pop}} = \frac{2^{3/2}\pi\hbar^2(e^{T_{\text{LO}}/T}-1)}{q(m^*)^{3/2}(kT_{\text{LO}})^{1/2}(\epsilon_\infty^{-1}-\epsilon_0^{-1})} \quad \text{for } T \leq 220\,\text{K}$$

$$\mu_{\text{pop}} = \frac{8\sqrt{2kT}}{3\sqrt{\pi m^*}E_0}\frac{e^{z_0}-1}{e^{z_0}+1} \quad \text{for } T \gg \theta_{\text{LO}}(\equiv T_{\text{LO}})$$

Alloy	$\dfrac{q\pi\hbar^4}{2^{1/2}V_c x(1-x)E_{AB}^2(m^*)^{5/2}}(kT)^{-1/2}$	$-5/2$
Potential barrier	$\dfrac{3q\hbar^4}{2^{5/2}\pi^{3/2}Na^6V_0^2(m^*)^{5/2}}(kT)^{-1/2}$	$-5/2$
Potential well	$\dfrac{q(m^*)^{1/2}\xi_B}{2^{3/2}\pi^{5/2}\hbar^2 N}(kT)^{-1/2}$	$1/2$
Space charge	$\dfrac{2^{3/2}q}{3\pi^{1/2}N\sigma_{sc}(m^*)^{1/2}}(kT)^{-1/2}$	$-1/2$
Dipole	$\dfrac{2^{9/2}\pi^{-1/2}\hbar^2\epsilon^2(kT)^{1/2}}{(m^*)^{3/2}qNq_d^2}$	$-3/2$

With $\xi \equiv kT$.

	$-1/2$
	$-1/2$
	$-1/2$
	$-1/2$
	$-1/2$

major parameters are tabulated. As indicated in Table 4.6, there are quite a few mechanisms with $T^{-1/2}$ mobility dependence. Consequently, if a few or all are present, using temperature dependence as the means of determining the scattering mechanisms involved becomes somewhat difficult.

Switching to scattering by charged dislocation, GaN represents a serendipitous exception among just about all the semiconductors in that despite the large densities of dislocations, highly efficient LEDs have been achieved. Dislocation densities as low as even $N_{dis} = 10^4 \, \text{cm}^{-2}$ are typically sufficient to degrade laser longevity. In fact, stacking fault density of about the same order has been attributed to insufficient operating lifetimes in ZnSe lasers, which are now the relics of the past. Motivated by somewhat incomplete and premature experimental observations of rather puzzling insensitivity of device operation to inordinate numbers of dislocations, theories have been advanced to explain this phenomenon by invoking the notion that threading dislocations in GaN do not induce electronic states in the bandgap (simply nonradiative recombination centers), which later has been attributed to small cell size used in initial calculations. Dislocations with edge component, such as mixed and perfect edge dislocations, have been associated with the formation of dangling bonds. Consequently, the consideration that these dislocations through their dangling bonds could trap electrons making them negatively charged evoked interest first in Ge many decades ago and now in GaN. The notion of negatively charged edge dislocations in GaN is also supported by another theoretical calculation.

Throughout this text, references are made to dislocations and their deleterious effects on a wide range of properties of GaN and related compounds. In terms of transport, dislocations are important only if they introduce charged centers, particularly in large densities, and to some effect in terms of their possible disordering effect on vibrational properties that among other processes can affect the thermal conductivity (the latter is discussed in Chapter 1). Consideration of the effect of dislocations on transport is not new to GaN. For example, it was pointed out decades ago that if the dislocations have an edge component, they introduce acceptor centers along the dislocation line that capture electrons from the conduction band in an n-type semiconductor, and are thus negatively charged. For example, for a dislocation density of $10^{10} \, \text{cm}^{-2}$, one expects $2 \times 10^{17} \, \text{cm}^{-3}$ of dangling bonds along the dislocation lines, assuming traps at intervals of the c-direction lattice constant of GaN. The negatively charged dislocation lines create a space charge region around them by which the electrons traveling across the dislocations are scattered.

The acceptor states that are empty are electrically neutral, which would not be effective in scattering the carriers. However, each filled trap carries one electronic charge and would scatter. In fact, Bonch-Bruevich and Glasko calculated the potential due to a vertical line charge as seen by electrons moving perpendicular to this line. Dislocation scattering in the context of n-type Ge has been investigated both theoretically and experimentally. Furthermore, Pödör reported an expression for the relaxation as

$$\tau_{dis}(k) = \frac{\hbar^3 \varepsilon_s^2 c^2}{N_{dis} f^2 m^* q^4} \frac{(1 + 4\lambda_D^2 k_\perp^2)_{3/2}}{\lambda_D^4} \tag{4.51}$$

and thus electron mobility average over some energy to yield mobility limited by dislocation scattering as albeit approximate. If we use $\mu_{dis} = q\tau_{dis}/m^*$ and approximate $k_\perp^2$ by $2m^*kT/\hbar^2$, we obtain for mobility

$$\mu_{dis} = \frac{\hbar^3 \varepsilon_s^2 c^2}{N_{dis} f^2 (m^*)^2 q^3} \frac{(1 + 4\lambda_D^2 2m^* kT/\hbar^2)^{3/2}}{\lambda_D^4} \approx \frac{8\varepsilon_s^2 2^{3/2} c^2}{N_{dis} f^2 q^3 \sqrt{m^*}} \frac{(kT)^{3/2}}{\lambda_D}. \quad (4.52)$$

If we take the Boltzmann distribution, then $\xi = (3/2)kT = \hbar^2 k_\perp^2/2m^*$, $k_\perp^2$ would become $3m^*kT/\hbar^2$, and mobility expression becomes

$$\mu_{dis} = \frac{\hbar^3 \varepsilon_s^2 c^2}{N_{dis} f^2 (m^*)^2 q^3} \frac{(1 + 4\lambda_D^2 2m^* kT/\hbar^2)^{3/2}}{\lambda_D^4} \approx \frac{8\varepsilon_s^2 2^{3/2} c^2}{N_{dis} f^2 q^3 \sqrt{m^*}} \frac{(kT)^{3/2}}{\lambda_D}. \quad (4.52)$$

where N_{dis} is the dislocation density, f is the occupation rate of the acceptor centers caused by the edge component of dislocations, and c is the separation of the acceptor centers taken to be the c-direction lattice parameter. As a reminder, the Debye screening length decreases with the square root of the electron concentration. Consequently, electron mobility defined by Equation 4.51 increases with increasing electron concentration (through a square root dependence) due to screening.

A more rigorous theory of charged dislocation line scattering as compared to the one described above has been developed within the framework of the BTE. This theory also utilizes the Bonch-Bruevich–Glasko and Pödör potential due to a line charge and a modified electron concentration, n', in conjunction with the effective screening in Equation 53, which may involve both free carriers and bound carriers.

The relaxation time in this case is expressed as

$$\tau_{dis}(k) = \frac{\hbar^3 \varepsilon^2 c^2}{N_{dis} m^* q^4} \frac{(1 + 4\lambda_D^2 k_\perp^2)^{3/2}}{\lambda_D^4}, \quad k_\perp^2 = 2m^* \xi/\hbar^2, \tau_{dis}(k)$$

$$= \frac{\hbar^3 \varepsilon^2 c^2}{N_{dis} m^* q^4} \frac{(1 + 8\lambda_D^2 2m^* \xi/\hbar^2)^{3/2}}{\lambda_D^4}. \quad (4.53)$$

The relaxation time expression of Equation 4.53 is consistent with the result obtained by Pödör, who then carried out an unspecified average over energy and obtained a drift mobility, $\mu_{dis} = C(kT)^{3/2}/\lambda_D$. However, to get an accurate power factor, a more rigorous treatment such as the BTE is needed.

4.3
Calculated Mobility of GaN

Calculations of mobility limited by piezoelectric processes and deformation potential associated with acoustic phonons, polar optical phonons, and ionized and neutral impurities are provided. The mobilities limited by various, but individual scattering mechanisms are included in the form of figures in the same section right after the discussion of a particular scattering mechanism. In this section, those methods will be applied to the specific case of GaN with all the scattering mechanisms combined. In order to be useful for the reader, the impurity scattering limited mobility is calculated for a range of ionized impurity concentrations beginning with very small and ending at degeneracy. The overall mobility is calculated using Matthiessen's rule that implicitly assumes that scattering events are independent of each other, meaning electrons undergo scattering events sequentially and all the scattering events can be described with relaxation time approximation that does not hold for polar optical phonon scattering. However, this method is applied even when it should not be, as alternatives require full numerical methods. In Table 4.7, the parameters needed for mobility calculations for GaN, InGaN, and AlGaN are tabulated. The values for the ternary alloys are deduced on the assumption that they can be represented by linear interpolations from the binary end points.

Experimental investigations of the transport in ternary and quaternary layers are relatively weak. Moreover, which of the models mentioned above is applicable is a matter of debate let alone alloy potentials not being known. Naturally, further discussion of the matter will continue for quite some time.

Ridley suggests that this piezoelectric scattering is notably strong when it is weakly screened. Look used piezoelectric scattering terms and acoustic deformation terms in conjunction with LO scattering to predict the maximum mobility in bulk GaN at room temperature. When other scattering terms were used in conjunction with polar optical phonon scattering, the maximum predicted mobility turned out to be about $1350\,\text{cm}^2/(\text{V s})$.

Table 4.7 Parameters used for mobility calculations for GaN, InGaN, and AlGaN.

Parameter	Symbol (units)	Fitted value (from mobility)	Literature value
High-frequency dielectric constant[a]	$\epsilon_\infty/\epsilon_0$		$5.47\varepsilon_0, 5.4\varepsilon_0, 5.37\varepsilon_0, 5.8\varepsilon_0, 5.43\varepsilon_0$
Low-frequency dielectric constant	ϵ_s/ϵ_0		9, 9.5, 9.7, 9.87, 10.4
$\frac{\epsilon_0}{\epsilon_\infty} - \frac{\epsilon_0}{\epsilon_s}$		0.113	
Polar phonon Debye temperature	θ_{LO} (K)		1057; 1060
Mass density	ρ (kg/m^3)		$6.15 \times 10^3, 6.10 \times 10^3$
Sound velocity[b]	s (m/s)		6.56×10^3 (LA), 2.68×10^3 (TA) 6.59×10^3
Piezoelectric constant	ε_{14} (C/m^2), h_{pz}		0.5 (0.375–0.6) 0.118 0.16 for GaAs
Acoustic deformation potential	E_{dp} or D_{ac} (eV)	13.2 (unscreened)	$8.3, 8.4, 8.54, 9.2, 9.1 \pm 0.7$
Effective mass	m^* (kg)		$0.218m_0, 0.22m_0, 0.23m_0$

a) From $\epsilon_s = \epsilon_\infty (\omega_{LO}/\omega_{TO})^2$.
b) From $C_L = \rho v_s^2$, where C_L is the spherically averaged elastic constant for longitudinal acoustic phonons and v_s is the average velocity of the particular acoustic mode.

4.3 Calculated Mobility of GaN

With the aid of Equation 4.17 for deformation potential scattering, Equation 4.22 for piezoelectric scattering, Equation 4.34 for polar optical phonon scattering (all of which so far represent lattice scattering), and Equation 4.11 for ionized impurity scattering, while taking the degeneracy into account for high impurity concentrations, the overall electron mobility can be calculated using Matthiessen's rule $\mu_{\text{Total}} = [1/\mu_{\text{dp}} + 1/\mu_{\text{pop}} + 1/\mu_{\text{I}} + 1/\mu_{\text{pz}}]^{-1}$ with its limitations; that is, the scattering events can be described with relaxation approximation that does not hold for polar optical phonon scattering. Using the parameters of $m^* = 0.22 m_n$, $N_D = 10^{15} - 10^{19}$ cm^{-3}, $\rho = 6.1 \times 10^3$ kg/m^3, $s = 6.59 \times 10^3$ m/s, $\varepsilon_s = 10.4 \varepsilon_0$, and $h_{\text{pz}} = 0.5$ C/m^2, the calculated mobility for each of scattering mechanisms as well as the overall mobility of GaN is shown in Figure 4.14 for ionized impurity densities of 10^{15} cm^{-3}.

For the ionized impurity scattering, the ionized donor concentration, N_I, was calculated using

$$N_I = \frac{N_D}{1 + g_D \exp[(E_F - E_D)/kT]} \approx n = N_C \cdot \frac{2}{\sqrt{\pi}} F_{1/2}\left(-\frac{E_C - E_F}{kT}\right), \quad (4.54)$$

with a donor binding energy of 30 meV and donor degeneracy factor, g_d, of 2. For degenerate case, simplifying approximations cannot be made and the Fermi integrals must be evaluated numerically, for example, with the aid of Matlab package. We should mention that the exponent in Equation 4.54 is also expressed

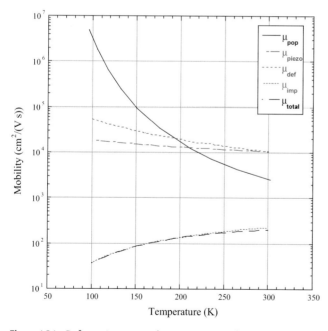

Figure 4.14 Deformation potential scattering, piezoelectric scattering, polar optical phonon scattering, and ionized impurity (10^{15} cm^{-3}) scattering limited mobility in GaN versus temperature along with the total mobility using Matthiessen's rule.

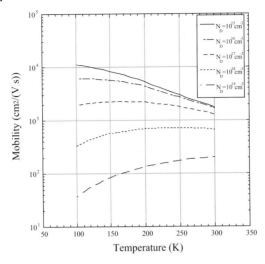

Figure 4.15 Calculated overall electron mobility in GaN versus temperature for ionized donor concentrations of 10^{15}, 10^{16}, 10^{17}, 10^{18}, and 10^{19} cm^{-3} using Matthiessen's rule.

with reverse sign owing to the assumptions made as to the reference to which the energies are defined (see Chapter 6 for a detailed discussion of ionized impurity concentration calculations and degeneracy factors for electrons and holes).

To cap the major scattering mechanisms commonly encountered in GaN, among which are the impurity, deformation potential, piezoelectric scattering, and polar optical phonon scattering, the temperature dependence of electron mobility in GaN limited by the aforementioned processes is shown in Figure 4.15. The donor concentrations used are 10^{15}–10^{19} cm^{-3} with increments of a factor of 10.

The calculated electron mobility of impurity-free GaN with the BTE is shown in Figure 4.16 for the directions parallel and perpendicular to the *c*-axis. Here the mobility determined by the iterative method of Rode is imbedded in the components of the mobility, which are limited by the various scattering processes. As expected, the electron mobility at room temperature and above is dominated by polar optical (LO phonon) scattering. Superimposed, as a thick line, is the overall mobility calculated with the analytical expressions for impurity-free GaN and Matthiessen's rule. The major disagreement between the iterative method and this one is that this one relies on the calculations of mobility components occurring at high temperatures. This indicates that the polar optical phonon scattering is the process causing the disagreement.

As the carrier concentration increases further to about 10^{18} cm^{-3}, the mobility remains essentially unchanged until the temperature reaches about 150 K, and then decreases rapidly with increase in temperature. This behavior at low temperatures arises from the dominance of piezoelectric scattering (at lower carrier concentrations) and from ionized impurity scattering (at higher carrier concentrations). Above 200 K, the polar optical phonon contribution is the most important scattering mechanism. Carrier–carrier scattering is also important particularly when the

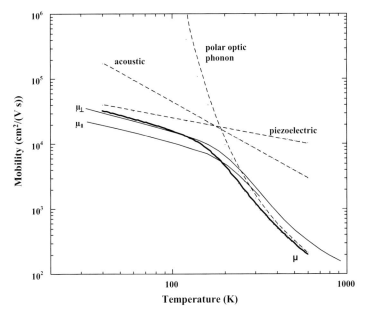

Figure 4.16 Theoretical electron drift mobility of pure GaN as calculated by Rode's iterative method for a transport transverse to the c-axis ($\mu_\parallel$) or in the c-plane and parallel to the c-axis ($\mu_\perp$) or out of the c-plane. The mobility as determined is imbedded in the mobility limited by the well-known scattering processes discussed in the text with the overall mobility being determined by Matthiessen's rule. The BTE results courtesy of D.L. Rode.

carrier concentration is high and other scattering mechanisms are suppressed as in 2DEG. Knowing the scattering rates paves the way for estimation of the impact of electron–electron scattering on electron mobility.

4.4
Scattering at High Fields

The behavior of carriers at high fields was first analyzed by Shockley using classical mechanics, which is not very accurate for complicated band structures and strong electron–phonon coupling. This is in turn further complicated by the lack of good understanding of the phonon generation and decay mechanisms and presence or absence of any mixed modes. Despite these shortcomings, his method provides a vivid image of how the carriers are scattered at high fields and why the velocity saturates in Si and Ge. Advanced treatment of hot carriers using the BTE and Monte Carlo calculations is more accurate, at the expense of complexity. When an electron is accelerated by an electric field, it gains energy from the field. The energy can be expressed as kT, assuming a one-dimensional model, where the electron temperature represents a higher temperature than the lattice temperature and energy transfer is from the hot electrons to the lattice, because their temperature is higher than that of the lattice. In a lattice wave, the movement of atoms can be either in the direction of wave propagation (longitudinal mode) or normal to it (transverse mode). In diamond and zincblende structures, there are two atoms per unit cell and these atoms can either move in phase or out of phase with each other. The out-of-phase component requires more energy and forms the optical branch of the phonon dispersion relationship. The in-phase motion requires less energy and thus forms the acoustic branch. In the simple treatment, there would be four modes, transverse and longitudinal optical and acoustical modes, termed as LA, TA, LO, and TO branches. In Si and Ge, the room-temperature low-field mobility is dominated by LA phonons, even though there are contributions from LO phonons. In GaAs and GaN, the room-temperature low-field mobility is determined by polar optical phonons, LO phonons.

In the case of Si and Ge, as the field is increased, the electrons give off their energy by emitting LA phonons at moderate fields where the electron temperature is not high enough to emit an optical phonon. The thermal acoustic velocity of electrons in semiconductors is generally in the range 10^5–10^6 cm/s. However, as the field is increased and as soon as the electrons gain sufficient energy (the threshold determined by the velocity needed to reach the optical phonon energy), the electron would emit optical phonons. Even though the reverse process, phonon absorption, can also occur, because the electron temperature is higher than the lattice temperature, the dominant process is phonon emission. Consider a case in which a moving electron collides with a phonon of mass M ($M \gg m$, the electron mass) at rest. In a 1D system, the electron reverses its direction, and the M particle moves in the direction of electron motion before the collision. This represents the case where a phonon is emitted by an electron. Considering this, the energy and momentum conservation would lead to a rate of energy change for each collision given as

$$\langle \partial \xi \rangle_1 = \frac{2s^2 p_1^2}{kT_1}, \tag{4.55}$$

where T_1 is the lattice temperature, s is the sound velocity, and p_1 is the momentum of the electron before collision.

In the reverse process, an electron at rest colliding with a phonon, the average energy exchange per collision is given by

$$\langle \partial \xi \rangle_2 = \frac{m^*}{2} \left(\frac{p_r + p'_r}{M} \right)^2 \cong 2m^* s^2, \tag{4.56}$$

where p_r and p'_r represent the momentum of phonon before and after the collision, respectively. When an electron collides with acoustic phonons, the energy transfer is very small. However, the direction of the electron motion is randomized, which cannot be obtained from a 1D treatment. Because the electrons gain energy from the field, but do not really dissipate that energy through LA phonon collisions, their thermal velocity increases above the thermal velocity when they were in equilibrium with the lattice. Mathematically, this can be represented by $v_e = \sqrt{kT_e/m^*} > \sqrt{kT_1/m^*} = v_{th}$. Here T_e and T_1 represent the electron and lattice temperatures, respectively. For a constant mean free length between collisions, which assumes no energy distribution for carrier velocity,

$$\mu = \frac{q\tau}{m^*} = \frac{ql}{m^* v_e} = \mu_0 (T_1/T_e)^{1/2}, \quad (4.57)$$

where μ_0 is the low-field mobility.

Power balance equation can be used to determine the dependence of mobility on electric field E. Shockley considered a 3D lattice and averaged velocity, determining the rate of energy gained by an electron from the field as

$$\left\langle \frac{d\xi}{dt} \right\rangle = \frac{8s^2}{\sqrt{\pi}} \left(\frac{(m^*)^2 v_e^2}{3kT_1} - m^* \right) \frac{v_e}{l} \quad (4.58)$$

and the energy given by an electron to lattice through phonon emission as

$$\left\langle \frac{d\xi}{dt} \right\rangle = qEv_d = q\mu E^2 = q\mu_0 E^2 \frac{v_{th}}{v_e}. \quad (4.59)$$

Under equilibrium, the energy gained and lost must balance, which after several steps yields

$$\mu = \mu_0 \left[1 - \frac{3\pi}{64} \left(\frac{\mu_0 E}{s} \right)^2 \right]. \quad (4.60)$$

At moderately high fields, when $\mu_0 E > s$ and $T_e \gg T_1$, Equation 4.60 reduces to

$$\mu = 1.36 \mu_0 \left(\frac{s}{\mu_0 E} \right)^{1/2} \quad (4.61)$$

and the drift velocity

$$v_d = \mu E = 1.36 \mu_0 (s\mu_0 E)^{1/2}. \quad (4.62)$$

So far, the treatment has been limited to acoustic phonons. Following a treatment similar to that of the acoustic phonon (meaning energy loss by an electron by optical phonon emission is equal to energy gained by lattice under equilibrium), one can arrive at an expression for the drift velocity. By making the assumption that a relaxation time can be defined for nonoptical phonons, which is not as severe as if it were done for polar optical phonons, one can write for the rate of change in energy:

$$\frac{d\xi}{dt} = qEv_s - \frac{\hbar \omega_{LO}}{\tau_e}, \quad (4.63)$$

where τ_e is the energy relaxation time. The first term on the right-hand side is the rate of energy gained from the field and the second one is the rate of energy loss due to optical phonon emission.

Other treatments lead to a rate of change in energy at high fields:

$$\left\langle \frac{d\xi}{dt} \right\rangle_{field} = \frac{qE^2 \tau_{op}}{m^*} = \left\langle \frac{d\xi}{dt} \right\rangle_{phonon} = \frac{\hbar \omega_{LO}}{\tau_{op}}, \quad (4.64)$$

where τ_{op} is the mean free time of electrons for optical phonon scattering and $\hbar\omega_{LO}$ is the optical phonon energy.

A corresponding rate expression for the momentum relaxation can be written as

$$\frac{d(m^*v_s)}{dt} = qE - \frac{m^*v_s}{\tau_m}, \tag{4.65}$$

where τ_m is the momentum relaxation time. If we assume that LO phonon scattering is the dominant process at high fields, the relaxation times τ_e and τ_m are equal to the optical phonon scattering time, τ_{op}. At steady state, $d\xi/dt = 0$ and $d(m^*v_s)/dt = 0$, and the steady-state saturation velocity v_s can be found as (recognizing Equation 4.64 as well)

$$v_s = \mu E = \frac{qE\tau_{op}}{m^*} = \left(\frac{\hbar\omega_{LO}}{m^*}\right)^{1/2}. \tag{4.66}$$

For Si, using 50 meV for $\hbar\omega_{LO}$ and $m^* = 0.27m_0$, v_s in the [001] direction is found to be 1.7×10^7 cm/s, which is very close to the experimental value of 1×10^7 cm/s.

For direct bandgap semiconductors with satellite valleys in the conduction band, the electron velocity initially increases with increasing field, as in the case of Si (indirect bandgap). Energy is gained by an electron from the field and lost by optical LO phonon emission to the lattice until a critical field is reached where the electrons can gain sufficient energy to be transferred to the low-lying satellite band, which in the case of GaAs is the L valley lying about 0.3 eV above the conduction band minimum. Because the density of states is very high, due to large effective mass, the scattering rate to the satellite valley is very efficient, and electron transfer to the satellite valley takes place in a small E field range. Consequently, the electron velocity drops from the peak value with further increase in field because of transfer to the band in which the mass is heavy. The average drift velocity in a two-valley semiconductor can be written as

$$v_d = \frac{n_1\mu_1 + n_2\mu_2}{n_1 + n_2} E. \tag{4.67}$$

If we assume that $\mu_2 \ll \mu_1$, the above equation can be approximated as

$$v_d = \mu_1 E \left(1 + \frac{n_2}{n_1}\right)^{-1}. \tag{4.68}$$

If the densities of states for two valleys are N_{C1} and N_{C2}, the electron concentrations in the two valleys are given by

$$n_1 = N_{C1} \exp\left(-\frac{\xi_{C1} - \xi_{F1}}{kT_{e1}}\right) \quad \text{and} \quad n_2 = N_{C2} \exp\left(-\frac{\xi_{C2} - \xi_{F1}}{kT_{e2}}\right), \tag{4.69}$$

and assuming that the electron temperature is the same for both valleys,

$$\frac{n_2}{n_1} = \frac{N_{C2}}{N_{C1}} \exp\left(-\frac{\xi_{C2} - \xi_{C1}}{kT_e}\right), \tag{4.70}$$

where ξ_{C1} and ξ_{C2} represent the bottom of the conduction band energy for the zone-center conduction band and the valley band, respectively. We also assumed that there is only one electron temperature involved. As in the case of Si and Ge, we additionally need the energy balance equation to get the field dependence of the electron velocity. Noting that the average kinetic energy of an electron obeying the Boltzmann distribution is $3kT_e/2$, and that there is no net energy exchange between the electron and lattice if $T_e = T_l$, the net rate of energy exchange can be expressed in a 3D system as

$$\left\langle \frac{d\xi}{dt} \right\rangle = \frac{3}{2\tau_e} k(T_e - T_l), \tag{4.71}$$

where τ_e represents the energy relaxation time. The interaction with the electric field is still given by Equation 4.59. Equating Equation 4.71 to Equation 4.59, and using Equation 4.68, we have

$$\frac{v_d}{v_0} = (E/E_0)^2 \left[1 + \frac{N_{C2}}{N_{C1}} \exp(-1/\theta_e)\right]^{-1} \quad \text{with} \quad v_0 = \mu E_0, \tag{4.72}$$

which can be used to calculate the velocity–field characteristics knowing the density of states in both bands, low-field conduction band mobility, energy difference between the conduction band and the satellite valley involved, and the electron temperature.

A simple analytical expression, empirical in nature, can be used to describe saturating velocity–field characteristics as follows:

$$v_d = \frac{\mu_0 E}{1 + (\mu_0 E/v_s)}, \tag{4.73}$$

which is a more simplified version of Equation 4.3, but expressed as the product of mobility and velocity. The terms μ_0 and E represent the low-field drift mobility and electric field, respectively. At high fields, the mobility field product is equal to v_s, the saturation velocity. This simple expression holds only for single-valley semiconductors under steady state without velocity overshoot effects and where the saturating field is large, such as Si. However, for semiconductors such as GaAs, the saturating field is much smaller. In that case, a modified version of Equation 4.73 in a fashion mimicking Equation 4.3 can be used:

$$v_d = \frac{\mu_0 E}{[1 + (\mu_0 E/v_s)^\beta]^{1/\beta}}. \tag{4.74}$$

The higher the mobility or the smaller the critical field causing saturation or intervalley scattering, the larger the β factor must be. It should be noted that this expression does not trace the peak and valley nature of the velocity–field characteristics. If the peak and negative differential resistance aspects are neglected, the fit is good.

In multiple-valley systems, as shown in Figure 4.17a, the electron velocity peaks about when the intervalley scattering causes population of the upper valley with relatively large electron effective mass. Figure 4.17b shows a schematic representation of the velocity–field characteristics for a single-band system that can be approximated by Equation 4.73, a two-piece model that allows a simpler analytical treatment of FET current–voltage characteristics (two-piece model), and a two-band (the lowest conduction band and the satellite valley) system.

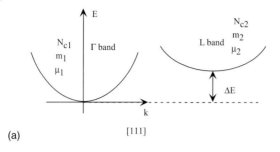

(a)

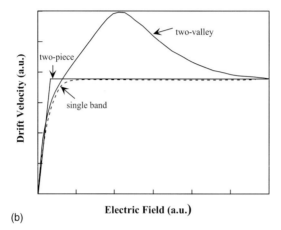

(b)

Figure 4.17 (a) A two-valley conduction band system that causes peaked velocity–field characteristics followed by reduction in velocity due to upper band population. (b) Schematic representation of velocity–field characteristics for a single-band system that can be approximated by Equation 4.73, a two-piece model that allows simpler analytical treatment of FET current–voltage characteristics (two-piece model), and a two-band (conduction band and satellite valley) system.

If the temperature in the upper valley is assumed to be equal to the lattice temperature ($T_{e2} = T_l$) because of typical heavy electron mass (low mobility), the energy transfer rate becomes

$$P_{12} = \frac{3}{2}r_{21}k(T_{e1} - T_l) = \frac{3k(T_{e1} - T_l)n_2}{2\tau_{21}}. \tag{4.75}$$

4.4.1
Transport at High Fields: Energy and Momentum Relaxation Times

When transport is caused by an external electric field, the electrons are supplied continuously with energy from the field at a rate determined by the dot product of the current density and the field vectors. This would leave the impression that the electrons can gain energy from the field indefinitely and attain very high velocities.

However, as the previous section indicated, the gain is balanced by a loss caused by scattering that transfers energy to the lattice through emission or absorption of phonons. It is convenient to view this process as an electron responding to an average electric field along with a series of independent momentum scattering events. The motion of an average electron can be described by the momentum balance equation and the energy balance equation:

$$\frac{d(m^* \langle \vec{v}_d \rangle)}{dt} = q\vec{E} - m^* \frac{\vec{v}_d}{\tau_m(\langle \xi \rangle)}, \tag{4.76}$$

$$\left\langle \frac{d\xi}{dt} \right\rangle = q \langle \vec{v}_d \rangle \cdot \vec{E} - \frac{\langle \xi - \xi_0 \rangle}{\tau_e(\langle \xi \rangle)}, \tag{4.77}$$

where $\tau_m(\langle \xi \rangle)$ and $\tau_e(\langle \xi \rangle)$ are the momentum and energy relaxation times, respectively, $\vec{E}$ is the electric field, and ξ_0 is the average electron energy at equilibrium in the absence of electric field. Electron interactions with lattice and defects, including impurities and other electrons, all of which are energy dependent, are included in the properly averaged momentum and energy scattering times.

When the electric field is small, the energy gained by the carriers from the field is low compared to the thermal energy that the carriers already have and the distribution function can be thought of as unchanged except for a small shift in momentum space. Such a displaced distribution function leads to ohmic behavior with the mobility independent of electric field. As the electric field is increased to a point that the energy gained from the field is comparable to or larger than the carrier average energy, the distribution function changes drastically signifying the hot carrier regime. The carrier mobility deviates from Ohm's law and becomes dependent on the electric field.

Under steady-state conditions, hot electron phenomena can be determined by energy balance between the power input from the electric field into the electronic system and the power loss from the inelastic collisions between the carriers and phonons. Under transient conditions, the evolution of the perturbed distribution function is also determined by carrier–carrier and carrier–phonon interactions. For a succinct discussion of electron–phonon interaction in a 3D system, and high-field transport in multivalley systems, including the GaN system, refer to Morkoç (2008).

4.4.2
Energy-Dependent Relaxation Time and Large B

In the limit of large magnetic field so that $\omega_c \tau \gg 1$, the elements of the conductivity matrix of Equation 92 can be written as (provided that $\hbar \omega_c \ll kT$)

$$\sigma_{xx} \equiv \frac{ne^2}{m^*} \left[\left\langle \frac{\tau}{1+\omega_c^2 \tau^2} \right\rangle \right] = \frac{ne^2}{m^*} \left\langle \frac{1/\omega_c^2 \tau}{1+1/\omega_c^2 \tau^2} \right\rangle \approx \frac{ne^2}{m^*} \frac{\langle \tau^{-1} \rangle}{\omega_c^2}, \tag{4.78}$$

$$\sigma_{xy} = -\frac{ne^2}{m^*} \left[\left\langle \frac{\omega_c \tau^2}{1+\omega_c^2 \tau^2} \right\rangle \right] = -\frac{ne^2}{m^*} \left\langle \frac{1/\omega_c}{1+1/\omega_c^2 \tau^2} \right\rangle \approx -\frac{ne^2}{m^*} \frac{1}{\omega_c}. \tag{4.79}$$

Again, setting the y component of the current to zero due to the Hall geometry and solving,

$$\langle j_x \rangle = \frac{\sigma_{xx}^2 + \sigma_{xy}^2}{\sigma_{xx}} E_x = \sigma E_x, \quad (4.80)$$

where

$$\sigma = \frac{\sigma_{xx}^2 + \sigma_{xy}^2}{\sigma_{xx}} \approx \frac{ne^2}{m^*} \left[\frac{(\langle \tau^{-1} \rangle / \omega_c^2)^2 + (1/\omega_c)^2}{(\langle \tau^{-1} \rangle / \omega_c^2)} \right] = \frac{ne^2}{m^*} \frac{1 + 1/(\omega_c \tau^{-1})^2}{\langle \tau^{-1} \rangle}$$

$$\approx \frac{ne^2}{m^*} \frac{1}{\langle \tau^{-1} \rangle}. \quad (4.81)$$

The Hall coefficient

$$R_H = \frac{1}{B} \frac{\sigma_{xy}}{\sigma_{xx}^2 + \sigma_{xy}^2} = -\frac{m^*}{Bne^2} \frac{1/\omega_c}{(\langle \tau^{-1} \rangle / \omega_c^2)^2 + 1/\omega_c^2} \approx -\frac{m^* \omega_c}{Bne^2} = -\frac{1}{ne}. \quad (4.82)$$

As for conductivity, if we use the nomenclature of σ_∞ for the high B field conductivity, we note that σ_∞ saturates at high B fields, which is not easy to observe experimentally. If we use σ_0 for the low B field value, the ratio of the low to high B field conductivities is given by

$$\sigma_0/\sigma_\infty = \langle \tau \rangle \langle \tau^{-1} \rangle. \quad (4.83)$$

For a nondegenerate semiconductor and with the help of Γ functions, Equation 4.83 can be written as

$$\sigma_0/\sigma_\infty = \gamma = \frac{\Gamma[(5/2) + s] + \Gamma[(5/2) - s]}{[\Gamma(5/2)]^2}. \quad (4.84)$$

For a two-carrier system as in the case of lightly doped samples where both electrons and holes must be considered, the relative change in resistivity can be expressed as

$$\frac{\Delta \rho}{\rho_0} = \left[\frac{\Gamma^2(5/2)\Gamma[(5/2) - 3s]}{\Gamma^3[(5/2) - s]} \left(\frac{n\mu_n^3 + p\mu_p^3}{n\mu_n + p\mu_p} \right) - \left(\frac{\Gamma(5/2)\Gamma[(5/2) - 2s]}{\Gamma^2[(5/2) - s]} \right)^2 \left(\frac{n\mu_n^2 + p\mu_p^2}{n\mu_n + p\mu_p} \right)^2 \right] B^2. \quad (4.85)$$

For an n-type sample ($n \gg p$), Equation 4.85 reduces to $\Delta\rho/\rho_0 \approx \mu_n^2 B^2$ and as before $\rho \approx \rho_0 (1 + \mu_n^2 B^2)$.

A very important implication of Equation 4.82 is that the Hall factor is unity and it holds for any band structure, any scattering mechanism, and degeneracy, provided that the material is homogeneous (isotropic) and spin-dependent scattering is excluded. This means that the electron concentration can be determined from the Hall measurements very accurately if they are made under conditions satisfying

$\omega_c \tau = \mu B$. In addition, this also means that if one performs Hall measurements for small magnetic fields for which $\omega_c \tau \ll 1$ and for sufficiently large magnetic fields for which $\omega_c \tau \gg 1$, the ratio of the low Hall coefficient determined in the low magnetic field measurement to that for the large magnetic field would result in the Hall factor. Mathematically,

$$r_H = \frac{R_H(\text{low } B \text{ field})}{R_H(\text{high } B \text{ field})}. \tag{4.86}$$

For geometric magnetoresistance, refer to Morkoç (2008).

4.4.3
Hall Factor

Even though the Hall factor was discussed in relation to many special cases of magnetoresistance measurements, a need for a summary discussion with emphasis on how it is dependent on the various types of scattering mechanisms is compelling. For the Boltzmann distribution for nondegenerate semiconductors with spherical constant energy surfaces, the average relaxation time

$$\langle \tau \rangle = \frac{\int \tau(\xi) f(\xi) d\xi}{\int f(\xi) d\xi} \tag{4.87}$$

in general, and with a distribution function of $f(\xi) = \xi^{3/2} \exp(-\xi/k_B T)$, the average relaxation time becomes

$$\langle \tau^m \rangle = \frac{\int_0^\infty [\tau(\xi)]^m \xi^{3/2} \exp(-\xi/kT) d\xi}{\int_0^\infty \xi^{3/2} \exp(-\xi/kT) d\xi}. \tag{4.88}$$

The general form describing the relaxation time for processes other than the inelastic ones such as optical phonon scattering is of the form $\tau \sim a\xi^{-s}$. Here a and s are constants. In semiconductors with spherical constant energy surfaces, the mean free time between collisions takes the form of $\tau \sim \xi^{-1/2}$ for deformation acoustic phonon scattering and alloy scattering, $\tau \sim \xi^{1/2}$ for piezoelectric scattering, and $\tau \sim \xi^{3/2}$ for ionized impurity scattering. Using the general form of $\tau \sim a\xi^{-s}$ for its energy dependence, $\langle \tau^2 \rangle$ and $\langle \tau \rangle^2$ can be obtained from Equation 4.87, the manipulation of which leads to expressions for the average of the square of the relaxation time and square of the average of the relaxation time.

Using $\Gamma(m) = \int_0^\infty x^{m-1} e^{-x} dx$ for the $\Gamma(x)$ integrals, expressions for r_H for all scattering events except the optical phonon scattering can be developed as

$$r_H = \frac{\langle \tau^2 \rangle}{\langle \tau \rangle^2} = \frac{\Gamma[(5/2) - 2s]\Gamma(5/2)}{[\Gamma[(5/2) - s]]^2}. \tag{4.89}$$

Calculation of the Hall factor requires $\Gamma(x)$ functions. The $\Gamma(x)$ functions for various values of x are given in Table 4.8. The Hall factors for various scattering mechanisms, when defined in a closed form that excludes polar optical phonon scattering,

Table 4.8 $\Gamma(x)$ functions.

x	$\Gamma(x)$
0.5	$\sqrt{\pi}$
1.0	1.0
1.25	0.906
1.5	0.886
1.75	0.919
2.0	1.0
2.5	1.33
3	2
3.5	3.325

Note that the values of $\Gamma(x)$ for $x<1$ and $x>2$ can be calculated using $\Gamma(m) = \int_0^\infty x^{m-1} e^{-x} dx$ or $\Gamma(x) = \Gamma(x+1)/x$, $\Gamma(x) = (x-1)\Gamma(x-1)$, and $\Gamma(1/2) = \sqrt{\pi}$.

are provided in Table 4.9. To reiterate, the polar optical phonon scattering cannot be described with the relaxation time approximation without loss of vital accuracy; as such, it causes r_H to vary with temperature and must be deduced from numerical calculations.

For a treatment of multiband effects and mixed conductivity, refer to Morkoç (2008).

4.5
Delineation of Multiple Conduction Layer Mobilities

Hall measurements give some average value for the electron concentration and mobility that is fine for uniform samples in both vertical and lateral directions. However, for multiple conduction layers, the experimental approach interpretation must be modified. For example, magnetoresistance measurements can be employed, with the aid of numerical fitting techniques, to discern the mobility for each of the layers contributing to current conduction. For simplicity, one can consider a two-layer stack in which case some of the parameters needed would have to be determined by other technique(s).

This parallel channel affects the behavior of the electron concentration as well as the mobility. This is conceptually similar to multiband conduction and mixed conductivity. Correct values of the bulk Hall mobility, μ, and bulk Hall electron concentration, n, can be extracted from the experimental Hall mobility, μ_{exp}, and electron concentration, n_{exp}, using a two-layer Hall analysis. The corrected values can be expressed as

$$\mu = \frac{\mu_{meas}^2 n_{meas} - \mu_{int}^2 n_{int}}{\mu_{meas} n_{meas} - \mu_{int} n_{int}}, \tag{4.90}$$

$$n = \frac{(\mu_{meas} n_{meas} - \mu_{int} n_{int})^2}{\mu_{meas}^2 n_{meas} - \mu_{int}^2 n_{int}}, \tag{4.91}$$

where μ_{int} and n_{int} are the mobility and electron concentration of the interfacial defective layer, respectively. The values of μ_{int} and n_{int} can be obtained from the low-temperature experimental uncorrected Hall data.

4.5 Delineation of Multiple Conduction Layer Mobilities

Table 4.9 Hall factors for various scattering mechanisms.

Scattering process	s factor	Relaxation time		Mobility		Hall factor, r_H
		Energy dependence	Mass dependence	Temperature dependence	Mass dependence	
Acoustic deformation	1/2	−1/2	−3/2	−3/2	−5/2	$3\pi/8 = 1.1781$
Acoustic piezoelectric	−1/2,	1/2	−1/2	−1/2	−3/2	$45\pi/128 = 1.1045$
Alloy scattering	1/2	−1/2	−3/2	−1/2	−5/2	$3\pi/8 = 1.1781$
Neutral impurity scattering	s = 0 (energy-independent scattering)	0	1	0	0	1
Ionized impurity scattering	−3/2	3/2	1/2	3/2	−1/2	$315\pi/512 = 1.9328$
Nonpolar optical phonon	Complicated but if $\tau \sim E^{1/2}$ is assumed, then s = −1/2					1.104
Polar optical phonon	Cannot be described with simple expression to infer s factor					Cannot be described with simple expression to infer Hall factor
Degenerate semiconductor	Sharply peaked at E_F and can be represented with a delta function					1

For energy and mass dependence of the relaxation time and mobility for various scattering mechanisms, see Tables 4.5 and 4.6.

As noted above, the magnetic field dependences of the conductivity and Hall coefficient can be used to deal with multilayer conduction. However, very high magnetic fields (>10 T) are needed for an accurate determination of the mobility and carrier concentration. Yet another method, "differential Hall measurement," can be used to tackle multilayer conduction. In this approach, thin layers of thickness δd are successively removed from the sample by reactive ion etching (RIE) coupled with Hall effect measurements after each RIE step. Needless to say, an accurate knowledge of the removed material or remaining thickness of the layer after each step is needed. This is attained by making profilometer measurements through a small hole, ion milled down to the Al_2O_3 substrate. (Note that the Al_2O_3 is not affected by the particular chemistry used in RIE.) One can then use the following expressions, in much the same way as Equations 4.90 and 4.91:

$$\mu_{\delta d} = \frac{\mu_d^2 n_d - \mu^2 n_{d-\delta d}}{\mu_d n_d - \mu_{d-\delta d} n_{d-\delta d}}, \qquad (4.92)$$

$$n_{\delta d} = \frac{(\mu_d n_d - \mu_{d-\delta d} n_{d-\delta d})^2}{\delta d [\mu_d^2 n_d - \mu_{d-\delta d}^2 n_{d-\delta d}]}. \qquad (4.93)$$

4.6
Carrier Transport in InN

As is the case for GaN, InN suffers from the lack of a suitable substrate material and, to a much larger extent, high native defect concentrations that really hinder its progress and analysis. Furthermore, the large disparity of the atomic radii of In and N is an additional contributing factor to the difficulty in obtaining InN of good quality. It is not clear as to the source of high electron concentration in unintentionally doped InN, and as such there is a substantial scatter in the electron mobilities obtained from various films.

Growth schemes for InN have been improved substantially in terms of the MBE approach and room-temperature mobilities of about 2000 $cm^2/(V s)$ have been obtained for both N-polar and Ga-polar surfaces. Although InN is under consideration, the term Ga-polar is used because the buffer layer on which InN is grown is GaN, which determines the polarity.

The main issues for OMVPE growth of InN include the temperature of growth, the flow rate of gases, the III/V ratio, the nature of the substrate and preparation of its surface prior to growth, and difficulty of growing thick films. Increased substrate temperatures would lead to more efficient pyrolysis/catalysis of NH_3 and thus more abundant active N. However, increased temperature would hasten InN decomposition. Experimentally, increased substrate temperatures while providing as much NH_3 as possible lead to reduced donor concentration, possibly due to decreased amounts of nitrogen vacancies that are thought to be the source of donors. Clearly, optimization of the NH_3 flow rate and substrate temperature requires a great deal of experimentation. It appears that a growth temperature of about 600 °C is the highest that can be tolerated, meaning the ultimate mechanical limitation is the flow rate of NH_3. The relatively lower carrier concentrations of $1-2 \times 10^{18}$ cm^{-3} in MBE-grown InN are possibly

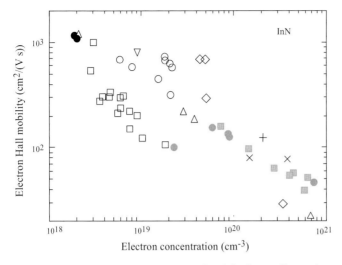

Figure 4.18 Carrier concentration versus Hall mobility for InN films inclusive of the data reported for MBE- and MOVPE-grown layers.

due to active nitrogen being supplied independent of the growth temperature. A compilation of OMVPE and MBE data on InN is shown in Figure 4.18.

Theoretical modeling, again a variational approach to solving the BTE carried out for temperatures of 77 and 300 K, demonstrated that the carrier concentration dependence of the Hall and drift mobilities in InN is a significant function of the compensation ratio. These calculated results (Figures 4.19 and 4.20) agree well with the available experimental data for the high compensation ratios, but not others in which the mobility is relatively high. Calculated results for the electron drift mobility as a function of temperature, compensation ratio, and carrier concentration yield peak electron mobilities of 25 000, 12 000, and 8000 cm^2/(V s) for doping densities of 10^{16}, 10^{17}, and 10^{19} cm^{-3}, respectively. Piezoelectric acoustic phonon scattering and ionized impurity scattering are the two dominant scattering mechanisms at temperatures $T \leq 200$ K, and the polar optical phonon scattering is the most significant scattering mechanism for temperatures $T \geq 200$ K.

As in the case of GaN, InN also exhibits characteristics reminiscent of multiple-layer conduction. The variable magnetic field measurements in the realm of quantitative mobility spectrum analysis (QMSA) have been applied to determine the spectra of electron mobilities throughout an InN layer grown by MBE. Pertinent details can be found in Morkoç (2008).

4.7
Carrier Transport in AlN

Electrical transport in AlN has not been studied extensively, at least in relation to GaN, contributed in part because of the erroneously perceived insulating nature. The basis for this lies in the large, 6.1 eV, energy bandgap and the defective nature of

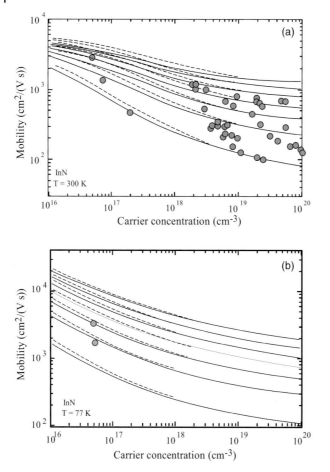

Figure 4.19 The electron drift mobility (solid curves) and the Hall mobility (dashed curves) for InN as a function of carrier concentration with the compensation ratios of 0.00, 0.15, 0.30, 0.45, 0.60, 0.75, and 0.90 at (a) 300 K (mobility decreases as the compensation ratio increases) and (b) 77 K. The horizontal axis represents the actual electron concentration for the drift mobility, but the Hall concentration for the Hall mobilities. Solid circles represent the experimental data. Courtesy of Chin and Tansley.

the material. In fact, the criterion should be dupability of the material not the bandgap. The effective electron mass for AlN is still relatively unknown but for numerical simulations of the electron mobility, estimates at 0.48 ± 0.05 and 0.27 have been employed. Using this and the energy bandgap $E_g = 6.0\,\mathrm{eV}$, the polar optical phonon limited drift mobility as a function of temperature has been calculated, with a value of about $2000\,\mathrm{cm^2/(V\,s)}$ at 77 K and $300\,\mathrm{cm^2/(V\,s)}$ at 300 K. Si is a donor in AlN, with $\sim 70\,\mathrm{meV}$ activation energy in the framework of the hydrogenic model. The data are consistent with optical measurement considering the donor-bound exciton. However, the Hall measurements show much higher activation energy that must be treated with extreme care as any compensation would affect the slope of the temperature dependence of the electron concentration.

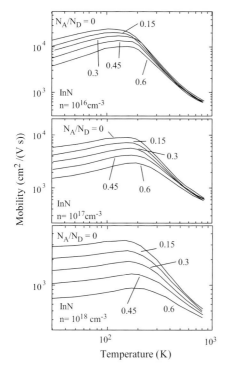

Figure 4.20 The electron drift mobility in InN as a function of temperature for electron concentrations of 10^{16}, 10^{17}, and 10^{18}, and for the compensation ratios of 0. 0, 0.3, 0.45, and 0.6. Courtesy of Chin and Tansley.

4.8 Carrier Transport in Alloys

Fundamentally, with all else being equal, the alloy scattering is the main difference between the binaries and alloys when it comes to carrier transport. As clearly seen in Equation 4.48, the parameter of most importance is the alloy scattering potential used.

If reliable experimental mobility data were to exist, this quadratic term could be treated as a fitting parameter, but in the absence it must be calculated. The reality is that there is no consensus on the alloy scattering potential. Phillip's electronegativity theory has been used to obtain these potentials. Significant bowing in the mobility of InGaN and InAlN versus the mole fraction occurs.

Alloy scattering is very dependent on the choice of the alloy potentials, as demonstrated quite clearly by Equation 4.48. If larger alloy potentials were chosen for AlGaN, significant bowing and a reduction in the mobility would occur. In fact, Figure 4.21 makes this point quite clear for AlGaN with significant bowing when alloy potential values of 0.5 and 1 eV are used for $V_A - V_B$. The donor and acceptor concentrations adopted for the calculations based on Equation 4.48

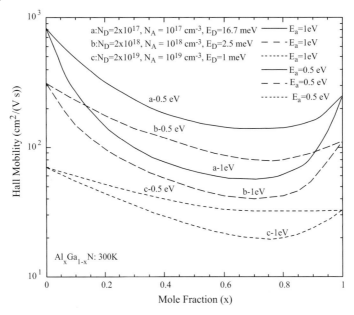

Figure 4.21 Calculated electron mobility in AlGaN with the following donor and acceptor concentrations: (1) $N_D = 2 \times 10^{17}$, $N_A = 1 \times 10^{17}$, (2) $N_D = 2 \times 10^{18}$, $N_A = 1 \times 10^{18}$, and (3) $N_D = 2 \times 10^{19}$, $N_A = 1 \times 10^{19}$ for $V_A - V_B = 0.5$ eV and $V_A - V_B = 1.0$ eV. In part, courtesy of D.C. Look.

are (1) $N_D = 2 \times 10^{17}$ cm^{-3}, $N_A = 1 \times 10^{17}$ cm^{-3}, (2) $N_D = 2 \times 10^{18}$ cm^{-3}, $N_A = 1 \times 10^{18}$ cm^{-3}, and (3) $N_D = 2 \times 10^{19}$ cm^{-3}, $N_A = 1 \times 10^{19}$ cm^{-3}.

Ridley calculated the mobility of bulk AlGaN. Although AlGaN was thought to be a single-mode alloy, it is, in fact, a two-mode alloy with the Fröhlich coupling associated with the lower frequency mode so weak that for LO phonon coupling AlGaN behaves as a single-mode alloy. The parameters used by Ridley for GaN (AlN) were as follows: piezoelectric constants (in C/m^2) $e_{33} = 0.7$ (1.5), $e_{31} = -0.4$ (−0.6), $e_{15} = -0.48$ (−0.3); deformation potential (in eV) 8.3 (9.5); density (in 10^3 kg/m^3) 6.15 (3.3); sound velocity (in 10^3 m/s) for LA modes 6.56 (9.06), TA modes 2.68 (3.70); dielectric constant at high frequency 5.4 (4.5), static dielectric constant 9.7 (8.5). A liner interpolation was used to determine the aforementioned parameters for the alloy. Figure 4.22 shows the compositional dependence of the various components of the mobility as well as the total mobility in AlGaN versus composition. Defect scattering and impurity (by assuming low impurity concentration) scattering are neglected in order to obtain a mobility of 1300 cm^2/(V s) for GaN and 877 cm^2/(V s) for Al$_{0.3}$Ga$_{0.7}$N with $n = 10^{16}$ cm^{-3}.

A similar treatment carried out for InGaN also reveals significant bowing in mobility. Figure 4.23 plots the electron mobility for an uncompensated donor concentration of 10^{16} cm^{-3} at 300 K as a function of the InN mole fraction for alloy potentials in the range 0–1.4 eV.

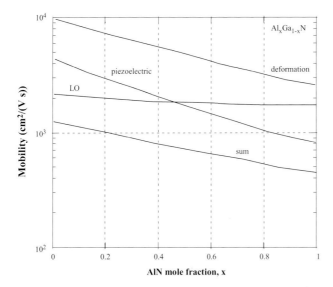

Figure 4.22 Calculated electron mobility in AlGaN at 300 K as a function of composition. Courtesy of B.K. Ridley.

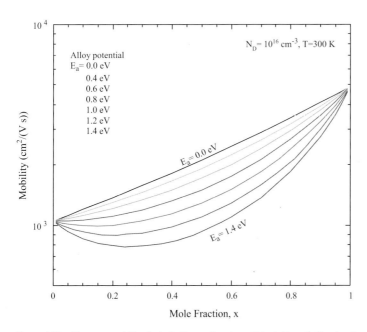

Figure 4.23 Electron mobility for InGaN as a function of the InN mole fraction for an uncompensated donor concentration of 10^{16} cm^{-3} for alloy potentials in the range from 0 to 1.4 eV at 300 K.

Table 4.10 Parameters derived and used in the mobility calculations in InGaN.

Parameter	Symbol (units)	InN	In$_x$Ga$_{1-x}$N
High-frequency dielectric constant[a]	ϵ_∞ (F/m)	$8.4\epsilon_0$	$5.47 + 2.93x$
Low-frequency dielectric constant	ϵ (F/m)	$15.3\epsilon_0$ estimated	$10.4 + 4.9x$
$\dfrac{\epsilon_0}{\epsilon_\infty} - \dfrac{\epsilon_0}{\epsilon_s}$		$\epsilon_0 \left(\dfrac{1}{8.4} - \dfrac{1}{10.5} \right)$	
Polar phonon Debye temperature (K)	θ_{LO} (K) or θ_{po} (K)	1033 or 1038	$1044 - 11x$ or $1044 - 6x$
Mass density	ρ (kg/m^3)	6.81×10^3	$(6.1 + 0.71x) \times 10^{-3}$
Sound velocity[b]	s (m/s)	6.24×10^3	$(6.59 - 0.35x) \times 10^{-3}$
Piezoelectric constant	ε_{14} (C/m^2), h_{pz}	0.375–0.5	0.375–0.5
Acoustic deformation potential	D_{ac} (eV)	7.1	$9.2 - 2.1x$, assuming 9.2 eV for GaN although 12–15 eV has been reported
Effective mass	m^* (kg)	0.115 or $0.15 m_0$	$0.22 - 0.105x$ or $0.22 - 0.07x$

InN parameters that are not listed above: a (lattice constant) = 3.548 Å; c (lattice constant) = 5.760 Å; m^* (effective mass) = $0.115 m_0$; ρ (mass density) = 6.8×10^3 kg/m^3; c_l (longitudinal elastic constant) = 2.65×10^{11} N/m^2; M (reduced mass of atoms) = 2.073×10^{-23} g; v_a (volume of the unit cell) = $\sqrt{3} a^2 c/4 = 3.140 \times 10^{-23}$ cm^3; ω (frequency of the polar optical phonon) = 1.352×10^{14} s^{-1}.
a) From $\epsilon_s = \epsilon_\infty (\omega_{LO}/\omega_{TO})^2$.
b) From $C_L = \rho s^2$.

In Tables 4.10 and 4.11, the parameters needed for the mobility calculations for InGaN and AlGaN are tabulated. The parametric values for the ternary alloys are deduced on the assumption that they can be represented by linear interpolations from the binary end points. Experimental investigations of the transport in ternary and quaternary layers are relatively weak compared to GaN and the controversial InN. Moreover, as mentioned often, the value of the alloy scattering potential is still a debated topic. Naturally, further discussion of the matter will continue for quite some time.

4.9
Two-Dimensional Transport in n-Type GaN

A 2DEG system is one in which the motion in one direction is ideally eliminated leaving behind only the in-plane motion. This can be accomplished by inducing an electron gas in the interface between two semiconductors with an energy discontinuity to the extent that the motion perpendicular to the interface is impeded. The characteristic dimension for confinement effect is that of the de Broglie wavelength (also called the thermal wavelength) of an electron. It should be

Table 4.11 Parameters pertinent to mobility calculations in AlGaN.

Parameter	Symbol (units)	AlN	Al$_x$Ga$_{1-x}$N
High-frequency dielectric constant[a]	ϵ_∞ (F/m)	$4.5\epsilon_0$, $4.68\epsilon_0$	$5.47 - 0.79x$
Low-frequency dielectric constant	ϵ (F/m)	$8.5\epsilon_0$	$10.4 - 1.9x$
$\dfrac{\epsilon_0}{\epsilon_\infty} - \dfrac{\epsilon_0}{\epsilon_s}$		$\epsilon_0\left(\dfrac{1}{4.68} - \dfrac{1}{8.5}\right)$	
Polar phonon Debye temperature	θ_{LO} (K)	1150	$1044 + 106x$
Mass density	ρ (kg/m^3)	3300, 3.23×10^3	$(6.1 - 2.87x) \times 10^{-3}$
Sound velocity[b]	s (m/s)	$6.59 \times 10^3 \sqrt{6.10/3.23} = 9.06 \times 10^3$; 9060 (LA), 3700 (TA)	$(6.59 - 2.47x) \times 10^{-3}$
Piezoelectric constant	ε_{14} (C/m^2), h_{pz}	0.566	$0.5 + 0.066x$
Acoustic deformation potential	E_{ds} (eV)	9.5	$9.2 + 0.3x$, assuming 9.2 eV for GaN although 12–15 eV has been reported
Effective mass	m^* (kg)	$0.35m_0$, $0.48m_0$	$0.22 + 0.26x$

a) From $\epsilon_s = \epsilon_\infty(\omega_{LO}/\omega_{TO})^2$.
b) From $C_L = \rho s^2$, where C_L is the longitudinal elastic constant from D.L. Rode.

mentioned that in the case of excitons, discussed in Chapter 6, the characteristic dimension is the Bohr radius. The de Broglie wavelength for thermalized electrons is given by

$$\lambda_{dB} = \frac{h}{p} = \sqrt{\frac{h^2}{2m^*\xi}} \quad \text{with} \quad p = mv \quad \text{and} \quad \frac{1}{2}mv^2 = \xi, \tag{4.94}$$

where ξ represents the kinetic energy of the electron. The de Broglie wavelength for 26 meV electrons (room temperature) is about 20 nm for GaN and about 14 nm for AlN. An artistic view of a model AlGaN/GaN single heterojunction is shown in Figure 4.24.

In the doped structures, the carriers donated by donors that are situated in the wider gap material diffuse to the one with the smaller bandgap, where they are confined due to the energy barrier on the one side and the charge-induced band bending on the other, as shown in Figure 4.24. In highly polar materials such as GaN and AlGaN, both spontaneous and piezoelectric polarization-induced screening charge causes the formation of free electron charge at the interface without any need for doping of the barrier layer. In the doped barrier case, the carriers and the ionized impurity centers are physically separated. Moreover, the large interface charge screens the local ionized impurities to some extent, leaving phonon scattering mechanisms to dominate. To summarize, the scattering mechanisms in 2DEG

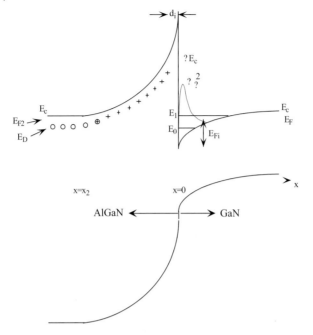

Figure 4.24 Schematic representation of the conduction band edge of an AlGaN/GaN heterostructure with an AlGaN layer (all or part away from the interface can be doped or no doping is introduced in which case the polarization charge-induced free carriers form the interfacial 2DEG). Electron probability, ψ_0^2, of the ground state of the confined system is also shown.

systems are optical and acoustic phonon scattering, alloy scattering due to any wavefunction overlap into the barrier, remote ionized impurity scattering due to charge centers in the barrier or at the surface, residual screened impurity scattering (background or intentionally doped channel impurities), and interface roughness scattering. Theory of scattering in a 2D system is beyond the scope of this book but the curious reader can refer to Morkoç (2008) and references cited therein, but a succinct description on the extent of the wavefunction, among others, is provided.

4.9.1
Scattering in 2D Systems

The electron density in the triangular well region depends on the shape of the wavefunction associated with the quantum states in the quasitriangular well shown in Figure 4.24. The wavefunction for the ith subband in such a triangular well can be described by Stern–Howard wavefunctions, developed initially for the SiO_2/Si system, as

$$\psi_{i,k_\parallel}(r_\parallel, z) = \varsigma_i(z) e^{i\vec{k}_\parallel \cdot \vec{r}_\parallel}, \tag{4.95}$$

where the confinement is in the z-direction with $\vec{k}_{\|} = \hat{i}k_x + \hat{j}k_y$ and $\vec{r}_{\|} = \hat{i}r_x + \hat{j}r_y$ and $\varsigma_i(z)$ is the quantized wavefunction and the solution of the Schrödinger equation describing one-dimensional bound motion as

$$\left[-\frac{\hbar^2}{2m_z^*}\frac{d^2}{dz^2} - eV(z)\right]\varsigma_i(z) = \xi_i\varsigma_i(z), \tag{4.96}$$

with boundary conditions $\varsigma_i(z) = 0$ for $z = \pm\infty$. The in-plane two-dimensional free motion of the electrons can be described by

$$\left[-\frac{\hbar^2}{2m_x^*}\frac{d^2}{dx^2} - \frac{\hbar^2}{2m_y^*}\frac{d^2}{dy^2}\right]e^{i\vec{k}_{\|}\cdot\vec{r}_{\|}} = \xi_{x,y}e^{i\vec{k}_{\|}\cdot\vec{r}_{\|}}. \tag{4.97}$$

Each eigenvalue ξ_i represents the bottom of the continuum (in-plane) of energy levels called the subband that can be grouped into ladders with respect to the bulk conduction band minimum from which they originate, as shown in Figure 4.24 for two subbands. Assuming isotropy for the in-plane mass, the solution of Equation 4.97 represents a parabolic band in the x and y directions (in-plane) and is given by

$$\xi_i(\vec{k}_{\|}) = \xi_i + \frac{\hbar^2 k_{\|}^2}{2m^*}, \tag{4.98}$$

the lowest energy of which is the subband energy represented by ξ_i, and can be found from a solution of Equation 4.96 as follows: Referring to Figure 4.24, the energy variation in the z-direction can be described as

$$-eV(z) = \begin{cases} eF_s z, & \text{for } z \geq 0, \\ \infty, & \text{for } z \leq 0, \end{cases} \tag{4.99}$$

where F_s is the effective interfacial layer electric field, assumed to be constant, and is to the first extent given by $e(n_s + n_{\text{depl}})/\varepsilon$. This equation can be solved by use of the Bohr–Sommerfeld quantization condition (described, for example, in Landau and Lifshitz (1977)):

$$\int_0^\infty (\xi_i - eF_s z)^{1/2}\,dz = \frac{\hbar\pi(i + (3/4))}{\sqrt{2m_z^*}}. \tag{4.100}$$

The solution to this equation leads to the energy levels quantized in the z-direction as

$$\xi_i = \left(\frac{\hbar^2}{2m_z^*}\right)^{1/3}\left(\frac{3\pi}{2}eF_s\right)^{2/3}[i + (3/4)]^{2/3}, \quad i = 0, 1, 2, \ldots. \tag{4.101}$$

Within this framework, the electron density in each of the subbands can be calculated.

Equation 4.101 represents an approximate solution with only 2% error. Alternatively, the variational method of the Schrödinger equation applied by Stern and Howard to an inversion layer can be applied to find the zeroth-order $\varsigma_0(z)$. This has also been expressed for a 2DEG at an AlGaAs/GaAs interface by Hirakawa and

Sakaki, which is called the Fang–Howard variational wavefunction, the normal component (z-direction) of which is expressed as

$$f(z) = \left(\frac{1}{2}b^3 z^2\right)^{1/2} \exp\left(-\frac{1}{2}bz\right). \tag{4.102}$$

The simplest approximation for the electron charge distribution, when only the lowest subband is occupied, for inversion layers is described by a trial function reported by Fang and Howard, which is the square of the above equation:

$$\varsigma_0(z) = \left(\frac{1}{2}b^3 z^2\right) \exp(-bz). \tag{4.103}$$

This expression is also the appropriate representation of the electron charge density in the lowest subband for image potential outside liquid helium if the thickness of the helium–vacuum interface is taken as zero.

In addition, the variational parameter b is given as, which includes the correction to the original treatment ($\varsigma_0(z)$ normalized, $\int_0^\infty \varsigma_0(z)dz = 1$):

$$\begin{aligned}b &= \left[\left(\frac{12m_z^* e^2}{\varepsilon \hbar^2}\right)\left(n_{\text{dep}} + \frac{11}{32}n_s\right)\right]^{1/3} \quad \text{in SI units} \quad \text{and} \quad b \\ &= \left[\left(\frac{48\pi m_z^* e^2}{\varepsilon \hbar^2}\right)\left(n_{\text{dep}} + \frac{11}{32}n_s\right)\right]^{1/3} \quad \text{in cgs units.} \end{aligned} \tag{4.104}$$

The average value of z weighted by the charge distribution of Equation 4.103 (for the $\varsigma_0(z)$) is $w = 3/b$, which is taken as the thickness of the inversion layer. Price takes the same as $w = 8/3b$.

The parameters n_s and n_{depl} represent the areal electron density of the induced interface charge and the equivalent areal depletion charge (as it applies to Si MOS and heterojunction FETs if the active region is actually p-type). The equivalent areal depletion charge for an n-type inversion layer would be the areal density of ionized acceptor density in the p-type bulk. In the AlGaN/GaN case, the GaN is typically n-type in which case any positive charge present in the region where the electron wavefunction is nonzero must be considered. Also, the parameter b represents the extent of the wavefunction penetration into the small-bandgap material.

4.9.1.1 Electron Mobility in AlGaN/GaN 2D System

High-quality heterostructures grown on bulk GaN or GaN templates paved the way for cleaner experiments, which allowed the determination of the conduction band deformation potential at $E_{\text{dp}} = 9.1 \pm 0.7$ eV from temperature and density dependence of the mobility, although values as high as 12–15 eV have been extracted (see Morkoç (2008) and references cited therein). Suffice it to state that the low-field electron mobility and density were then measured in the range 1.7–300 K. The inverse electron mobility followed clearly a linear relationship with respect to temperature as shown in Figure 4.25 for two sheet electron densities of 6.6×10^{11} and 1.67×10^{12} cm^{-2}.

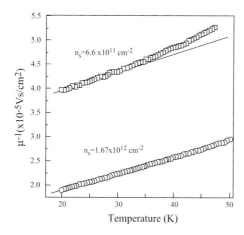

Figure 4.25 The inverse 2DEG electron mobility versus temperature in the range 20–50 K for electron densities of 6.6×10^{11} and 1.67×10^{12} cm^{-2} measured in a FET-like structure that allowed varying the electron density by the gate potential. The solid lines are a guide to the eye. Courtesy of H.L. Störmer.

Keeping in mind that $\mu^{-1} = \mu_0^{-1} + \mu_{ac}^{-1}$. The acoustic phonon limited mobility in the linear temperature dependence region can be characterized with $\mu_{ac}^{-1} = \alpha T$ with $\alpha = \mathrm{d}(\mu_{ac}^{-1})/\mathrm{d}T$. The component of the mobility depicted by μ_0 that represents the extrapolated mobility at $T=0$ takes into account all the temperature-independent scattering mechanisms such as those by remote ionized impurities and charged extended defects. Figure 4.26 displays the slope α as a function of electron density, n_s.

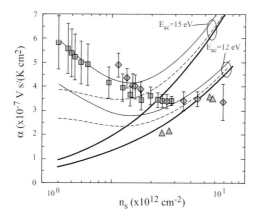

Figure 4.26 The α parameter versus the electron density. Data shown are from the FET (squares), fixed density samples (diamonds), and literature (triangles). Calculated fits are for two values of the deformation potential ($E_{dp} = D_{ac} = 12$ and 15 eV): thick solid lines for a degenerate 2DEG system for $D_{ac} = 12$ and 15 eV; broken lines for a nondegenerate 2DEG system for $D_{ac} = 12$ and 15 eV, with $n_{depl} = 0$; and thin lines for a nondegenerate 2DES system for $D_{ac} = 12$ and 15 eV, with $n_{depl} = 6 \times 10^{11}$ cm^{-2}. Courtesy of H.L. Störmer.

Of importance is the proportionality of α to the acoustic phonon scattering rate, $1/\tau_{ac} = \alpha(q/m^*)T$. Therefore, α is a measure of the density dependence of the scattering rates solely by acoustic phonon scattering. The upward trend of α at low n_s values seen in Figure 4.26 is of significance. It departs noticeably from predictions based on acoustic phonon scattering in a screened, degenerate 2DEG system.

It should be noted that both deformation potential and piezoelectric acoustic phonon scattering limit the mobility at low temperatures in a 2DEG system, the former becoming more important at low sheet carrier densities. However, the piezoelectric constants for wurtzite GaN are not well known. Relating to the well-known GaAs case, it has been noted that in GaAs the rise in the α parameter at low densities could be reproduced by increasing the model value of h_{14} piezoelectric constant (in effect in a cubic system) by 1.5. Therefore, the rising values of the α parameter might in part be due to the piezoelectric constant being larger than that reported so far.

4.9.1.2 Numerical Two-Dimensional Electron Gas Mobility Calculations

Calculations of the electron mobility at the AlGaN/GaN interface have been performed in the context of acoustic phonon scattering, Coulomb scattering from both the donor-like defects on the AlGaN barrier surface and unintentional dopants in the GaN, and alloy disorder scattering as well as the interface roughness scattering. Figure 4.27 shows calculated 2DEG mobilities at low temperature as a function of barrier width for several heterostructures with different Al compositions. Calculations show three distinct regions in that for very small barrier widths, the mobility is quite low and increases slowly with increasing barrier width. The maximum mobilities are obtained for 2DEG densities in the range of about $3.5 \times 10^{12}\,\text{cm}^{-2}$ (for $x = 0.25$) to $5 \times 10^{11}\,\text{cm}^{-2}$ (for $x = 0.05$).

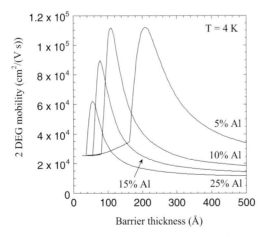

Figure 4.27 Low-temperature 2DEG mobilities as a function of barrier thickness for four different AlGaN/GaN heterostructures with different Al barrier compositions. Courtesy of W. Walukiewicz.

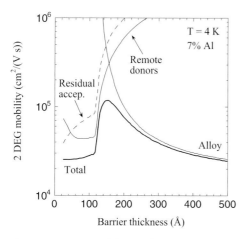

Figure 4.28 2DEG mobility as a function of AlGaN barrier thickness for an $Al_{0.07}Ga_{0.93}N$/GaN heterostructure. Courtesy of W. Walukiewicz.

The overall and individual mobilities as a function of barrier thickness for a 2DEG at the interface of an $Al_{0.07}Ga_{0.93}N$/GaN heterostructure are shown in Figure 4.28. This is the Al alloy fraction that would produce the highest possible low-temperature mobilities for the indicated doping levels. The component mobility curves illustrate the dominant scattering mechanisms in each of the three different regions mentioned above.

The calculated mobilities agree well with the experimental values for barrier Al compositions of less than about 15%, as shown in Figure 4.29. However, for higher Al fractions, the measured mobilities are a nearly constant factor of 2–2.5 lower. Two possible reasons can be suggested. First, the electron density exceeds 6×10^{12} cm^{-2} for $x > 0.15$. This corresponds roughly to the electron density at which the higher subbands in the GaN quantum well begin to be occupied, which causes increased scattering. Second, the lattice mismatch at the interface increases with increasing x, which leads possibly to increased roughness and thereby reduced electron mobility at the interface. In the end, technological issues, other than interface roughness, may come to be the culprit as was the case in the early development of the GaAs/AlGaAs heterointerface.

The effect of interface roughness scattering on the total mobility has been estimated using 15 Å as the parameter for the characteristic lateral extent of the islands and adjusting the height of the islands in order to obtain the best fit with experimental data. A constant island height of 0.9 Å produced the fit shown in Figure 4.29a (for a constant barrier thickness and variable barrier composition) and Figure 4.29b (for a constant barrier composition and variable barrier thickness). The points represent experimental data. The solid line is the calculated mobility neglecting interface roughness scattering, whereas the dashed line includes interface roughness scattering.

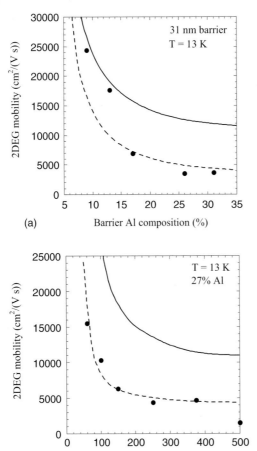

Figure 4.29 (a) 2DEG mobilities as a function of Al composition of the barrier. The points represent experimental data. The solid line represents the calculated mobility neglecting interface roughness scattering. The dashed line represents the calculated mobility with interface roughness scattering. (b) 2DEG mobilities as a function of thickness of an $Al_{0.27}Ga_{0.73}N$ barrier along with the experimental data. The solid line represents the calculated mobility neglecting interface roughness scattering. The dashed line represents the calculated mobility including interface roughness scattering. Courtesy of W. Walukiewicz.

Above arguments indicate that the highest electron mobilities are expected from heterostructures with 2DEG densities in the range of a few $10^{12}\,cm^{-2}$. Because the polarization-induced charge transfer increases with increasing width and Al composition of the barrier, the highest mobilities are limited to heterostructures with relatively thin barriers with low Al content for undoped structures. A maximum electron mobility of about $10^5\,cm^2/(V\,s)$ at 4 K can be expected from heterostructures with a 15 nm thick $Al_{0.07}Ga_{0.93}N$ barrier.

4.9.1.3 Magnetotransport and Mobility Spectrum

Hall measurements, particularly the conductivity matrix elements measured as a function of magnetic field up to sufficiently large values, the extent of which depends on the mobility with higher mobilities requiring smaller fields, can be made to delineate the mobility of various layers in the system. The conductivity matrix element σ_{xy} is given by

$$\sigma_{xy} \equiv -\frac{ne^2}{m^*}\frac{\omega_c\tau^2}{1+\omega_c^2\tau^2} = \frac{\sigma(B)\mu B}{1+\mu^2 B^2} \tag{4.105}$$

and assumes its maximum value for $\mu B = 1$. By simply measuring the conductivity element as a function of magnetic field and noting the magnetic field at which it is maximized, one can easily determine the mobility. The difficulty lies in the experiments in that high-mobility samples coupled with high magnetic fields are required. To reiterate, having variable temperature and variable magnetic field conductivity data allows one, using analyses such as QMSA, to determine the mobility of each of the layers contributing to current conduction. Doing so, in one experiment, led to electron mobilities and carrier densities in the bulk GaN buffer layer and for the 2DEG (sheet density) of $\mu_{GaN} \sim 880\,cm^2/(V\,s)$ (80 K) and $\mu_{2DEG} \sim 7100\,cm^2/(V\,s)$ (80 K), and $n_{GaN} \sim 8 \times 10^{14}\,cm^{-3}$ and $n_{2DEG} \sim 7 \times 10^{11}\,cm^{-2}$, respectively. For these measurements, a six-contact Hall bar sample was fabricated using standard photolithography and electron beam evaporated Ti/Al/Ti/Au ohmic contacts. Variable-field Hall measurements were carried out in a superconducting cryogenic physical parameter measurement system (PPMS). With the PPMS system, measurements were performed under constant current of 0.1 mA and varying magnetic field (0.01–7.0 T) that was applied perpendicular to the sample surface. Measurements were performed in a temperature range 5–300 K.

Application of QMSA to the variable-field data uncovered dominantly a two-carrier electron transport scheme with one high-mobility electron and one low-mobility electron. Somewhat surprising (most likely an artifact), a moderate mobility hole conduction was also observed in this spectrum for the best fit. Figure 4.30 shows the multicarrier mobility spectra for the $Al_{0.08}Ga_{0.92}N/GaN$ sample at 80 K. The high-mobility carrier corresponds to the 2DEG electrons within the quantum well channel layer at the $Al_{0.08}Ga_{0.92}N/GaN$ interface and is consistent with the results of standard variable temperature and low magnetic field Hall measurements. It should be reiterated that the sample was not intentionally doped and the 2DEG was formed due to polarization. The $880\,cm^2/(V\,s)$ (80 K) low-mobility carrier represents the bulk electrons within the HVPE-grown GaN template on which the heterostructure was grown with MBE, and is consistent with independent measurements performed in HVPE layers only. The ghost hole mobility of $2000\,cm^2/(V\,s)$ is too high for any known carrier of the type in the GaN/AlGaN material system, which is curious because the observation of positive Hall coefficients unambiguously confirms its presence. At this point, the origin of the hole peak remains unknown.

For an extensive discussion of the two-dimensional hole gas, refer to Morkoç (2008). Likewise discussion of interface roughness scattering and quantum transport can be found in the same source.

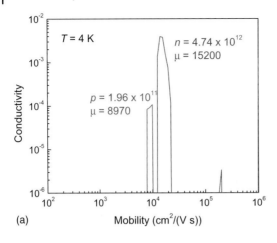

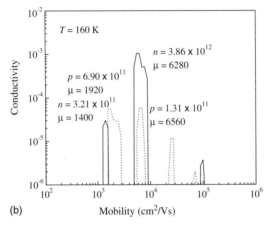

Figure 4.30 QMSA spectrum for an Al$_{0.25}$Ga$_{0.75}$N/GaN sample at 4 K (a) and 180 K (b). At 4 K, the Hall bar data and magnetotransport measurements indicate the carrier conduction to be entirely due to 2DEG. Therefore, the hole-type conduction seen in (a) is most likely not real and might have its genesis in experimental measurement errors (although all attempts were made to ensure reliable data collection) and/or some other nonideality. However, at 180 K, there is participation from the bulk electrons in the GaN buffer layer. The 1400 cm^2/(V s) contribution is therefore attributed to the bulk conduction. Again, there are hole transport contributions at 180 K that are most likely due to measurement errors. The 2DEG mobility figures at both temperatures are consistent with Hall measurements. The QMSA fits are provided by J. Meyer and I. Vurgaftman.

Further Reading

Blakemore, J.S. (1982) *J. Appl. Phys.*, 53 (10), R123–R181.

Bonch-Bruevich, V.L. and Glasko, V.B. (1966) *Fiz. Tverd. Tela*, 3, 36.

Callen, H. (1949) *Phys. Rev.*, 76, 1394.

Chin, V.W.L., Tansley, T.L., and Ostockton, T. (1994) *J. Appl. Phys.*, 75, 7365–7372.

Chin, V.W.L., Zhou, B., Tansley, T.L., and Li, X. (1995) *J. Appl. Phys.*, 77, 6064.

Conwell, E.M. (1967) *High Field Transport in Semiconductors*, in *Solid State Physics Supplements* 9, 108 (eds F. Seitz, D. Turnbull, and H. Ehrenreich), Academic Press, New York.

Datta, S. (1995) *Electronic Transport in Mesoscopic Systems*, Cambridge University Press, Cambridge.

DeMey, G. (1983) Potential calculations in Hall plates, in *Advances in Electronics and Electron Physics*, vol. 61 (eds L. Martin, and C. Marton), Academic Press, pp. 1–61.

Ehrenreich, H. (1959) *J. Phys. Chem. Solids*, 8, 130.

Fang, F.F. and Howard, W.E. (1966) *Phys. Rev. Lett.*, 16, 797–799.

Hamaguchi, C. (2001) *Basic Semiconductor Physics*, Springer.

Hammar, C. and Magnusson, B. (1972) *Phys. Scr.*, 6, 206.

Hirakawa, K. and Sakaki, H. (1986) *Phys. Rev. B*, 33, 8291–8303.

Howarth, D.J. and Sondheimer, E.H. (1953) *Proc. R. Soc. Lond. A*, 219, 53–74.

Hsu, L. and Walukiewicz, W. (2002) *Appl. Phys. Lett.*, 80 (14), 2508–2510.

Isenberg, I., Russel, B.R., and Green, R.F. (1948) *Rev. Sci. Instrum.*, 45, 1479–1480.

Landau, L.D. and Lifshitz, E.M. (1977) *Quantum Mechanics: Non-Relativistic Theory*, Pergamon Press, Oxford.

Look, D.C. (1989) *Electrical Characterization of GaAs Materials and Devices*, John Wiley & Sons, Ltd, Chichester.

Lundstrom, M. (2000) *Fundamentals of Carrier Transport*, 2nd edn, Cambridge University Press.

Lyddane, R.H., Sachs, R.G., and Teller, E. (1941) *Phys. Rev.*, 59, 673.

Meyer, J.R. and Bartoli, F.J. (1987) *Phys. Rev. B*, 36, 5989–6000.

Morkoç, H. (2008) *Handbook of Nitride Semiconductors and Devices*, Wiley-VCH Verlag GmbH, Weinheim.

Nag, B.R. (1980) *Electron Transport in Compound Semiconductors*, Springer Series in Solid-State Sciences, vol. 11, Springer, Berlin.

Phillips, J.C. (1970) *Rev. Mod. Phys.*, 42, 317.

Pödör, B. (1966) *Phys. Status Solidi B*, 16, K167.

Price, P.J. (1981a) *Ann. Phys. NY*, 133, 217–219.

Price, P.J. (1981b) *J. Vac. Sci. Technol.*, 19, 599–603.

Price, P.J. and Stern, F. (1983) *Surf. Sci.*, 132, 577–593.

Ridley, B.K. (1999) *Phys. Status Solidi A*, 176 (1), 359–362.

Rode, D.L. (1975) Low field electron transport, in *Semiconductors and Semimetals*, vol. 10 (eds R.K. Willardson and A.C. Beer), Academic Press, New York, p. 1.

Seeger, K. (1991) *Semiconductor Physics*, 5th edn, Springer Series in Solid State Sciences, vol. 40, Springer, Berlin.

Shockley, W. (1950) *Electrons and Holes in Semiconductors*, Van Nostrand, Princeton, NJ.

Stern, F. (1972) *Phys. Rev. B*, 5, 4991–4899.

Stern, F. (1980) *Phys. Rev. Lett.*, 44, 1469.

Stern, F. and Howard, W.E. (1967) *Phys. Rev.*, 163, 816–835.

van der Pauw, L.J. (1958) *Philips Res. Rep.*, 13, 1.

Vurgaftman, I., Meyer, J.R., Hoffman, C.A., Redfern, D., Antoszewski, J., and Faraone, L. (1998) *J. Appl. Phys.*, 84 (9), 4966–4973.

Wasscher, J.D. (1969) *Electrical transport phenomena in MnTe5, an antiferromagnetic semiconductor.* Dissertation, Philips Gloeilampenfabriken, Eindhoven, The Netherlands, reproduced in Wieder, H.H. (1979) *Laboratory Notes on Electrical and Galvanometric Measurements*, Elsevier, Amsterdam.

5
The p–n Junction

5.1
Introduction

A p–n junction depicts the stacking of two semiconductors having n- and p-type conductivities (polarities). The two regions can be of the same (homojunction) or structurally similar (heterojunction) semiconductors. When two similar semiconductors with different bandgaps are conjoined, the conduction and the valence bands at the junction do not match precisely giving way to types *I*, *II*, and *broken* band alignment. In *type I* the conduction and valence bands of the larger bandgap semiconductor straddle those of the other. In *type II* the conduction and valence bands are staggered, and in the *broken* variety the valence band of one is above the conduction band of the other. The sum of discontinuities at the conduction and valence bands in *type I* add up to the difference in bandgaps of the two sides if there is no compositional gradient. The topic of band alignment is a complex matter and has been the topic of many reports. Additional details of the band alignment in general and those pertinent to the GaN family can be found in Morkoç (2008) and references therein.

5.2
Band Alignment

Various types of band alignments are often encountered at semiconductor interfaces depending on the relative positioning of the energy bands with respect to each other. Figure 5.1 depicts two types of possible alignments, which occur most commonly in semiconductor heterojunctions. Type I alignment, in which the bandgap of one semiconductor lies completely within the bandgap of the other (straddling type), is the most useful one for optoelectronic devices. A type II alignment occurs when bandgaps of the two materials overlap, but one does not completely enclose the other (staggering type) (Figure 5.1). Type II alignment has been studied in the GaSb-based systems for long-wavelength applications. A more obscure application has to do with type II ZnSe/ZnTe heterojunctions that have been used to overcome crucial

Nitride Semiconductor Devices: Fundamentals and Applications, First Edition. Hadis Morkoç.
© 2013 Wiley-VCH Verlag GmbH & Co. KGaA. Published 2013 by Wiley-VCH Verlag GmbH & Co. KGaA.

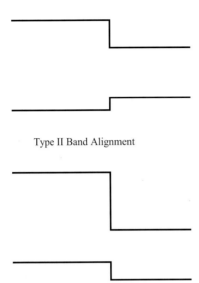

Figure 5.1 Schematic description of type I and type II band alignments.

problems related to difficulties in performing p-ohmic contacts for blue ZnSe-based lasers.

A practical guide to band discontinuities in the GaN system, including SiC, is shown in Figure 5.2 that also indicates both the absolute values and percentile

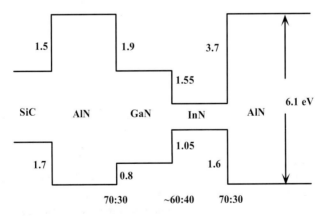

Figure 5.2 Band discontinuities in nitride semiconductors including SiC that obey transitivity. The InN data have been shifted to take into account its 0.8 eV bandgap by maintaining the previously determined valence band discontinuity. It should also be pointed out that the bandgap of AlN has been refined to be 6.1 and about 6 eV at low temperatures and room temperature, respectively. Note that the stack of layers is constructed for the purpose of showing the transitivity nature.

offsets to the extent known. One note of caution is that the piezoelectric effect has been evoked to explain the asymmetry of the band discontinuities. The dangling bonds unique to one polarity of the interface may also induce such an asymmetry, refer to Chapter 1 for details regarding bonds. Though this concept accounts for the major features of the observed data, deterministic investigations must be undertaken before a more definitive statement can be made (see Morkoç (2008) for more details).

5.3
Electrostatic Characteristics of p–n Heterojunctions

When n- and p-type semiconductors are joined, free carriers near the interface diffuse across the junction until the electric field caused by the depletion charge so created balances this motion, which causes the Fermi levels on both sides to align and remain constant if no external current flows. To segue into the discussion of heterojunctions, Figure 5.3a depicts the case where the larger

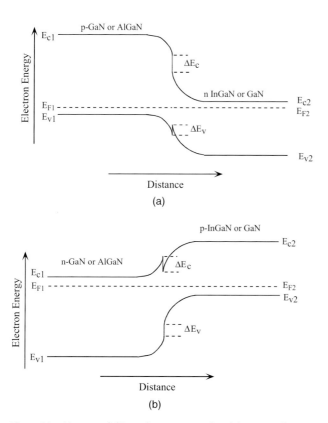

Figure 5.3 (a) p–n and (b) n–p heterojunction band diagrams after contact at equilibrium.

bandgap semiconductor is the p-type and Figure 5.3b represents that of n-type semiconductor. The total band bending is called the built-in potential. The potential that develops upon contact is comparable to the bandgap when the p- and n-layers are reasonably doped. When the junction is forward biased, the current increases rapidly as the internal voltage approaches the built-in voltage. The built-in voltage, which is the separation of the Fermi levels on either side, can be written as

$$E_{g1} - (E_{F1} - E_{v1}) - \Delta E_c - (E_{c21} - E_{F2}) - qV_{bi} = 0, \quad (5.1)$$

where V_{bi} is the built-in potential and the other parameters are as defined in Figure 5.3. If the p- and n-type semiconductors were doped at the onset of degeneracy, meaning the Fermi levels align with the valence and conduction band edges in p- and n-type semiconductors, respectively, the built-in voltage would simply be

$$qV_{bi} = E_{g1} - \Delta \xi_c. \quad (5.2)$$

Shown in Figure 5.4 are the charge, electric field, and potential distributions in the space charge region of an abrupt p–n heterojunction in the depletion approximation. Charge balance dictates that the positive and negative charges on either side of the junction equal with the net total charge being zero. The field discontinuity at the heterointerface is due to the larger bandgap semiconductor having a relatively smaller dielectric constant.

Poisson equation can be written as

$$-\frac{\partial^2 V}{\partial x^2} = \frac{\partial E}{\partial x} = \frac{\rho(x)}{\varepsilon} = \frac{q}{\varepsilon}[p(x) - n(x) + N_D^+ - N_A^-], \quad (5.3)$$

where $p(x)$ and $n(x)$ represent the hole and electron concentrations, respectively, and N_D^+, N_A^- represent the ionized donor and acceptor concentrations, respectively, all within the space charge region. Relying on the depletion approximation leads to equating $p(x)$ and $n(x)$ to 0 in the space charge region. Furthermore, assuming that all the donors and acceptors are ionized allows replacing N_D^+, N_A^+ with just the donor and acceptor concentrations. Doing so reduces the charge density given by Equation 5.3, for the case of abrupt junction and with constant doping on both sides, to

$$\begin{aligned}\rho &= q(N_D), \quad \text{for} \quad 0 < x \leq x_n, \\ \rho &= -q(N_D), \quad \text{for} \quad -x_p \leq x < 0.\end{aligned} \quad (5.4)$$

Equation 5.3 (Poisson equation) for the case of depletion approximation can then be rewritten for both sides of the space charge region as

$$\begin{aligned}dE/dx &= q\varepsilon_n(N_D), \quad \text{for} \quad 0 < x \leq x_n, \\ dE/dx &= -q\varepsilon_p(N_A), \quad \text{for} \quad -x_p \leq x < 0.\end{aligned} \quad (5.5)$$

In the depletion approximation, and with the additional assumptions that the semiconductor is not compensated and all the dopants are ionized,

(a)

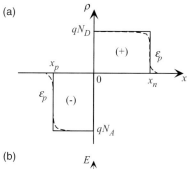

(b)

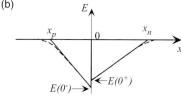

(c)

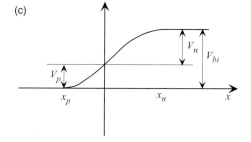

Figure 5.4 An abrupt heterojunction under equilibrium. The edge of the space charge region on the p- and n-sides are represented by x_p and x_n, respectively. The charge distribution is shown in part (a) with broken lines and within the depletion approximation with solid lines, the electric field is shown in part (b) with broken lines and within the depletion approximation with solid lines, and the potential is shown in part (c). Charge (a), field (b), and voltage (c) profiles of a nitride p–n heterojunction.

Poisson's equation is reduced to

$$dE/dx = -\frac{d^2 V}{dx^2} = \frac{q}{\varepsilon_n}(N_D), \quad \text{for the n-side, meaning } 0 < x \leq x_n, \quad (5.6)$$

and

$$dE/dx = -\frac{d^2 V}{dx^2} = -\frac{q}{\varepsilon_p}(N_A), \quad \text{for the p-side, meaning } -x_p \leq x < 0, \quad (5.7)$$

where subscripts n and p attached to the dielectric constant depict that the parameter is for the n- and p-sides, respectively. In a homojunction they are equal.

Equation 5.3 (or Equations 5.6 and 5.7) can be solved with the boundary conditions that the field is zero at the edges of the depletion region ($x = -x_p$ and $x = x_n$), and that the normal component of the displacement vector is continuous at the interface between the p- and n-layers, that is, $\varepsilon_n E(0^+) = \varepsilon_p E(0^-)$, which leads to

$$E = -\frac{qN_D}{\varepsilon_n}(x_n - x), \quad \text{for} \quad 0 < x \leq x_n,$$
$$E = -\frac{qN_A}{\varepsilon_n}(x_p + x), \quad \text{for} \quad -x_p \leq x < 0. \tag{5.8}$$

By integrating Equation 5.8, the voltage drop across the space charge regions on the n- and p-sides can be obtained as

$$V_n = (qN_D/2\varepsilon_n)x_n^2,$$
$$V_p = (qN_A/2\varepsilon_p)x_p^2. \tag{5.9}$$

And by definition, the built-in potential V_{bi} is expressed as

$$V_{bi} = V_n + V_p. \tag{5.10}$$

Utilizing Equation 5.9, the charge neutrality equation ($N_A x_p = N_D x_n$), and Equation 5.10, the expressions for the depletion region widths on the n- and p-sides of the junctions with an applied voltage V (changes the internal junction voltage from V_{bi} to $V_{bi} - V$ with $V > 0$ for forward bias and $V < 0$ for reverse bias) can be obtained as

$$x_n^2 = \frac{2\varepsilon_n \varepsilon_p N_A (V_{bi} - V)}{qN_D(\varepsilon_n N_D + \varepsilon_p N_A)}. \tag{5.11}$$

Similarly,

$$x_p^2 = \frac{2\varepsilon_n \varepsilon_p N_D (V_{bi} - V)}{qN_A(\varepsilon_n N_D + \varepsilon_p N_A)}. \tag{5.12}$$

The total depletion depth W is simply some of the extent of depletion regions into the n- and p-regions:

$$W = x_n + x_p = \left[\frac{2\varepsilon_n \varepsilon_p (N_D + N_A)^2 (V_{bi} - V)}{qN_A N_D(\varepsilon_n N_D + \varepsilon_p N_A)}\right]^{1/2}, \tag{5.13}$$

which in the case of homojunction simplifies to

$$W = \left[\frac{2\varepsilon(N_D + N_A)(V_{bi} - V)}{q(N_D N_A)}\right]^{1/2} = \left[\frac{2\varepsilon(N_A^{-1} + N_D^{-1})}{q}(V_{bi} - V)\right]^{1/2}. \tag{5.14}$$

A more accurate expression can be obtained by retaining the majority carrier tails near the edges of the depletion region (as opposed to assuming a unit function like distribution) that is $p(x)$ for the p-side and $n(x)$ for the n-side in Equations 5.4–5.7. A correction factor of kT/q must be subtracted from both V_n and V_p leading to a $2kT/q$, in which case the depletion depth expression for a homojunction is given as

$$W^2 = \frac{2\varepsilon(N_D + N_A)(V_{bi} - V - 2kT/q)}{q(N_D N_A)}$$
$$= \frac{2\varepsilon[(1/N_D) + (1/N_A)]}{q}(V_{bi} - V - 2kT/q). \quad (5.15)$$

For a one-sided abrupt junction in which either N_D or N_A is much larger than the other, the depletion depth is given by

$$W^2 = \frac{2\varepsilon}{qN_B}(V_{bi} - V - 2kT/q), \quad (5.16)$$

with N_B being the doping concentration of the side with the lower doping level. Equation 5.16 is often presented in the form of

$$W = L_D \sqrt{2[\beta(V_{bi} - V) - 2]}, \quad (5.17)$$

where L_D is the Debye length that is given by $L_D = \sqrt{(\varepsilon kT)/(q\beta N_B)}$ and is an important characteristic length and determines the extent of the sharpness of the depletion edge as seen in capacitance–voltage measurements and doping profile deduced from such measurements, and $\beta = q/kT$. The one-sided junction approximation applies directly to Schottky barriers.

The built-in voltage can also be calculated by the band diagram of the p–n junction, an example of which is shown in Figure 5.5 for heterojunction with the p-section formed by the larger bandgap material. By geometrical means, the built-in potential can be obtained using either the conduction band or the valence band.

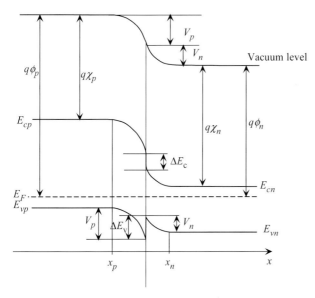

Figure 5.5 Energy band diagram of an ideal p–n heterojunction in equilibrium. The vacuum level, electron affinities (χ) for both polarities, and the work function (ϕ) for both polarities are also shown.

Using the valence band leads to the following equality:

$$(E_F - E_{vn}) - qV_n + \Delta E_v - qV_p - (E_F - E_{vp}) = 0. \tag{5.18}$$

The above equation together with Equation 5.10 leads to

$$qV_{bi} = qV_n + qV_p = \Delta E_v + (E_F - E_{vn}) - (E_F - E_{vp})$$

$$= \Delta E_v + kT \ln\left(\frac{N_{vn}}{p_n}\right) - kT \ln\left(\frac{N_{vp}}{p_p}\right)$$

$$= \Delta E_v + kT \ln\left(\frac{N_{vn}}{N_{vp}}\right) + kT \ln\left(\frac{p_p}{p_n}\right)$$

or

$$qV_{bi} = E_g|_n + \Delta E_v - (E_{cn} - E_F) - (E_F - E_{vp}), \tag{5.19}$$

where N_{vn} and N_{vp} represent the valence band density of states for the semiconductors constituting the n- and p-sides of the heterojunction, respectively. Likewise, p_n, p_p represent the hole concentrations for the n- and p-sides of the heterojunction, respectively.

Working instead with the conduction band across the junction, we obtain

$$(E_{cn} - E_F) + qV_n + \Delta E_c + qV_p - (E_{cp} - E_F) = 0 \tag{5.20}$$

and

$$qV_{bi} = qV_n + qV_p = -\Delta E_c - (E_{cn} - E_F) + (E_{cp} - E_F)$$

$$= -\Delta E_c + kT \ln\left(\frac{N_{cn}}{n_n}\right) - kT \ln\left(\frac{N_{cp}}{n_p}\right)$$

$$= \Delta E_v + kT \ln\left(\frac{N_{cn}}{N_{cp}}\right) + kT \ln\left(\frac{n_p}{n_n}\right)$$

or

$$qV_{bi} = E_g|_p - \Delta E_c - (E_{cn} - E_F) - (E_F - E_{vp}).$$

The built-in potential can also be determined from the knowledge that the electron and hole currents, individually and in aggregate, vanish in equilibrium, as shown in Figure 5.5. Taking the electron diffusion and drift current and equating it to zero, we obtain

$$J_n = q\mu n E + qD_n \frac{dn}{dx} e^{-\Delta E_c/kT} = 0. \tag{5.21}$$

Solving for the E field,

$$E = -\frac{D_n}{\mu} \frac{dn}{ndx} e^{-\Delta E_c/kT}. \tag{5.22}$$

Integrating the E field across the junction within the depletion region leads to the built-in voltage as

$$V_{bi} = -\int_{n(x_p)}^{n(x_n)} E dx = -\frac{kT}{q}\int_{n(x_p)}^{n(x_n)} \frac{D_n}{\mu}\frac{dn}{n} e^{-\Delta E_c/kT}$$

$$= \frac{kT}{q}\ln\frac{n(x_n)}{n(x_p)} + \frac{\Delta E_c}{q} \qquad (5.23)$$

$$= \frac{\Delta E_c}{q} + \frac{kT}{q}\ln\frac{n_n}{n_p} = \frac{\Delta E_c}{q} + \frac{kT}{q}\ln\frac{N_D N_A}{n_{ip}^2},$$

where n_{ip} depicts the intrinsic carrier concentration in the p-type material.

5.4
Current–Voltage Characteristics of p–n Junctions

In p–n junctions formed in wide-bandgap semiconductors, the diffusion current component in reverse bias and for small applied voltages in forward bias is small, because of the small minority carrier concentrations. Consequently, the generation–recombination current takes on a prime importance. As the name suggests, the generation–recombination current is caused by capture of free carriers by traps (defect centers) near the center of the gap. Generally speaking, if these centers are located in the upper half of the bandgap, they capture and emit electrons from and to the conduction band. Conversely, if the centers are in the lower half of the energy gap, they capture and emit holes from and to the valence band. A trap can capture an electron from the conduction band if the trap has not already captured one or already emitted it. The electron captured by a trap can later be emitted back to the conduction band. Similar processes occur involving hole traps and holes. Centers close to the middle of the gap are very efficient in the capture and emission processes, and can capture electrons and holes. To recognize the pioneers' contribution, this GR current is referred to as *Shockley–Read–Hall* (SRH) generation–recombination current. Here, only the final expression following a myriad of approximations is given leaving the curious reader to refer to Morkoç (2008) and references therein or books and scientific papers. The surface effects often complicate the treatment of SRH current, but that can be handled and expressed in a compact manner by invoking the concept of surface recombination velocity, which is again discussed in the above-mentioned reference. As the term suggests, a large surface recombination velocity explicitly implies a notable surface recombination activity, which is detrimental to devices such as solar cells and bipolar transistors, among others. The generation rate can be expressed as

$$U \approx \frac{\sigma_p v_{th} N_t (pn - n_i^2)}{n} = \sigma_p v_{th} N_t (p_n - p_{n0}) = \frac{p_n - p_{n0}}{\tau_p}, \qquad (5.24)$$

where $p \equiv p_n$ and $n_i^2/n = p_{n0}$ represent the minority carrier hole and equilibrium hole concentrations, respectively. Here, we replaced the term τ_{nr} with τ_p to indicate specifically that we are dealing with minority carrier hole lifetime in an n-type semiconductor. Because $U = (p_n - p_{n0})/\tau_p$, the minority carrier recombination lifetime has the familiar form of $\tau_p^{-1} = \sigma_p v_{th} N_t$, where σ_p, v_{th}, and N_t represent the capture cross section, thermal velocity, and the trap center concentration, respectively.

Similarly, for a p-type semiconductor,

$$U \approx \frac{\sigma_n v_{th} N_t (p_p n_p - n_i^2)}{p_p} = \sigma_n v_{th} N_t (n_p - n_{p0}) = \frac{n_p - n_{p0}}{\tau_n}, \quad \text{with } \tau_n^{-1}$$
$$= \sigma_n v_{th} N_t.$$

(5.25)

Recombination rate for surface effects, if the centers are located at or near the middle of the gap ($\xi_t = \xi_i$) and the electron and hole capture cross sections are equal, can be expressed as

$$U_s = s_0 \frac{(p_s n_s - n_i^2)}{n_s + p_s + 2n_i},$$

(5.26)

where s_0 denotes the surface recombination velocity for a surface without a space charge layer and defined as $s_0 = \sigma v_{th} N_{ts}$, where N_{ts} is the surface defect density.

5.4.1
Diode Current under Reverse Bias

In order to arrive at a compact equation, the following popular approximations are made: $\sigma_n = \sigma_p = \sigma$, and the generation–recombination center is at or very near the intrinsic level, so $n_1 = p_1 = n_i$. In a reverse-biased junction, the free electron and hole concentrations are very small and can be neglected. In addition, the electron–hole density product is $pn - n_i^2 \approx -n_i^2$ due to a large reverse bias. The net recombination rate is given by

$$U = (\sigma v_{th} N_t) \frac{(pn - n_i^2)}{(n + n_1) + (p + p_1)} = (\sigma v_{th} N_t) \frac{(-n_i^2)}{(n_i) + (n_i)} = -\frac{n_i}{2\tau_{nr}} = -\frac{n_i}{\tau_e},$$

(5.27)

where $2\tau_{nr}$ has been taken as τ_e, which is an *effective recombination lifetime* in the depletion region. Within the above assumptions, the generation–recombination current density in a reverse-biased junction with a depletion depth of W is then give by

$$J_{GR} = -q \int_0^W |U| dx = -\frac{q n_i}{\tau_D} W,$$

(5.28)

where W is the width of depletion region. Here, the term τ_e is replaced with τ_D to emphasize the realm of depletion.

5.4.1.1 Poole–Frenkel and Schottky Effects

Electric field impacts the current–voltage characteristics through modification of the effective trap barrier, a process termed as *Poole–Frenkel* current, named after the pioneers who worked independently. To distinguish this process from the Schottky effect, the trap must be positively charged when empty and neutral when it traps an electron. A trap that is neutral and charged after capturing an electron will not experience this effect. In both the *Schottky* and *Poole–Frenkel* effects, the barrier can be attributed to a Coulombic interaction between the escaping electron and the positive charge left behind. The difference, which will give a factor of 2 dissimilarity between the two cases, is that in the case of Poole–Frenkel emission, the positive charge left behind is fixed in the lattice, while the positive charge in the case of the *Schottky* effect is a mirror image, moving away from the emission site as the electron moves away. Hence, as the electron moves away, at a distance r, the distance from the electron and its image (only in the case of Schottky emission) is $2r$. Note here that although our discussion is for the case of electron emission, the same arguments can also apply to hole emission.

While the details of the *Poole–Frenkel* current derivation can be found in Morkoç (2008) and references therein, the final form of the expressions signifying *Poole–Frenkel* and *Schottky* effects are provided here for convenience:

$$J_{pf} = q\mu n_0 E \exp\frac{-q\varphi_{pf}}{\varsigma kT} \exp\left(\frac{\beta_{PF} E^{1/2}}{\varsigma kT}\right), \quad \text{for Poole} - \text{Frenkel}, \tag{5.29}$$

$$J_s = A \times T^2 \exp\frac{-q\varphi_s}{\varsigma kT} \exp\frac{\beta_s E^{1/2}}{\varsigma kT}, \quad \text{for Schottky}, \tag{5.30}$$

where ς is a parameter ranging between 1 and 2 that depends on the position of the Fermi level, μ is the mobility, and n_0 is the electron density. In addition, ϕ_{pf} represents the barrier height in the absence of electric field for the case of *Poole–Frenkel*. Similarly, ϕ_s represents the barrier height in the absence of electric field for the case of *Schottky* emission. The *Poole–Frenkel* term can be conveniently expressed as

$$j = j_0 \exp\left(\frac{\beta_{pf} E^{1/2}}{\varsigma kT}\right), \tag{5.31}$$

$$\beta_{PF} = \sqrt{\frac{q^3}{\pi \varepsilon}} \quad \text{and} \quad J_0 = q\mu n_0 E \exp\frac{-q\phi_{pf}}{\varsigma kT}, \tag{5.32}$$

where j_0 is the low-field current density, β_{PF} is the Poole–Frenkel constant, and ε is the high-frequency dielectric constant of the semiconductor. For GaN, taking $\varepsilon(\infty) = 5.35\varepsilon_0$, the value of $\beta_{PF} = 3.3 \times 10^{-4}\,\text{eV}\,\text{cm}^{1/2}/\text{V}^{-1/2}$. The E term in these expressions would apply to the cases where the current is not fully limited by the generation rate. When limited by generation rate, the current is independent of the E field applied. For example, in samples where the in-plane current is under investigation, the carrier rate affects the carrier concentration that is available and the E field determines the carrier velocity along the direction of carrier motion.

5.4.1.2 Avalanching

In the reverse bias case and with large electric fields and, when a critical field is reached or exceeded, avalanche multiplication would occur giving rise to a sudden rise in current. Basically, carriers entering a high-field region gain sufficient energy to create new electron–hole pairs while losing their energy and thermalizing to the lowest energy of the band. In addition to the primary carriers, secondary carriers generated in this way can also be accelerated by the field creating their own electron–hole pairs and so on.

Let us assume that $J_n(0)$ and $J_p(0)$ are the electron and hole current densities entering the depletion region at $x=0$. In presence of avalanching, or multiplication, taking the hole current density alone, it will reach a value of $M_p J_{p0}$ at $x=W$. Here, M_p is the multiplication factor for holes and is defined as $M_p = J_p(W)/J_p(0)$. The spatial variation of the electron current in the presence of an optical generation at a rate of $G(x)$ in the units of carriers/(cm^3 s) can be expressed as

$$\frac{d}{dx}J_n(x) = \alpha_n(x)J_n(x) + \beta_p(x)J_p(x) + qG(x). \tag{5.33}$$

Similarly, for hole current,

$$-\frac{d}{dx}J_p(x) = \alpha_n(x)J_n(x) + \beta_p(x)J_p(x) + qG(x). \tag{5.34}$$

Using the total current being equal to the sum of electron and hole currents and expressing the electron current as the hole current subtracted from the total current, Equation 5.34 can be rewritten, in the case of no optical generation of carriers, that is, $G(x)=0$, as

$$-\frac{d}{dx}J_p(x) = \alpha_n(x)J(x) + \left[\beta_p(x) - \alpha_n(x)\right]J_p(x). \tag{5.35}$$

Equation 5.35 can be solved with the boundary condition that $J = J_p(W) = M_p J_p(0)$:

$$J_p(x) = J\left\{1/M_p + \int_0^x \alpha_n \exp\left[-\int_0^x (\beta_p - \alpha_n)dx'\right]dx\right\} \Big/ \exp\left[-\int_0^x (\beta_p - \alpha_n)dx'\right]. \tag{5.36}$$

Equation 5.36 can be expressed as

$$1 - \frac{1}{M_p} = \int_0^W \beta_p \exp\left[-\int_0^x (\beta_p - \alpha_n)dx'\right]dx. \tag{5.37}$$

The breakdown voltage is reached when the multiplication coefficient approaches ∞, in which case the left-hand side in Equation 5.37 becomes equal to 1. For equal electron and hole ionization rates, Equation 5.37 reduces to the ionization integral:

$$\int_0^x \alpha_n dx = 1. \tag{5.38}$$

Using the expressions above with the field dependence of ionization rates, the breakdown voltage, the maximum field, and the depletion depth can be calculated. The electric field and the potential distribution in the depletion region can be calculated using Poisson's equation, which is covered in all semiconductor device texts. Utilizing Equation 5.37 at breakdown with infinitely large multiplication factor and iterative methods, depletion layer boundaries can be obtained. Knowing those, one can calculate the breakdown voltage as

$$V_B = \frac{E_m W}{2} = \frac{\varepsilon_s E_m^2}{2q}(N_B)^{-1}. \tag{5.39}$$

It is clear that the breakdown voltage is inversely proportional to the donor concentration, N_B, in the depletion region, and E_m is the maximum field, or the field needed for impact ionization.

5.4.2
Diffusion Current

The total diffusion current is the sum of the diffusion currents on the n- and p-sides owing to minority carriers generated by both thermal generation and other means within, on the average, a diffusion length of the depletion region such as optical excitation, that is,

$$J_{\text{diff}} = J_{\text{diff n}} + J_{\text{diff p}} = qD_n \frac{n_{p0} + \tau_n G_L}{L_n} + qD_p \frac{p_{n0} + \tau_p G_L}{L_p}, \tag{5.40}$$

where G_L is the optical generation rate. In the dark, this expression with $L_{p,n'} \equiv (D_{p,n'}\tau_{p,n})^{1/2}$ reduces to the more familiar relation:

$$J_{\text{diff}} = qn_{p0}\sqrt{\frac{D_n}{\tau_n}} + qp_{n0}\sqrt{\frac{D_p}{\tau_p}} \quad \text{or} \quad J_{\text{diff}} = q\frac{n_i^2}{p_{p0}}\sqrt{\frac{D_n}{\tau_n}} + q\frac{n_i^2}{n_{n0}}\sqrt{\frac{D_p}{\tau_p}}. \tag{5.41}$$

The diffusion constant for holes and electrons can be estimated from the mobilities, but the hole and electron minority carrier lifetimes would have to be determined.

5.4.2.1 Diffusion Current under Reverse Bias

The total current under reverse bias due to diffusion and generation–recombination is

$$J_{\text{diff}} = q\left[\frac{n_i^2}{N_A^-}\sqrt{\frac{D_n}{\tau_n}} + \frac{n_i^2}{N_D^+}\sqrt{\frac{D_p}{\tau_p}} + \frac{n_i}{\tau_e}\left[\frac{2\varepsilon_s(V_{\text{bi}} - V)}{q}\left(\frac{1}{N_A^-} + \frac{1}{N_D^+}\right)\right]^{1/2}\right]$$

$$= q\left[\frac{n_i^2}{N_A^-}\sqrt{\frac{D_n}{\tau_n}} + \frac{n_i^2}{N_D^+}\sqrt{\frac{D_p}{\tau_p}} + \frac{n_i(x_n + x_p)}{\tau_e}\right]. \tag{5.42}$$

Here, we assume that the forward bias and the reverse bias are represented by a positive and a negative voltage, respectively. For a heterojunction, this expression should be modified. For a p–n heterojunction,[1] the first term in the diffusion-limited current should contain the parameters of the wider bandgap material. The second term should contain the smaller bandgap parameters. The generation–recombination current is more complex, but can be separated into two parts – each dealing with the respective semiconductor. This requires some knowledge of the effective recombination lifetime in addition to the dielectric constants. In wide-bandgap materials such as GaN, the n_i^2 term is very small, and thus the reverse current is dominated by the generation–recombination current component when neglecting the surface recombination current.

If we assume that the surface recombination velocity is large enough to bring the carrier concentration to zero, the surface recombination velocity for minority holes and electrons is the same, and the region of surface recombination is limited to the depletion depth $W = x_n + x_p$, then the surface recombination current density can be expressed as

$$J_s = \int_0^W q|U_s|dx = sq\left[\frac{n_i^2}{N_A^-}x_n + \frac{n_i^2}{N_D^+}x_p\right]. \tag{5.43}$$

The total current, neglecting the diffusion current, is the sum of the generation and surface recombination currents and is given by

$$J_r = \frac{qn_i(x_n + x_p)}{\tau_e} + sq\left[\frac{n_i^2}{N_A^-}x_n + \frac{n_i^2}{N_D^+}x_p\right]. \tag{5.44}$$

The small intrinsic carrier concentration appears to make the second term negligible, but only if the surface recombination velocity ($s = \sigma v_{\text{th}} N_{\text{st}}$) is small.

5.4.2.2 Diffusion Current under Forward Bias

In a forward-biased p–n junction, the minority carriers are injected across the junction and pave the way for the diffusion current. Generation–recombination centers in the depletion region give rise to an additional current component: *generation–recombination current*. The latter is said to cause the diode current to deviate from being ideal. In addition, as in the reverse bias case, surface

---
1) The italic capital letter depicts the wider bandgap semiconductor.

recombination current too may be noticeable if the junction area is small and the surface recombination velocity is high.

$$J_{\text{diff}} = q\left[\frac{n_i^2}{N_A^-}\sqrt{\frac{D_n}{\tau_n}} + \frac{n_i^2}{N_D^+}\sqrt{\frac{D_p}{\tau_p}}\right]\left[\exp\left(\frac{qV}{kT}\right) - 1\right]. \quad (5.45)$$

This expression can also be modified to give a first-order representation of the current–voltage relationship for a heterojunction. This is accomplished by using the parameters of the p-type layer for the first term in the parentheses and of the n-type layer for the second term.

The forward-biased recombination current, for the case where trapping centers are at the intrinsic level and hole and electron capture cross sections that are equal, is again obtained in exactly the same manner as in the reverse bias case. Through a modification of Equation 5.25, the generation rate for the forward bias case can be expressed as

$$U = \sigma v_{\text{th}} N_t \frac{n_i^2\{\exp[(qV)/kT] - 1\}}{n + p + 2n_i}, \quad (5.46)$$

where we made use of the quasi-equilibrium expression

$$pn = n_i^2\left[\exp\left(\frac{qV}{kT}\right)\right]. \quad (5.47)$$

Recognizing that the p–n product is constant throughout the junction, U would be maximized when $p+n$ is minimized, which occurs where $n=p$. Therefore, at the point where U is maximum,

$$p = n = n_i\left[\exp\left(\frac{qV}{2kT}\right)\right], \quad (5.48)$$

$$U_{\max} = \sigma v_{\text{th}} N_t \frac{n_i[\exp(qV/kT) - 1]}{2[\exp(qV/2kT) + 1]}. \quad (5.49)$$

When $V \gg kT/q$, $U_{\max} = \sigma v_{\text{th}} N_t [n_i \exp(qV/2kT)]/\tau_e$, and $\tau_e = 1/(2\sigma v_{\text{th}} N_t)$ is the effective lifetime.

The recombination current expression can be found by integration:

$$J_{\text{rec}} = -\frac{qn_i}{\tau_e} W \exp\left(\frac{qV}{2kT}\right) = -\frac{qn_i}{\tau_e}\left[\frac{2\varepsilon_s(V_{\text{bi}} - V)}{q}\left(\frac{1}{N_A^-} + \frac{1}{N_D^+}\right)\right]^{1/2} \exp\left(\frac{qV}{2kT}\right). \quad (5.50)$$

The right-hand side of the equation holds true for an abrupt junction. The voltage dependence of the recombination current is such that it is called $2kT$ current.

Further Reading

Blakemore, J.S. (1987) *Semiconductor Statistics*, Dover.

Frenkel, J. (1938) *Tech. Phys. USSR*, **5**, 685; Frenkel, J. (1938) *Phys. Rev.*, **54**, 647.

Grove, A.S. (1967) *Physics and Technology of Semiconductor Devices*, John Wiley & Sons, Inc., New York.

Hall, R.N. (1952) *Phys. Rev.*, **87**, 387.

Milnes, A.G. and Feucht, D.L. (1972) *Heterojunctions and Metal Semiconductor Junctions*, Academic Press, New York.

Morkoç, H. (2008) *Handbook on Nitride Semiconductors and Devices*, Wiley-VCH Verlag GmbH, Weinheim.

Sah, C.T., Noyes, R.N., and Schokley, W. (1957) *Proc. IRE*, **45**, 1228.

Shockley, W. and Read, W.T. (1952) *Phys. Rev.*, **87**, 835.

Sze, S.M. (1981) *Physics of Semiconductor Devices*, John Wiley & Sons, Inc., New York.

6
Optical Processes

6.1
Introduction

An astonishing property of direct bandgap semiconductors is the efficient light emission that transformed our lives. Light emission through electrical injection of minority carriers, *electroluminescence* (EL), has seen the most practical applications. When an external voltage is applied across a forward-biased p–n junction, electrons and holes that are injected into the medium from their respective ends recombine. This recombination results in the emission of a photon whose energy is mainly equal to the difference in the energies of states occupied by electrons and holes prior to recombination. Typically, the emission spectrum is rich in emission associated with intrinsic processes, meaning those not involving defects and impurities of any kind, and extrinsic processes, meaning those involving impurities and defects either in the simple or in the complex form.

Not all recombination results in the emission of a photon. The recombination process resulting in photon emission is termed *radiative recombination*. When this process does not involve an electromagnetic field, such as in photoluminescence (PL) experiments and light-emitting diodes (LEDs), it is called *spontaneous emission*. In other words, electron–hole (e–h) pairs are annihilated followed by photon emission. When an electromagnetic field of appropriate frequency is involved in the process, this emission is termed *stimulated emission* (provided that the frequency, polarization, phase, and direction of propagation are consistent) such as that found in semiconductor lasers (see Chapter 8 for definition of stimulated emission).

Among the most important optical processes taking place in semiconductors that directly influence device operation are the absorption and emission of photons. Considering absorption, if $I(\nu)$ represents the optical intensity at point x in a semiconductor, the spatial rate of change of the intensity at the same point is proportional to the intensity, and is given by

$$\frac{dI}{dx} = -\alpha I, \tag{6.1}$$

Nitride Semiconductor Devices: Fundamentals and Applications, First Edition. Hadis Morkoç.
© 2013 Wiley-VCH Verlag GmbH & Co. KGaA. Published 2013 by Wiley-VCH Verlag GmbH & Co. KGaA.

where α is the absorption coefficient with a unit of cm^{-1}. In an absorptive medium, the dielectric function is complex and can be expressed as

$$\varepsilon = \varepsilon' + j\varepsilon'' = \varepsilon_0(n + j\kappa)^2 = \varepsilon_0(n^2 - \kappa^2 + j2n\kappa), \tag{6.2}$$

where ε' and ε'' are the real and imaginary components of the dielectric constant, respectively, and n and κ represent the refractive index and the extinction coefficient, respectively.

The power of an electromagnetic field propagating along the x-direction is proportional to

$$\exp[-2(j(n + j\kappa)k_0 x)], \tag{6.3}$$

where $k_0 = 2\pi/\lambda_0$ is the free space wave vector, and the real part of the exponent is that associated with the absorption coefficient as

$$\alpha = 2\kappa k_0, \tag{6.4a}$$

where

$$\kappa = \frac{\varepsilon''}{2\varepsilon_0 n}. \tag{6.4b}$$

It is clear that the imaginary part of the dielectric constant is responsible for loss. Direct bandgap semiconductors, nitrides included, have large absorption coefficients with near-bandgap values in excess of 10^5 cm^{-1}, which bode very well for LEDs, lasers, and detectors. In the case of population inversion in a cavity, as in lasers, the absorption coefficient changes its sign, effectively becoming negative leading to gain as the electromagnetic wave traverses through the medium. When the gain exceeds the losses in the system, lasing oscillations would ensue.

6.2
Einstein's A and B Coefficients

Here, the first-order treatment for calculating the PL spectral distribution in a semiconductor is presented. Relying on the treatment of Planck, Einstein described absorption and stimulated emission constant per unit electromagnetic energy with energies between $h\nu$ and $h(\nu + \Delta\nu)$. In this nomenclature, the transition of an electron from a higher lying level 2 to a lower lying level 1 is depicted by the coefficient B_{21} for stimulated emission, which is referred to as the transition probability from state 2 to state 1. The spontaneous emission from level 2 to level 1 is depicted as A_{21}. The transition from level 1 to level 2 is called absorption and depicted by coefficient B_{12}. The A and B coefficients are called the Einstein's A and B coefficients.

In a two-level system with 1 representing the lower and 2 the upper level, the rates of upward and downward transitions for a system at thermal equilibrium at the temperature T were expressed by Einstein as

$$R_{21} = B_{21}\rho(\nu) \quad \text{and} \quad R_{12} = B_{12}\rho(\nu), \tag{6.5}$$

where the term $\rho(\nu)d\nu$ is the volume density of the electromagnetic modes in the frequency range ν and $\nu + d\nu$. If N_2 and N_1 represent the population levels (or photon occupation numbers) of levels 2 and 1, respectively, under thermodynamic equilibrium we can write

$$R_{21} = [A_{21} + B_{21}\rho(\nu)]N_2 \quad \text{and} \quad R_{12} = B_{21}\rho(\nu)N_1. \tag{6.6}$$

The product $\rho(\nu)N_1$ represents the photon energy density in the frequency range of ν and $\nu + d\nu$. Let us now define the expression governing $\rho(\nu)$. The number of modes between ν and $\nu + d\nu$ in a cubic cavity of side L is given by

$$\frac{2(1/8)4\pi\nu^2\,d\nu\,n^3}{(c/2L)^3} = \frac{8\pi V n^3}{c^3}\nu^2\,d\nu \quad \text{with} \quad V = L^3, \tag{6.7}$$

where V is the total volume, c denotes the velocity of light in vacuum, n is the refractive index, and $4\pi\nu^2\,d\nu$ is the volume of thin spherical shell. It is convenient to express this density (in terms of frequency and also energy) and also per unit volume as

$$\rho(\nu)d\nu = \frac{8\pi\nu^2 n^3}{c^3}d\nu \quad \text{and in terms of energy } \rho(E)dE = \frac{8\pi E^2 n^3}{(hc)^3}dE. \tag{6.8}$$

Boltzmann statistics imply that the probability of a given cavity mode lying between $h\nu$ and $h\nu + h\,d\nu$ is proportional to $\exp(-h\nu/kT)h\,d\nu$. The average energy per mode using the aforementioned distribution function is given by

$$\langle E \rangle = \frac{h\nu}{e^{h\nu/kT} - 1}. \tag{6.9}$$

The average number of photons for each mode is obtained from Equation 6.9 by dividing it with $h\nu$, which leads to

$$\langle \rho \rangle = \frac{1}{e^{h\nu/kT} - 1}. \tag{6.10}$$

The volume photon energy mode density in a frequency interval between ν and $\nu + d\nu$ (in a frequency interval $d\nu$) can be expressed as the product of Equations 6.8 and 6.9:

$$\rho(\nu)d\nu = \frac{8\pi\nu^2 n^3}{c^3}\frac{h\nu}{e^{h\nu/kT} - 1}d\nu, \tag{6.11}$$

which is called the Planck's formula.

Returning to Einstein coefficients, $A_{21}N_2$ represents the spontaneous emission process in which the electromagnetic radiation does not participate and $B_{21}\rho(\nu)N_2$ denotes the stimulated process with the electromagnetic radiation field participating. Because the spectral width of the radiation is finite, the unit of the beam intensity is W/m² per unit frequency interval.

Energy balance requires that R_{12} and R_{21} be equal, which determines the *spectral distribution*. In other words, the upward transition rate must be equal to the total downward transition rate at thermal equilibrium. Equating the temperature-independent components of this equality leads to

$$A_{21} = \frac{8\pi h \nu^3 n^3}{c^3} B_{21}. \tag{6.12}$$

The term relating the A and B coefficients in Equation 6.12 represents the density of the electromagnetic waves with the frequency between $h\nu$ and $h\Delta\nu$ inside the medium times $h\nu$; in other words, the density of the electromagnetic wave energy with the frequency between $h\nu$ and $h(\nu + \Delta\nu)$ inside the medium.

Equating the temperature-dependent terms of the same balance equation leads to

$$B_{12} = B_{21}. \tag{6.13}$$

The above treatment can be extended to a semiconductor with the additional conditions that momentum conservation and Pauli's exclusion principle must hold. The population difference between levels 2 and 1 is

$$N_2 - N_1 = V \frac{8\pi k^2 \, dk}{(2\pi)^3} [f_c(1 - f_v) - f_v(1 - f_c)], \tag{6.14}$$

where f_c and f_v denote the electron occupancy factors for the conduction and the valence bands in a semiconductor with excess carriers. Needless to say, $1 - f_c$ and $1 - f_v$ depict the probability of a corresponding state in the conduction and the valence band, respectively, being empty to satisfy the Pauli's exclusion principle. As mentioned just above, the term $8\pi k^2 \, dk/(2\pi)^3$ accounts for the density of the electromagnetic waves in k-space and V is the volume. If $N_2 - N_1 > 0$, the semiconducting medium would amplify as opposed to attenuating. See Chapter 8 for continuation of the above treatment in the context of lasers.

6.3
Absorption and Emission

For a semiconductor with parabolic bands, the upper states correspond to the conduction band and the lower to the valence band. We can write, by assuming one participating valence band only,

$$E_2 = E_c + \frac{\hbar^2 k^2}{2m_n^*} \quad \text{and} \quad E_1 = E_v - \frac{\hbar^2 k^2}{2m_p^*}. \tag{6.15}$$

With the aid of the reduced effective mass, $m_r^{*-1} = m_n^{*-1} + m_p^{*-1}$, where m_n^* and m_p^* are the electron and hole effective masses, respectively,

$$E_2 - E_1 = h\nu = E_g + \frac{\hbar^2 k^2}{2m_r^*}. \tag{6.16}$$

By solving for k and inserting it in the expression for the density of the electromagnetic waves in the k-space, Equation 6.14, or the density of directly associated states, we have

$$N(h\nu)d(h\nu) = \frac{(2m_r^*)^{3/2}}{2\pi^2\hbar^3}(h\nu - E_g)^{1/2}d(h\nu). \tag{6.17}$$

The absorption coefficient for a given $h\nu$ is proportional to the probability for a transition from the initial state to the final state, and to the density of available electrons in the first state and the density of empty states in the excited state, as depicted with Equation 6.14. Equation 6.17 accounts for the dependence of the absorption coefficient on energy in a direct bandgap semiconductor. In other words, the absorption coefficient is proportional to the square root of the energy above the gap energy. Below the gap energy and in this ideal picture, the absorption coefficient tends to zero.

In an absorption measurement on a high-purity sample, the probabilities of having an electron in the lower state (valence band) and that for the higher state (conduction band) can be taken as 1 and 0, respectively. Hence, the absorption coefficient reduces to

$$\alpha(h\nu) = A^*(h\nu - E_g)^{1/2}, \tag{6.18}$$

with

$$A^* \approx \frac{q^2(2m_r^*)^{3/2}}{nch^2 m_n^*}, \tag{6.19}$$

where c is the velocity of light in vacuum. Once the absorption coefficient versus photon energy is either calculated or measured, the spontaneous emission spectrum can be calculated from

$$I(h\nu) = (h\nu)r(h\nu) = (h\nu)g(h\nu), \tag{6.20}$$

where $r(h\nu)$ and $g(h\nu)$ are the recombination and generation rates that are equal at thermal equilibrium. In a photoluminescence experiment, this assumption can be used provided that the exciting light intensity is very low:

$$I(h\nu) = \frac{8\pi\nu^2}{c^2}h\nu\alpha(h\nu)[f_c(1-f_v)], \tag{6.21a}$$

where

$$f_c = \left[1 + \exp\left(\frac{E_c - F_n}{kT}\right)\right]^{-1} \tag{6.21b}$$

and

$$f_v = \left[1 + \exp\left(\frac{F_p - E_v}{kT}\right)\right]^{-1}. \tag{6.21c}$$

The term f_c represents the occupation probability of being in the upper (conduction) state and f_v represents the occupation probability of being in the lower (valence) state. F_n and F_p are the quasi-Fermi levels for electrons and holes, respectively. Simplifying these probabilities for a nondegenerate semiconductor, which means replacing them with their Boltzmann factors, we get for $I(h\nu)$:

$$I(h\nu) = \frac{8\pi n^2 h\nu^3}{c^2} \alpha(h\nu) \exp\left(\frac{F_n - F_p}{kT}\right) \exp\left(\frac{-h\nu}{kT}\right). \tag{6.22}$$

The spectral emission response is proportional to the product of the absorption coefficient and $\exp(-h\nu/kT)$. On the lower energy side, the emission spectrum has the spectral dependence of the absorption coefficients, and above the bandgap it roughly declines exponentially.

6.4
Band-to-Band Transitions and Efficiency

The average lifetime of carriers before radiative recombination is called the *radiative lifetime* τ_r. The rate of emission of photons by recombining electrons, n, and holes, p, is a bimolecular process and is given by

$$R = Bnp, \tag{6.23}$$

where B is the radiative recombination probability. The terms R and B have the units of $cm^{-3} s^{-1}$ and cm^3/s, respectively. For a p-type (or n-type) semiconductor where the excess carrier concentration is much less than the equilibrium hole (or electron) concentration, radiative recombination lifetime reduces to $\tau_r = (pB)^{-1}$. For thermalized electrons and holes, the recombination time depends on the electron and hole energies, which means that it will depend on the photon energy. Consequently, it is customary to define an average lifetime as $\langle \tau_r \rangle$, which depends on the k selection rules as is the case for perfect or nearly perfect semiconductors. In heavily excited semiconductors, this does not hold. In electroluminescent devices, such as in LEDs and lasers, with p-type active regions, when an electron is injected in thermal equilibrium, in terms of electron and hole distributions, we define a lifetime called the *minority carrier radiative lifetime* τ_{rad}, also commonly termed as τ_r, to depict the radiative recombination process. This is the time it takes for an extra minority carrier to be annihilated radiatively by a majority (hole) carrier. In intrinsic and/or near-intrinsic semiconductors with very low electron/hole concentrations, the minority radiative recombination time is rather long as in the case of indirect semiconductors because the probability of these processes is very small. The radiative recombination time can be made smaller with increased doping up to a certain limit as the more the doping is increased the less the above expression is valid. The stimulated emission lifetime does not follow this rule as the stimulated emission rate depends on the photon density as well.

In addition to the radiative processes, there are nonradiative processes in semiconductors because of imperfections that act as nonradiative centers. We should mention some defects as radiative recombination centers that in a PL experiment can shed light on the energy levels of defect states. For a semiconductor containing nonradiative traps or recombination centers, in an experiment such as time-dependent PL, the decay in the intensity of the integrated PL intensity versus temperature is related to the low-temperature integrated PL intensity as

$$I_{PL}(T) = \eta_{PL}(T) I_{PL}(0), \tag{6.24}$$

where $\eta_{PL}(T)$ is the temperature-dependent PL efficiency, and $I_{PL}(T)$ and $I_{PL}(0)$ are the integrated PL intensities at a temperature T and zero, respectively.

The measured PL decay time can be expressed in terms of radiative and nonradiative lifetimes as

$$\tau_{PL}(T) = [1/\tau_r(T) + 1/\tau_{nr}(T)]^{-1}, \tag{6.25}$$

where $\tau_{PL}(T)$, also referred to as $\tau_{total}(T)$ or $\tau_{eff}(T)$, is total or effective recombination lifetime and is also the quantity that is experimentally measured. The magnitude of $\tau_{PL}(T)$ represents the average length of time a photoexcited carrier can remain in the conduction (or valence) band before recombination and is thus directly correlated to material quality, purity, and doping level approaching $\tau_r(T)$ or also referred to as $\tau_{rad}(T)$ in pure and defect-free materials. The term $\tau_{nr}(T)$ or also referred to as τ_{nrad} is the lifetime for all the nonradiative recombination channels combined. For an intrinsic material, the total recombination rate is

$$R_T = (1/\tau_{eff})(np/2n_i), \tag{6.26}$$

where n_i is the intrinsic carrier concentration, and n and p are the injected electron and hole concentrations, respectively. The radiative recombination rate is

$$R = (1/\tau_{rad})(np/2n_i). \tag{6.27}$$

The radiative efficiency is then

$$\eta = R/R_T = \tau_{nonrad}/(\tau_{rad} + \tau_{nonrad}) \quad \text{or} \quad \eta = \tau_{nr}/(\tau_r + \tau_{nr}). \tag{6.28}$$

The expression for radiative recombination efficiency can also be found by taking advantage of Equations 6.24 and 6.25:

$$\eta_{PL}(T) = \tau_{PL}/\tau_r. \tag{6.29}$$

In the limit where there is no nonradiative recombination, which means that $1/\tau_{nonrad}$ is zero, the radiative efficiency becomes unity. The lifetime of excess minority carriers can be obtained by measuring the dynamic behaviors of optical emissions involved using time-resolved photoluminescence (TR-PL).

In a PL experiment, the band-to-band emission lineshape is determined by the joint density of states and the probability of participating states being available for recombination. The former has the form $(\hbar\omega - E_g)^{1/2}$ and the latter $\exp(-E/kT) = \exp(-\hbar\omega/kT)$. When the semiconductor is excited by the above-bandgap photon radiation, the two lineshapes put together lead to a lineshape of the form

$$(\hbar\omega - E_g)^{1/2}\left[\exp\left(\frac{-\hbar\omega}{kT}\right)\right]. \tag{6.30}$$

Lower energy photons do not excite electrons into the conduction band; therefore, the band-to-band emission will be zero in "linear" experiments.

6.5
Optical Transitions in GaN

Optical transitions in semiconductors, GaN is no exception, can be grouped into two categories. The *intrinsic transitions* are those that are associated with semiconductors void of impurities and defects. The *extrinsic transitions* have their genesis in impurities and defects. Free excitons and their phonon replicas, if any, and free-to-free transitions represent the intrinsic transitions. The impurity-bound excitons, transitions involving impurities such as free-to-bound and bound-to-bound, and defects constitute extrinsic transitions. A collage of the intrinsic and extrinsic transitions is sketched in Figure 6.1.

6.5.1
Excitonic Transitions in GaN

Excitons are first classified into free and bound excitons. Owing to the band structure of GaN, there are three free excitons (A, B, and C) in GaN. The bound excitons can be bound to shallow donors (SDBE), deep donors (DDBE), and structural defects (StDBE), as shown in Figure 6.2. In high-quality samples with low impurity concentrations, the free excitons can also exhibit excited states, in addition to their ground-state transitions. Wurtzite (Wz) structures are more interesting due to the splitting of the valence band by crystal field and spin–orbit interactions, as will be described below.

Excitons in GaN take on a special meaning in that the valence band is not degenerate due to the crystal field and spin–orbit interactions at the Γ point. The three emerging states are termed Γ_9^v, upper Γ_7^v, and lower Γ_7^v. The related free exciton transitions from the conduction band to these three valence bands are termed A, B, and C excitons. In terms of symbols, they are $A \equiv \Gamma_7^c \to \Gamma_9^v$ (also referred to as the *heavy hole state*), $B \equiv \Gamma_7^c \to \Gamma_7^v$ the upper one (also referred to as the *light hole state*), and $C \equiv \Gamma_7^c \to \Gamma_7^v$ the lower one (also referred to as the *crystal field split band*). In ideal Wz crystals, that is, strain free, they have the following symmetries: Excitons associated with all three bands are allowed in the α polarization ($\mathbf{E} \perp c$ and $\mathbf{k} \parallel c$). In the σ polarization ($\mathbf{E} \perp c$ and $\mathbf{k} \perp c$), A and B excitons are observed, with the C exciton being very weak. In the π polarization ($\mathbf{E} \parallel c$ and $\mathbf{k} \perp c$), A exciton is forbidden, and C exciton is strong with the B exciton being weak. Here $\mathbf{E}$ and $\mathbf{k}$ are the electric field and momentum vectors,

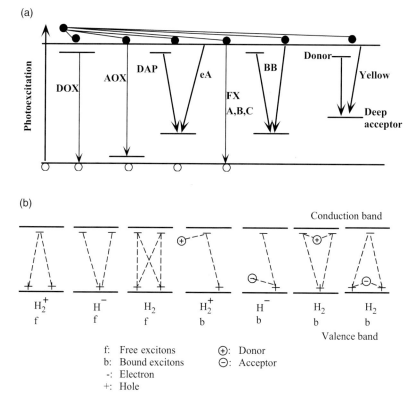

Figure 6.1 (a) Intrinsic and extrinsic optical transitions that occur in response to an above-bandgap excitation in GaN. (b) Various free and bound exciton complexes that can form in a semiconductor.

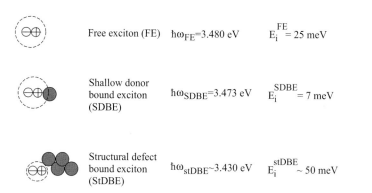

Figure 6.2 Examples of free and bound excitons in GaN along with their graphical description, absolute energies at low temperatures, and binding energies to the extent that is known.

Table 6.1 Excitonic transitions, whether they allowed or not, for α polarization ($\mathbf{E} \perp c$ and $\mathbf{k} \parallel c$), σ polarization ($\mathbf{E} \perp c$ and $\mathbf{k} \perp c$), and π polarization ($\mathbf{E} \parallel c$ and $\mathbf{k} \perp c$).

Polarization	A exciton	B exciton	C exciton
α polarization ($\mathbf{E} \perp c$ and $\mathbf{k} \parallel c$)	Allowed	Allowed	Allowed
σ polarization ($\mathbf{E} \perp c$ and $\mathbf{k} \perp c$)	Allowed	Allowed	Weak
π polarization ($\mathbf{E} \parallel c$ and $\mathbf{k} \perp c$)	Forbidden	Weak	Strong

respectively, and c denotes the c-axis of the crystal (see Table 6.1). Excitonic states are also expected to have fine structure due to short-range exchange interaction (responsible for splitting the A state into a dipole-active Γ_5 and a dipole-forbidden Γ_6 state, separated by the energy of the short-range exchange interaction Δ_{56}) and dipole–dipole interaction (responsible for splitting the Γ_5 state into a transverse Γ_5 and a longitudinal Γ_5 state, separated by the energy of the longitudinal–transverse splitting Δ_{LT}). These splittings are expected to be on the order of 1–2 meV and only upon drastic improvements in the GaN quality, the fine structure of exciton emission has been observed and identified.

As an example, PL spectra taken in a high-quality freestanding GaN substrate and MBE-overgrown layer at 15 K are shown in Figure 6.3. The main features of the PL spectra are similar, yet a few distinctions stand out. A peak at 3.4673 eV is attributed to the A exciton bound to unidentified shallow acceptor ($A^0, X_A^{n=1}$). The most intense peaks at 3.4720 and 3.4728 eV are attributed to the A exciton bound to two neutral shallow donors, $D_1 = vO_N$ and $D_2 = Si_{Ga}$: ($D_1^0, X_A^{n=1}$) and ($D_2^0, X_A^{n=1}$). Positions of

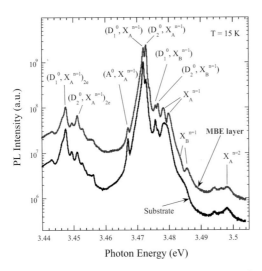

Figure 6.3 Excitonic PL spectrum for GaN in the area of substrate and in the MBE-overgrown part. Excitation density is 100 W/cm².

6.5 Optical Transitions in GaN

these peaks are the same with an accuracy of 0.2 meV for the substrate and different points of the overgrown layer. The ratio between the intensities of the $(D_1^0, X_A^{n=1})$ and $(D_2^0, X_A^{n=1})$ peaks is different in the substrate and overgrown layer as can be deduced from Figure 6.3. This feature helps to identify four other peaks related to the excitons bound to the same shallow donors. Namely, the peaks at 3.4758 and 3.4766 eV, having the same intensity ratio and energy separation as the $(D_1^0, X_A^{n=1})$ and $(D_2^0, X_A^{n=1})$ peaks, are attributed to the B exciton bound to the D_1 and D_2 donors: $(D_1^0, X_B^{n=1})$ and $(D_2^0, X_B^{n=1})$.

6.5.1.1 Strain Effects

Strain-induced modifications of the exciton energies can be obtained through the use of the strain Hamiltonian for a Wz crystal such as that reported by Pikus. The energies of the three free excitons are related to the strain-free energies by

$$\begin{aligned} E_A &= E_{A0} + a_z \varepsilon_{zz} + a_{xy}(\varepsilon_{xx} + \varepsilon_{yy}) + b_z \varepsilon_{zz} + b_{xy}(\varepsilon_{xx} + \varepsilon_{yy}), \\ E_B &= E_{B0} + a_z \varepsilon_{zz} + a_{xy}(\varepsilon_{xx} + \varepsilon_{yy}) + \Delta_+ [b_z \varepsilon_{zz} + b_{xy}(\varepsilon_{xx} + \varepsilon_{yy})], \\ E_C &= E_{C0} + a_z \varepsilon_{zz} + a_{xy}(\varepsilon_{xx} + \varepsilon_{yy}) + \Delta_- [b_z \varepsilon_{zz} + b_{xy}(\varepsilon_{xx} + \varepsilon_{yy})], \end{aligned} \quad (6.31)$$

where E_{k0} ($k = A, B, C$) and a_{ij} and b_{ij} ($i, j = x, y, z$) represent, in order, the strain-free exciton energies, combined hydrostatic deformation potentials, and uniaxial deformation potentials, and ε_{ij} is the particular component of the strain tensor. We limit our discussion to c-plane GaN that represents the majority of GaN films and means that the parallel or the in-plane component is in the x–y plane and the c-direction is in the z-direction. The strain components are

$$\varepsilon_{xx} = \varepsilon_{yy} = \varepsilon_{//} \frac{a_s - a_0}{a_0} \quad (6.32a)$$

and

$$\varepsilon_{zz} = \varepsilon_\perp = \frac{c_s - c_0}{c_0}, \quad (6.32b)$$

where a and c represent the in-plane and out-of-plane lattice constants, respectively, with the subscripts s and 0 indicating strained and relaxed parameters. Under biaxial strain conditions, the in- and out-of-plane strains are related through the stiffness (elastic) coefficients as

$$\varepsilon_\perp = \frac{-2C_{13}}{C_{33}} \varepsilon_{//}. \quad (6.33)$$

The coefficients Δ_+ and Δ_- in Equation 6.31 account for the valence band mixing through spin–orbit interaction and are given as

$$\Delta_\pm = \frac{1}{2} \left\{ 1 \pm [1 + 8(\Delta_3/(\Delta_1 - \Delta_2)^2)]^{1/2} \right\}, \quad (6.34)$$

where Δ_1 is the splitting of the Γ_9 and Γ_7 valence bands due to the crystal field. The symbols Δ_2 and Δ_3 represent the spin–orbit coupling. By plotting the measured excitonic transition energies as a function of strain, as determined from X-ray diffraction, one can obtain a value of about 0.531 for $\Delta_3/(\Delta_1 - \Delta_2)$. Utilizing the experimentally observed excitonic transition energies together with Equation 6.31, one can separate out the uniaxial component of the strain-induced shift. Doing so leads to

$$b_z - \frac{C_{33}}{C_{13}} b_{xy} = 15.2 \text{ eV}. \tag{6.35}$$

A similar approach for the in-plane strain leads to

$$a_z - \frac{C_{33}}{C_{13}} a_{xy} = 37.9 \text{ eV}. \tag{6.36}$$

Utilizing the quasicubic approximation, that is, the $C_{11} \approx C_{33}$ assumption that is justified by the small strain-induced shift relative to the total excitonic energies, leads to $b_z \approx -2b_{xy}$ and $a_z - a_{xy} \approx 2b_{xy}$. One then obtains $b_z \approx -5.3$ eV and $b_{xy} \approx 2.7$ eV by using 106 and 398 GPa for C_{13} and C_{33}, respectively.

6.5.1.2 Bound Excitons

Excitons may be bound to neutral and ionized donors and acceptors, as well as to isoelectronic defects. Not all of the excitons may be observed in a given semiconductor as only some of them are stable. Because the bound excitons have much smaller kinetic energies than free excitons, the spectral width of the bound exciton lines is generally narrower than those for the free excitons. The A, B, and C excitons discussed earlier represent intrinsic processes as they do not involve pathways requiring extrinsic centers. Available semiconductors contain impurities such as donors and acceptors and shallow donor- and acceptor-like defects. In theory, excitons could be bound to neutral and ionized donors and acceptors. Because the bound excitons have much smaller kinetic energies, as compared to free excitons, when bound excitons recombine their emission is characterized by a spectral line at a lower photon energy and narrower linewidth than those of the free excitons. The energy of the photon emitted through bound exciton recombination is given by

$$h\nu = E_g - E_x - E_{bx}, \tag{6.37}$$

where E_x is the free exciton binding energy and E_{bx} is the additional energy binding the free exciton to the impurity center. The radiative recombination lifetime of bound excitons increases with binding energy E_{bx}. When acceptors are present, the exciton can be bound to a neutral acceptor (A^0), and the corresponding PL emission line is called A^0X transition (or I_1). In n-type GaN materials, a free exciton can be bound to a neutral donor (D^0) or an ionized donor (D^+), with resulting PL emission lines labeled as D^0X and D^+X transitions (or I_2 and I_3), respectively. The nomenclature follows that established for hexagonal II–VI semiconductors, such as ZnO. In excitons bound to neutral donors and acceptors, the like particles predominantly

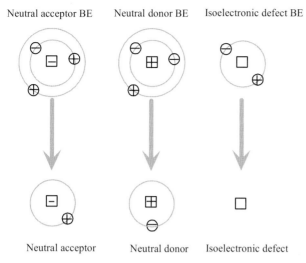

Figure 6.4 A schematic representation of the electronic structure of bound excitons. The bound exciton states as well as the corresponding defect ground states are shown, for neutral donors, neutral acceptors, and neutral "isoelectronic" defects. Courtesy of B. Monemar.

couple. The particles with opposite signs are weakly bound by the attractive Coulomb potential of the two coupled particles. Because for shallow dopants all three particles involved are shallow, with a substantial wavefunction overlap, other interactions do exist, which may lead to additional excited configurations.

In addition, excitons can also be bound to structural defects. The bound exciton (BE) transitions occur between the excited state, that is, the donor BE (DBE) state, and the neutral donor ground state for the case of donors, as shown in Figure 6.4. As the name implies, a dominant coupling of the like particles in the BE states is assumed to describe the bound exciton states for neutral donors and acceptors. In the case of a shallow neutral donor, the two electrons in the BE state are assumed to pair off into a two-electron state with zero spin. The additional bound hole is assumed weakly bound in the net hole attractive Coulomb potential set up by this bound two-electron complex. In this picture, the excited states would be the higher shallow bound hole states. Because for shallow dopants all three particles are shallow, with a substantial wavefunction overlap, other interactions do exist, which may lead to additional excited configurations, as observed in most direct bandgap semiconductors.

6.6
Free-to-Bound Transitions

At low temperatures, the carriers in nondegenerate semiconductor are frozen on the impurities because the thermal energy required for their ionization is no longer available. In n-type semiconductor, electrons excited into the conduction band in PL experiment can recombine radiatively with photogenerated holes after the latter are captured by acceptors at the acceptor energy levels. Thus, the resultant emitted photon energy would represent the difference between the conduction band and acceptor level or the bandgap minus the acceptor binding energy, $E_g - E_A$. At elevated temperatures, the average kinetic energy ($kT/2$) of free electrons should be added. These transitions involving free carriers and bound charges constitute *free-to-bound transitions* (e.g., e–A transitions in case of free electrons and bound holes). If the impurity concentration is increased, donors or acceptors become closer to one another and their wavefunctions begin to overlap more and more with an associated broadening of the level, which is called the impurity band. With continued increase in impurity concentration, the impurity band may widen so much that it would overlap with the nearest band, the conduction band in the case of donors and the valence band in the case of acceptors. As a result, the carriers are freed and this delocalization is called the *Mott transition*, which can be observed even before such overlap because the impurity band is half filled owing to spin degeneracy. The higher portion of the photoluminescence spectrum deviates from a simple exponential by spreading out to higher and lower energies, and also changing because of bandgap renormalization (*redshift*). Deep impurities and defects causing radiative decay of electrons to their ground states can be probed with luminescence experiments, and GaN is no exception.

In unintentionally doped GaN, there are no free holes and very low concentration of free electrons at low temperature (below 15 K) in dark due to relatively large binding energies of the shallowest donors and acceptors. Consequently, only *bound-to-bound*, specifically donor–acceptor pair (DAP), transitions are expected, involving transitions from the shallow donors (mostly filled with electrons in n-type GaN) to different acceptor levels (filled with photogenerated holes according to their hole capture cross sections). With increasing temperature, electrons thermally first escape to the conduction band. When their concentration increases, the e–A transitions are expected to dominate over the DAP transitions.

6.7
Donor–Acceptor Transitions

Though some semiconductors, such as Ge, represent the purest materials available, all contain both donors and acceptors to varying degrees, and are known by the term *compensated*. The nomenclature results from acceptors capturing electrons from the donor states. Consequently, a compensated semiconductor contains both ionized donors and acceptors. Carriers generated by optical excitation can be trapped at the donor and acceptor sites causing them to be neutral. In returning toward equilibrium, some electrons on the neutral donor sites will recombine with holes on the neutral acceptors, a process termed DAP transition. The DAP transition energy is given by

$$h\nu = E_g - E_D - E_A + \frac{q^2}{\varepsilon R}, \tag{6.38}$$

where E_D and E_A are the donor and acceptor binding energies, and the last term on the right-hand side is the Coulomb interaction contribution resulting from the interaction of ionized donors and acceptors. R is the distance between such donors and acceptors, and is assumed much larger than the lattice constant. This is easily satisfied in semiconductors that are not highly doped. In high-quality samples, many DAP transitions can be observed for many values of R. The Coulomb interaction causes the energy of the final state to be lowered. The pair spectra have attracted much discussion in the semiconductors GaP and GaAs. Unfortunately, the term pair spectrum is commonly used very loosely, and is attributed to any donor–acceptor transition observed in GaN.

Further Reading

Bir, G.L. and Pikus, G.E. (1974) *Symmetry and Strain-Induced Effects in Semiconductors*, John Wiley & Sons, Inc., New York.

Casey, H.C. and Panish, M.B. (1978) *Heterostructure Lasers*, Academic Press, New York.

Einstein, A. (1917) *Phys. Z.*, **18**, 121.

Hopfield, J.J. (1958) *Phys. Rev.*, **112**, 1555.

Kittel, C. (1958) *Elementary Statistical Physics*, John Wiley & Sons, Inc., New York, p. 175.

Monemar, B. (2001) *J. Phys.: Condens. Matter*, **13**, 7011–7026.

Monemar, J.J., Bergman, J.P., and Buyanova, I.A. (1997) Optical characterization of GaN and related material, in *GaN and Related Materials* (ed. S.J. Pearton), Gordon and Breach, Amsterdam.

Morkoç, H. (2008) Chapter 5, in *Handbook on Nitride Semiconductors and Devices*, vol. 2, Wiley-VCH Verlag GmbH, Weinheim.

Pankove, J.I. (1971) *Optical Processes in Semiconductors*, Prentice Hall, Englewood Cliffs, NJ.

Pikus, G., (1962) *Sov. Phys. JETP*, **14**, 1075; Ivchenko, E.L. and Pikus, G. E. (1997) *Superlattices and Other Heterostructures*, 2nd edn, Springer Series in Solid-State Sciences, vol. 110, Springer, Berlin.

Svelto, O. and Hanna, D.C. (1976) *Principles of Lasers*, Plenum Press.

Thomas, D.G., Hopfield, J.J., and Augustyniak, W.M. (1965) *Phys. Rev.*, **140**, A202.

Voos, M., Leheny, R.F., and Shah, J. (1986) *Handbook on Semiconductors: Optical Properties of Solids* (ed. M. Balkanski), North-Holland, Amsterdam.

Yu, P.Y. and Cardona, M. (1995) *Fundamentals of Semiconductors*, Springer, Berlin.

7
Light-Emitting Diodes and Lighting

7.1
Introduction

Owing primarily to nitride semiconductors with blue and green emission, light-emitting diodes (LEDs) morphed from simple indicators to high-tech marvels with applications far and wide in every aspect of modern life. LEDs are simply p–n junction devices and convert electrical power into generally visible light through spontaneous emission, the wavelength of which is determined by the bandgap of the semiconductor. Generally, unlike the semiconductor lasers, the junction is not biased to and beyond the transparency. In the absence of transparency, self-absorption occurs in the medium, which is why this region thickness is kept to an optimum, where the photons are generated and they are emitted in random directions. A modern LED is generally of a double-heterojunction type with the active layer being the only absorbing layer in the entire structure inclusive of the substrate. LEDs gravitated from simply being indicator lamps replacing nixie signs to highly efficient light sources featuring modern technologies with high quantum and extraction efficiencies making packaging all too important. The standard 5 mm plastic dome to focus the emitted light gave way to sophisticated packaging and epitaxial design strategies, as will be discussed. Furthermore, the area of the device as well as the shape of the chip is designed for maximum étendue, a measure of the optical area of the device. Strategies had to be adopted not only to remove the heat generated but also to deal with the thermal mismatch between the chip and the heat sink.

Elaborating further, as LEDs became brighter and white light generating varieties became available, the role of LEDs shifted from being simply indicator lights to illuminators. The mobile electronics devices, high-efficiency TVs, signs, displays, automotive and aircraft industries, and general lighting are the beneficiaries of nitride-based LEDs. In 2011 with nearly $12.5B LEDs in sales worldwide, about 27.2% accounted for mobile electronics, 11.2% for signs, 8% for automotive, 24% for TV/monitor, and 14.4% for lighting. The lighting component would be expected to grow considerably when the free market conditions fully prevail. The goals for a 35 A/cm^2 injection current density are 90% photon extraction efficiency, 100% internal quantum efficiency (IQE), near 250 lm/W efficacy, and only 5% carrier loss

Nitride Semiconductor Devices: Fundamentals and Applications, First Edition. Hadis Morkoç.
© 2013 Wiley-VCH Verlag GmbH & Co. KGaA. Published 2013 by Wiley-VCH Verlag GmbH & Co. KGaA.

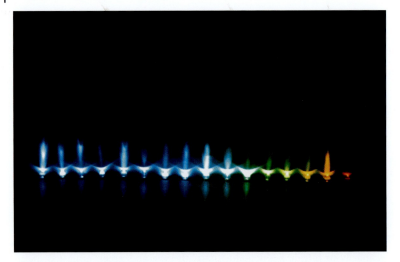

Figure 7.1 InGaN LEDs spanning the spectral range from violet to orange. Courtesy of S. Nakamura, then with Nichia Chemical Co. Ltd.

to recombination outside the active region. Nitride-based LEDs with InGaN-active regions span the visible spectrum from yellow to violet, as illustrated in Figure 7.1.

While saturated color red LEDs can be produced using semiconductors such as GaP, AlGaAs, AlGaInP, the green and blue commercial LEDs with sufficient brightness to be of use for outdoor applications have so far been manufactured with nitride semiconductors. Figure 7.2 exhibits the various ternary and quaternary materials used for LEDs with the wavelength ranges indicated. The color bar corresponds to the visible portion of the spectrum. We should also mention that another wide-bandgap semiconductor, ZnO, with its related alloys is being pursued for light emission as it is a very efficient light emitter. However, lack of convincingly high p-type doping in high concentration has kept this approach from reaching its potential so far.

Even though there is still some discussion of the fundamentals of radiative recombination in InGaN LEDs, the basics of LEDs, assuming that the semiconductors of interest are well behaved, will be treated first. This will be followed by the

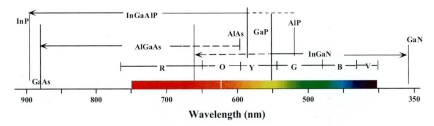

Figure 7.2 The LED materials and range of wavelength of the emission associated with them. The color band indicates the visible region of the spectrum.

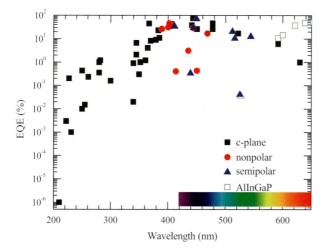

Figure 7.3 The available data for LEDs spanning the visible and UV wavelengths. Courtesy of Prof. S. Chichibu.

performance of available nitride LEDs and their characteristics. The discussion is completed with succinct treatments of the reliability of nitride-based LEDs, and of organic LEDs, which have progressed to the point that indoor applications are being considered.

External quantum efficiency (EQE) is an exceedingly valuable parameter to gauge the performance of an LED. In the visible region, violet and blue LEDs are the domain of nitride semiconductors. The red is in the domain of another III–V material, AlInGaP at least for the time being. The green color and its variants are challenging, to say the least, in that both nitride and AlInGaP system performance is subpar; the latter is a fundamental one because of its limited bandgap. The problem with the nitride variety is most likely not a fundamental one and is technological in nature, which over time can be mitigated. For relatively shorter wavelengths, the active region eventually give way to GaN and then AlGaN for even shorter wavelength varieties and the EQE plummets due to a large extent the technology and limitation of p-type doping, which become acute for large AlN mole fraction AlGaN alloy. The data available in the literature are presented in Figure 7.3 to illustrate the point.

7.2
Current Conduction Mechanism in LED-Like Structures

Consider an AlGaN(p)/GaN(p)/AlGaN(n) double-heterojunction device, which is forward biased for which the carrier and light distributions in the active layer are depicted schematically in Figure 7.4. For simplicity, assume that all the carriers recombine in the smaller bandgap-active region. In reality, some fraction of the recombination is nonradiative as well as carriers leaking out of the active region and

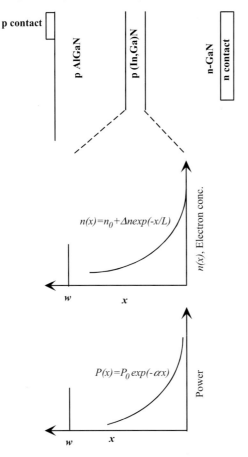

Figure 7.4 The spatial carrier and light distribution in a double heterostructure LED.

recombine elsewhere in the structure, which does not contribute to the wavelength of interest.

Because the active layer is chosen p-type (for n-type which is the case in GaN-based LEDs, one can simply replace n with p), we will be dealing with minority electron carriers. The continuity equation for electrons can be written as

$$D\frac{\partial^2 n}{\partial x^2} - \frac{n - n_o}{\tau} + G = \frac{\partial n}{\partial t}, \qquad (7.1)$$

where n and n_o represent the minority-carrier concentration and the equilibrium minority-carrier concentration, respectively. The terms D, G, and τ represent the electron diffusion length, the generation rate, and the carrier lifetime, respectively, and x is the distance. If the active layer were n-type, the same equations would apply with the minority electron parameters replaced with the minority hole parameters.

7.2 Current Conduction Mechanism in LED-Like Structures

Under steady-state conditions and large injection levels, such as is the case for high-brightness LEDs, the time dependence vanishes, and the generation rate and the equilibrium minority carrier concentration can be neglected, and the continuity expression reduces to

$$D\frac{d^2 n}{dx^2} - \frac{n}{\tau} = 0. \tag{7.2}$$

This second-order differential equation can be solved with appropriate boundary conditions, which can be arrived at by considering the rate of change in the carrier concentration at each side of the active p-layer. The general solution of the continuity equation is given by

$$n(x) = A\exp\left(\frac{-x}{L}\right) + B\exp\left(\frac{x}{L}\right) \tag{7.3}$$

or in the p-active region of interest

$$n(x) = A\sinh\left(\frac{w-x}{L}\right) + B\cosh\left(\frac{w-x}{L}\right). \tag{7.4}$$

Here, L is the diffusion length $L = (D\tau)^{1/2}$, w is the active layer thickness, and the constants A and B can be found subject to the boundary conditions as described below.

The rate of change in carrier concentration at $x = 0$ is the difference between the injection rate and the interface recombination rate. The rate of change in carrier concentration at $x = w$ is the difference between the injection rate at $x = w$ and the interface recombination rate at $x = w$.

$$-\left.\frac{dn}{dx}\right|_{x=0} = \frac{J_{\text{diff}}(0)}{qD} - \frac{v_s n(0)}{D}, \quad \text{at } x = w, \tag{7.5}$$

and

$$-\left.\frac{dn}{dx}\right|_{x=w} = \frac{J_{\text{diff}}(w)}{qD} - \frac{v_s n(w)}{D}, \quad \text{at } x = w, \tag{7.6}$$

where q is the electronic charge, J_{diff} is the diffusion current density, and v_s (cm/s) is the interface recombination velocity. It is assumed that $J_{\text{diff}}(w)$ is negligible in the case when the p-layer is thicker than the diffusion length. One must keep in mind that the rate of change in the minority carrier is always negative.

The solution to the continuity equation subject to the above boundary conditions is

$$n(x) = \frac{J_{\text{diff}}(x=0)}{q}\sqrt{\frac{\tau}{D}}\left\{\frac{\cosh[(w-x)/L] + v_s\sqrt{(\tau/D)}\sinh[(w-x)/L]}{[v_s^2(\tau/D)+1]\sinh(w/L) + [2v_s\sqrt{(\tau/D)}\cosh(w/L)]}\right\}. \tag{7.7}$$

Here $J_{\text{diff}}(x=0)$ can be assumed to be the terminal current as the hole injection is negligible, given the very small intrinsic carrier concentration.

The average electron concentration in the active region can then be calculated from the integral:

$$n_{\text{ave}} = \frac{1}{w}\int_0^w n(x)\,dx = \frac{J\tau_{\text{eff}}}{qw}. \tag{7.8}$$

Substitution of the electron concentration (Equation 7.7) into Equation 7.8 leads to an effective carrier lifetime, which reduces to

$$\tau_{\text{eff}}^{-1} = \tau^{-1} + 2\frac{v_s}{w} = \tau_{\text{rad}}^{-1} + \tau_{\text{nrad}}^{-1} + 2\frac{v_s}{w}, \tag{7.9}$$

if $w/L < 1$, and $v_s^2(\tau/D) \ll 1$. In addition, in the absence of interface recombination, the effective lifetime would reduce to τ, which is related to radiative and nonradiative recombination times through $\tau^{-1} = \tau_{\text{rad}}^{-1} + \tau_{\text{nrad}}^{-1}$.

It would be instructive to note that in case an LED is illuminated as in photodiodes and photodetectors, the photo current would be a reverse current. The electron–hole pairs, preferably generated in the depletion region, would be swept by the built-in field causing the electrons to traverse to the n-region and holes to the p-region. In contrast, when the diode is forward biased the electrons are injected from the n-electrode and traverse toward the p-side while the holes take the opposite journey. Therefore, when photocurrent measurements are made to draw conclusions about electron overflow, it is prudent to recall that carrier flow is in opposite direction to that observed in LEDs.

7.3
Optical Output Power and Efficiency

Before we delve into optical power, it is imperative to define the visible light terminology and intricate details of light extraction from an LED. The photons generated in the active layer are emitted in all directions with only a fraction of them escaping the device to reach the human eye unless special precautions are taken. It is, therefore, imperative that photon extraction efficiency be increased. These include photons heading toward the surface anyway and the others that have many opportunities to do the same. This means that the entire structure with the exception of the active region where the photons are generated must be transparent to the emitted photon wavelength and a backside reflector must be in place. Furthermore, the phosphor conversion efficiency used for white light should be very high. We must also make sure that voltage charge product to photon energy ratio is as close to unity as possible that requires minimal series resistances and heterojunction discontinuities (the latter cannot be made infinitely small as discontinuities are needed for carrier confinement – typically $3kT$ is acceptable). One must also be cognizant of the photon down conversion energy loss from, for example, blue photons to yellow photons when blue photons are used to pump a yellow dye as is the case in phosphor conversion-based white light sources. Also to be considered is the eye response that peaks at 555 nm but drops for both red and blue. Below is a succinct discussion of the commonly used efficiency terms along with their determination.

7.3.1
Efficiency and Other LED Relevant Terms

Internal quantum efficiency: Ratio of the photons emitted from the active region to the number of minority carriers, that is, electrons or holes (in the case of undoped quantum wells either electrons or holes as attempts are made to inject equal quantities) *injected into* the LED p–n junction.

Extraction efficiency, χ: Ratio of photons emitted from the packaged chip into space to the photons generated in the chip. This includes the effect of photons reflected back into the chip due to index of refraction difference between air and the device surface, but excludes losses related to phosphor conversion.

External quantum efficiency: Ratio of extracted photons to the injected electrons, as such it is the product of the internal quantum efficiency, IQE, and the extraction efficiency χ.

Scattering efficiency: Ratio of the photons emitted from the LED to the number of photons emitted from the chip, which accounts for the scattering losses in the encapsulant of the lamp.

Electrical efficiency, η_v: Represents the conversion of electrical energy to photon energy and is defined by photon energy divided by forward voltage multiplied by electronic charge, qV_{appl} ($\%_{el}$). The forward voltage applied is determined by the diode characteristics and should be as low as possible for the photon-emitting medium, which ideally is the bandgap of the medium if radiative recombination is through conduction band electrons with holes in the valence band. Resistive losses and electrode injection barriers add to the forward voltage.

Wall plug efficiency (power efficiency): Optical power output divided by the electrical power provided to the device, irrespective of the spectrum of that output. The device electrical *efficacy* is the product of the wall plug efficiency and the spectral or optical efficacy. This is the most appropriate term in terms of figuring out the energy usage.

Color-mixing efficiency, η_{color}: Losses incurred during mixing the discrete colors for white light generation (not the spectral efficacy but just optical losses only).

Brightness: A subjective term used to describe the perception of the human eye, such as very dim, on the one hand, to blinding, on the other. The relationship between brightness and luminance is very nonlinear.

Luminance: The luminous intensity per unit area projected in a certain direction in SI units (cd/m^2).

Luminous efficiency: The power in photometric terms, measured in lumens per watt, divided by the electric power that generates it. In short it is lumens out divided by electric power in. To avoid confusion, *luminous efficacy* is used in the display field. In the LED literature one finds *luminous performance* for this term (lm/W).

Luminous flux: Power of visible light in photometric terms.

Luminous intensity: The luminous flux per solid angle emitted from a point. The unit is lumens per steradian, or candelas. This term is dependent on the package and the angle of measurement, and as such is not reliable.

The specified total luminous flux Φ of an LED is determined for photopic vision and can be calculated through the relation

$$\Phi = S(\lambda) K(\lambda) d\lambda, \tag{7.10}$$

where $S(\lambda)$ is the spectral power output of the LED and $K(\lambda)$ is the luminous efficacy of monochromatic radiation at wavelength λ. The luminous efficiency of light sources involves the efficiency of energy conversion from electrical power (W) to optical power (radiant flux in watts), followed by conversion by the eye sensitivity over the spectral distribution of light. The conversion by eye also accompanies a conversion of units, from radiant flux (W) to luminous flux (lm), and is called *luminous efficacy of radiation* having units of lumen per watt. The luminous efficacy of monochromatic radiation $K(\lambda)$ at wavelength λ is established and is defined by $K(\lambda) = K_m V(\lambda)$, where $K_m = 683$ lm/W (representing the peak at 555 nm) and is the *maximum luminous efficacy of radiation*, $V(\lambda)$ is the spectral luminous efficiency of photopic vision defined by the *Commission internationale de l'Eclairage* or *International Commission on Illumination* (CIE) and is the basis of photometric units. The values of $K(\lambda)$ are the theoretical limits of light source efficacy at each wavelength. For example, monochromatic light at 450 nm has a luminous efficacy of only 26 lm/W (theoretical limit). For real light sources, including LEDs, the overall luminous efficacy of radiation, K, is calculated from its spectral power distribution $S(\lambda)$ by

$$K = K_m \frac{\int_0^\infty S_\lambda(\lambda) V(\lambda) d\lambda}{\int_0^\infty S_\lambda(\lambda) d\lambda}. \tag{7.11}$$

The loss associated with the reflection at the semiconductor–air interface is called the Fresnel loss. When light passes from a medium of refractive index n_2, which is the active layer here, to a medium with the refractive index n_1, being air in this case, a portion of the radiation is reflected at the interface. Assuming a smooth interface, this loss is given in the case of normal incidence by

$$R = \left(\frac{n_2 - n_1}{n_2 + n_1}\right)^2. \tag{7.12}$$

The *Fresnel loss efficiency* can thus be defined as $\eta_F = (1 - R)$ representing the fraction of photons escaping the semiconductor surface. We should keep in mind that the surface can be roughened and/or epoxy dome can be placed to increase this efficiency.

Photons incident on the air–semiconductor interface from the semiconductor side at an angle larger than the *critical angle* θ_c suffer total internal reflection. This critical angle is determined, again assuming smooth interface, by Snell's law

$$\theta_c = \sin^{-1}\left(\frac{n_1}{n_2}\right). \tag{7.13}$$

The *critical loss efficiency*, therefore, can be expressed as $\eta_C = \sin^2 \theta_c$.

If the efficiency term associated with internal losses including interface recombination and self-absorption is denoted by η_A, then with what learned so far, the overall efficiency, would be $\eta_{opt} = \eta_F \eta_C \eta_A$.

For comparison and relevant knowledge, the total efficacy (lumens per electrical power, including ballast losses) of traditional light sources can be found elsewhere (Rea, 2000). Within a lamp type, the higher wattage sources are generally more efficient than the lower wattage sources, a property that holds for incandescent light bulbs as well. High-pressure sodium, metal halide, and fluorescent lamps are the most efficient white light sources outside of LEDs. Obviously, the spectral power distribution of white light-producing LEDs should be designed to have high luminous efficacy.

7.3.2
Optical Power and External Efficiency

The photon flux density can be approximated by a Gaussian function of the form

$$S(\lambda) = S_0 \exp\left[\frac{-4(\lambda - \lambda_0)^2}{(\Delta\lambda)^2}\right], \tag{7.14}$$

where S is the number of photons per unit time per unit volume with S_0 representing the same in the center of the spectrum. At a given point, x, in the active layer, $S_0 = \Delta n(x)/\tau_{rad} \approx n(x)/\tau_{rad}$, with τ_{rad} being the radiative lifetime. Recognizing that the photon energy equals $h\nu = hc/\lambda$ the power is given by

$$P = Ahc \int_0^\infty \frac{S(\lambda)}{\lambda} d\lambda, \tag{7.15}$$

where A is the cross-sectional area emitting photons.

With further manipulation and substitutions connecting the photon density to the carrier density in the form of $S(\lambda) = \tau_{rad}^{-1} \int_0^w n(x)\exp[-\alpha(\lambda)x]dx$, we obtain

$$P = Ahc \int_0^\infty \frac{d\lambda}{\lambda} \int_0^w \frac{n(x)\exp[-\alpha(\lambda)x]}{\tau_{rad}} dx, \tag{7.16}$$

with $\alpha(\lambda)$ being the absorption coefficient, which is of course a function of wavelength, λ.

Assuming wavelength-independent absorption coefficient, α_0, within the spectral width of emission, and narrow emission around a central wavelength λ_0, the power equation is reduced to

$$P_0 = \frac{Ahc}{\lambda_0 \tau_{rad}} \int_0^W n(x)\exp(-\alpha_0 x)dx = \frac{Ahc}{q\lambda_0 \tau_{rad}} J\tau_{eff}, \tag{7.17}$$

where h is Planck's constant, c is the speed of light, $n(x)$ is the electron concentration as function of position x, α_0 is the absorption constant at the central wavelength, A is the cross-sectional area of the active region, W is the active region width (thickness of the recombination region), q is the electron charge, J is the current density, τ_{rad} is the radiative recombination time constant, and $\tau_{\text{eff}} = (\tau_{\text{rad}}^{-1} + \tau_{\text{nrad}}^{-1})^{-1}$ where τ_{nrad} is the nonradiative recombination time constant. From here on we will refer to these time constants as τ_{rad} and τ_{nrad} for simplicity. Recognizing that hc/λ_0 represents the photon energy and if the photon energy is E_{ph} in terms of electronvolts, the internal quantum efficiency can be defined as

$$\eta_{\text{int}} = \frac{P_0/E_{\text{ph}}}{I/q}, \qquad (7.18)$$

where I is the current. Utilizing Equation 7.17 for the power, we obtain

$$\eta_{\text{int}} = \left(\frac{hc}{q\lambda_0}\right)\left(\frac{I}{\tau_r \tau_{\text{eff}}^{-1}}\right)\left(\frac{1}{IE_{\text{ph}}}\right) = \frac{\tau_{\text{eff}}}{\tau_r}. \qquad (7.19)$$

Part of the current may flow out of the intended recombination medium, the effect of which could be represented by an additional reduction τ_{eff} term or the overflow (spillover, leakage) current component. This can be treated separately as will be discussed shortly as part of the discussion of the internal quantum efficiency. Multiplying the internal quantum efficiency with the combined loss and efficiency factors, the EQE becomes

$$\eta_{\text{ext}} = \eta_{\text{opt}} \frac{\tau_{\text{eff}}}{\tau_r}, \qquad (7.20)$$

which is about 10% for ultraviolet (UV) and blue GaN-based diodes unless special precautions are taken to enhance this figure. The internal quantum efficiency, η_{int}, will be discussed in detail shortly. In the case of Ohmic losses, the term E_{ph} must be replaced by the energy corresponding to the applied voltage qV_{app}. Then, the EQE will assume the form

$$\eta_{\text{ext}} = \eta_{\text{opt}} \left(\frac{\tau_{\text{eff}}}{\tau_{\text{rad}}}\right)\left(\frac{E_{\text{ph}}}{qV_{\text{app}}}\right). \qquad (7.21)$$

The third term on the right-hand side represents the electrical conversion efficiency, or *electrical efficiency*.

7.3.3
Internal Quantum Efficiency

The internal quantum efficiency defines the fraction of electron–hole pairs that are converted into band edge photons within the active region. Some are lost to nonradiative recombination processes. Although many defect states can capture carriers nonradiatively, when the energy of that defect falls nearly at the middle of the bandgap the process is called the Shockley–Read–Hall process or *SRH generation/recombination process*. The other form of nonradiative recombination is the Auger process in which the energy given off during a band-to-band recombination is used to excite another carrier to a higher energy, which in turn is thermalized by phonon emission. Another process, which seems more dominant and is not counted against the internal quantum efficiency, is the escape of carriers from the active region with ensuing recombination outside of the active region either radiatively, in which case at an undesired wavelength, or nonradiatively, or collected by the contact. This is referred to as the leakage or carrier spillover.

7.3.3.1 Auger Recombination

The *Auger* process has been recognized for many decades and theoretically treated by Landsberg and colleagues. Because Auger recombination encompasses carrier recombination across the band and also carrier excitation to higher energies, it becomes more important at high carrier densities and in materials with relatively small bandgaps. Below a brief discussion of Auger recombination is given and how it is quantified for instructional purposes at the very least. In brief and focusing on electron initiated variety, in Auger recombination the energy given off when an electron in the conduction band recombines with a hole in the valence bands is used to excite another electron to a higher energy level in the conduction band instead of radiation. The process naturally requires that both the energy and momentum be conserved, necessitating indirect transitions to occur. The process can also involve phonon participation with more complicated routes. For example, if the process follows the recombination of an electron excitation to a higher energy, the process goes by the depiction of CCCH and so on.

The dominant temperature and energy dependence of the Auger recombination-specific carrier lifetime (the inverse of Auger recombination rate) to a first extent is given by a modified expression of Beattie and Landsberg by

$$\tau \propto \left[\frac{E_g(T)}{kT}\right]^{3/2} \exp\left[\frac{1+2M}{1+M}\frac{E_g(T)}{kT}\right] \tag{7.22}$$

for a nondegenerate and intrinsic semiconductor, where k is the Boltzmann constant. The terms $E_g(T)$ and M represent the temperature (T)-dependent bandgap and electron-to-hole mass ratio, respectively. A disparately smaller electron mass, as in GaN, would ensure the lifetime is determined by electron–electron collisions, namely, electron recombination with a hole followed by another electron excitation to a higher energy or electron excitation from the valence band to the conduction band coupled with hot electron relaxation to near the bottom of the conduction band, with the entire process conserving energy and momentum. In the case of electron excitation to a higher level in the conduction band, the Auger recombination rate can be expressed as

$$\begin{aligned} U_{\text{Auger}} &= C_n[n(pn - n_i^2)] \\ &= C_n(n_0 + \Delta n)[((p_0 + \Delta p)(n_0 + \Delta n) - n_i^2)], \end{aligned} \tag{7.23}$$

where C_n represents the Auger recombination coefficient for electrons, p is the hole concentration, n_0 and p_0 represent the equilibrium, and Δn and Δp the excess carrier concentrations. Recognizing that $p_0 \ll n_0$ and $n_i^2 \ll (p_0 + \Delta p)(n_0 + \Delta n)$ for GaN, which is normally n-type and the intrinsic carrier concentration n_i is very small, and

$\Delta p = \Delta n$, Equation 7.23 can be rewritten as

$$U_{\text{Auger}} = C_n \Delta p (n_0 + \Delta n)^2 = C_n \Delta n (n_0 + \Delta n)^2. \tag{7.24}$$

For high injection levels, the excess carrier concentration dominates over the equilibrium electron concentration making the latter negligible and the Auger rate can be written as

$$C_n (\Delta n)^3 \approx C_n n^3. \tag{7.25}$$

In the case of hole excited to a higher energy in the valence band, the Auger recombination rate is given by

$$U_{\text{Auger}} = C_p \left[p(pn - n_i^2) \right], \tag{7.26}$$

where C_p represents the Auger recombination coefficient for holes. If both of electron and hole initiated Auger processes take place, then the sum of Equations 7.23 and 7.26 must be used.

Typically Auger recombination rate is relatively higher or the associated lifetime is shorter for relatively small bandgap materials as compared to wide-bandgap semiconductors. By the same token, Auger recombination is more important at high temperatures. To a first order, the traps do not participate in the process, although their taking part could be important in some cases. In large bandgap semiconductors, the Auger process would depend on the doping level and become important in degenerate cases. At high doping concentrations, the electron wavefunction of adjacent impurities would overlap and delocalize the electrons (or holes), which increase the likelihood of Auger recombination. To get a relative appreciation for Auger recombination in a wide-bandgap material such as GaN, the Auger recombination lifetime is some 50 times longer than it is in GaAs. Shown in Figure 7.5 is a collage of Auger recombination lifetime data reported for Si, Ge, and various III–V semiconductors including InGaN. The overall trend of reduction in lifetime with increasing bandgap is noted with InGaN data deduced from lasers (tend to be more reliable) and LEDs, which do not yet uncover the bandgap dependence and calls for further investigations.

7.3.3.2 SRH Recombination

The generation–recombination process described by Read and Sah et al., which is now commonly referred to as the *SRH generation/recombination process*, entails capture and release of free carriers by deep defect centers. The rate of electron capture is proportional to the product of electron concentration in the conduction band and the concentration of traps not occupied by electrons. Unlike the Auger effect, SRH recombination is more noticeable at low injection levels. The recombination rate for SRH in a p-type semiconductor is given by

$$U = \frac{\sigma v_{\text{th}} N_t (pn - n_i^2)}{n + p + 2n_i \cosh[(E_t - E_i)/kT]} = \frac{1}{\tau_{\text{nr}}} \frac{(pn - n_i^2)}{n + p + 2n_i \cosh[(E_t - E_i)/kT]}, \tag{7.27}$$

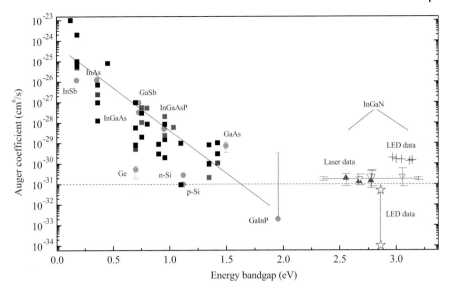

Figure 7.5 Reported Auger recombination lifetime versus bandgap for various semiconductors including conventional III–Vs and InGaN (deduced from LEDs and lasers, the latter being more reliable. The InGaN data do not yet follow bandgap dependence. In part courtesy of Drs. J. Piprek and A. Hangleiter.

with the nonradiative rate or the inverse of nonradiative recombination time due to this event $\tau_{nr}^{-1} = \sigma v_{th} N_t$, which depicts the average lifetime of the minority carriers, where N_t is the trap concentration, σ is the capture cross section of the trap, and v_{th} is the thermal velocity. If we further assume that the defect centers are at or near the intrinsic level (i.e., $E_t = E_i$), Equation 7.27 would reduce to

$$U = \frac{1}{\tau_{nr}} \frac{(pn - n_i^2)}{n + p + 2n_i}. \tag{7.28}$$

If we now express the carrier concentrations in terms of the equilibrium and excess carrier concentrations, Equation 7.28 can be rewritten as

$$U = \frac{1}{\tau_{nr}} \frac{[(p_0 + \Delta p)(n_0 + \Delta n) - n_i^2]}{p_0 + \Delta p + n_0 + \Delta n + 2n_i}. \tag{7.29}$$

Recognizing that $p_0 \ll n_0$ and $n_i^2 \ll (p_0 + \Delta p)(n_0 + \Delta n)$ for GaN, which is normally n-type, the intrinsic carrier concentration n_i is very small, and $\Delta p = \Delta n$ Equation 7.28 reduces to

$$U = \frac{1}{\tau_{nr}} \frac{\Delta n(n_0 + \Delta n)}{2\Delta n + n_0}. \tag{7.30}$$

For high injection levels and/or relatively low equilibrium electron concentrations $\Delta n > n_0$ Equation 7.29 becomes

$$U = \frac{\Delta n}{2\tau_{nr}} = A\Delta n \quad \text{with} \quad A = (2\tau_{nr})^{-1}. \tag{7.31}$$

7.3.3.3 Radiative Recombination

Following the treatment of the balance equation by van Roosbroeck and Shockley, the equilibrium emission intensity $\phi(E)$ at a photon energy E and under thermal equilibrium for a bulk semiconductor is given by

$$\phi(E) = \frac{8\pi n_R^2 E^2 \alpha(E)}{h^3 c^2} \frac{1}{e^{E/kT} - 1}, \quad (7.32)$$

where $\alpha(E)$ is absorption coefficient and expressed for a direct bandgap semiconductor as

$$\alpha(E) = A^*(E - E_g)^{1/2}, \quad (7.33)$$

where $A^* \approx q^2 (2m_r^*)^{3/2}/(n_R c h^2 m_n^*)$. m_r^* is the reduced mass given by $1/m_r^* = 1/m_n^* + 1/m_p^*$ in terms of the electron and hole effective masses, m_n^* and m_p^*, respectively, and n_R, represents the refractive index (the energy dependent one can take the value at the band edge as a first-order approximation).

The generation rate per unit volume is the integral of emission intensity at a given energy over all possible energies:

$$G = \int_0^\infty \phi(E) dE = \int_0^\infty \frac{8\pi n_R^2 E^2 \alpha(E)}{h^3 c^2} \frac{dE}{e^{E/kT} - 1}$$

$$= A^* \int_0^\infty \frac{8\pi n_R^2 E^2 (E - E_g)^{1/2}}{h^3 c^2} \frac{dE}{e^{E/kT} - 1}. \quad (7.34)$$

With external excitation, the total generation rate is given by the material relationship

$$R = \frac{G n p}{n_0 p_0} \equiv B n p, \quad (7.35)$$

where $B \equiv G/(n_0 p_0)$. For low excitation levels in a p-type semiconductor $p \approx p_0$ and the radiative lifetime is given by

$$\tau_r = \frac{n - n_0}{R} = \frac{n - n_0}{G/(n_0 p_0) np} \approx \frac{1}{Bp} \approx \frac{1}{Bp_0},$$

or $\frac{dn}{dt} = Bp_0 \Delta n.$ (7.36)

As mentioned already, B is the radiative recombination coefficient and varies between 10^{-9} and 10^{-11} cm^3/s for direct and 10^{-13} and 10^{-15} cm^3/s for indirect bandgap semiconductors. Similar to the p-type semiconductor for low excitation levels in an n-type semiconductor $n \approx n_0$ and the radiative lifetime is given by

$$\tau_r = \frac{p - p_0}{R} = \frac{p - p_0}{G/(n_0 p_0) np} \approx \frac{1}{Bn} \approx \frac{1}{Bn_0},$$

or $\frac{dp}{dt} = Bn_0 \Delta p.$ (7.37)

We can surmise from the above equations that under low-level injection, the radiative lifetime of minority carriers is inversely proportional to the majority carrier concentration. Under high excitation levels, the rate of radiative recombination is given by (using Equation 7.35 and $n = n_0 + \Delta n$, $p = p_0 + \Delta p$, $\Delta n = \Delta p \gg n_0, p_0$)

$$R = B\Delta n^2 \approx Bn^2 \tag{7.38}$$

At these high injection levels, the radiative intensity depends on the product of excess electron and hole concentration or the square of the excess electron concentration assuming the radiative recombination coefficient remains the same. The quadratic dependence is a result of diminishing radiative lifetime with increasing injection level.

7.3.3.4 Continuity or Rate Equations as Pertained to Efficiency

As a preamble, be forewarned that LED structures and the particulars of the InGaN-active layer are too complex for all the well-established and overly simplified treatments, such as the use of A, B, and C coefficients. The layer quality issues and active layer integrity under injection make a compelling case against the use of a constant A coefficient. Similarly, active layer design and the any carrier confinement, either due to localization or by design of the active layer, would cause the B coefficient to be injection dependent in addition to its temperature dependence. As long as the A and B coefficients are injection dependent, one cannot pull out a reliable C coefficient from polynomial fits. Having made the aforementioned case, it is still very valuable to treat the problem with constant coefficients in order to gain some level of appreciation of the processes, *albeit* limited, taking place and develop a sense for the overall problem. This must, however, be complemented with a discussion as the scope and nature of changing A, B and C, coefficients.

In the case of high level of injection, the electron and hole concentrations, n and p, are nearly equal to each other and both are much larger than n_0 and thus $n = \Delta n + n_0 \approx \Delta n$. Consequently, the Auger recombination rate would be proportional to n^3. For comparison, the band-to-band radiative recombination rate $[U_{b\text{-}b} = B(pn - n_i^2)]$ at high injection levels would be proportional to Bn^2, and the recombination by traps, SRH recombination, $[U_{SRH} \approx \tau_n^{-1}(n_p - n_{p0})]$ would be proportional to n. In simple terms, the total rate of recombination at high injection levels due to deep levels, band-to-band radiative recombination and nonradiative Auger recombination combined can be expressed as (assuming that excess carrier concentrations dominate $n = \Delta n + n_0 \approx \Delta n$)

$$U = -\frac{dn}{dt} = B_{dl}n + B_{b\text{-}b}n^2 + B_A n^3 = An + Bn^2 + Cn^3, \tag{7.39}$$

where $B_{dl}(\equiv A)$, $B_{b\text{-}b}(\equiv B)$, and $B_A(\equiv C)$ represent the deep level recombination, band-to-band recombination (B in Equation 7.35), and Auger recombination (C in Equation 7.25) coefficients, respectively, in the units of s^{-1}, cm^3/s, and cm^6/s, respectively. If there is generation, for example, by photoexcitation or carrier injection by current, then a generation term G or injection rate term $\nabla J/(q)$ (with J being the current density in the recombination region) or both depending on the situation must be added to the right-hand side of Equation 7.39 which leads to

$$-\frac{dn}{dt} = B_{dl}n + B_{b-b}n^2 + B_A n^3 - \left(G \text{ and or } \frac{\nabla J}{q}\right). \quad (7.40)$$

The implicit assumption is that all the carriers that are injected by current density J recombine either radiatively or nonradiatively in the active region. We should point out that in many structures some of the carriers spillover the barrier and recombine in regions of the semiconductor of no interest to the user. In that case the term $\nabla J/q$ would have to be modified. Carrier spillover and its effect on the efficiency versus current characteristic of the LED will be discussed shortly.

It is very instructive to rewrite Equation 7.39 in terms of the lifetimes associated with each of the three processes

$$-\frac{dn}{dt} = \frac{n}{\tau_{nr}} + \frac{n}{\tau_r} + \frac{n}{\tau_{Au}} \quad \text{with} \quad \tau_{nr} = \frac{1}{A}, \quad \tau_r = \frac{1}{Bn}, \quad \tau_{Au} = \frac{1}{Cn^2}, \quad (7.41)$$

where $\tau_{nr}, \tau_n, \tau_{Au}$, represent the nonradiative, radiative, and Auger recombination lifetimes. While on the topic the internal quantum efficiency, assuming no carrier overflow, is defined as

$$\eta_{int} = \frac{\tau_{eff}}{\tau_r} = \frac{Bn^2}{An + Bn^2 + Cn^3} \quad \text{with} \quad \tau_{eff}^{-1} = \tau_{nr}^{-1} + \tau_r^{-1} + \tau_{Au}^{-1}. \quad (7.42)$$

It is extremely important to note that A, B, and C coefficients in Equation 7.39 should not be taken as constants when the injection level is changed not to mention that the same equation is good only under the various assumptions leading to its derivation, as outlined above, and in particular the assumption that injected carrier concentration is larger than the equilibrium concentration in the material. Beginning with the A coefficient, it can change due to, for example, defect generation and/or increased access to existing defects with increased injection.

As for the B coefficient, it is a reflection of the likelihood of radiative recombination of an electron and a hole. This probability depends on the band structure of the material and also on the spatial separation of electrons and holes. Furthermore, at very low injection levels ($\Delta n \ll n_0$) and in materials with sizeable unintentional background concentration, as in the case of InGaN, the rate of radiative recombination would follow a linear dependence on injection ($\tau_r = [B(n_0 + \Delta n)\Delta p]^{-1}$). With increasing injection the electron hole spatial separation becomes smaller and the radiative recombination lifetime becomes inversely proportional to injection level as represented with $\tau_r = (Bn)^{-1}$. However, at much higher injection levels where the already small electron hole spatial separation does not increase the radiative recombination likelihood by much, the slope of the recombination rate with respect to n begins to get smaller.

In summary, with constant A and B coefficients the slope of the light output versus current optical excitation power (depending on the nature of carrier injection) would initially be about 2 in log scale when the nonradiative recombination is dominant, which gives way to 1 as the radiative processes become dominant. The shape and nature of the change in the slope depends on the background concentration and the values of A and B coefficients. The picture becomes more complex if and when A and B coefficient are injection dependent and Auger and/or Auger-like processes, n^3 dependence, are present. The A coefficient can change with injection due to defect generation and localization-induced problems. The B coefficient can change due to Stark effect and associated screening, and also limitation of total available states. These points will shortly be discussed in some detail below.

The transition from the low to high injection regime naturally depends on the unintentional background carrier concentration in the active region. When nonradiative centers are present, that is, $A \neq 0$, the overall slope of the emission versus excitation would be affected in that steeper slope during low and intermediate excitation levels may be observed. One can appreciate the complexity of the problem when injection dependence of the A and B coefficients is considered. Not only new defect generation due to injection could be the source of the changing A coefficient, but also carrier localization would be responsible in that the manifold to other defects that are already present in the material could become available as the injection level is increased. One can conclude that the problem is much more complex and that the assumption of the constancy of A and B coefficients as well as the applicability of Equation 7.39 regardless of the injection level could be misleading. This being the case, it is not always prudent to simply fit a polynomial with constant A, B, and C coefficients to the light output dependence on injection with the ensuing extraction of constant A, B, and C coefficients and use them as the basis to draw far-reaching fundamental conclusions about the nature of optical processes. For example, ignoring any contribution of Auger or Auger-like processes with n^3 dependence, the emission intensity can be calculated with respect to injected carrier concentration or laser excitation power in the case of optical excitation. Initially when the injected minority concentration is much smaller than the background concentration, one would expect a linear slope versus n of the emission intensity. With increasing injection the radiative recombination lifetime gets shorter, which causes a square dependence on the injected carrier density. This reduction in radiative recombination lifetime should eventually cease, which should return the dependence to a linear one. The above discussion assumes no degradation and ignores carrier leakage, thermal issues, and any Auger-like processes.

Considering the emission dependence on laser excitation power, in case of optical excitation, requires manipulation of the rate equation (because $\Delta p = \Delta n$, Δn is chosen here to represent the minority carrier concentration), Equation 7.35.

Consider the steady-state rate equation with optical excitation and negligible C coefficient

$$A\Delta n + B(n_0 + \Delta n)\Delta n - G = 0, \tag{7.43}$$

which when solved leads to

$$\Delta n = \frac{-A - Bn_0 + \sqrt{A^2 + 2ABn_0 + B^2 n_0^2 + 4GB}}{2B}. \tag{7.44}$$

The optical emission in a resonant photoluminescence (PL) experiment can be expressed as

$$I_{PL} = \eta_{co} B(n_0 + \Delta n)\Delta n, \qquad (7.45)$$

with η_{co} depicting the collection efficiency of photons in the PL experiment. With the aid of Equation 7.44, this can be rewritten as

$$I_{PL} = \frac{\eta_{co}A^2 + 2BG\eta_{co} + ABn_0\eta_{co} - \eta_{co}A\sqrt{A^2 + 2n_0 AB + n_0^2 B^2 + 4GB}}{2B}. \qquad (7.46)$$

Equation 7.46 allows us to plot the light emission intensity versus the generation rate, or the optical excitation density ($G \propto P_{exc}$) for a given set of A and B coefficients and background electron concentration as shown in Figures 7.6 and 7.7. Equation 7.46 can be solved with the approximation of $R_{SRH} = A\Delta n$ or with the full SRH term $R_{SRH} = A(\Delta n n_0 + \Delta n^2)/(n_0 + 2\Delta n)$ in the rate equation. There are very minor variations between the full and approximated solutions. Therefore, for illustration purposes the approximate expression can be used. For the calculations, resonant excitation at 3.22 eV, a coupling efficiency of $\eta_{co} = 1$, an active region absorption coefficient of $\alpha = 10^5$ cm^{-1}, and a front surface reflection of $R = 12\%$ have been assumed.

To illustrate the roles of A and B coefficients, the output power versus the excitation power is shown in Figure 7.8 wherein the contribution due to radiative recombination is indicated as positive, whereas the effect of the nonradiative recombination characterized by the A coefficient is indicated as negative or loss. Subtraction of the negative term from the positive one gives a measure of the light output. Clearly the slope of light emission with respect to the excitation power is dependent on the value of the A coefficient used, with slope of 2 indicating sizeable effect of nonradiative recombination to the process. For a very small A coefficient, the data are dominated by radiative recombination with a slope of nearly 1 for the injection levels considered.

A treatment similar to the optical excitation can be undertaken for electrical injection as well. In this case we again begin with the rate equation given by Equation 7.40, but for the case of steady state, ignoring the third power term, assuming linear minority carrier distribution in the active layer, and no carrier spillover,

$$\frac{dn}{dt} = An + Bn^2 - \frac{\nabla J}{q} = 0 \quad \text{and} \quad \nabla J = \frac{J}{d}, \qquad (7.47)$$

with d is the active region thickness, q is the electron charge, J is the current density, and I_{EL} is the electroluminescence (EL) intensity.

The electron concentration can be deduced from Equation 7.47 as

$$\Delta n = \frac{-Aqd + \sqrt{A^2 d^2 q^2 + 2ABn_0 d^2 q^2 + B^2 d^2 n_0^2 q^2 + 4JBqd} - Bqdn_0}{2Bqd}. \qquad (7.48)$$

7.3 Optical Output Power and Efficiency

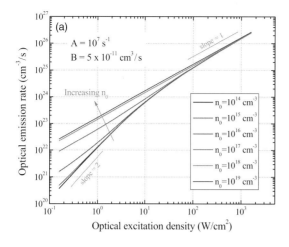

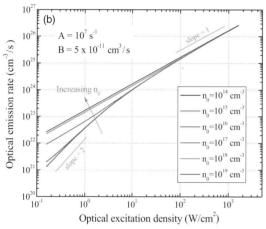

Figure 7.6 Calculated optical emission rate versus the excitation power for a fixed set of A and B coefficients with background electron concentration varied from 10^{14} to 10^{19} cm^{-3} to represent the cases with virtually undoped active layer, 10^{15} cm^{-3}, to what is expected in practical device structures, 10^{18} cm^{-3} and beyond. Note that for very high injection levels ($\Delta n \gg n_0$) the optical emission rate increases linearly with increasing excitation power, whereas for very low levels the slope is close to 2, and all curves converge, (a) with $R_{SRH} = A\Delta n$ approximation and (b) with full $R_{SRH} = A(\Delta n n_0 + \Delta n^2)/(n_0 + 2\Delta n)$. The difference between the two sets is sufficiently small.

Recalling that $I_{EL} = \eta_{co} B n^2$, we can isolate and plot I_{EL} as a function of current density, J.

$$I_{EL} = \eta_{co} B \Delta n (n_0 + \Delta n). \tag{7.49}$$

An example in the form of calculated electroluminescence intensity versus the injection current for a fixed B coefficient and a set of A coefficients is illustrated in Figure 7.9.

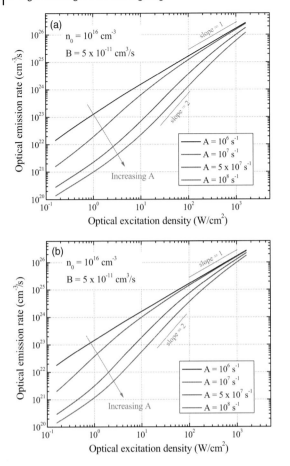

Figure 7.7 Calculated optical emission rate versus the excitation power for a fixed set of background electron concentration and B coefficient with varying A coefficient. The point is that value of the A coefficient has an effect on the slope of light output versus the excitation power in the relatively large. The point is that value of the A coefficient has an effect on the slope of light output versus the excitation power: the relatively large A coefficients give rise to larger slopes at medium excitation levels, which should not be confused with high quality: (a) with $R_{SRH} = A\Delta n$ approximation and (b) with full $R_{SRH} = A(\Delta n n_0 + \Delta n^2)/(n_0 + 2\Delta n)$. The difference between the two sets is sufficiently small.

A common procedure for determining IQE makes use of the ratio of the PL intensity (resonant variety is much better as it illuminates only the active regions) at room temperature to that obtained at low temperature with the inherent assumption of 100% IQE at low temperature, which is not always correct. In fact, in many cases the low-temperature quantum efficiency deviates from unity considerably. A more dependable approach relies on excitation-dependent PL measurements and plotting the optical power versus the excitation power. If the efficiency were unity, the slope would be 1 for all excitation levels within reason. A larger slope would indicate

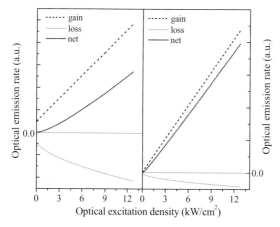

Figure 7.8 Optical emission rate due to B coefficient and loss to nonradiative recombination due to A coefficient and the difference between the two which represent the expected light output for a set of A and B coefficients and a background electron concentration.

participation of nonradiative centers. This concept can be implemented by performing optical power measurements at, for example, 10 and 295 K using 385 nm excitation from a frequency-doubled Ti:Sapphire laser ensuring photogeneration of carriers only in the LED-active regions. The highest excitation density used in the particular experiment under discussion corresponds to an average carrier concentration of $\sim 10^{18}\,\text{cm}^{-3}$ in a single double heterostructure (DH) LED structure. As the collected PL intensity is proportional to excitation intensity, $L_{PL} \propto I_{exc}^{m}$, the linear dependence ($m \approx 1$) DH structures with 1, 2, 4, 6, and 8 active regions at 10 K (see Figure 7.10a) indicate that the radiative

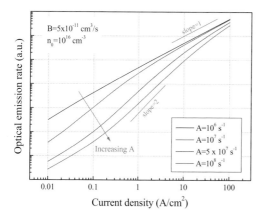

Figure 7.9 Calculated optical emission rate versus the injection current density with a fixed B coefficient and the A parameter as a variable assuming no carrier spillover. Note the similarity to Figure 7.7 which depicts the same but versus excitation laser power.

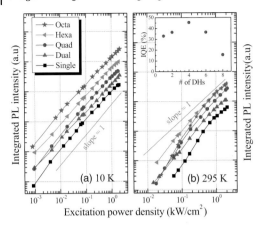

Figure 7.10 Integrated PL intensity as a function of excitation power density (a) at 10 K and (b) at 295 K for LED structures with 1, 2, 4, 6, and 83 nm-thick $In_{0.15}Ga_{0.85}N$ active layers separated by 3 nm $In_{0.06}Ga_{0.94}N$ barriers and emitting at ∼425 nm. The inset of (b) displays the PL-IQE versus the number of 3 nm DHs in the active region. The gray-solid lines indicate slope of 1. Scaling of integrated intensity, particularly at low temperature, with the number of active layers (active layer volume) is indicative of reasonably uniform resonant excitation of all the active layers. Lagging of intensity at room temperature with respect to the number of active layers at room temperature is indicative of somewhat increased contribution of nonradiative centers when the volume is increased, which should be treated as being specific to these particular layers.

recombination is dominant with an associated efficiency of unity except for very low excitation power levels. The consistent increase in optical power detected with increasing number of active regions (increased pumped volume) versus the optical excitation is indicative of nearly uniform excitation.

The room temperature data (Figure 7.10b), however, show a super-linear dependence ($m ≈ 1.4$–1.95) at low excitation, which is attributed to the sizeable effect of nonradiative recombination ($m = 2$ in case of injection independent τ_{nonRad}). As the excitation density is increased, the slope gradually approaches to $m = 1$ ($I_{PL} \propto I_{exc}$). The gradually decreasing slope in the intermediate excitation regime indicates strong competition between the nonradiative and radiative processes and can be attributed to a decreasing radiative lifetime at moderate injections. It is plausible that with increasing excitation saturation of localized states and ensuing delocalization of carriers (particularly holes, as the electron density in the wells is in mid-10^{17} cm^{-3}) may allow access to additional nonradiative centers and result in enhanced recombination rate with respect to low excitation. Furthermore, Coulomb screening of the quantum-confined Stark effect (QCSE) with increasing excitation leads to an increased interband recombination rate of $1/\tau_{Rad}(N) \propto BN$, where B is the bimolecular recombination coefficient and N is injected carrier density. Despite its shortcomings, the IQE values deduced from comparison of PL intensities at 10 and 295 K at the highest excitation density employed are shown in the inset of Figure 7.10b.

7.3.3.5 Carrier Overflow (Spillover, Flyover, Leakage)

At the onset with no *a-priori*-predilection, the electron overflow might have its origin either in (i) thermionic emission of equilibrium electrons from the bottom of the active region over the barrier into the p-layer (the large band discontinuities preclude it), or (ii) ballistic and quasiballistic transport of the injected electrons aided by field. The thermally activated one is in which it is assumed that the onset of carrier spillover begins when the Fermi level in the active region rises all the way up to the conduction band edge of the barrier layer, which is negligible in the nitride system due to the large barriers involved. In terms of analytical treatment, one notes that the electron concentration in a degenerate semiconductor is given in terms of the Fermi level in reference to the conduction band edge and density of states by

$$n \approx \frac{4}{3\sqrt{\pi}} N_C \left(\frac{E_F - E_C}{kT} \right)^{3/2}. \tag{7.50}$$

If now one assumes that the carrier spillover in noticeable quantity would occur when the Fermi level aligns with the conduction band edge of the barrier layer and that the said barrier is 0.2 eV, the electron concentration when the spillover occurs is calculated to be 4.2×10^{19} cm^{-3} in GaN-active region and 3.5×10^{19} cm^{-3} in In$_{0.2}$Ga$_{0.8}$N-active regions. In GaN and In$_{0.2}$Ga$_{0.8}$N, this electron concentration is reached at an injection current density of approximately 5.6 and 4 kA/cm^2, respectively, if one further assumes an active region thickness of 10 nm. Obviously, this current level is highly dependent on the barrier height, as shown in Figure 7.11. Any small increase in the barrier height increases this current considerably to levels that are practically out of reach. We should keep in mind that the presence of carriers and scattering processes that might take place in the quantum well (QW) will have a bearing on at what current level the carrier spillover begins at noticeable levels.

As a way to segue into the calculation of hot electron-induced electron overflow, Figure 7.12 displays the experimental relative EQE as a function of the current

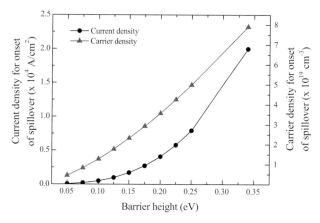

Figure 7.11 Carrier density calculated from Equation 7.50 and the corresponding current levels for barrier heights in the range 0.05–0.35 eV in In$_{0.2}$Ga$_{0.8}$N for the onset of carrier spillover.

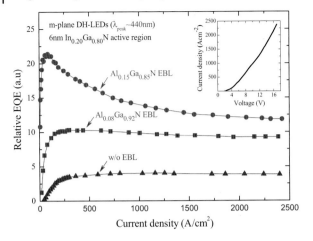

Figure 7.12 Relative EQE of m-plane In$_{0.20}$Ga$_{0.80}$N LEDs with Al$_x$Ga$_{1-x}$N EBLs having Al compositions $x = 15\%$, 8%, and 0%, measured under pulsed current, 1 μs pulse width and 0.1% duty cycle. Emission wavelength is $\lambda_{peak} \sim 440$ nm. The inset shows the current–voltage dependence for the LED with 15% Al in EBL.

density for three m-plane (nonpolar) LEDs with varying electron blocking layer (EBL) barrier heights. Because the IQE is independent of the composition of the EBL barrier height (confirmed by resonant PL experiments), the observed difference in the relative EQE can safely be ascribed to carrier injection and overflow. The relative EQE for the m-plane LED with 15% Al in EBL shows a pronounced peak at approximately 80 A/cm^2, which drops by ~45% with increasing current. A negligible efficiency degradation (~3–5%) is observed for the structure without EBL (represented by 0% Al in the EBL), albeit its relative EQE is approximately three to five times lower than that with 15% Al EBL. Intermediate values for EQE and efficiency degradation (~10%) are obtained for the LED with 8% Al within the EBL.

It is assumed that the LED with 15% Al EBL has negligible electron overflow at the current density of 80 A/cm^2 where the EQE peaks. Thus, in the LED without EBL less than one-fifth of the overall current is contributing to the EL and approximately four-fifths of the current passes the active region without contributing to light emission. After refuting the thermionic emission route, in our opinion, we focus our attention on hot electron transport as the cause for carrier overflow.

In terms of numerical treatment, in addition to one's own, commercial products such as Silvaco Atlas software are available. One such simulation for a p-GaN/In$_{0.20}$Ga$_{0.80}$N/n-GaN LED structure without an EBL takes into account thermionic emission of equilibrium electrons and tunneling within the Wentzel–Kramers–Brillouin approximation. Using the commonly accepted parameters for an In$_{0.20}$Ga$_{0.80}$N-active layer, such as $1 \times 10^7 \, s^{-1}$ for the SRH recombination coefficient of $1 \times 10^7 \, s^{-1}$, and $1 \times 10^{-11} \, cm^3/s$ for the radiative recombination coefficient, $1 \times 10^{-30} \, cm^6/s$ for the Auger recombination coefficient, and $\Delta E_c = 0.5$ eV (representing 70% of ΔE), the simulations show that even at an uncharacteristically elevated junction temperature of 1000 K and at an unreasonably high current density of $1 \times 10^4 \, \text{A/cm}^2$, the thermionic emission-driven overflow electron current into the p-GaN region due to electrons in thermal equilibrium with the lattice is only ~11% of the total current density. This means that the hot electron process is a dominant one and should be pursued.

Available experimental data indicate that the optical power and/or the efficiency degrade without introducing a barrier layer, which acts to prevent electron overflow. The holes are relatively heavy and thus the process is driven by electrons. This in and of itself is indicative of the fact that electron overflow does indeed takes place, but calculations indicate this not to be a thermally driven one which paves the way for consideration of hot electrons.

In the hot electron scenario, electrons injected into the active region gain additional kinetic energy equal to the conduction band offset between the n-GaN and the $\text{In}_{0.20}\text{Ga}_{0.80}\text{N}$-active layer ($\Delta E_c, \sim 0.5 \, \text{eV}$). The hot electrons either undergo thermalization by losing their excess energy (mainly through their interaction with LO-phonons) or leave the active region. Let us consider the electrons that experience ballistic motion (no scattering in the active region) and quasiballistic motion (scattering events with LO phonons). Calculations show that LEDs without EBL or electron cooler prior to injection suffer from serious overflow accounting for some ~67% of the total electron current under flat-band conditions for a 6 nm-thick active region. Because only 1% of the injected electrons enter the p-GaN layer after two scattering events, this process does not need to be included.

The probability of ballistic transport for a particular energy follows an $\exp(-t/\tau_{sc})$ dependence where t is the transit time, and τ_{sc} is the electron–LO-phonon scattering time, which is given by $\tau_{sc} = 1/(1/\tau_{abs} + 1/\tau_{em})$ with τ_{abs} and τ_{em} representing the LO-phonon absorption and emission times, respectively. If the acceleration and deceleration of the electrons in the active region are neglected, the percentage of the overflow electrons can be expressed as

$$P_1 = \frac{\int_{\max\{0,(\phi_{EBL}-qV)\}}^{+\infty} f(E) N(E) \exp\left(\frac{-L/v(E+\Delta E_c)}{\tau_{sc}}\right) dE}{\int_0^{+\infty} f(E) N(E) dE}, \qquad (7.51)$$

where E depicts the excess electron energy with respect to the bottom of the conduction band of the n-GaN layer, L is the active region thickness (6 nm in our case), V is the net potential drop across the InGaN region, ϕ_{EBL} is the barrier height of EBL (i.e., the conduction band offset between the EBL and the p-GaN which is 0 when EBL is absent), $N(E)$ is the conduction band density of states, $f(E)$ is the Fermi–Dirac distribution function, $v(E + \Delta E_c) = \sqrt{2(E + 0.5 \, \text{eV})/m_e}$ is the electron velocity with an initial energy of E plus the gained energy due to the 0.5 eV band offset, and m_e is the electron effective mass. For a first-order approximation the electrons can be assumed to traverse only in the direction normal to the heterointerface, which turns out to insignificantly overestimate the spillover.

The percentile of the overflow electrons following one scattering event is given by

$$P_2 = \left[\int_0^{+\infty} f(E)N(E)\mathrm{d}E\right]^{-1} \int_{\max\{0,(\phi_{EBL}-qV\mp\hbar\omega_{LO})\}}^{+\infty} f(E)N(E) \int_0^L \frac{1}{v(E+\Delta E_c)\cdot \tau_{ph}}$$
$$\times \exp\left(-\frac{x/v(E+\Delta E_c)}{\tau_{sc}}\right) \cdot \exp\left[-\frac{(L-x)/v(E+\Delta E_c \mp \hbar\omega_{LO})}{\tau_{sc}}\right] \cdot \mathrm{d}x \cdot \mathrm{d}E.$$

(7.52)

where $+\hbar\omega_{LO}$ and $-\hbar\omega_{LO}$ in the lower integration limit correspond to phonon emission ($\tau_{ph} = \tau_{em}$) and absorption ($\tau_{ph} = \tau_{abs}$), and $\hbar\omega_{LO} = 88$ meV is the LO phonon energy in InGaN. Equation 7.52 takes into account energy conservation and the probability to reach position x without being scattered, to be scattered near x, and to exit the active region without being scattered between x and L. The integration over $\mathrm{d}x$ and $\mathrm{d}E$ accounts for all possible paths and all suitable electrons.

Figure 7.13 depicts the ratio of the overflow electron current to the total current calculated for an *m*-plane (nonpolar) LED within the framework of Equations 7.51 and 7.52 as a function of the EBL barrier height for flat-band (in the active region) reached through compensation of the built-in electric field of the p–n junction by the applied voltage. The phonon emission time (τ_{em}), the phonon absorption time (τ_{abs}), and the phonon scattering time (τ_{sc}) of 0.01, 0.1, 0.009 ps, respectively, have been used.

For the LED without EBL, a significant portion (~67%) of the total electron current is due to the overflow electrons, while almost no electron overflow was found to occur when an EBL with 15% Al is used. The calculations are in reasonable agreement with experimental observations for the LED without EBL: approximately

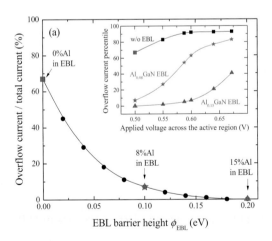

Figure 7.13 Calculated ratio of overflow electron current to total current as a function of EBL barrier height (ϕ_{EBL}) under flat-band condition in the active region (the applied external voltage compensates the built-in potential, ~0.5 V) for *m*-plane LEDs with EBLs: 0% Al, 8% Al, and 15% Al. The inset shows the same as a function of the applied voltage (forward direction) across the InGaN-active region.

four-fifth of the current is the overflow current while only one-fifth of the electron current is captured by the InGaN-active region as evident from Figure 7.12.

The inset of Figure 7.13 shows the ratio of calculated overflow electron current to total current as a function of the forward bias across the active region for the *m*-plane LEDs with different EBLs. Note that 0.5 V applied voltage corresponds to the flat-band condition owing to the compensation of the built-in voltage inside the active region, and larger biases would lower the effective EBL barrier height, which causes the overflow current to increase. With 15% Al in EBL, the overflow current is nearly zero under the flat-band condition and gradually increases with increasing bias. The increase in the overflow current component with decreasing EBL barrier height, and its weaker dependence on the applied bias for lower EBL barrier heights are consistent with both theory and experiment.

7.3.3.5.1 Staircase Electron Injector (SEI) to Mitigate Electron Overflow
To restate *a priori*, SEI is an effective substitution for EBL but without the detrimental hole blocking effect. SEI is simply designed to cool the electrons before being injected into the active region for recombination. Each time an electron is injected over a down a step, its potential energy is converted into kinetic energy with an associated velocity. Replacing one potential step into several smaller ones reduces the kinetic energy while at each of the ledge. If the ledge length is chosen appropriately, electrons would rid of some of their potential energy as they enter the active region and thus traverse with increased residence time, which increases recombination probability (Figure 7.14a). This thesis is of course predicated on the presence of holes, the concentration of which, in the end, is the ultimate limiting factor for recombination. However, under forward bias, particularly high bias and on polar surface, an optimum design for a particular LED-active layer structure is needed. Figure 7.14a shows that the relative EQE values of the two *m*-plane LEDs with SEI (5 nm with >100 meV steps), one with and the other without EBL, are essentially the same for the same current injection levels, contrasting that observed in LEDs without the SEI. It is, therefore, reasonable to suggest convincingly that the InGaN SEI reduces or electron overflow due to ballistic or quasiballistic electron transport by efficiently thermalizing injected electrons as illustrated in Figure 7.14b. The data suggest that the particular SEI under consideration is an effective substitution for the EBL, which aggravates already meager hole transport. However, with electric field, induced by either bias or polarization when present, even with SEI spillover of electrons could occur particularly for large biases in which case a thicker SEI region should be employed as we will discuss shortly.

To account for the reduced, if not totally eliminated, electron overflow in LEDs with a SEI (one-layer SEI with one intermediate composition of $In_{0.10}Ga_{0.90}N$ for simplicity), first-order calculations of the electron overflow have been performed. The SEI considered is 15 nm thick – the same as the total thickness of the three-layer SEI of the LEDs investigated experimentally. The percentage of overflow electrons due to ballistic transport in the LED with SEI is given by

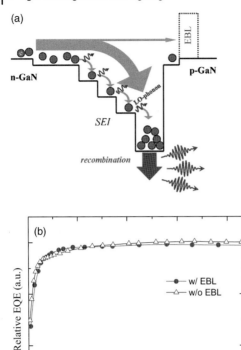

Figure 7.14 (a) A schematic depiction of the conduction band structure and the enhancement of electron thermalization in the presence of a three-layer SEI; (b) Relative EQE of two m-plane LEDs grown on freestanding m-plane ($1\bar{1}00$) GaN substrates with a three-layer SEI: one with and one without the EBL.

$$P_0 = \frac{\int_{\max\{0,(\phi_{EBL}-qV)\}}^{+\infty} f(E)N(E) \cdot \exp(-[d/v_1(E+\Delta E_{c1}) + L/v_2(E+\Delta E_{c2})]/\tau_{sc})dE}{\int_0^{+\infty} f(E)N(E)dE},$$

(7.53)

where $d = 15$ nm and $L = 6$ nm are the thicknesses, and $v_1(E + \Delta E_{c1}) = \sqrt{2(E + 0.25 \text{ eV})/m_e}$ and $v_2(E + \Delta E_{c2}) = \sqrt{2(E + 0.5 \text{ eV})/m_e}$ are the electron velocities in the SEI region ($\text{In}_{0.10}\text{Ga}_{0.90}\text{N}$) and the active region under flat-band conditions, respectively. The energies, $\Delta E_{c1} = 0.25$ eV $\Delta E_{c2} = 0.5$ eV in the velocity expressions represent the excess energy gained by electrons from the conduction band discontinuities upon injection into the $\text{In}_{0.10}\text{Ga}_{0.90}\text{N}$ SEI layer and into the $\text{In}_{0.20}\text{Ga}_{0.80}\text{N}$-active layer, respectively. Again, for simplicity electron transport normal to the heterointerfaces is assumed.

For electrons experiencing one scattering event, a total of four different cases corresponding to emission or absorption of only one LO phonon in the SEI and the active regions can be envisioned. The probability of overflow resulting from only one phonon emission in the SEI is given by

$$P_3 = \left[\int_0^{+\infty} f(E)N(E)dE\right]^{-1} \int_{\max\{0,(\phi_{EBL}-qV+\hbar\omega_{LO})\}}^{+\infty} f(E)N(E) \int_0^d \frac{\exp(-[x/v_1(E+\Delta E_{c1})]/\tau_{sc})}{v_1(E+\Delta E_{c1}) \cdot \tau_{em}}$$
$$\times \exp(-[(d-x)/v_1(E+\Delta E_{c1}-\hbar\omega_{LO})+L/v_2(E+\Delta E_{c2}-\hbar\omega_{LO})]/\tau_{sc}) \cdot dx \cdot dE. \quad (7.54)$$

Equation 7.54 takes into account the probabilities for a suitable electron (e.g., with energy $\hbar\omega_{LO}$ higher than the bottom of conduction band of n-GaN for the case of no EBL and flat bands) to reach position x ($0 \leq x \leq d = 15$ nm) without being scattered, to emit a phonon near x, and to exit the active region without being scattered between x and L. The probability of overflow due to only one phonon emission in the active region is given by

$$P_4 = \left[\int_0^{+\infty} f(E)N(E)dE\right]^{-1} \int_{\max\{0,(\phi_{EBL}-qV+\hbar\omega_{LO})\}}^{+\infty} f(E)N(E) \int_0^d \frac{\exp(-[d/v_1(E+\Delta E_{c1})]/\tau_{sc})}{v_2(E+\Delta E_{c2}) \cdot \tau_{em}}$$
$$\times \exp(-[(x-d)/v_2(E+\Delta E_{c2})+(d+L-x)/v_2(E+\Delta E_{c2}-\hbar\omega_{LO})]/\tau_{sc}) \cdot dx \cdot dE. \quad (7.55)$$

Similarly, we can obtain the probability for one-phonon absorption. Again, the contribution of two scattering events to the overflow electrons is negligible for the SEI considered (<1%). The total electron overflow is then obtained by summing the ballistic and quasiballistic contributions. Figure 7.15 shows the percentile overflow electron current for the two LEDs without EBL ($\phi_{EBL} = 0$): one with and that without

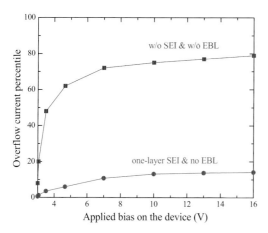

Figure 7.15 Calculated overflow electron current/total electron current as a function of applied forward voltage across the 6 nm-thick $In_{0.20}Ga_{0.80}N$ active region for the LED without a SEI and without an EBL, and the other LED with a one-layer $In_{0.10}Ga_{0.90}N$ SEI and without an EBL. Solid lines are guides to the eye. The inset shows the band diagrams for the LEDs with and without (dashed line) the SEI.

SEI, calculated using $\tau_{em} = 10$ fs, $\tau_{abs} = 100$ fs, and $\tau_{sc} = 9$ fs. For flat band, a significant portion (~62%) of the total electron current is the overflow electron current for the LED without SEI. However, for the LED with the SEI, the electron overflow is substantially reduced to ~8%. An SEI causes the electrons to have a longer transit time owing to lower velocity resulting from lower kinetic energy gained from a relatively small band discontinuity.

7.3.3.5.2 Polarization Effects on Electron Overflow

Because of the dominance of the polar *c*-plane orientation, polarization-induced field in the form of the band bending must be considered for completeness. The polarization-induced charge densities of 4×10^{12} and 1×10^{13} cm^{-2} along with 15% Al EBL are assumed. Figure 7.16 shows the so simulated conduction band edge profile for a *c*-plane DH LED with 6 nm active region and 15% Al in EBL, along with its *m*-plane counterpart under the same injection current (1200 A/cm²) for comparison. Due to the polarization-induced field in the AlGaN EBL as well as the InGaN-active region in the *c*-plane case, the conduction band edge of EBL is "pulled down" at the active-region side (due to the immobile positive polarization charge at the active region–EBL interface) even at a high current density of 1200 A/cm². When compared to its *m*-plane counterpart (broken lines), the polarization field in the *c*-plane LED (solid line) reduces the effective barrier of the EBL for the electrons.

The calculated percentile of electron overflow (i.e., overflow current/total current) for *c*-plane 6 nm DH LEDs, in the framework of hot electrons, is shown in Figure 7.17. The LED without EBL and SEI shows a much higher electron overflow percentile than the one with 15% Al in the EBL. Moreover, the overflow in the sample LED without EBL saturates at 90% once the applied voltage exceeds 5 V, while for the LED with 15% Al in EBL the overflow increases with the applied voltage: from 0% to 50% when the voltage increases from 3 to 14 V (the voltage

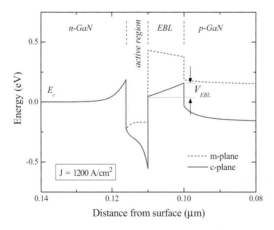

Figure 7.16 Calculated conduction band edge of *c*-plane (solid line) and *m*-plane (dotted line) InGaN LEDs with a 6 nm In$_{0.15}$Ga$_{0.85}$N active region and a 10 nm Al$_{0.15}$Ga$_{0.85}$N EBL. The injection current density for both LEDs is 1200 A/cm².

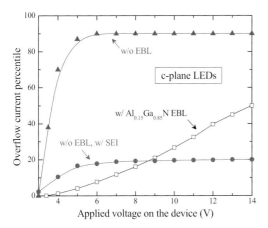

Figure 7.17 Calculated electron overflow current as a function of the voltage applied across the device for c-plane DH LEDs (6 nm In$_{0.15}$Ga$_{0.85}$N active region) with different EBL heights (i.e., $x = 0\%$ Al and 15% Al in Al$_x$Ga$_{(1-x)}$N) and without SEI as well as a c-plane DH LED with SEI and without any EBL.

here attempts to represent all the expected contributions from series resistances, etc., for the particular LEDs used in the laboratory – even then the magnitude of the latter voltage figure is larger than it should be and is most likely an artifact of the simulator, but the trend holds). For the c-plane LED, which has a one-layer SEI and no EBL, the calculated electron overflow is greatly reduced (~4.5 times lower which is consistent with experiments) compared to that having no SEI and no EBL. Moreover, the c-plane LEDs (shown in Figure 7.17) have higher electron overflow than their m-plane counterparts (shown in Figure 7.14) with similar structure. It is worth noting that in the c-plane case, the probability for electron to traverse the active region and reach the EBL (before surpassing the EBL) is already ~90% (in both 0% Al and 15% Al EBL cases due to the lowering of the barrier in the p-region by the polarization-induced electric field. Marginal (3–5%) enhancement of the ballistic or quasiballistic electron transport is gained by adding the electric field acceleration effect on the electron velocity.

7.3.3.5.3 Optimization of SEI For simplicity for the time being let us consider the one-step SEI whose schematic is shown in Figure 7.18 in which we aim to optimize the step height (ΔE_c) and thickness (d) for minimizing the electron overflow. The contribution by electrons undergoing two or more scattering events to overflow is negligible for low fields/biases. However, they should be included for accuracy particularly at high applied voltages.

For this first-order optimization, the electrons not experiencing ballistic (no scattering) and the quasiballistic transport (i.e., only one phonon emission or absorption) within the SEI region are assumed thermalized to the bottom of the conduction band of the SEI layer. The percentile of these partially thermalized electrons (see process "b" in the schematic in Figure 7.18) can be described as

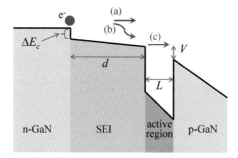

Figure 7.18 A schematic for the conduction band of an LED with a one-intermediate layer SEI (of thickness d and step height ΔE_c). After being injected into the SEI from the n-GaN region, some electrons will have ballistic and quasiballistic (only one phonon emission or absorption) transport through the SEI [process (a)], while the others (experiencing two or more scattering events) are considered to be thermalized in the SEI [process (b)]. Under high bias condition and with a small step height ΔE_c, the thermalized electrons from process (b) might contribute to the overflow current through ballistic or quasiballistic transport [process (c)]. For the calculations, the conduction band discontinuity between the active region and p-GaN is assumed to be 0.5 eV, and no EBL is employed.

$(1 - P_{bal\_1} - P_{abs\_1} - P_{em\_1})$, where P_{bal_1}, P_{abs_1}, and P_{em_1} are the probabilities of ballistic transport, and quasiballistic transport with one phonon absorption or one phonon emission within the SEI region, respectively. Having been injected from the SEI into the active region, these electrons will have a certain probability of ballistic or quasiballistic transport the active region, depending on the conduction band discontinuity between the SEI and the active region and also the voltage drop across the active region [see process "c)" in the schematic in Figure 7.18]. This can be described as $(P_{bal\_2} + P_{abs\_2} + P_{em\_2})$, where P_{bal_2}, P_{abs_2}, and P_{em_2} are the probabilities of ballistic transport and quasiballistic transport with one phonon absorption and one phonon emission within the active region, respectively. Therefore, the amount of overflow due to the thermalized electrons from the SEI can then be described as $(1 - P_{bal\_1} - P_{abs\_1} - P_{em\_1}) \times (P_{bal\_2} + P_{abs\_2} + P_{em\_2})$, which is added to the overflow percentile calculated with the SEI theoretical model to obtain the total electron overflow percentile.

The calculated percentiles of electron overflow for the one-step SEI with varying step height (ΔE_c) and thickness (d) are shown in Table 7.1. Under flat-band condition (net potential drop across the active region $V = 0$), the 15 nm-thick SEIs with step heights of 0.1, 0.2, and 0.3 eV all result in an overflow percentile of 11% for the cases enumerated in the table. For a voltage drop of $V = 0.1$ V, the overflow percentile for the 15 nm-thick SEI having a step height of 0.1 eV increases to 61% due to a significant increase of overflow contribution from the electrons thermalized in SEI but traversing the active region ballistically and/or quasiballistically without recombination due to lowering of the conduction band of p-GaN to the same level as that of the SEI. For the same bias condition corresponding to $V = 0.1$ V, the SEI structures with the step heights of 0.2 or 0.3 eV and a thickness of 15 nm have comparable electron overflow percentiles (21% versus 22%). Clearly multiple step

7.3 Optical Output Power and Efficiency

Table 7.1 Calculated electron overflow percentiles for a one-intermediate layer SEI (see the schematic in Figure 7.18), with varying SEI step height (ΔE_c) and SEI thickness (d) under flat-band conditions ($V = 0$) for an LED with a 6 nm-thick InGaN-active region.

ΔE_c (eV)	$d = 3$ nm	$d = 9$ nm	$d = 15$ nm
0.1	36% (70%)	18% (66%)	11% (61%)
0.2	37% (52%)	18% (38%)	11% (21%)
0.3	39% (51%)	21% (36%)	11% (22%)
0.4	41% (52%)	24% (39%)	13% (26%)
0.5	42% (53%)	26% (42%)	18% (29%)

The values in parentheses are for $V = 0.1$ V.

SEI with larger ledge widths should be considered particularly for biased active region due to either external voltage or polarization charge, which is discussed next.

We now turn our attention to the case of multiple-layer SEI. Specifically a two-layer SEI structure (see Figure 7.19) under the flat-band conditions with a total thickness of 9 nm (4.5 + 4.5 nm), the same as that for the one-step variety is considered. The conduction band discontinuity between the active region and the SEI is 0.2 eV (step height for the one-layer SEI is $\Delta E_c = 0.3$ eV). The calculated overflow percentile is 16% for the two-layer SEI compared to 21% (Table 7.1) associated with one-step SEI.

In total 19% of the electrons are thermalized by the first SEI layer and the probability of ballistic and quasiballistic transport of these thermalized electrons from point B to point C (refer to Figure 7.19) is 0.8%. The product of the two probabilities (19% × 0.8%), representing the overflow, gives a relatively small electron overflow percentile of ~0.2%. Naturally, SEI with more steps in (with potential steps no less than one LO phonon energy) would result in smaller electron overflow percentile. But the effect SEI on the material quality and strain must be considered. Moreover, the probability to emit photons with longer wavelength than

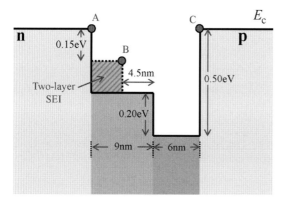

Figure 7.19 Schematics for one-layer and two-layer SEIs. For the one-layer SEI, the total SEI thickness is 9 nm and the energy step height is 0.3 eV. For the two-layer SEI, each energy step height is 0.15 eV and each layer thickness is 4.5 nm.

desired will increase if the energy level of the bottom SEI step will be too close to the bottom of the conduction band of the active region.

In order to study the impact of carrier overflow experimentally a baseline is established by resonant PL investigations as this method does not allow carrier overflow, the results of which are exhibited in Figure 7.10. When compared and contrasted to EL experiments, which are affected by carrier overflow, one can deduce the carrier overflow component qualitatively. Along these lines, the integrated EL intensities versus injection current for a set of DH LEDs with varying number of 3 nm active layers, for example, mono, duo, trio, quad, hexa, and octa are shown in Figure 7.20. The integrated EL intensity, L_{EL}, can be described by a power dependence on the injection current density as $L_{EL} \propto J^m$, where the power index m, as in the case of optical excitation, reflects the degree of radiative and nonradiative recombination processes, with $m = 1$ the representing fully radiative recombination process and $m > 1$ indicating participation by nonradiative processes. The superlinear growth of EL intensity ($m \sim 1.4$ for single, ~ 1.3 for dual and quad, and ~ 1.6 for hexa and octa DH LEDs) at low current densities can be attributed to the impact of nonradiative recombination, a larger number indicating more nonradiative recombination centers. The EL intensity changes nearly linearly at high current levels; therefore, EL efficiency tends to be constant (Figure 7.20 inset).

What is even more striking is that scaling with number of active layer up to the hexa set is indicative of electron overflow, indicative of the fact that the two-step SEI composed of 5 nm $In_{0.04}Ga_{0.96}N$ and 5 nm 8% $In_{0.08}Ga_{0.92}N$, while an efficient substitute for electron blocking layer, is insufficient on polar c-plane devices to prevent overflow. It is, therefore, imperative that SEI is optimized for a particular active layer design, which is discussed next. In LEDs using multiple quantum well (MQW)-active

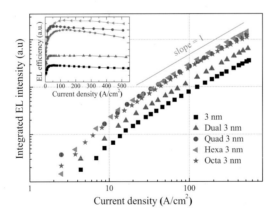

Figure 7.20 Integrated EL intensity dependence on current density (the gray solid line indicates slope of 1); the inset shows EL efficiencies of multi-3 nm DHs versus injected current density. Under normal conditions, the EL intensity with respect to injection current should be the same for all structures if carrier overflow is absent. Scaling up to six active layers indicates the 5 nm/5 nm two-step SEI is not sufficient for mono up to hexa 3 nm active layers. This particular SEI is shown to be an effective substitute for EBL in nonpolar LEDs, but on polar c-plane it is not sufficient to eliminate the electron overflow altogether. Instead, the SEI layer should be optimized for a specific active layer design.

regions, some of the MQWs on the *n*-layer side automatically act as electron coolers already.

As can be discerned from Figure 7.20 and as alluded to earlier while 5 nm $In_{0.04}Ga_{0.96}N$ and 5 nm 8% $In_{0.08}Ga_{0.92}N$ SEI successfully replaces 15% AlGaN EBL layer, it is not sufficient to prevent electron overflow in its entirety. The obvious remedy is to optimize the SEI structure for a given active layer design. This would be an opportune time to mention that in MQW-active layer designs a portion of the MQW on the *n*-side would act as electron cooler. One optimization scheme involves just simply increasing the thickness of the staircase region, for example, 20 nm $In_{0.04}Ga_{0.96}N$ and 20 nm 8% $In_{0.08}Ga_{0.92}N$. While the electron overflow is not detected in resonant PL measurements, modification by the SEI design would be discernible in the EL experiments in that if SEI is fully successful there would be no overflow and thus active layer design, inclusive of its total thickness (provided that one is in a realm wherein self-absorption is not notable), should have no bearing on light emission intensity predicated on the assumption that there is no damage to the active region under operation. Note that in 3 nm active region there is no quantum confinement and that the recombination occurs in the lowest energy state, because if the number of states is limited even in the 3D case of a thin active region, blueshift in emission may occur at high injection. In one such experiment, active layer designs consisting of stacks of 3 nm active regions [numbering 1–4 (indexed as Quad)] with 5 nm $In_{0.04}Ga_{0.96}N$ and 5 nm 8% $In_{0.08}Ga_{0.92}N$ SEI, or 20 nm $In_{0.04}Ga_{0.96}N$ and 20 nm 8% $In_{0.08}Ga_{0.92}N$ SEI, or 30 nm $In_{0.04}Ga_{0.96}N$ and 30 nm 8% $In_{0.08}Ga_{0.92}N$ SEI have been examined.

Salient features of the data displayed in Figure 7.21 are the mono 3 nm DH LED with 5 nm $In_{0.04}Ga_{0.96}N$ and 5 nm 8% $In_{0.08}Ga_{0.92}N$ SEI shows much lower EL

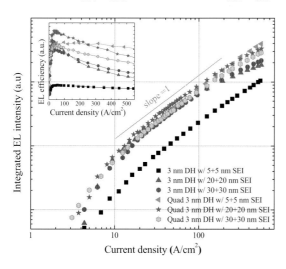

Figure 7.21 The integrated EL efficiency dependence on current density (gray solid lines represent a slope of 1) for LEDs with varying SEI thickness. The active layers depicted are 1 × 3 nm DH and 4 × 3 nm DH, while the SEI structure involved is 5 nm $In_{0.04}Ga_{0.96}N$ and 5 nm 8% $In_{0.08}Ga_{0.92}N$, or 20 nm $In_{0.04}Ga_{0.96}N$ and 20 nm 8% $In_{0.08}Ga_{0.92}N$ or 30 nm $In_{0.04}Ga_{0.96}N$ and 30 nm 8% $In_{0.08}Ga_{0.92}N$. The inset displays the corresponding EL efficiency data in linear scale as this particular format is ubiquitously used in the literature.

emission as compared to the quad variety with the same SEI or the mono variety with 20 nm $In_{0.04}Ga_{0.96}N$ and 20 nm 8% $In_{0.08}Ga_{0.92}N$ SEI. The dispersion for the mono active layer with 20 nm $In_{0.04}Ga_{0.96}N$ and 20 nm 8% $In_{0.08}Ga_{0.92}N$ SEI and 30 nm $In_{0.04}Ga_{0.96}N$ and 30 nm 8% $In_{0.08}Ga_{0.92}N$ SEI is negligible indicating the effectiveness of the 20 nm $In_{0.04}Ga_{0.96}N$ and 20 nm 8% $In_{0.08}Ga_{0.92}N$ SEI design. Consistent with the above discussion, the effect of SEI design is diminished in quad 3 nm DH layer (each active region is separated by a barrier) as a portion of the quad design on the n-side acts as an electron cooler. Data for all quad active layer designs, regardless of the particulars of SEI, are more or less comparable. Another observation that can be drawn from an examination of Figure 7.21 is that the EL intensity varies linearly, in logarithmic scale, with injection above 20 A/cm^2, which clearly indicates the dominance of radiative recombination.

Ultimately no matter how optimized the device is, the decisive factor is the hole concentration that can be obtained in the p-GaN layer. One cannot escape the reality that holes must be available for raditative recombination after all said and done. Therefore, stating the obvious any effort to increase the hole concentration would be well worthwhile.

In order to ascertain whether structural properties play any role in the observation, scanning transmission electron microscopy (STEM) investigation of the LED structures were undertaken. Atomic resolution images displayed in Figure 7.22, taken from a 6 × 3 nm active layer device structure show that the interfaces are well defined, sharp and InGaN channels to be of uniform composition with no evidence of compositional inhomegeities.

7.4
Effect of Surface Recombination

Surface recombination has not been the dominant narrative in InGaN-based LEDs, but the concept deserves some discussion when and if it becomes a critical issues. For a double-heterojunction LED, where the active layer is the only absorbing layer in the entire structure on a transparent substrate, the internal absorption term, η_A, including interface recombination, has been determined to be

$$\eta_A = \left\{2\left[(V_s^2 + 1)\sinh\left(\frac{w}{L}\right) + 2V_s \cosh\left(\frac{w}{L}\right)\right]\right\}^{-1}$$

$$\left\{\left[\frac{1+V_s}{1+\alpha L}\right]\left[1 - \exp w\left(\frac{1+\alpha L}{L}\right)\exp\frac{w}{L}\right] - \left[\frac{1-V_s}{1-\alpha L}\right]\left[1 - \exp w\left(\frac{1-\alpha L}{L}\right)\exp\frac{-w}{L}\right]\right\},$$

(7.56)

with $V_s = v_s L/D$.

Figure 7.23 exhibits η_A as a function of the active layer thickness for two surface recombination velocities (100 and 1000 cm/s). The other parameters used are for GaN, even though all the LEDs are made of InGaN [center wavelength: 450 nm, electron mobility: 600 $cm^2/(Vs)$]. The effective carrier lifetime is as indicated

7.4 *Effect of Surface Recombination* | 245

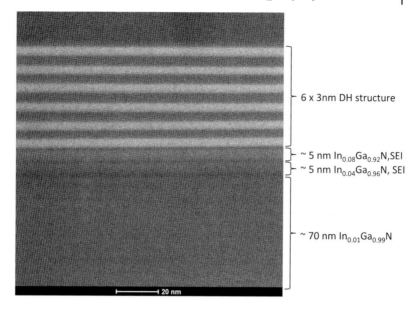

- 6 × 3nm DH structure
- ~ 5 nm $In_{0.08}Ga_{0.92}N$, SEI
- ~ 5 nm $In_{0.04}Ga_{0.96}N$, SEI
- ~ 70 nm $In_{0.01}Ga_{0.99}N$

(a)

Top 3 $In_{0.15}Ga_{0.85N}$ layers   Bottom 3 $In_{0.15}Ga_{0.85N}$ layers

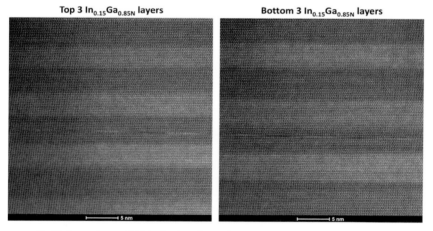

- 13-15 atom columns thick in the $In_{0.15}Ga_{0.85N}$ layers, corresponds to ~3.5 nm.
- $In_{0.06}Ga_{0.94N}$ layers are ~4-4.5 nm thick.

(b)

Figure 7.22 Atomic resolution image of a DH LED with 6 × 3 nm active layer, (a) lower (b) higher resolution. Note smooth interfaces and absence of any compositional inhomogeneity. Courtesy of A. Yankovich and P. Voyles.

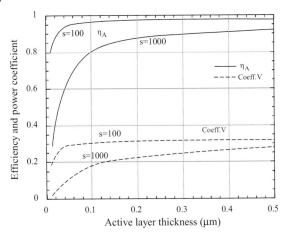

Figure 7.23 The efficiency reduction term caused by interface recombination and bulk absorption in an otherwise ideal GaN-based LED for surface recombination velocities of 100 and 1000 cm/s. The coefficient term relates the output power to the injection current.

(radiative lifetime $= 2 \times 10^{-9}$ s, absorption coefficient $\alpha = 10^5$ cm^{-1}, and refractive index $= 2.6$). Moreover, the coefficient in front of the injection current in Equation 7.17 relating the power to the injection is also plotted. In the absence of available data, what would be plausible was chosen based on the assertion that the GaN surface is reasonably inert. Further consideration was given to the observation that the Schottky barrier height seems to become higher with an increased work function of the metal. Figure 7.24 displays the same parameters as a function of the surface recombination velocity in the range 1 to 10 000 cm/s for several thicknesses of the active layer ranging from 3.5 to 20 nm.

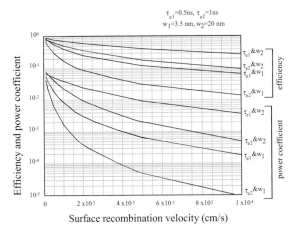

Figure 7.24 The parameters of Figure 7.23 as a function of the surface-recombination velocity for active layer thicknesses ranging from 3.5 to 20 nm. Here τ_e and w represent the lifetime and active layer thickness, respectively.

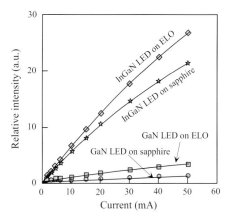

Figure 7.25 Relative output power of UV InGaN and GaN LEDs as functions of forward current for LED chip size is as large as 350 μm × 350 μm, which covers many ELO stripes covering both higher and lower dislocation density regions. Courtesy of T. Mukai, Nichia Chemical Co.

7.5
Effect of Threading Dislocation on LEDs

Nonradiative recombination center density is some ways related to threading dislocation density. Unmistakably, the smaller the dislocation density, the better the quantum efficiency of LEDs. This is more so true as the wavelength of emission is made to decrease from visible by first reducing the InN content in the active region and then increasing the AlN mole fraction in the active region for even shorter wavelengths. Two examples will be used to illustrate the case, one compares 380 nm LEDs fabricated on standard and ELO buffer layers, latter having lower dislocation density. The other example relates the internal quantum efficiency of 280 nm UV LED with the dislocation density. Figure 7.25 shows the relative output power of UV (380 nm) InGaN and GaN LEDs produced using sapphire and ELO substrates as functions of forward current: the ELO GaN (average dislocation density $7 \times 10^6 \, \text{cm}^{-2}$) and GaN on sapphire (dislocation density $1 \times 10^{10} \, \text{cm}^{-2}$). As the figure indicates, the both InGaN and UV GaN LED (380 nm) on ELO provide better output power than their counterparts on sapphire. The shorter wavelength variety seems to be more sensitive to the dislocation density. This sensitivity increase on dislocation density increases at short wavelengths is clearly demonstrated in Figure 7.26 in conjunction with UV LEDs emitting at 280 nm.

7.6
Current Crowding

Current crowding, nonuniform current distribution, is associated with LEDs with both anode and cathode electrodes on top of the chip such as in high voltage varieties. It is also in effect in other devices including bipolar transistors and those with interdigitated layout. Current crowding robs off the device of its optimum performance as well as causing premature breakdown, not to mention performance

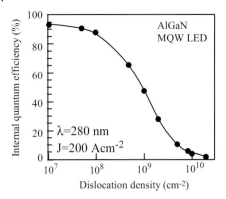

Figure 7.26 Dependence of internal quantum efficiency on the dislocation density for UV LEDs emitting at 280 nm, representing the most stringent (most sensitive) case. Courtesy of M. Kneissl.

loss with increasing current. Reasonably uniform current in such mesa structures can be expected if the p- and n-cladding layers are thick and highly conductive. However, the growth of thick and highly conductive p-type (Al, In, Ga)N films is not an option, leading to crowding near the edge p-type contact.

Current crowding can be calculated by devising an equivalent circuit model shown in Figure 7.27, as has been developed for bipolar transistors earlier. It is inclusive of the resistances of the p-type contact, and the n-type and p-type cladding layers with the assumption that the p-type metal contact has the same electrostatic potential at every point. The p–n junction region is characterized by an ideal diode.

Figure 7.27 shows the schematic LED structure with lateral injection geometry. In this structure, important distributed components of the total series resistance can be categorized into the lateral resistance component of the n layer, r_n, and the transparent electrode, r_t, and the vertical component of the p layer, r_p, and the p contact, r_c. Assuming that the p-pad is grounded, the current continuity equation for the circuit in Figure 7.27 leads to the following two basic equations:

$$\frac{d^2 V_n}{dx^2} = \frac{\rho_n}{t_n} J, \quad \frac{d^2 V_t}{dx^2} = \frac{\rho_t}{t_n} J. \tag{7.57}$$

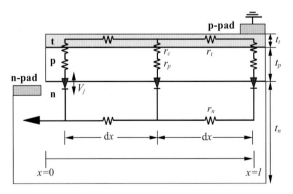

Figure 7.27 Equivalent LED circuit, with a p-pad as the physical ground that can be used to model current crowding. Courtesy of H. Kim, S.J. Park, H. Hwang, and N.M. Park.

7.6 Current Crowding

The relation between V_n and V_t can be also expressed as follows:

$$V_n = V_j + R_v I_0 \exp\left(\frac{eV_j}{kT}\right) + V_t. \tag{7.58}$$

The parameters V_n, V_t, ρ_n, and ρ_t are the lateral voltage drops and the electrical resistivities of the n layer and p transparent electrode, respectively, and J is the current density across the p–n junction region. V_j is the junction voltage drop, I_0 is the reverse saturation current, and R_v is the vertical resistance of the area element wdx.

Using the above three equations, the current distribution can be expressed as

$$J(x) = J(0)\exp\left[\mp x \Big/ \left(\sqrt{(\rho_c + \rho_p t_p)\left|\frac{\rho_n}{t_n} - \frac{\rho_t}{t_t}\right|^{-1}}\right)\right], \quad -\text{ for } p \text{ pad}, + \text{for } n \text{ pad}, \tag{7.59}$$

where $J(0)$ is the reverse saturation current density at the mesa edge, the $(-)$ sign holds for the p-pad and the $(+)$ sign for the n-pad as physical grounds, respectively. As a result, the current spreading length, L_s, can be expressed as follows:

$$L_s = \sqrt{(\rho_c + \rho_p t_p)|\rho_n/t_n - \rho_t/t_t|^{-1}}. \tag{7.60}$$

Figure 7.28 shows the calculated current distribution of the LED, indicating that it is possible to achieve a nearly uniform current distribution at the critical condition of $\rho_t/t_t = \rho_n/t_n$.

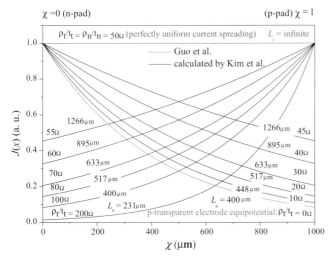

Figure 7.28 Calculated current distributions versus the lateral length x in an LED. The parameters used in the calculation: $\rho_n = 0.01\,\Omega$ cm, $t_n = 2\,\mu$m. Courtesy of H. Kim, S.J. Park, H. Hwang, and N.M. Park.

7.7
Perception of Color

Vision encompasses simultaneous interaction of the eyes and the brain through a network of neurons, receptors, and other specialized cells. The process commences with the stimulation of light receptors in the retina, which lines the back of the eyeball, and transmission of electrical signals to the brain through the optic nerves. This information is processed in several stages ending at the visual cortex of the brain.

The human eye is equipped with a variety of optical elements including the cornea, iris, pupil, aqueous and vitreous fluids, variable-focus lens, and the retina with the bodies of nerve cells arrayed in three rows separated by two layers packed with synaptic connections. The back of the retina contains the photoreceptive sensory cells. The retina has two types of photoreceptors, rods, and cones with the former dominating. Rods are for low-light vision and cones for daylight. At dusk, dawn and in dimly lit places, rods provide gray vision without color. The cones are responsible for bright-colored vision. Most mammals have two types of cones, green-sensitive and blue-sensitive, but primates have three types including red-sensitive as well as the green and blue.

Rod sensitivity peaks at green ($\sim$550–555 nm). Rod vision is commonly referred to as *scotopic* or twilight vision. When all three types of cone cells are stimulated equally, we perceive the light as being *achromatic* or white. In daylight, the human eye is most sensitive to the wavelength of 555 nm with a maximum sensitivity of 683 lm/W. This is called *photopic vision*. In low light and night situations, *scotopic vision*, the peak sensitivity blue shifts to 507 nm. The maximum sensitivity for the scotopic vision is 1754 lm/W. At red and blue extremes, the sensitivity of human eye drops dramatically. Figure 7.29 presents the *luminous efficacy*, $K(\lambda)$, which represents the effectiveness of the radiant power of a monochromatic light source in stimulating the visual response for the daylight (*photopic vision*) and night visions (*scotopic vision*). As seen in Figure 7.29, the efficacy curve falls drastically at both ends of the visual spectrum. This increases optical power requirements for emitters in the blue and red regions to achieve the same brightness or luminous performance offered by green-light sources.

The optical power generated by an LED must excite the human eye. This brings into the discussion the color perception of the human eye, which has been standardized by the CIE. This commission produces charts used by the display society to define colors. Detection and measurement of radiant electromagnetic energy is called *radiometry*, which when applied to the visible portion of the spectrum involving the human eye, is termed *photometry*. The nomenclature for the latter delineates itself from the former by adding the adjective luminous to those terms used for the former. For example, energy in the former is called the *luminous energy* in the latter. The former can be converted into the latter, and vice versa, if the perception of color by the human eye is known.

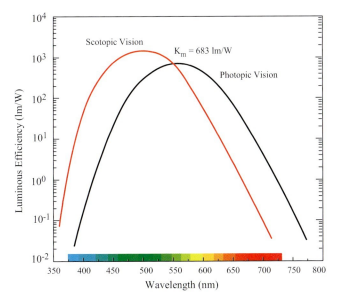

Figure 7.29 Luminous efficacy of monochromatic radiation, $K(\lambda)$, for the human eye under light, photopic vision, and dim, scotopic vision, conditions. The band indicates the color at visible wavelengths. The maximum luminous efficacy, K_m, for photopic vision occurs at 555 nm and is 683 lm/W.

7.8
Chromaticity Coordinates and Color Temperature

Visible light in its entire spectra can be represented in a number of different three-dimensional spaces. To convert the light spectrum to color space, the color matching function of CIE colorimetric observer functions are used, which are shown in Figure 7.30. These plots, or functions, X, Y, and Z are representative of the sensitivity of the human eye to a given wavelength. The common conversion splits the color space into two variables for color and one for luminance. This reduces the color space into a three-dimensional representation such as CIE 1931 *xy* chromaticity diagram (or space) and CIE 1976 UCS (uniform color space) chromaticity diagram (or space).

The spectrum of a given light source is weighted by the *XYZ color matching functions*, as shown in Figure 7.30. From the resultant three weighted integral values (called *tristimulus values* $\overline{X}$, $\overline{Y}$, $\overline{Z}$) the *chromaticity color coordinates* x, y are then calculated by

$$x = \frac{\overline{X}}{\overline{X}+\overline{Y}+\overline{Z}}, \quad y = \frac{\overline{Y}}{\overline{X}+\overline{Y}+\overline{Z}}. \tag{7.61}$$

The boundaries of this horseshoe-shaped CIE diagram represent the plots of monochromatic light, called the *Spectrum Locus*, which represents saturation color for that wavelength. Also plotted near the center of the diagram is the so-called *Planckian locus*, which is the trace of the chromaticity coordinate of a blackbody with

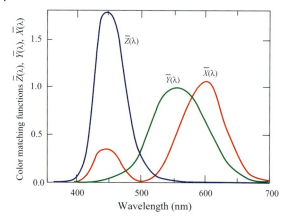

Figure 7.30 CIE 1931 XYZ color matching functions (standard colorimetric observer sensitivity plots for red (x), green (y), and blue (z) color perception cones of human eye. Courtesy of Lumileds/Philips.

its temperature ranging from 1000 to 20 000 K. The CIE (x, y) chromaticity (color coordinate) diagram, commonly known color domain, Planckian blackbody radiation locus with its temperatures, several available white light illuminants, and the wavelength range achievable with various LED semiconductor materials are shown in Figure 7.31.

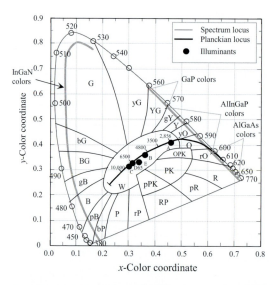

Figure 7.31 CIE (x, y) chromaticity (color coordinate) diagram, commonly known color domain, Planckian blackbody radiation locus with its temperatures, several available white light illuminants, and the wavelength range achievable with various LED semiconductor materials. Courtesy of Lumileds/Philips.

The color temperature is not used for color coordinates (x, y) off the Planckian locus. In this case, the *correlated color temperature* (CCT) is used. CCT is the temperature of the blackbody whose perceived color most resembles that of the light source in question. Due to the nonuniformity of the (x, y) diagram, the ISO-CCT lines are not perpendicular to the Planckian locus on the (x, y) diagram. To calculate CCT, an improved chromaticity diagram, namely the CIE 1960 (eV) diagram – later replaced by 1976 (u', v') diagram is used where the ISO-CCT lines are perpendicular to the Planckian locus by definition.

An important characteristic of the chromaticity diagram is that light stimuli on the diagram are additive. A mixture of two colors will produce a chromaticity coordinate falling on the line between their respective chromaticity coordinates. For example, the mixture of two sources of light with wavelengths at 485 nm (blue) and 583 nm (yellow-orange) each with a half-bandwidth of 20 nm results in an (x, y) value of (0.38, 0.38). This value is very close to the spectrum locus, producing a soft white color with a temperature of about 4000 K. This idea is the basis of the phosphor conversion LED (pcLED) in which a yellow-emitting yttrium aluminum garnet (YAG) phosphor is pumped with a blue LED, generating white light discussed in Section 7.11.

7.9
LED Degradation

Prior to the debut of GaN, experience with all LED materials had been that acceptable quantum efficiencies can be obtained only when the defect concentration in the semiconductor is well below about $10^4 \, \text{cm}^{-2}$. GaN, with defect concentrations about six orders of magnitude higher than the aforementioned figure at the time of GaN-based LED introduction, was not consistent with the trend at the time and thus GaN was discounted by many LED manufactures. Even with large defect concentrations, in the vicinity of $10^9 \, \text{cm}^{-2}$, GaN LEDs, with longevity well over the minimum 10 000 hours, required by the display society (CIE), were marketed in early 1994. Characteristic of semiconductor industry, the effort turned to understanding the degradation mechanisms and developing means to increase lifetime.

Degradation can be thermally activated and/or hot electron induced, the former being an activated process such as that involving metal semiconductor reaction. In addition, photon extraction can be hampered, which would manifest itself as degradation of the device, if the transparency of semitransparent contacts, used extensively in earlier varieties, and the transparent conducting oxide suffers degradation. As the overall efficiency of LEDs, the wallplug efficiency, which is the rate of conversion of electrical power to optical power, enhances the heat to be dissipated would be reduced, which would mitigate the thermally driven, by which equilibrium is meant, processes. This would require high-quality material, which also has the benefit of withstanding defect, both point defect and extended varieties, generation during operation either thermally driven or hot electron driven. It should be noted that hot electron-driven processes, which would tend to generate nonradiative centers, could occur prematurely with use.

A classical technique for investigation of degradation is to subject the device under test to current stress (e.g., $4\,kA/cm^2$, upward of hundreds of hours) meaning passing current through the device at room temperature or at elevated temperatures, the latter for accelerated measurements particularly used for long-lived devices. In early varieties LEDs so tested, the degradation rate was found to be more prominent at low currents, which is attributable to generation of point defects during the stress. One can assume that a very high-quality GaN-based material system should handle this type of current stress. Not doing so would imply that the defective nature of the starting material is at least in part responsible for this degradation. Hot electrons present in the active region could trigger defect generation either directly or most likely through generation of hot phonons because electron phonon coupling in GaN is very high – see the chapter of field effect transistors (FETs), and group velocity of LO phonon is miniscule. As the material quality gets better, stress measurements would be done at higher ambient temperature to reduce the time required for experiments.

After certain set limit of degradation, reduction in optical output power, that is, down to 80%, the device is considered nonusable and thus a lifetime can be determined. If the lifetime of the device is longer than the stress time that can practically be employed, accelerated lifetime measurements can be undertaken at three different elevated temperatures. If the degradation is an activated process, meaning exponential, extrapolation to room temperature would yield the lifetime of the device. However, if the degradation is not an activated process and/or has other components, such as that induced by hot electron/hot phonons, the aforementioned technique would not be applicable. In this case premature degradation would be observed.

As for the current crowding leading to emission crowding and increased series resistance, pending confirmation, the degradation process has been attributed to the presence of hydrogen, which can diffuse in the p-layer and generate Mg—H bonds with the acceptor atoms, thus compensating the overall active already low hole concentration and reducing device performance. The likely source of hydrogen could be from the passivation of, for example, Si_3N_4, layer typically deposited by plasma enhanced chemical vapor deposition (PECVD) on the LEDs for chip encapsulation and surface leakage current reduction.

In phosphor conversion LEDs (Section 7.11.3), any degradation caused by the reliability of YAG and any encapsulant used enter into the picture in addition to degradation of LEDs themselves, at least in theory. The YAG:Ce phosphor has been proven to be extremely stable under most adverse conditions in that an ensemble of YAG-pcLEDs has been put into a 60% relative humidity/85 °C test chamber and driven at $50\,A/cm^2$. After 3000 h, no significant degradation of the lumen-output was observed. The caveat is that the aforementioned studies have been conducted on pcLED in a special package, which did not contain an epoxy encapsulant. Tests on 5 mm LEDs with epoxy dome lenses, which are still used in some applications, failed often because of "browning" (oxidation) of the epoxy, reducing the transmission.

LEDs fabricated in high-quality materials with high quantum efficiency coupled with high photon collection efficiency would consume less electrical power. Well-developed packaging technologies can effectively remove the wasted power, which is converted into heat and thus keep the device temperature lower and increase the operation lifetime. However, even with 100% internal quantum efficiency, owing to electron overflow, and less than unity voltage conversion and phosphor conversion efficiencies, there will be heat dissipation, which makes good packaging imperative discussed next.

7.10
Packaging

Packaging is pivotal in any LEDs and fills in multiple roles in that it mechanically transitions the LED chip to its operating interface, inclusive of electrical connections, in many cases focuses and increases the light extracted from the LED, dissipates the heat generated (result of less than unity power conversion), and provides protection, both in physical terms and against stray fields. LEDs nowadays are utilized in many different areas and the old standard bearer 5 mm LED package is in many cases not appropriate.

For high-power LEDs flip-chip mounting combined with metal-based back reflector techniques are utilized, as shown in Figure 7.32. The high-power and low-voltage devices with active dimensions of $1 \times 1 \,\text{mm}^2$ require the current spreading, heat removal and efficient photon extraction features. There are many advantages to the flip-chip LED design among which is the fact that light exits through the polished transparent sapphire substrate instead of the semitransparent (somewhat absorbing Ni/Au contact layer unless transparent ohmic contacts are employed, which is becoming the norm). This particular package is capable for electrical input power levels up to 5 W.

For enhancing the light extraction efficiency further, a thin film LED structure, in which the substrate is completely removed through a combination of laser lift-off (LLO) and photoelectrochemical etching to produce a roughened surface (nitrogen polarity n-GaN) with cone-like features, from which the light emerges. In the realm of thin film LEDs, once the chip is removed from the substrates, it can be mounted on a metal alloy substrate with a high thermal conductivity and thermal match in terms of expansion and contraction, for example, $400 \,\text{W}/(\text{m K})$, which allows high current operation, see Figure 7.33 for a schematic. The chip has a patterned surface with "photon-injecting nozzle" microstructures to enhance light extraction in the forward direction. It should also be noted that for high luminance the optical size of a component, étendue, must be increased.

In an effort to assure the reliability of the LEDs especially under the high temperature ($>100\,°\text{C}$) and extreme photon fluence ($>50 \,\text{W/cm}^2$), several improvements in base material and chip attachment technologies have been undertaken. Among them is a package that can allow $150\,°\text{C}$ (white LED) or higher ($185\,°\text{C}$, non-white LED) junction temperature operation, as provided by the LUXEON K2

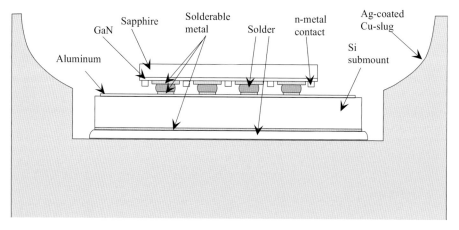

(a)

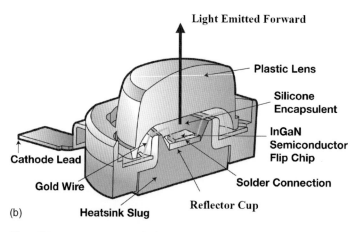

(b)

Figure 7.32 (a) Cross-sectional schematic flip-chip-mounted high-luminance LED package with Ag back reflector. Electro static discharge protection is integrated into the Si submount (circa 1997). The package is able to handle power dissipation associated with 350 mA current injection with resulting LED lumens of 20–40 lm. (b) Artistic rendition of the package inclusive of plastic lens. In this flip-chip model, the substrate can be removed and the exposed N polarity GaN can be made dark by chemical etching or a polymeric photonic crystal can be placed on what is now the top surface for better photon collection. Polymeric photonic crystal can be produced by laser lithography (holography) for reduced cost of fabrication. Courtesy of Lumileds/Philips.

package, an artistic view of which is shown in Figure 7.34 for junction temperatures $T < 120\,°C$ and $T \geq 150\,°C$. While on the topic, it should be mentioned that the epoxy used in some packages suffers from yellowing, a degradation mechanism that can be mitigated by substituting it with silicone resins or silicone-epoxy resins with thermal stability and UV resistance for encapsulating LED.

7.11 Luminescence Conversion and White Light Generation

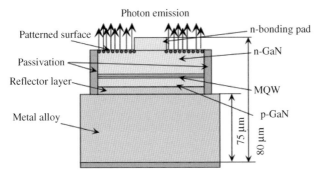

Figure 7.33 A Metal Vertical Photon LED (MVP-LED) structure from SemiLEDs. Courtesy of C.A. Tran, Semileds.

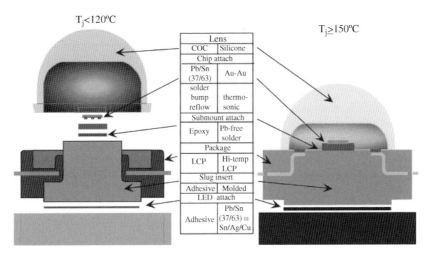

Figure 7.34 High-power LED packages for lighting applications. (a) LUXEON about 1998. (b) LUXEON K2 about 2006. Courtesy of LumilEDs/Philips.

7.11
Luminescence Conversion and White Light Generation

Availability of violet, particularly blue in the context of this chapter, compact LED emitters has paved the way for alternative approaches to generate blue, green, and red primary colors. A blue or a violet LED can be used to pump a medium containing the desired color centers, dyes in organic and phosphors in inorganic materials to generate the color(s) desired including white. For white LEDs to be accepted for indoor illumination, the *color-rendering index* (CRI), must conform to the CIE standards, at least what the standards being followed dictate, which might be changed. While the color temperature requirements can be met with phosphor pumped LEDs, the strict CRI together with very high luminous efficacies might be unattainable. The former is due to very close adherence to produce the Planckian spectrum and the latter is due to the Stokes shift (loss in this case) associated with phosphor down conversion. To meet the above mentioned requirements of CRI and color temperature, the multicolor LED approach is proposed, which is very cumbersome to begin with, in which the bandwidth (must be narrow) and in particular the wavelength of emission must be precisely controlled, which is very stringent.

With broad bandwidth associated with green LEDs (in addition to the much reduced power available in the so-called green gap) and to a lesser extent for blue LEDs, it may not be quite feasible to use three or four LED approach to achieve white light meeting the CIE indoor illumination standards. If so, an inexpensive and attainable method for white light generation may be to pump tricolor phosphors, which are known to have sharp emission linewidths, with LEDs. Alternatively, a blue source could pump two-color or three-color phosphors for white light generation. However, if one were to limit the phosphors to what has traditionally been available, one needs deep UV LED source with high power (which would not lead to high luminous efficacy because of losses associated with Stokes shift during down conversion) or develop efficient phosphors, which can be excited efficiently with blue LEDs and the less desirable UV LED approach that are accessible by the GaN system. The phosphors that can be pumped by GaN-based LEDs are under development and seem to be progressing well. As a matter of practicality and changing customer preferences may allow modifications in CRI in which case a blue pump coupled with yellow dye might be all that is needed for white light generation.

Since the introduction of the incandescent lamp in 1879, there has been a drive for brighter, cheaper, smaller, and more reliable light sources. In the USA, about 30% of all generated electricity is used for lighting. This is representative of the global trend; consequently, significant improvements in lighting efficiency would have a major impact on worldwide energy consumption. Unfortunately, none of the conventional light sources (incandescent, halogen and fluorescent) have improved significantly in the past several decades in terms of efficiency. Because an average of about 70% of the energy consumed by these conventional light sources is wasted as heat, which in many cases ends up only increasing the cooling required – consuming additional energy.

7.11.1
Color-Rendering Index

Color rendering of a light source is characterized by comparing the appearance of various object colors under illumination by the particular light source versus that under reference illumination, day light for CCT > 5000 K, and Planckian radiation for CCT < 5000 K. The standardized method, the CRI, is very well outlined by the CIE. In this method, 14 Munsell samples of various different colors, spectra for eight are given in Figure 7.35 including several saturated colors, are carefully selected and the color differences, denoted as ΔE_i, of these color samples under the test

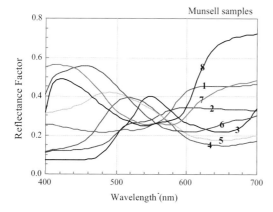

Figure 7.35 Munsell samples spectra for determining the CRI. Courtesy of M.E. Coltrin of Sandia National Laboratory.

illumination and under the reference illumination are calculated on the 1964 $W^*U^*V^*$ uniform color space. The process incorporates corrections for chromatic adaptation. Then the *Special Color Rendering Index*, R_i, for each color sample is calculated using $R_i = 100 - 4.6\Delta E_i$. Here a figure of 100 represents the best CRI.

To gain an insight about the extent of color saturation, the LED output is generally shown on the chromaticity diagram, as depicted in Figure 7.36. The oval near the center indicates various grades of "white light." The line through the white light region indicates the color diagram for white light with the accompanying color temperature (Planckian locus). The narrower the output spectrum of an LED, the closer its color is to the outer periphery. As the spectrum gets wider, the corresponding color on the chromaticity diagram is pulled toward the center reducing the range of colors that can be obtained by the color-mixing scheme. Moreover, the output of a commercially available white LED constructed from a blue InGaN LED overcoated with a yellow light under blue photoexcitation emitting cerium (Ce)-doped yttrium aluminum garnet (YAG:Ce) [$Y_3Al_5O_{12}$: Ce^{3+} (4f)] inorganic phosphor (YAG) is marked with data points indicative of various Gd concentrations.

The blue and red LEDs available commercially are almost saturated while the same cannot yet be said for green ones. The spectral broadening observed for green LEDs is attributed to compositional inhomogeneities, which get larger with increasing InN mole fraction. When used in conjunction with the available red and blue LEDs, present InGaN green LEDs provide the means for achieving some 70–80% of all the color possible.

7.11.2
White Light from Multichip LEDs

The three primary colors, red, green, and blue with identical intensities perceived by the eye, can be mixed together to generate white light. With currently available LEDs,

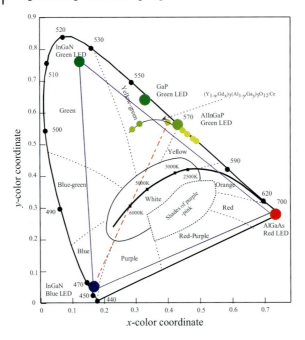

Figure 7.36 The chromaticity diagram along with available commercial LED performance data. Clearly, blue and green InGaN LEDs constitute the two important legs of the triad, the three primary colors that are needed for full-color displays. Moreover, the output of an optically pumped YAG medium doped for yellow emission is shown with data points indicative of various Gd concentrations. The broken line that connects the blue LED to a one particular composition of the YAG medium indicates the range of warm white colors that can be obtained.

the generation of white light can have luminous efficacies of approximately 45 lm/W. An example of a white light spectrum produced by combining the outputs from a three-color multiple LED is shown in Figure 7.37. Note that because the human eye's reduced response to the red color, the intensity of the red-color LED must be higher to generate white light with acceptable CRI. LED-based white light sources have been in commercial production and consist of collectively housed LED chips, or arrays of different-colored LED lamps, that is, multiple-chip LEDs. This three-color approach is potentially very efficient, high-quality white lighting approach, but the cost is expected to be high as it involves InGaN and AlInGaP technologies with each LED requiring different drives.

This particular approach has a few problems. The perceived color may change with viewing angle due to the discrete wavelengths of light used. Multiple LED chip requirements make this approach relatively expensive. Obtaining a consistent color across an array of such white pixels could also be a source of problem because the light intensity of LEDs and driving voltages tend to vary from diode to diode as well as color-tuning individual diodes is likely to be difficult. Temperature

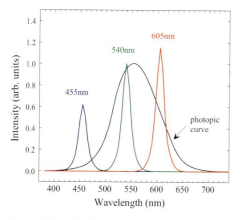

Figure 7.37 White light output emission spectrum from a three-color multichip LED. Courtesy of M.E. Coltrin of Sandia National Laboratory.

dependence of wavelength shift for each of the three diodes used may be different causing color variation with temperature. Another consideration is the variation in operating life and/or degradation rate of different color LEDs. Because the intensity variations also lead to overall color change, uneven degradation of the three LEDs would lead to color change over time. For example, the light output level of AlGaAs-based LEDs is found to decrease by about 50% after 15 000 to 40 000 hours of operation.

The chromaticity coordinates (x, y), CCT, CRI, and the luminous efficacy of radiation can be calculated if the spectral power distribution of a light source is known. Standard two-chip LEDs with any combination of wavelengths cannot produce white light with an R_a value that may not be acceptable in some cultures for general lighting applications. The three LED chip approach and in particular the four LED approach are both capable of producing much better color rendering, but the selection of peak wavelengths is critical in order to produce a CRI acceptable for general lighting, not to mention the very stringent wavelength and very narrow bandwidths control required. Simulations of three LEDs having peak wavelengths of 450, 550, and 650 nm, with their relative power adjusted to create white light with a color temperature of 4000 K results in a CRI value that is not acceptable. Each LED is modeled using a Gaussian line shape function, with a half-bandwidth of 20 nm. In this case, CRI (R_a) is only 37 with luminous efficacy of 228 lm/W (theoretical maximum). Again, a value of $R_a = 37$ is not acceptable for use in general lighting, except for limited outdoor use.

In terms of the three LED solution, when the wavelengths of each of the three LEDs in the three LED chip set are optimized and controlled precisely, general CRI can be improved substantially. Simulations of a three-chip LED set with peak wavelengths of 459.7, 542.4, and 607.3 nm lead to $R_a = 80$ and a luminous efficacy of 400 lm/W, which represents the theoretical maximum. This combination certainly meet the criteria for general lighting applications, but the wavelength accuracy assumed in the simulations is unrealistic and unattainable.

7.11.3
Combining LEDs and Phosphor(s)

This method for white light generation from multiple-chip LEDs involves the use of one or more phosphors, such as the combined use of a blue LED and a yellow phosphor, which represents the least expensive solution and has seen considerable activity. This method is favored in that the technology already exists; it is low cost, and requires small space. Already cool white with a color temperature of 5500 K and CRI of 70–85, and warm white with a color temperature of 3200 K and CRI of 90 have been reported to be available. These data alone out the difficult road ahead for the multi-LED approach. Let us now discuss various LED/phosphor approaches.

In the two-LED/one-phosphor approach called the RG_BB, the red and blue primary semiconductor sources are used in conjunction with a green phosphor pumped by blue. An efficiency of 95% is assumed for the green phosphor (less the 15.4% Stokes loss), which is a very challenging goal. Narrower range of red wavelengths can be used with efficiencies of 80% (615 nm red) or 90% (626 nm red) to reach 286 lm/W. Broad linewidths needed for the green phosphor (50–75 nm) as the red wavelength increases, the CRI improves to make up for the "missing" short wavelength red. The broad green phosphor used in this approach replaces the green and yellow LEDs in the four-LED approach.

In the one-LED/two-phosphor, R_BG_BB, approach, the blue primary semiconductor source is used to pump the green and red phosphors. An efficiency of 95% is typically assumed for both phosphors with Stokes losses: 24.2% (red), 15.4% (green). Owing to the properties of the red phosphor, very narrow range of red wavelengths are allowed for >286 lm/W. However, one needs a blue efficiency of 90% (if 615 nm red is used) or 100% (if 625 nm red is used) to reach the 286 lm/W luminous efficacy mark. The broad linewidths are needed for green phosphor (50–70 nm) as red wavelength increases. Narrow linewidths (1–20 nm) are needed for red phosphor pumped by blue. The irony is that there is no phosphor system meeting these specifications is available with developments reportedly in place. Figure 7.38 shows the interrelationship between the linewidth, wavelength, and luminous efficacy, underscoring again the importance of the red wavelength. We discuss the details, particularly in terms of the phosphors, later in this section following the discussion of the BY_Y approach in which a blue LED source is used to pump a yellow phosphor.

A more simplified approach, one-LED/one-phosphor, wherein a blue LED source is used to pump a yellow phosphor (as opposed to green and red phosphors shown in Figure 7.38, termed the BY_B approach, has gained considerable interest despite its low CRI for outdoor lighting in particular. Commercially available white LEDs are constructed from a blue InGaN LED over coated with a yellow light-emitting (under blue photoexcitation) cerium (Ce)-doped yttrium aluminum garnet (YAG:Ce) [$Y_3Al_5O_{12}:Ce^{3+}$ (4f′)] inorganic phosphor. This is called the white phosphor conversion LEDs (pcLEDs). In this approach, the InGaN LED generates blue light at a

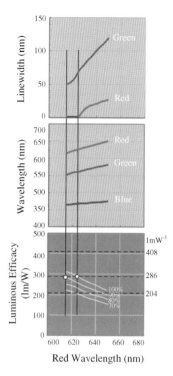

Figure 7.38 The relationship between the linewidth (top), wavelength (center), and luminous efficacy (bottom) in the BG$_G$R$_R$ approach wherein a blue LED source is used to pump green and red phosphors to achieve white light. Courtesy of M.E. Coltrin, Sandia National Laboratories.

peak wavelength of about 460–470 nm, which excites the trivalent cerium Ce^{3+}:YAG phosphor, which emits pale-yellow light, centered at about 580 nm with a full-width-at-half-maximum linewidth of 160 nm.

The combination of the blue light from the LED, which is transmitted through the phosphor, and the pale-yellow light from the Ce^{3+}:YAG results in soft white light with a color temperature of, for example, 4600 K, as shown in Figure 7.39 for several drive currents. The emission spectrum of the YAG phosphor can be modified (tuned) by substituting some or all the yttrium sites with other rare earth (RE) elements such as gadolinium (Gd), or terbium (Tb). The RE^{3+}:YAG emission and absorption spectrum can be further engineered by replacing some or all of the aluminum sites with gallium. Instead of illuminating inorganic phosphors such as RE^{3+}:YAG, the blue light emission from the InGaN LED can also be used to generate luminescence from organic polymers, which are coated on the epoxy encapsulate of an LED, but practicality of this approach is questionable.

The efficacy of the properly packaged (for heat removal as well as light extraction) production phosphor-white LEDs was around 60 lm/W with half-life time of 100 000 hours in 2005 (the figures improved considerably by a factor of nearly 3

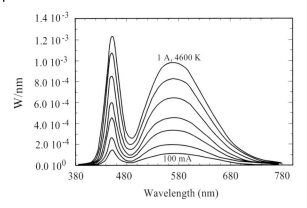

Figure 7.39 Emission spectrum of a white single-phosphor conversion LED, pcLED (blue + YAG:Ce) showing the blue peak of light leaking through the phosphor and the broader yellow peak from the phosphor. In a well-designed pcLED the spectrum changes negligibly with drive current. The color temperature (CCT) stays at 4600 K perfectly dimmable. Courtesy of Lumileds/Philips.

for efficacy by the year 2008). When compared to multiple chip LEDs for red-, green-, and blue-color outputs, an advantage of the phosphor white or hybrid-white phosphor conversion LED, is that it only requires one blue LED (or UV), in which case the blue light must be emitted by another phosphor). Also, the conversion efficiencies of about 90% are possible in inorganic YAG-based converters without the bounds imposed on the active layer composition. In addition, white-light LEDs based on phosphors have been shown to be relatively insensitive to temperature, which is very desirable.

While simple, there are several technological problems with this approach, at least with early designs. Among them is a halo effect of blue/yellow color separation due to the different emission characteristics of the LED (directional) and the phosphor (isotropic). Moreover, as mentioned above the CRI is low, only about 75–85 for the cool white, and broad color "bins" are necessary to ensure reasonable product yields. Finally, most lamps have color points that do not lie on the blackbody curve, which is undesirable and eventually a color shift from blue to yellow with aging and variation in drive current is noted. In addition to the aforementioned challenges, which mainly deal with the excitation source, there are other challenges associated with the phosphor material.

The flux that can be obtained in the pcLED combination is given as

$$\phi = \eta_{int} \cdot \eta_{extr} \cdot \eta_v \cdot \varepsilon_{o,ph} \cdot \eta_{QD} \cdot \eta_{ph} \cdot \eta_{pkg} \cdot P, \qquad (7.62)$$

where η_{int} is the internal quantum efficiency, η_{extr} is the photon extraction efficiency (% photons extracted per photon generated), η_v is the electrical efficiency (photon energy divided by the injection energy), $\varepsilon_{o,ph}$ is the luminous efficacy of phosphor/LED combination in terms of lm/W, η_{QD} is the quantum deficit due to Stokes shift (in terms of %), η_{ph} is the phosphor quantum efficiency, η_{pkg} is the

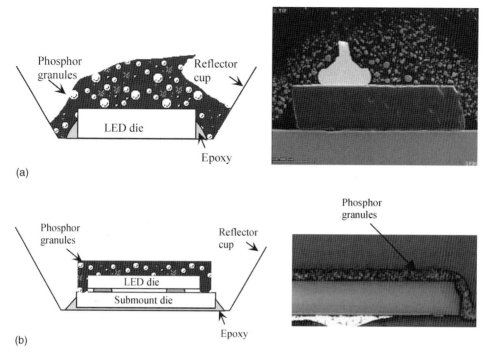

Figure 7.40 (a) Conformal coating of the yellow die (b) versus the standard coating to avoid color tint dependence on the viewing angle. Courtesy of Lumileds/Philips.

package photon extraction efficiency, and P is the electrical power applied (W). Improvement of the internal quantum efficiency depends on the materials quality and quantum well design. The extraction efficiency depends on the use of generated photons and can be improved by chip and package design. The phosphor quantum efficiency is dependent on the phosphor material and can be improved by progress in phosphor science. Finally, the level of electrical power that can be applied depends on the chip and package design). All of these components are on the table for improvement for increasing the flux obtainable from white LEDs based on the phosphor/LED combination.

In standard one LED/one phosphor solution (pcLED), the thickness of the yellow phosphor through which the blue LED light travels changes the tint by changing absorption and therefore the yellow emission. This means that the tint depends on the viewing angle for the construct show in Figure 7.40a. Conformal coating of the LED die with the yellow die shown in Figure 7.40b is designed to mitigate the "tint" change problem. The viewing angle dependence of the normalized CCT, with respect to on axis, for the phosphor slurry method and conformal phosphor coating method indicates a change of as large as 700 K in the CCT for the slurry case as opposed to 80 K for the conformal coating case is noted.

Further Reading

Berg, A.A. and Dean, P.J. (1976) *Light-Emitting Diodes*, Clarendon Press, Oxford.

CIE (1987) *International Lighting Vocabulary*, CIE 17.4/IEC 50 (845).

CIE (1995) *Method of measuring and specifying colour rendering properties of light sources*, CIE Publication 13.3.

Jones, E.D. (2000) *Inorganic Light Emitting Diodes for General Illumination. An Optoelectronics Industry Development Association Technology Roadmap*, Optoelectronics Industry Development Association, October 26–27.

Jones, C.F. and Ohno, Y. (1999) Colorimetric accuracies and concerns in spectroradiometry of LEDs. *Proceedings of the CIE Symposium'99 – 75 Years of CIE Photometry, Budapest*, pp. 173–177.

Landsberg, P.T. and Adams, M.J. (1973) Theory of donor-acceptor radiative and Auger recombination in simple semiconductors. *Proc. R. Soc. Lond. A.*, **334**, 523–539.

Munsell, A.H. (1905) *A Color Notation*, 1st edn, Munsell Color Company, Baltimore, MD.

Pankove, J.I. (1971) *Optical Processes in Semiconductors*, Prentice Hall.

Piprek, J. (2003) *Semiconductor Optoelectronic Devices: Introduction to Physics and Simulations*, Academic Press, San Diego.

Rea, M.S. (2000) *IESNA (Illuminating Engineering Society of North America) Lighting Handbook*, 9th edn, IESNA, pp. 26–33.

Shim, J.-I., Kim, H., Shin, S., and Yoo, H.-Y. (2011) An explanation of efficiency droop in InGaN-based light emitting diodes: saturated radiative recombination rate at randomly distributed in-rich active areas. *J. Korean Phys. Soc.*, **58** (3), 503–508.

van Roosbroeck, W. and Shockley, W. (1954) Photon-radiative recombination of electrons and holes in germanium. *Phys. Rev.*, **94**, 1558–1560.

8
Semiconductor Lasers: Light Amplification by Stimulated Emission of Radiation

8.1
Introduction

Semiconductor lasers cover a respectable wavelength range of about 0.3–11 μm (up to about 100 μm considering quantum cascade lasers relying on intraband transitions). Semiconductor lasers have many applications, for example, in optical communications as carrier and pumping sources, sources for projection TV, theater, the state-of-the-art electromechanical systems, and mundane applications such as pointers. One of the most salient applications of GaN-based lasers is in digital versatile disks (DVDs) that made it into game platforms and Blu-ray.

Wide-bandgap semiconductors based on ZnSe were considered as prime candidates for efficient compact light sources in the early 1960s, but it was after nearly three decades that the green light-emitting diodes (LEDs) and diode lasers were finally demonstrated in the early 1990s. New epitaxial techniques coupled with postgrowth processes proved to be crucial for the synthesis of wide-bandgap quantum well (QW) heterostructures. They also had a profound role in solving nearly intractable and long-standing defects and doping-related problems such as p-type doping. Major improvements have been made in the performance and device durability since the first commercially available laser diodes (LDs) in 2000. It is a foregone conclusion that 405–415 nm InGaN lasers, or preferable shorter wavelength devices utilizing GaN or (Al, In)GaN layers, form the basis of the new generation of optical disk technology. Last but not least, the gain in GaN is calculated to be high (without saturation up to $25\,000\,\text{cm}^{-1}$ and perhaps beyond) based on the large density of states (DOS), absorption coefficient ($10^5\,\text{cm}^{-1}$), and matrix element considerations. Now that high-quality GaN templates, typically prepared by hydride vapor phase epitaxy (HVPE), are available and laser structures must be of the highest quality, much of the laser work is done with heterostructures prepared on GaN templates/substrates, a representative of which is shown in Figure 8.1.

Nitride Semiconductor Devices: Fundamentals and Applications, First Edition. Hadis Morkoç.
© 2013 Wiley-VCH Verlag GmbH & Co. KGaA. Published 2013 by Wiley-VCH Verlag GmbH & Co. KGaA.

8 Semiconductor Lasers: Light Amplification by Stimulated Emission of Radiation

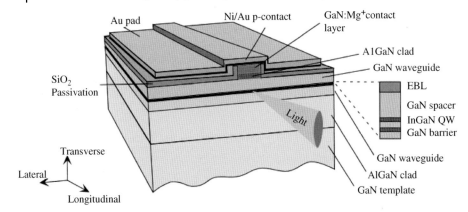

Figure 8.1 Schematic of a representative edge-emitting laser on GaN template. Adapted from W. Scheibenzuber and U. Schwarz.

8.2
A Primer to the Principles of Lasers

Semiconductors were not considered for lasers for a long time, as they are normally absorptive. It is, however, now well known that a semiconductor can be made transparent by injecting (generating) large concentrations of electrons and holes. In a laser, several processes take place simultaneously with important consequences. The injected electrons can recombine with holes to give off photons in a process termed *spontaneous emission* (the probability of which does not depend on the presence of other photons); a photon can be absorbed by a valence electron, which then gets excited into the conduction band in a process termed *absorption*. Likewise, when a conduction band electron recombines with a hole to generate a photon in the presence of another photon of the same energy, which is in phase with the initial photon, this is termed *stimulated emission*. All three processes are shown schematically in Figure 8.2.

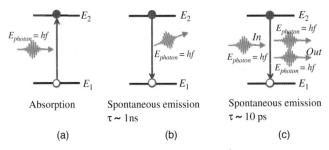

Figure 8.2 Schematic diagram of (a) stimulated absorption, (b) spontaneous emission, and (c) stimulated emission in a semiconductor. In stimulated processes, the probability of occurrence depends on the density of other photon with the same energy. Note that a photon generated through stimulated emission shares the same phase, direction, energy, and polarization as the photons present and goes on to amplify the electromagnetic wave. Courtesy of U. Ozgur.

With stimulated emission, which occurs beyond transparency, gain is obtained and it sustains the lasing oscillations. In calculating the conditions to yield lasing, all three processes that are strong functions of the conduction and valence band structures must be considered. In order for lasing to occur, the downward stimulated emission must dominate over the absorption process. This translates in the upper level, the conduction band in this case, having a higher population of electrons, which goes counter to standard distributions. This condition represents population inversion and at the onset the semiconductor becomes transparent to the stimulated emission. Stimulated emission can be understood by examining Figure 8.3, which schematically shows the upward and downward transitions across the bandgap. Optical excitation commences by absorbing a photon (step 1

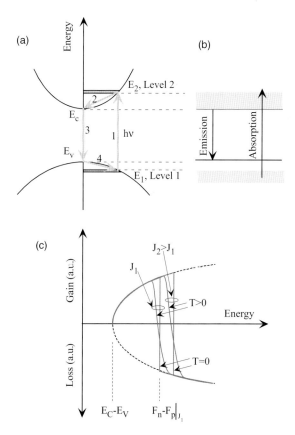

Figure 8.3 (a) Transitions involved in excitation (process 1), followed by relaxation to the bottom of the conduction band (process 2 on the order of 10^{-12} s) and recombination with resultant photon emission (with a carrier lifetime – time interval before recombination – of about 10^{-9} s) (process 3), and electron occupying a hole state created by process 1 (process 4). (b) Schematic representation of absorption and emission processes in a semiconductor beyond transparency. (c) Gain/loss diagram for a semiconductor for $T = 0$ and $T > 0$ K for two different pumping levels (J) beyond transparency.

in Figure 8.3a). To conserve momentum, as photon momentum is negligible, the excited electron must originate deep in the valence band. Nearly simultaneously (determined by the rate of dominant scattering mechanism that is polar optic phonon scattering in compound semiconductors taking place at picosecond timescale) the excited electron drops to the bottom of the conduction band (step 2 in Figure 8.3a). This is route to population inversion in that an electron exists at the bottom of the conduction band and a hole at the top of the valence band. This situation is retained for a time period on the order of 1 ns, which is the spontaneous emission (recombination) time. In a time period equal to the lifetime, on the average, an electron recombines with a hole at the top of the valence band (step 3 in Figure 8.3a). We should mention that as soon as a photon is absorbed (for excitation depicted with step 1 in Figure 8.3a), an electron leaves the top of the valence band to fill the hole created in the process (step 4 in Figure 8.3a). The notion of transparency can be understood by examining Figure 8.3b where the separation between the energy levels E_2 and E_1 (defining absorption) is larger than the photon energy produced through recombination from the bottom of the conduction band to the top of the valence band.

When the medium is made transparent, optical electromagnetic wave can be amplified as the medium converts from being an attenuating one to an amplifying one. Leaving the details for later discussions, before transparency, the loss in the semiconductor is a function of energy, as shown with lower dashed line in Figure 8.3c. As the transparency is approached, by optical or electrical pumping, the loss would be reduced giving way to zero at the onset of transparency and negative (gain) beyond the onset of transparency (upper solid lines in Figure 8.3c). First, the spectral dependence of the gain follows that of the *joint density of states* (JDOS). The gain would then convert to loss again for photon energies exceeding the separation of quasi-Fermi levels. The gain deviates from the functional dependence of the JDOS due to statistical spread as shown. As the pumping level is increased ($J_2 > J_1$), so is the separation between the quasi-Fermi levels ($F_n - F_p$) and so is the extension of the gain distribution to higher energies. A critical point is that high photon densities needed for laser operation cannot be maintained except in a waveguide cavity.

8.2.1
Waveguiding

Continuous-wave (CW) semiconductor lasers cannot exist without a proper waveguide. The refractive index of the narrower bandgap active layer should be larger than the refractive index of the wide-bandgap material (cladding layer) straddling it for confining and guiding. Waves undergoing total internal reflections are supported by the waveguide and are called the *waveguide modes*. The transverse field distribution of such a mode, vertical to the growth direction, can be described by a cosine or a sine function inside the layer. It decays exponentially in the outer, low refractive index media. The refractive index and its dispersion are imperative for accurate design. As for wurtzite (WZ) GaN and related nitrides, they lack cubic symmetry and therefore are anisotropic, which results in uniaxial birefringence, meaning different refractive indices for polarization perpendicular (n_o ordinary) and parallel (n_e extraordinary) to the c-axis. Refractive index dispersion has been measured for GaN and AlGaN in a variety of ways (interferometry, ellipsometry, and prism coupling) with notable spread in data.

The transmission (reflection) interferometry is the simplest of all, where the transmission (reflection) is measured as a function of wavelength. Oscillations are observed in the transparent wavelength region of the sample due to interference of multiple reflections at the surface/interfaces. When half-integer multiples of the wavelength equal the optical path length in the sample, maxima in the spectrum occur. Because the product of the thickness and the index is measured by this technique, thickness values from other measurements are required to determine the refractive index.

Spectroscopic ellipsometry detects the polarization change in the light beam reflected from the sample. The measured ratio of the reflected light for the transverse electric (TE) and transverse magnetic (TM) polarizations allows determination of the refractive index with a higher accuracy compared to the interferometric technique. However, the thickness of the film should be on the order of the wavelength to avoid multiple solutions. Consequently, thick films are not suitable for the ellipsometry method if the thickness is not known from other measurements and does not fall into the region causing multiple solutions.

The prism coupling method is a very accurate means for measuring the refractive indices and thickness of thin films. The measurement technique is based on phase matching the light in the prism to the modes allowed in the waveguide formed by the higher index film sandwiched between two lower index layers. It produces a higher accuracy compared to the other two techniques mentioned above; however, it requires the use of laser beams, limiting the measurement to the available laser wavelengths.

Using the highly accurate prism coupling technique, the dispersion of the ordinary ($n_o(\lambda)$) and extraordinary ($n_e(\lambda)$) refractive indices has been reported for two different sets of $Al_xGa_{1-x}N$ with $0 \leq x \leq 0.2$ and $0 \leq x \leq 1.0$, respectively, in the wavelength region of $457 < \lambda < 980$ nm. The ordinary and extraordinary indices measured at various laser lines for the MBE-grown samples are shown in Figure 8.4. The data for each sample were fit to the first-order Sellmeier dispersion formula. The resulting values of A_0 and λ_0 were fit to polynomials in x:

$$n^2(\lambda, x) = 1 + \frac{A_0(x)\lambda^2}{\lambda^2 - \lambda_0^2(x)},$$
$$A_0(x) = B_0 + B_1 x,$$
$$\lambda_0(x) = C_0 + C_1 x + C_2 x^2.$$

(8.1)

The B_i and C_i ($i=0, 1, \ldots$) coefficients are listed in Table 8.1.

A widely used method to obtain the InGaN refractive index relies on the shift of the GaN refractive index data with the assumption that the refractive index is identical at the band edge for both materials. Highly simplified calculations suggest that this extraction is reasonable for InGaN with low In compositions.

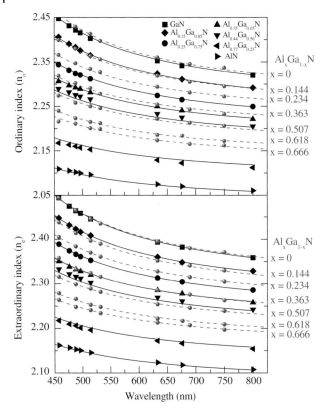

Figure 8.4 Measured ordinary and extraordinary index dispersion for MBE-grown AlGaN samples with different Al compositions. The lines are a result of the Sellmeier parameterization with bandgap energy. The solid spheres represent data measured for MOCVD- and HVPE-grown samples and the dashed lines are the Sellmeier fits to the data. Courtesy of U. Ozgur.

Table 8.1 B and C coefficients for the adjustable parameters in the Sellmeier dispersion formula for the MBE samples parameterized in x for the ordinary and extraordinary refractive indices.

Coefficient	n_o	n_e
B_0	4.1446 ± 0.0146	4.2957 ± 0.0165
B_1	-1.0021 ± 0.0273	-0.9817 ± 0.0310
C_0 (nm)	190.72 ± 2.48	191.71 ± 2.23
C_1 (nm)	-82.999 ± 12.363	-76.363 ± 11.142
C_3 (nm)	-27.521 ± 11.619	-23.427 ± 10.471

8.2.2
Analytical Solution to the Waveguide Problem

An ideal waveguide would fully confine the electromagnetic wave while incurring no loss. Waveguiding is treated in many electromagnetics texts. An electromagnetic wave with the incidence angle with respect to the normal equal to or larger than the critical angle θ_c,

$$\theta_c = \sin^{-1}\frac{n_2}{n_1}, \tag{8.2}$$

will undergo total internal reflection provided that $n_1 > n_2$. The complex dielectric function ε can be related to the refractive index n (n_r and $\bar{n}$ are also used throughout this book for the same) and the extinction coefficient K according to

$$\varepsilon = \varepsilon' + j\varepsilon'' = \varepsilon_0(n+j\kappa)^2 = \varepsilon_0\left(n^2 - \kappa^2 + j2n\kappa\right). \tag{8.3}$$

Because the power of an electromagnetic field propagating along the z-direction is proportional to

$$\exp[-2(j(n+j\kappa)k_0 x)], \tag{8.4a}$$

where

$$k_0 = \frac{2\pi}{\lambda_0} \tag{8.4b}$$

is the free space wave vector. The absorption coefficient is then given by

$$\alpha = 2\kappa k_0, \tag{8.5a}$$

$$\kappa = \frac{\varepsilon''}{2\varepsilon_0 n}. \tag{8.5b}$$

It is clear that the imaginary part of the dielectric constant is responsible for the loss term.

To segue into the waveguide problem, the concept of wave guidance for TE (no E field component in the direction of propagation) and TM (no H field component in the direction of propagation) modes is shown schematically in Figure 8.5. Internal reflection occurs when $n_1 > n_2$ or $n = (n_2/n_1) < 1$ with n being the relative refractive index. External reflection occurs when $n_1 < n_2$ or $n = (n_2/n_1) > 1$.

The process is characterized by a reflection coefficient r with an associated reflectance of R and by transmission coefficient t with an associated transmittance of T ($R+T \equiv 1$). The reflectance and transmittance represent the fraction of power in reflected and transmitted waves, respectively:

$$R = \frac{P_r}{P_i} = r^2 = \left(\frac{E_r}{E_i}\right)^2 \quad \text{and} \quad T = \frac{P_t}{P_i} = n\left(\frac{\cos\theta_t}{\cos\theta_i}\right)t^2, \tag{8.6}$$

where the subscripts i, r, and t indicate the incident, reflected, and transmitted waves, respectively.

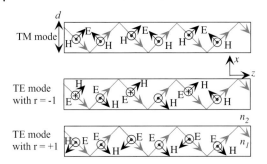

Figure 8.5 Wave guidance for TE (no E field component in the direction of propagation) and TM (no H field component in the direction of propagation) modes. Negative values of r indicate phase change.

8.2.2.1 TE Mode

A TE-mode plane wave propagating along the z-direction within the guide has even and odd solutions. The even solution field components for TE mode in phasor notation are obtained from Maxwell's equations:

$$E_y = \bar{E}\cos(\kappa x)\exp(-j\beta z), \tag{8.7a}$$

$$H_x = \frac{-j}{\omega\mu_0}\frac{\partial}{\partial z}E_y = -\frac{\beta}{\omega\mu_0}\bar{E}\cos(\kappa x)\exp(-j\beta z), \tag{8.7b}$$

$$H_z = \frac{j}{\omega\mu_0}\frac{\partial}{\partial x}E_y = -\frac{j\kappa}{\omega\mu_0}\bar{E}\sin(\kappa x)\exp(-j\beta z). \tag{8.7c}$$

Here, bar above the symbol means that the electric field amplitude E is used in phasor notation and the angular frequency ω is related to the frequency by $\omega = 2\pi\nu$, $\beta^2 = \omega^2\mu\varepsilon$.

The notion of guidance within the guide explicitly implies that the wave should decay outside the waveguide. For this to happen, the exponent in the x-direction should be real. Then,

$$E_{y_2} = \bar{E}\cos\left(\frac{\kappa d}{2}\right)\exp\left[-\gamma\left(x-\frac{d}{2}\right)\right]\exp(-j\beta z), \tag{8.8}$$

where $\gamma^2 = \beta^2 - n^2 k_0^2$, $k_0 = 2\pi/\lambda_0$ and $\lambda_0 = c/\nu$, with c being the speed of light in vacuum.

For the wave not to propagate or attenuate outside the core of the waveguide, γ should be positive that leads to the condition $\beta^2 > n_2^2 k_0^2$, which brings us back to the total internal reflection condition of Equation 8.2.

A similar treatment for the magnetic field components outside the core provides

$$H_{x_2} = \frac{-j}{\omega\mu_0}\frac{\partial}{\partial z}E_y = -\frac{\beta}{\omega\mu_0}\bar{E}\cos\left(\frac{\kappa d}{2}\right)\exp\left[-\gamma\left(x-\frac{d}{2}\right)\right]\exp(-j\beta z), \tag{8.9a}$$

$$H_{z_2} = \frac{j}{\omega\mu_0}\frac{\partial}{\partial x}E_y = -\frac{j\gamma}{\omega\mu_0}\bar{E}\cos\left(\frac{\kappa d}{2}\right)\exp\left[-\gamma\left(x-\frac{d}{2}\right)\right]\exp(-j\beta z). \tag{8.9b}$$

8.2 A Primer to the Principles of Lasers

Applying the boundary conditions that the y-component of the E field and the z-component of the H field must be continuous at $x = d/2$, we get

$$\frac{\kappa d}{2}\tan\frac{\kappa d}{2} = \frac{\gamma d}{2} \quad \text{or} \quad \tan\frac{\kappa d}{2} = \frac{\gamma}{\kappa} = \left[\frac{\beta^2 - n_2^2 k_0^2}{n_1^2 k_0^2 - \beta^2}\right]^{1/2}. \tag{8.10}$$

Here, β is the unknown and its solution can be found from the transcendental equation either numerically or graphically.

Recognizing that $\kappa^2 = n_1^2 k_0^2 - \beta^2$ and $\gamma^2 = \beta^2 - n_2^2 k_0^2$ and adding them lead to the elimination of β^2:

$$(n_1^2 - n_2^2)\left(\frac{k_0 d}{2}\right)^2 = \left(\frac{\kappa d}{2}\right)^2 + \left(\frac{\gamma d}{2}\right)^2. \tag{8.11}$$

The joint solution of Equation 8.10 is the one that satisfies the guidance condition. Equation 8.11 represents a circle with radius $r = (n_1^2 - n_2^2)^{1/2}[(k_0 d)/2]$ in the $\kappa d/2$, $\gamma d/2$ plane. Intersections of the graphical representations of Equations 8.10 and 8.11 are then the desired solutions, as illustrated in Figure 8.6. The left-hand side of

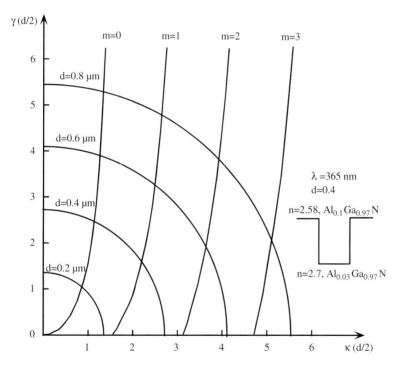

Figure 8.6 Graphical approach to determine the waveguide modes. The quarter circles represent the solution for Equation 8.10. The solution for Equation 8.11, with m as the mode parameter, is also shown. The intersections of the two sets of plots are then the solutions for modes supported by the waveguide.

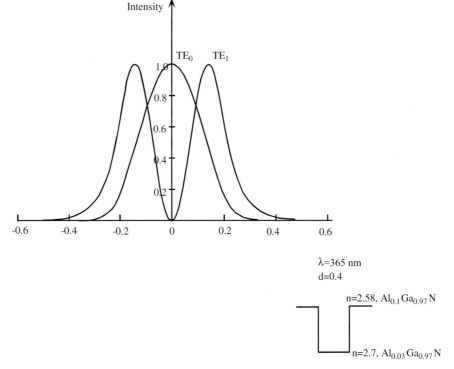

Figure 8.7 Optical field profile of fundamental and second-order modes for TE radiation obtained from Equation 8.7.

Equation 8.10, $\kappa d/2 \tan(\kappa d/2)$, is plotted as $\gamma d/2$ to give the quarter circles in Figure 8.6 for waveguide thicknesses in the range 0.2–0.8 µm.

It is clear that for small values of d, only the fundamental mode solution, $m = 0$, exists. Figure 8.7 depicts the field distribution of the fundamental and second-order modes. It is evident that the fundamental mode is necessary for sufficient confinement.

8.2.2.2 TM Mode

As for the TM polarization, a TM-mode plane wave propagating along the z-direction within the guide also has even and odd solutions. The even solution field components for TM mode, in phasor notation, is obtained from Maxwell's equations:

$$H_y = \bar{H}\cos(\kappa x)\exp(-j\beta z), \tag{8.12a}$$

$$E_x = \frac{-j}{\omega\mu_0}\frac{\partial}{\partial z}H_y = -\frac{\beta}{\omega\mu_0}\bar{H}\cos(\kappa x)\exp(-j\beta z), \tag{8.12b}$$

$$E_z = \frac{j}{\omega\mu_0}\frac{\partial}{\partial x}H_y = -\frac{j\kappa}{\omega\mu_0}\bar{H}\sin(\kappa x)\exp(-j\beta z). \tag{8.12c}$$

The rest of the treatment is identical to that depicting the TE mode case.

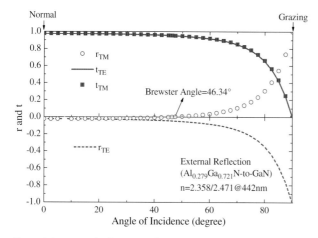

Figure 8.8 External reflection and transmission coefficients for both TE and TM modes as a function of angle of incidence for a model case representing AlGaN (~28%) to GaN interface.

At normal incidence,

$$R = r^2 = \left(\frac{1-n}{1+n}\right)^2. \tag{8.13}$$

The external reflection case with transmission and reflection coefficients as a function of angle of incidence θ for both TE and TM modes and for AlGaN–GaN interface is shown in Figure 8.8. Note that $\theta = 0$ and $\theta = 90°$ indicate normal and grazing incidence angles. Also note that when the TM mode reflection is zero, at $\theta = 46.34°$ the wave is TE polarized.

Internal reflection for TE and TM modes as a function of angle of incidence is shown in Figure 8.9 – again for the same model system as that depicted in Figure 8.8. The angle of incidence where the TM mode reflectance goes to zero, $\theta_p = 43.66°$, given by $\theta_p = \tan^{-1}(n) = \tan^{-1}(n_2/n_1)$, is called the polarizing angle or more popularly the Brewster angle that occurs for the TM polarized light only. At this angle, TM mode is perfectly transmitted with no reflection.

Recall that phase changes upon reflection. For example, negative reflection coefficient indicates the pertinent wave will have its vector switch direction upon reflection. As can be seen in Figure 8.8 depicting the external reflection case, the reflection coefficient is negative for all incidence angles for the TE mode and for $\theta < \theta_p$ for the TM mode. In the internal reflection case, Figure 8.9, and for incidence angles $\theta < \theta_c$, the reflection coefficient is positive and that for TM is also positive except for $\theta > \theta_p$. However, the angle of incidence values for $\theta > \theta_c$ lead to complex reflection coefficients given by

$$r_{TE} = \frac{\cos\theta - i\sqrt{\sin^2\theta - n^2}}{\cos\theta + i\sqrt{\sin^2\theta - n^2}} \quad \text{and} \quad r_{TM}$$

$$= \frac{-n^2\cos\theta + i\sqrt{\sin^2\theta - n^2}}{n^2\cos\theta + i\sqrt{\sin^2\theta - n^2}} \quad \text{or simply} \quad r = |r|e^{i\varphi}, \tag{8.14}$$

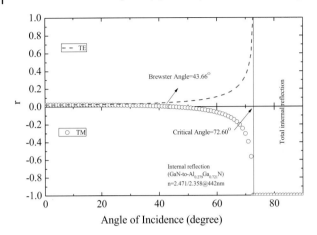

Figure 8.9 Internal reflection for TE and TM modes as a function of angle of incidence for the GaN to AlGaN (~28) model system. The critical angle for total reflection is $\theta_c = \sin^{-1}(n) = \sin^{-1}(n_2/n_1) = 72.6°$ and the angle at which the TM reflection goes to 0 is $\theta_p = 43.66°$. Also note that at normal incidence $\theta' = 90°$, the reflection coefficients for both TE and TM modes go to unity.

with ϕ representing the phase shift. The phase shift dependence on the angle of incidence for TE and TM modes is given by

$$\tan\left(\frac{\phi_{TE}}{2}\right) = -\frac{\sqrt{\sin^2\theta - n^2}}{\cos\theta}, \quad \tan\left(\frac{\phi_{TM} - \pi}{2}\right) = -\frac{\sqrt{\sin^2\theta - n^2}}{n^2\cos\theta}. \quad (8.15)$$

Another important parameter in waveguides is the *confinement factor* Γ that is defined as the ratio of the optical power overlapping with the active layer to the total optical power in that mode. Mathematically,

$$\Gamma = \frac{\int_{-d/2}^{d/2} E(x)^2 dx}{\int_{-\infty}^{\infty} E(x)^2 dx}, \quad (8.16)$$

where $E(x)$ is the distribution of the optical field amplitude in the direction x. The squared exponent is indicative of the need to use optical power. Figure 8.10 exhibits the optical field (E field) distribution for a multiple quantum well (MQW) InGaN active layer (GaN waveguide) and an AlGaN cladding layer containing 15% Al. The variable parameter is the thickness W of the GaN film on each side of the active layer. The active layer consists of four wells made of InGaN with 15% In and barriers made of InGaN with 5% In having a total MQW region thickness of 400 Å; it replicates a Nichia laser structure. Here, W is allowed to have the values of 0.08, 0.1, and 0.12 μm. Figure 8.11 shows the field and refractive index distribution calculated for the same waveguide (Figure 8.10). The effect of the 200 Å $Al_{0.2}Ga_{0.8}N$ employed to keep the InGaN MQW from dissociating during the high-temperature growth of the top waveguide has been neglected.

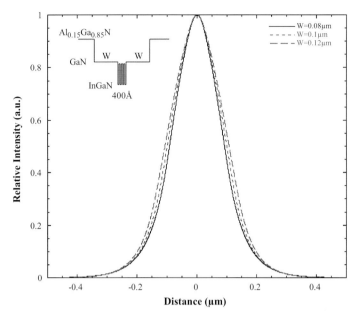

Figure 8.10 The optical field (E field) distribution for a four-well MQW InGaN active layer, GaN waveguide, and AlGaN cladding layer containing 15% Al. The variable parameter is the thickness W of the GaN on each side of the active layer and is half the total thickness of the waveguide.

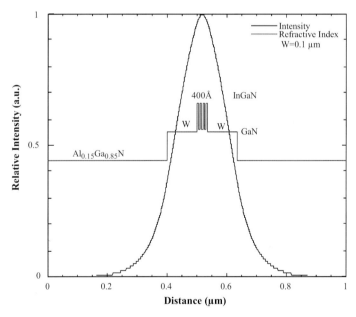

Figure 8.11 The field and refractive index distributions calculated for the waveguide as in Figure 8.10. Parameters used are 2.536 for the effective index, amplification region 6.46%.

8.2.3
Far-Field Pattern

Knowledge of the radiation characteristics, that is, the far-field pattern, of a semiconductor laser is necessary for proper collection of the radiated power. It is also a measure of the waveguiding properties that gives a reality check on the calculations of the field distribution and the validity of the parameters used such as the refractive indices that are functions of not only the composition but also the doping and the temperature (in an expanded view, the refractive index would also be a function of strain and presence of any externally applied electric field as in some optical, mainly passive, devices). Diffraction expressions must be used for a given mode or sets of modes to calculate the far-field pattern, which is beyond the scope of this text. The form of the radiation pattern emanating from a semiconductor laser is exhibited in Figure 8.12 where $\theta_\perp$ and $\theta_{//}$ represent the full angles at half power in

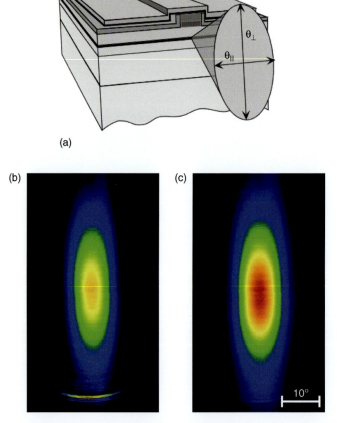

Figure 8.12 Schematic representation of the far-field characteristics of an edge-emitting injection laser (a) with (b) and without (c) leakage into the GaN buffer layer below the AlGaN clad (substrate mode peak). Adapted from W. Scheibenzuber and U. Schwarz.

the directions perpendicular and parallel to the plane of the p–n junction, respectively. Because the dimension perpendicular to the junction plane is small, considerable divergence of the beam occurs in this direction.

To a first approximation, a uniform aperture radiates a single lobe of which the full width at half maximum (FWHM) power $\theta_\perp$ changes inversely with the thickness W of the emitting region, as given by

$$R_{21}|_{st} \tag{8.17}$$

Similarly, for a diode with a strong confinement in the lateral direction such as in index-guided laser of width Z, the beam width in the plane of the junction is again for a uniform aperture,

$$\theta_{//} \approx \frac{1.2\lambda}{Z}. \tag{8.18}$$

There exist higher order modes as well that give rise to complex radiation patterns, which cannot be represented by Equations 8.16 and 8.17. The semiconductor is not impartial to which mode propagates, and the mode that has the largest fraction of its intensity distribution overlap the gain region will have the largest gain. Moreover, the facet reflectivity has an effect, but it is mainly manifested in the polarization. For example, a plane wave propagating in a waveguide whose E field lies in the facet experiences a lower facet loss than other polarizations and culminates in the TE mode that is supported.

The far-field pattern can be obtained by considering the TE waves, as we have done so in conjunction with waveguiding and solving the wave equation in free space, the details of which can be found in many textbooks:

$$\frac{I(\theta_\perp)}{I(0)} = \frac{\cos^2 \theta_\perp \left| \int_{-\infty}^{+\infty} E_y(x,0) \exp(j \sin \theta_\perp k_0 x) dx \right|^2}{\left| \int_{-\infty}^{+\infty} E_y(x,0) dx \right|^2}, \tag{8.19}$$

where k_0 is the magnitude of the propagation vector in free space and E_y is the y component of the electric field, it is the only component because the field is of TE mode.

8.3 Loss, Threshold, and Cavity Modes

In addition to losses germane to the semiconductor and the waveguide properties, the end losses must also be compensated for by the amplification inside the laser cavity. There are also other losses distributed inside the cavity, such as *free carrier losses*. In general, the absorption law $I(z) = I_0 \exp(-\alpha z)$, with I_0 being the incident light intensity, α the absorption coefficient, and z the distance, comes into play.

Recognizing that the distributed losses in the active layer and in the cladding layers (with absorption coefficients α_a and α_c, respectively) are not identical, we can write for the internal loss,

$$\alpha_i = \Gamma \alpha_{tot} = \Gamma \alpha_a + (1-\Gamma)\alpha_c, \tag{8.20}$$

where α_{tot} represents the total loss and Γ denotes the total confinement factor, as defined by Equation 8.16. Qualitative conditions for the onset of laser action or the threshold conditions are obtained when the gain or negative absorption, at the population inversion condition, compensates for all losses sustained during a round-trip in the laser cavity. This is given by

$$I_0 R_1 R_2 \exp[2(g_{th} - \alpha_i)L] = I_0. \tag{8.21}$$

The amplitude portion of which leads to

$$g_{th} = \alpha_i + \frac{1}{2L} \ln \frac{1}{R_1 R_2}, \tag{8.22}$$

where g_{th} is the threshold gain, α_i is the distributed internal losses, R_1, R_2 are the reflection coefficients of the laser facets (which are generally made unequal as the light extraction from one facet is favored over the other), and L is the cavity length. The first term in Equation 8.22 is the internal loss term, while the second term represents the end losses from the facets.

The phase portion of Equation 8.21, that is, the condition that a wave making a full round-trip in the cavity is in phase for sustained oscillations, leads to

$$\frac{4\pi n_r L}{\lambda_m} = 2m\pi \quad \text{or simply} \quad m\left(\frac{\lambda_m}{2n_r}\right) = L, \tag{8.23}$$

where m is an integer taking the values 1, 2, 3, ..., n is the refractive index, and λ_m and λ_m/n_r are the free space and inside-the-cavity wavelengths, respectively.

The longitudinal mode spacing near the fundamental mode λ_0 (corresponding to $m=1$) can be obtained from the derivative of the phase condition as

$$dm\lambda_0 + md\lambda_0 = 2Ldn_r. \tag{8.24}$$

By setting $dm = -1$ and substituting m from Equation 8.23, one can find the adjacent mode spacing:

$$\Delta \lambda_0 = \frac{\lambda_0^2}{2L[n_r - \lambda_0(dn_r/d\lambda_0)]}. \tag{8.25}$$

Polynomial fits to the data presented in Figure 8.4 can ameliorate the calculation of the longitudinal modes determined by Equation 8.25.

In a semiconductor laser, the propagating mode is not entirely within the active layer in which case Equation 8.22 must be modified to take the confinement factor into account:

$$\Gamma g_{th} = \alpha_i + \frac{1}{2L} \ln \frac{1}{R_1 R_2}, \tag{8.26}$$

where $\Gamma g_{th} = g_{n_{th}}$ is the modal gain with g_{th} being the intrinsic or material gain at the threshold (intrinsic gain). It is customary to lump all the waveguide losses other than the end losses into internal loss $\alpha_i = \Gamma\alpha_a + (1-\Gamma)\alpha_c$, where α_a and α_c represent the losses in the active region and the cladding, respectively.

8.4 Optical Gain

Calculation of the gain is rather complicated and requires a very accurate knowledge of the genesis of the lasing mechanisms as well as the semiconductor band structure, which includes effective masses. The gain can be calculated from the spontaneous emission spectrum or the absorption spectrum if we know the recombination mechanisms, the electron–hole plasma and the excitonic origin. The gain expression is directly related to the band structure, the transition probabilities, and the occupation probabilities. Under high injection, which is the case in semiconductor lasers, Coulomb interactions are screened completely in small-bandgap semiconductors with low exciton binding energies and excitons dissociate easily. Thus, electrons and holes can be treated independently as plasma, except for the many-body effects such as bandgap renormalization. As a first approximation, additional assumptions are made in the framework of which the electrons are assumed to interact with the electromagnetic field but not with other electrons and phonons; this is the basis for the so-called *single particle model*.

Let us now discuss the simultaneous processes, that is, spontaneous emission, stimulated emission, and absorption as they are eventually related to gas, solid, and semiconductor lasers. Relying on the treatment of Planck, *Einstein* treated stimulated emission per unit electromagnetic energy with energies between $h\nu$ and $h(\nu + \Delta\nu)$ due to the transition from level 2 to level 1 by the coefficient B_{21} that describes the transition probability from state 2 to 1. Transitions from level 1 to level 2, absorption, were described by the transition probability B_{12}. The spontaneous emission from level 2 to level 1 is depicted as A_{21}. These A and B coefficients are called the *Einstein's A and B coefficients*.

For a two-level system in thermal equilibrium at temperature T, the rate of upward ($1 \rightarrow 2$) and downward ($2 \rightarrow 1$) transitions were expressed by Einstein (see Chapter 6 for more details):

$$r_{21} = B_{21}\rho_e(\nu) \quad \text{and} \quad r_{12} = B_{12}\rho_e(\nu). \tag{8.27}$$

The term $\rho_e(\nu)d\nu$ is the volume density of the electromagnetic energy in the frequency range ν and $\nu + d\nu$ or in the energy range $h\nu$ and $h\nu + dh\nu$. The coefficients B_{21} and B_{12} are called the Einstein's B coefficient as alluded to above.

In a semiconductor, occupation probabilities of levels 1 and 2 must also be considered in which case Equation 8.27 would be modified:

$$r_{21} = B_{21}f_2(1-f_1)\rho(\nu) \quad \text{and} \quad r_{12} = B_{12}f_1(1-f_2)\rho(\nu), \tag{8.28}$$

where f_1 and f_2 represent the occupation probabilities of levels 1 and 2, respectively. Therefore, $(1-f_1)$ and $(1-f_2)$ represent the probabilities of levels 1 and 2 being empty, respectively. Essentially, if a transition is to occur between level 1 and level 2, there must be electrons available in level 1, indicated by f_1, and there must be empty states available in level 2, indicated by $(1-f_2)$, to receive that electron making the transition from level 1 to level 2. If N_2 and N_1 represent the population (or photon occupation numbers) of levels 2 and 1, respectively, the rate of change in the population of level 2 through spontaneous emission (represents decay in population, which is why the – sign) can be written as

$$\left.\frac{dN_2}{dt}\right|_{spon} = -A_{21}N_2. \tag{8.29}$$

Note that spontaneous emission is not coupled to the optical field and therefore does not depend on the photon density in the system having the same energy. The above rate equation simply indicates that the photon population in level 2 would decrease with a rate of A_{21} by which we can argue that the spontaneous emission lifetime is $\tau_{sp} = (A_{21})^{-1}$. Typically, this lifetime is on the order of 10^{-9} s.

Turning our attention to stimulated emission, which again involves transitions from level 2 to level 1, we can write the rate equation as

$$\left.\frac{dN_2}{dt}\right|_{st} = -B_{21}N_2\rho_e(\nu). \tag{8.30}$$

Note that this process is coupled to the photons having the same energy in the system, which is the genesis for the $\rho_e(\nu)$ term. The product $\rho_e(\nu)N_2 d\nu$ represents the photon energy density in the frequency range of ν and $\nu + d\nu$. A decrease in the N_2 population due to transition from level 2 to level 1 would be accompanied by an equal increase in N_1, which means that Equation 8.30 can also be written as

$$\left.\frac{dN_1}{dt}\right|_{st} = B_{21}N_2\rho_e(\nu). \tag{8.31}$$

Doing the same for stimulated absorption whose proportionality constant is B_{12} (as absorption involves excitation of an electron from level 1 to level 2),

$$\left.\frac{dN_2}{dt}\right|_{absorp} = R_{12} = B_{12}N_1\rho_e(\nu) \quad \text{or} \quad \left.\frac{dN_1}{dt}\right|_{absorp} = -R_{12} = -B_{12}N_1\rho_e(\nu). \tag{8.32}$$

Note that the stimulated absorption, which is involved here, is dependent on the photon density, having the same energy, in the system. Again, the product $\rho_e(\nu)N_1 d\nu$ represents the photon energy density in the frequency range of ν and $\nu + d\nu$. Let us now define the expression governing $\rho_e(\nu)$.

The sum of spontaneous emission and stimulated emission represents the total downward transitions from level 2 to level 1. The total rate of these two processes can then be expressed:

$$R_{21} = [A_{21} + B_{21}\rho_e(\nu)]N_2. \tag{8.33}$$

Under thermodynamic equilibrium, each upward transition must be balanced by a downward transition, in which case

$$R_{21} = R_{12} \Rightarrow A_{21}N_2 + B_{21}N_2\rho_e(\nu) = B_{12}N_1\rho_e(\nu). \tag{8.34}$$

Manipulation of Equation 8.34 leads to

$$\frac{N_2}{N_1} = \frac{B_{12}\rho_e(\nu)}{A_{21} + B_{21}\rho_e(\nu)}. \tag{8.35}$$

Determination of the photon density requires knowledge of the number of modes that can be accommodated in a cavity, which is treated in texts including Morkoç (2009). Recognizing that each point represents two modes, TE and TM, the mode density in a frequency interval between ν and $\nu + d\nu$ (in a frequency interval $d\nu$) and for a unit volume can be formulated by

$$\rho(\nu)d\nu = \frac{1}{V}\frac{dN}{d\nu}d\nu = \frac{8\pi n_r^3}{c^3}\nu^2 d\nu. \tag{8.36}$$

Note that $\rho(\nu)$ denotes the photon mode density in units of $1/m^3$, whereas $\rho_e(\nu)$ depicts the energy density in units of J/m^3. Again, skipping the details, which can be found in standard texts, the number of modes in a volume V in a frequency interval of $d\nu$ around a central frequency ν or energy E is given by

$$dN'(\nu) = \frac{8\pi n_r^3 \nu^2 V}{c^3} d\nu \quad \text{or in terms of energy,} \quad dN'(E) = \frac{8\pi E^2 n_r^3 V}{h^3 c^3} dE. \tag{8.37}$$

The number of modes (photon density) per unit volume and per unit frequency (spectral density) is then the derivative of N' with respect to ν divided by the volume that represents the spectral density:

$$\rho(\nu) = \frac{1}{V}\frac{dN'(\nu)}{d\nu} = \frac{8\pi \nu^2 n_r^3}{c^3} \quad \text{or in terms of energy,} \quad \rho(E) = \frac{1}{V}\frac{dN'(E)}{dE} = \frac{8\pi E^2 n_r^3}{h^3 c^3}. \tag{8.38}$$

Boltzmann statistics imply that the probability of a given cavity mode that lies between $h\nu$ and $h\nu + hd\nu$ is proportional to $\exp(-h\nu/kT)hd\nu$. The average energy per mode using the aforementioned distribution function is given by

$$\langle E \rangle = \frac{h\nu}{e^{h\nu/kT} - 1} = \frac{E}{e^{E/kT} - 1} \quad \text{or average number of photons/mode}$$
$$\langle \rho \rangle = \frac{1}{e^{h\nu/kT} - 1} = \frac{1}{e^{E/kT} - 1}. \tag{8.39}$$

The photon mode density can be expressed as the product of Equations 8.38 and 8.39:

$$\rho(\nu) = \frac{8\pi \nu^2 n_r^3}{c^3} \frac{1}{e^{h\nu/kT} - 1} \quad \text{or in terms of energy,} \quad \rho(E) = \frac{8\pi E^2 n_r^3}{h^3 c^3} \frac{1}{e^{E/kT} - 1}. \tag{8.40}$$

With dispersion of the refractive index, the same can be written as

$$\rho(\nu) = \frac{8\pi \nu^2 n_r^3}{c^3} \frac{1 + [(\nu/n_r)(dn_r/d\nu)]}{e^{h\nu/kT} - 1} \quad \text{or in terms of energy,}$$
$$\rho(E) = \frac{8\pi n_r^3 E^2}{h^3 c^3} \frac{1 + [(E/n_r)(dn_r/dE)]}{\exp(h\nu/kT) - 1}, \tag{8.41}$$

which represents the Planck's formula (Planck's blackbody radiation distribution law). The details of the photon density and photon energy density are given in Appendix 8.A.

Using the Planck's formula together with Einstein expansion, one can determine the gain in terms of population density at threshold, the details of which can be found in many texts including Chapter 2 of Morkoç (2009):

$$g_{th} = \frac{c^2 \Delta N_{th}}{8\pi v^2 n_r^2 \Delta v \tau_{sp}}. \tag{8.42}$$

Noting that the threshold carrier density can be related to threshold current density through (with d being the thickness of the pumped region)

$$J_{th} = \frac{q \Delta N_{th} d}{\tau_{sp}}, \tag{8.43}$$

Equation 8.42 can be written as

$$g_{th} = \frac{c^2 J_{th}}{8\pi q d v^2 n_r^2 \Delta v}. \tag{8.44}$$

Because the quantum efficiency is less than unity, Equation 8.44 must be rewritten as

$$g_{th} = \frac{c^2 J_{th} \eta}{8\pi q d v^2 n_r^2 \Delta v}. \tag{8.45}$$

with η being the quantum efficiency. The threshold current can then be expressed as

$$J_{th} = g_{th} \frac{8\pi q d v^2 n_r^2 \Delta v}{c^2 \eta}. \tag{8.46}$$

Equation 8.46 is quite telling in that the narrower the gain curve (small Δv), the thinner the region to be pumped (small d afforded, for example, by quantum wells), and the smaller the threshold current becomes. The gain curve lineshape in InGaN is of a special concern as due to homogeneous and inhomogeneous broadening, the linewidth is wide. If lasing occurred due to excitonic transitions, the gain curve would be very narrow, which is the attraction of ZnO with its 60 meV exciton binding energy, as opposed to slightly over 20 meV for GaN, which persists even above room temperature.

8.5
A Glossary for Semiconductor Lasers

In semiconductors, the lower level at an energy of E_1 and the upper level at an energy of E_2 are represented by bands associated with valence and conductions bands, respectively, as shown in Figure 8.3, and with more needed description in Figure 8.13.

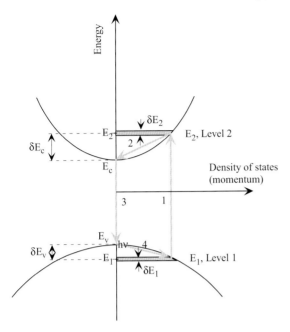

Figure 8.13 Optical transitions in a direct bandgap semiconductor in the energy versus momentum (which also represents energy versus DOS though the functional forms deviate) diagram that is pumped beyond transparency. The transitions $i = 1, 2$, and so on represent excitation, relaxation to the bottom of the conduction band, emission, and filling the hole state vacated by electron excitation to the conduction band (process 1), respectively.

Recall that momentum and energy as well as spin must be conserved. Because photons have negligible momentum transitions 1, 3 are straight up and down in the zone center. However, transitions 2 and 4 involve phonons for conserving momentum, which changes considerably between states before and after the transitions involved.

Let us designate the wavenumbers in the conduction and valence bands associated with transition 1 as k_c and k_v. The dispersion relationship in the parabolic band approximation can be written for conduction and valence bands as

$$\delta E_C = E_2 - E_C = \frac{\hbar^2 k_C^2}{2m_n^*} \text{ in the conduction band for transition 1}$$
$$\delta E_V = E_1 - E_V = \frac{\hbar^2 k_V^2}{2m_p^*} \text{ in the valence band for transition 1} \quad (8.47)$$

with k_c and k_v representing the wavenumbers in the conduction and valence bands, respectively. Because photon momentum is negligible, $k_c = k_v$ and with $h\nu = E_2 - E_1$, Equation 8.47 can be rewritten as

$$\delta E_c = E_2 - E_c = \frac{m_p^*}{m_n^* + m_p^*}(h\nu - E_g),$$
$$\delta E_v = E_v - E_1 = \frac{m_n^*}{m_n^* + m_p^*}(h\nu - E_g), \quad (8.48)$$

where E_g is the bandgap of the semiconductor and m_n^* and m_p^* are the electron and hole effective masses (m_e^* and m_h^* are also common). The extension of the quasi-Fermi levels into the conduction and valence bands are not the same ($\delta E_c \neq \delta E_v$) because of the inequality of the effective masses and the said extensions relate to each other by

$$\delta E_2 = \frac{m_p^*}{m_n^*}\delta E_1 \tag{8.49}$$

The occupation probabilities for energy levels E_1 and E_2 in the frame of Fermi–Dirac statistics are

$$f(E_i) = f_i = \frac{1}{1+\exp[(E_i - E_{F_i})/k_B T]}, \quad \text{with } i = 1, 2, \tag{8.50}$$

where E_{Fi} represents the Fermi level for level i. These occupation probabilities subtracted from unity would lead to the probability of those states being unoccupied or empty. Expanding f_1 and f_2 represents the occupation probabilities of level 1 and level 2, respectively. Therefore, $(1-f_1)$ and $(1-f_2)$ represent the probability of levels 1 and 2 being empty. Later on the Fermi levels for levels $i = 1$ and 2 will be replaced with quasi-Fermi levels (F_p and F_n for levels 1 and 2, respectively).

The rate equation for absorption in semiconductors is

$$R_{12}|_{abs} = B_{12}f_1(1-f_2)\rho(E_{21}). \tag{8.51}$$

It is the same for stimulated emission:

$$R_{21}|_{st} = B_{21}f_2(1-f_1)\rho(E_{21}). \tag{8.52}$$

(R_{ij}) terms are the transition rates per unit volume.

Electron can also make the transition from level 2 to level 1 via spontaneous emission, which can be formulated as

$$R_{21}|_{sp} = A_{21}f_2(1-f_1). \tag{8.53}$$

Again, note the involvement of the optical field in the absorption and stimulated emission by the term $\rho(E_{21})$, but not in the spontaneous emission process. Under thermal equilibrium, as in the case of Equation 8.34 but modified for a semiconductor, the downward transition rate must be equal to the upward transition rate:

$$\begin{aligned}R_{21} &= R_{21}|_{st} + R_{21}|_{sp} = R_{12}|_{abs} \Rightarrow A_{21}f_2(1-f_1) + B_{21}f_2(1-f_1)\rho(E_{21})\\ &= B_{12}f_1(1-f_2)\rho(E_{21}).\end{aligned} \tag{8.54}$$

Solving for $\rho(E_{21})$ from Equation 8.54 (by substituting E_{21} for E) while ignoring the dispersion in the refractive index, we obtain

$$\rho(E_{21}) = \frac{A_{21}f_2(1-f_1)}{B_{12}f_1(1-f_2) - B_{21}f_2(1-f_1)} = \frac{8\pi E_{21}^2 n_r^3}{h^3 c^3}\frac{1}{e^{E_{21}/kT}-1}. \tag{8.55}$$

Setting $B_{21} = B_{12}$ allows us to derive an expression for relating Einstein's A and B coefficients:

$$A_{21} = \frac{8\pi E_{21}^2 n_r^3}{h^3 c^3} B_{21}. \tag{8.56}$$

The necessary condition for population inversion (or gain) dictates that the downward transition rate represented by stimulated emission should be larger than the upward transition rate represented by absorption, which is expressed as

$$B_{21}f_2(1-f_1)\rho(E_{21}) > B_{12}f_1(1-f_2)\rho(E_{21}) \tag{8.57}$$

and because $B_{21} = B_{12}$, Equation 8.57 implies that

$$f_2(1-f_1) > f_1(1-f_2). \tag{8.58}$$

Substituting Equation 8.50 into Equation 8.58 leads to the population inversion condition being expressed in terms of the Fermi levels associated with levels 2 and 1:

$$F_2 - F_1 > E_2 - E_1 \quad \text{or in the semiconductor terms } F_n - F_p > E_c - E_v = E_g. \tag{8.59}$$

This means that the quasi-Fermi level separation must be larger than the bandgap for population inversion and thus stimulated emission and gain as articulated by *Bernard* and *Duraffourg*. Because carrier injection is through a diffusion process, the concentration of injected carriers in a homojunction laser cannot exceed the carrier concentrations in the n- or p-emitters. Consequently, to satisfy the condition (Equation 8.59), the equilibrium Fermi level in the emitter must also be shifted toward the corresponding band, as shown in Figures 8.3 and 8.13. Before the advent of heterostructures, the thickness of the pumped region needed to be comparable to the wavelength of the radiation, so the light traveling along this region diffracted severely into absorptive passive regions.

8.5.1
Optical Gain in Bulk Layers: a Semiconductor Approach

To develop expressions for gain in a semiconductor laser, we refer to Equation 8.54, which allows the net stimulated emission rate to be written as

$$R_{21}|_{st} = B_{21}f_2(1-f_1)\rho(E_{21}) - B_{12}f_1(1-f_2)\rho(E_{21}) = B_{12}(f_2 - f_1)\rho(E_{21}). \tag{8.60}$$

Using Equations 8.55 and 8.56, Equation 8.60 can be rewritten as

$$R_{21}|_{st} = A_{21}(f_2 - f_1)\frac{h^3 c^3}{8\pi E_{21}^2 n_r^3} \frac{8\pi E_{21}^2 n_r^3}{h^3 c^3} \frac{1}{e^{E_{21}/kT} - 1} = \frac{A_{21}(f_2 - f_1)}{e^{E_{21}/kT} - 1}. \tag{8.61}$$

For its similarity to Equation 8.53 depicting the spontaneous emission rate $(R_{21}|_{sp})$, $R_{21}(\text{stim})$ that popularly has assumed the nomenclature of "stimulated emission rate" is defined as

$$R_{21}(\text{stim}) = R_{21}|_{\text{st}}\left(e^{E_{21}/kT} - 1\right) = A_{21}(f_2 - f_1). \tag{8.62}$$

Again, the stimulated emission rate of Equation 8.62 [$R_{21}(\text{stim})$] describes the downward transition rate for stimulation of a similar photon. When this is multiplied by $(e^{E_{21}/kT} - 1)^{-1}$, $R_{21}|_{\text{st}}$ is arrived that represents the net stimulated emission rate. Notwithstanding customary use, to avoid confusion in this text, $R_{21}|_{\text{st}}$ is used for rate equations.

Solving the rate balance equation of Equation 8.54 for the absorption rate and recalling that $B_{21} = B_{12}$, we obtain

$$R_{12}|_{\text{abs}} = B_{12}(f_1 - f_2)\rho(E_{21}). \tag{8.63}$$

8.5.1.1 Relating Absorption Rate to Absorption Coefficient

Let us now embark on relating the net absorption rate $R_{12}|_{\text{abs}}$ to the absorption coefficient (when negative, it is called the gain coefficient). Consider a plane wave propagating along the z-direction (length of the waveguide) represented by $I_0 \exp(-\alpha z)$. The loss parameter α can then be expressed as

$$-\alpha(= g) = \frac{dI/dz}{I}. \tag{8.64}$$

The numerator in Equation 8.64 represents the net power emitted per unit volume, while the denominator represents the power per unit area. The term in the denominator is simply the photon density distribution per unit frequency (spectral photon density) multiplied by the group velocity, c/n_r. The term in the numerator can be represented by $R_{12}|_{\text{abs}}$ given in Equation 8.63. Rewriting Equation 8.64 in light of the aforementioned discussion, we obtain for α (now recognizing its energy dependence),

$$\alpha(E_{21}) = \frac{B_{12}(f_1 - f_2)\rho(E_{21})}{(c/n_r)\rho(E_{21})E_{21}} = \frac{B_{12}(f_1 - f_2)}{(c/n_r)E_{21}} = -g(E_{21}). \tag{8.65}$$

8.5.1.2 Relating Stimulated Emission Rate to Absorption Coefficient

The denominator in Equation 8.65 can be expressed in terms of the stimulated emission rate by utilizing Equations 8.56 and 8.62:

$$\alpha(E_{21}) = -g(E_{21}) = \frac{A_{21}(f_1 - f_2)}{(c/n_r)E_{21}} \frac{h^3 c^3}{8\pi E_{21}^3 n_r^3} = -\frac{h^3 c^2}{8\pi E_{21}^3 n_r^2} R_{21}(E_{21}, \text{stim}). \tag{8.66}$$

8.5.1.3 Relating Spontaneous Emission Rate to Absorption Coefficient

The absorption coefficient $\alpha(E_{21})$ can also be related to the spontaneous emission rate. From Equation 8.65, we obtain

$$B_{12} = B_{21} = \alpha(E_{21})\frac{E_{21}(c/n_r)}{(f_1 - f_2)} \quad \text{or} \quad \alpha(E_{21}) = B_{12}(f_1 - f_2)(n_r/c)(1/E_{21}). \tag{8.67}$$

By utilizing Equation 8.53 $[R_{21}|_{\text{sp}} = A_{21}f_2(1-f_1)]$ and Equation 8.56 $[A_{21} = B_{21}(8\pi E_{21}^3 n_r^3)/(h^3 c^3)]$, the spontaneous emission rate can be rewritten as

$$R_{21}(E_{21})|_{\text{sp}} = \alpha(E_{21}) \frac{f_2(1-f_1)}{f_1 - f_2} \frac{8\pi E_{21}^3 n_r^2}{h^3 c^2}. \tag{8.68}$$

Using the Fermi–Dirac statistics given in Equation 8.50 for f_i $(i = 1, 2)$ and taking E_{F_i} as F_1 and F_2 for levels 1 and 2 (later on these will be replaced with F_p and F_n, the quasi-Fermi levels for holes and electrons, respectively),

$$R_{21}(E_{21})|_{\text{sp}} = \alpha(E_{21}) \frac{8\pi E_{21}^3 n_r^2}{h^3 c^2} \frac{1}{\exp[(E_{21} - F_2 + F_1)/kT] - 1}. \tag{8.69}$$

Recalling Equation 8.56 $[A_{21} = (8\pi E_{21}^2 n_r^3)/(h^3 c^3) B_{21}]$, Einstein's A coefficient can now be written in terms of the momentum matrix element M as

$$A_{21} = \frac{8\pi E_{21}^2 n_r^3}{h^3 c^3} \left(\frac{\pi q^2 \hbar}{V m_0^2 \varepsilon_0 n_r^2 \hbar \omega} \right) |M|^2 = \frac{4\pi q^2 E_{21} n_r}{V m_0^2 \varepsilon_0 h^2 c^3} |M|^2. \tag{8.70}$$

Note that we utilized $\hbar \omega = E_{21}$. Also note that A_{21} has the units $1/(\text{cm}^3)$.

Note that in a semiconductor, the available states in the valence band (representing level 1) and in the conduction band (representing level 2) are distributed over energy. Therefore, an extension of the previous treatment must be undertaken to include all the eligible states in the conduction and valence bands.

8.5.2
Semiconductor Realm

To determine the DOS for electrons, we take an approach similar to the density of modes in a cavity (see, for example, Morkoç (2009) and references therein for details). Briefly, the DOS is the number of states between the momentum values of k and $k + dk$. The unit volume in k-space confined within the boundaries of the k-vector is $V = (2\pi)^3/a^3$ and the volume of a spherical shell of thickness dk is $4\pi k^2 dk$. The density of allowed values of k in a volume V is the number of cubes of face $2\pi/a$ that can be fit in that volume in k-space. Therefore, the number of states is the product of volume in k-space and the number of states divided by the volume in real space ($V = a^3$). Doing so leads to

$$\rho(k)dk = \rho(E)dE = \left[2(4\pi k^2 dk)/(2\pi/a)^3\right]/V = (k^2/\pi^2)dk. \tag{8.71}$$

The factor of 2 is picked up due to the spin-up and spin-down polarization of electrons (spin degeneracy). If spin conservation is applied, the factor 2 would not be included, halving the final value. Recalling that for a parabolic band, we have

$$(\hbar k)^2 = 2m_n^*(E - E_c) \tag{8.72}$$

The DOS in a semiconductor can be expressed as

$$\rho_c(E-E_c)|_{3D} = \frac{1}{2\pi^2}\left(\frac{2m_n^*}{\hbar^2}\right)^{3/2}(E-E_c)^{1/2}, \quad (E \geq E_c) \text{ for 3D}$$

$$\rho_c(E-E_n)|_{2D} = \frac{m_n^*}{\pi\hbar^2}, \quad (E \geq E_n) \text{ for 2D}$$

$$\rho_c(E)|_{2D} = \frac{m_n^*}{\pi\hbar^2}\sum_i u(E-E_i), \quad \text{for 2D with multiple quantum states}$$

$$\rho_c(E-E_n)|_{1D} = \frac{2}{\pi\hbar}\sqrt{\frac{m_n^*}{2(E-E_n)}}, \quad (E \geq E_n) \text{ for 1D}$$

(8.73)

where m_n^* is the conduction band effective mass (DOS effective mass) and $u(E-E_i)$ is the step function that is zero, except when $E > E_i$, and i represents the ith confined level. The factor of 2 in the 1D case accounts for the spin degeneracy. The 1D system is characterized by a spiked dependence at energy $E = E_n$ that repeats itself at every quantum state.

As for a 0D system, the DOS can be represented by $N\delta(E-E_n)$ at each of the quantum states. The coefficient N contains the spin degeneracy factor, any accidental degeneracy of the bound state involved, and the number of quantum dots per unit volume.

Likewise, the DOS for a 3D system in the valence band is given by

$$\rho_v(E_v-E) = \frac{1}{2\pi^2}\left(\frac{2m_p^*}{\hbar^2}\right)^{3/2}(E_v-E)^{1/2} \quad \text{(good for } E \leq E_v\text{)}, \qquad (8.74)$$

where m_p^* is the hole effective mass (DOS effective mass) in the valence band. We are assuming that only one of the bands in the valence band is participating. The equations for the valence band and for 2D and 1D systems can be written following the pattern seen in Equation 8.73. The derivations are available in many texts, including Morkoç (2009).

In optical devices relying on band-to-band transitions, both valence and conduction bands are involved. Therefore, it is often convenient to define JDOS, which can be used to calculate the emission and/or absorption rates. We can define $\rho_{\text{joint}}(E)$ as

$$\rho_{\text{joint}}(E) = (k^2/\pi^2)\frac{dk}{dE}, \quad \text{which is evaluated at } E = hv = E_2 - E_1. \qquad (8.75)$$

Considering a parabolic band, we can write

$$dE = dE_2 - dE_1 = \hbar^2\left(\frac{1}{m_n^*} + \frac{1}{m_p^*}\right)kdk. \qquad (8.76)$$

Solving for dk/dE and substituting it into Equation 8.75, we obtain

$$\rho_{\text{joint}}(hv) \equiv \rho_{cv}(hv) = \frac{k}{\pi^2\hbar^2}\frac{m_n^* m_p^*}{m_n^* + m_p^*}. \qquad (8.77)$$

8.5 A Glossary for Semiconductor Lasers

Using the band parabolicity described by Equation 8.47 for both the conduction and valence bands, which relate the momentum to energy, together with Equation 8.48, we can rewrite

$$\hbar k = \left[\frac{2m_n^* m_p^*}{m_n^* + m_p^*}(h\nu - E_g)\right]^{1/2} = [2m_r^*(h\nu - E_g)]^{1/2}, \qquad (8.78)$$

where $m_r^* = (m_n^* m_p^*)/(m_n^* + m_p^*)$ is the reduced effective mass. Then,

$$\rho_{cv}(h\nu) = \frac{1}{2\pi^2}\left(\frac{2m_r^*}{\hbar^2}\right)^{3/2}(h\nu - E_g)^{1/2} \quad \text{or} \qquad (8.79)$$

$$\rho_{cv}(E) = \frac{1}{2\pi^2}\left(\frac{2m_r^*}{\hbar^2}\right)^{3/2}(E - E_g)^{1/2}.$$

Let us now derive the expression for the absorption coefficient. The rate of excitation from the valence band to the conduction band depends on the availability of states with electrons in the valence band, which is $\rho_v(E_v - E)f_1$, and on the availability of empty states in the conduction band, which is $\rho_c(E - E_c)(1 - f_2)$. The absorption coefficient at energy $\hbar\omega$ must then be the integral of Equation 8.65 over all the possible energies corresponding to the energy difference $\hbar\omega$, the energy difference between level 1 and level 2:

$$\alpha(\hbar\omega) = \int_{-\infty}^{\infty} \frac{B_{12}(f_1 - f_2)V^2}{(c/n_r)(E_2 - E_1)}\rho_v(E_v - E_1)\rho_c(E_2 - E_c)\delta(E_2 - E_1 - \hbar\omega)dE_1\,dE_2, \qquad (8.80)$$

where V is the crystal volume. Note that B_{12} has the units J/s. The term $\rho(E)dE$ has the units $1/m^3$; therefore, the integral over volume is needed twice, once over E_1 and another over E_2, justifying the volume squared (V^2) term.

Following the treatment by Lasher and Stern (1964), the absorption coefficient can be expressed by defining new energy parameters variable, E', with a reference point of $E' = 0$ at the conduction band minimum (CBM). Making the substitutions $E_2 \to E'$, $E_1 \to E' - E$, $f_1 \to f_1(E' - E)$, and $f_2 \to f_2(E')$,

$$\alpha(E) = \frac{\pi q^2 \hbar}{m_0^2 \varepsilon_0 n_r c E^2}\int_{-\infty}^{\infty}\rho_{cv}(E')|M(E', E' - E)|^2[f_1(E' - E) - f_2(E')]dE'. \qquad (8.81)$$

The coefficient together with the momentum matrix element has the units m^2. Here, as we defined $E' = 0$ at the CBM as the reference, there is only one energy to integrate over, E'. The term $\rho_{cv}(E')dE'$ from Equation 8.79 has the units $1/m^3$, making the overall units for the right-hand side $1/m$. As shown in Lasher and Stern (1964), the volume term V does not appear when JDOS is used. But when separate DOS are used, we need to involve the crystal volume. From here and on, we define the rates as per unit energy per unit volume, in units $1/(cm^3 \text{ J})$.

To avoid any possible confusion and making additional nomenclature changes germane to semiconductors such as quasi-Fermi levels and breaking the integral into two, one for the valence band and the other for the conduction band, Equation 8.81 can be represented by a double integral:

$$\alpha(E) = \frac{\pi q^2 \hbar V}{m_0^2 \varepsilon_0 n_r c E^2} \int_0^\infty \int_0^\infty \rho_c(E_n) \rho_v(E_p) \left[f_p(F_p - E_p) - f_n(E_n - F_n) \right] |M(E_n, E_p)|^2 dE_n \, dE_p, \tag{8.82}$$

where F_n and F_p represent the quasi-Fermi levels for electrons and holes, respectively. We should point out that the Fermi functions in Equation 8.82 represent the electron occupation probabilities. In addition, the nomenclature $(E_n - F_n)$ and $(F_p - E_p)$ are used to indicate that the Fermi statistics for electrons and holes are to consider the electron quasi-Fermi level (F_n) and hole quasi-Fermi level (F_p), respectively. In some cases, a Dirac delta function is used to explicitly indicate that the equation under question has nonzero values only when the energy between the upper and lower levels (E_n and E_p) is near the energy of interest. In addition, instead of the electron occupation probabilities for the upper and lower levels, the statistics for the upper level for electron occupancy and statistics for the lower level for hole occupancy are sometimes used. In other words, $f(E_1)$ depicts the hole occupation factor and $[1 - f(E_1)]$ represents the missing hole or electron occupation factor in level 1. This is interchangeably done in the literature, which causes some confusion, but doing so allows the easily recognizable Fermi–Dirac equation for holes using the quasi-Fermi level for hole.

The knowledge of the matrix element would then pave the way for calculating the absorption coefficient and pertinent transition rates at a given energy. We can write for the spontaneous emission rate in the case of a single valence band (with the aid of Equation 8.53 and with JDOS within the direct transition selection rule):

$$R_{21}(E)|_{sp} = \frac{4\pi q^2 E n_r}{m_0^2 \varepsilon_0 h^2 c^3} \rho_{cv}(E) |M(E)|^2 f_2(E) [1 - f_1(E)]. \tag{8.83}$$

Note that in Lasher and Stern (1964), the lower case rates are defined per unit volume per unit energy. Hence, here there is an extra 1/J in units. The coefficients above together with the matrix element have the units m^3/s. When using the JDOS, the integral and the volume V should be removed.

If the matrix element for all the initial and final states is the same, it can be taken out of the integral in Equation 8.83 and we would then have (including indirect transitions as well)

$$R_{21}(E)|_{sp} = \frac{4\pi q^2 E n_r}{m_0^2 \varepsilon_0 h^2 c^3} \langle |M|^2 \rangle_{av} V \int_{-\infty}^{\infty} \rho_c(E') \rho_v(E' - E)$$
$$\times f_2(E') [1 - f_1(E' - E)] dE'. \tag{8.84}$$

With the definition of a recombination coefficient $B \equiv (4\pi q^2 E n_r)/(m_0^2 \varepsilon_0 h^2 c^3) \langle |M|^2 \rangle_{av} V$, Equation 8.84 can be rewritten as (B has the units m^3/s as also stated in Lasher and Stern (1964)

$$R_{21}(E)|_{sp} = B \int_{-\infty}^{\infty} \rho_c(E')\rho_v(E' - E)f_2(E')[1 - f_1(E' - E)]dE'. \quad (8.85)$$

Note that this rate also has the units $1/(m^3 s\, J)$. The integral should be over energy to get the overall rate per unit volume.

Doing similarly for spontaneous emission and shortly for stimulated emission rates,

$$R_{21}(\text{stim}) = \frac{4\pi q^2 E n_r V}{m_0^2 \varepsilon_0 h^2 c^3} \int_{-\infty}^{\infty} \rho_c(E')\rho_v(E' - E)|M(E', E' - E)|^2 [f_2(E') - f_1(E' - E)]dE',$$

$$R_{21}(\text{stim}) = \frac{4\pi q^2 E n_r}{m_0^2 \varepsilon_0 h^2 c^3} \rho_{cv}(E)|M(E)|^2 [f_2(E) - f_1(E)] \quad \text{(with the joint density of states).}$$

(8.86)

Referring to Equation 8.62, the net stimulated emission rate $R_{21}|_{st}$ can be found from $R_{21}|_{st} = R_{21}(\text{stim})(e^{E/kT} - 1)^{-1}$ and doing so leads to

$$R_{21}(E)|_{st} = \frac{4\pi q^2 E n_r V (e^{E/kT} - 1)^{-1}}{m_0^2 \varepsilon_0 h^2 c^3} \int_{-\infty}^{\infty} \rho_c(E')\rho_v(E' - E)$$

$$\times |M(E', E' - E)|^2 [f_2(E') - f_1(E' - E)]dE'. \quad (8.87)$$

As in the case of spontaneous emission, if the matrix element for all the initial and final states is the same, it can be taken out of the integral and we then would have

$$R_{21}(E)|_{st} = \frac{4\pi q^2 E n_r (e^{E/kT} - 1)^{-1}}{m_0^2 \varepsilon_0 h^2 c^3} \rho_{cv}(E)|M(E)|^2 [f_2(E) - f_1(E)]. \quad (8.88)$$

If there were no selection rules, one can integrate over the conduction and valence bands independently and doing so would lead to the rate at which photons are spontaneously emitted in unit volume:

$$R_{sp} = \int R_{21}(E)|_{sp} dE = Bnp, \quad (8.89)$$

where n and p are electron and hole concentrations, respectively. This total rate is now in units of $1/(m^3 s)$. If we further assume that the electrons in the conduction band (band 2) and holes in the valence band (band 1) are in equilibrium and can be characterized by quasi-Fermi levels, F_n and F_p, respectively (as in the case of the literature, they are also defined as E_{fn} and E_{fp}), then the Fermi–Dirac distribution function in the conduction band, f_c, for electrons is given by

$$f_c = \frac{1}{1 + \exp[(E_c - F_n)/kT]} \quad (8.90a)$$

and that for holes (*not electrons*) in the valence band by

$$f_v = \frac{1}{1 + \exp[(E_v - F_p)/kT]}. \quad (8.90b)$$

Scattering, heretofore neglected, causes a broadening in the absorption coefficient, and thus the gain coefficient. This is typically accounted for by using a broadening factor in the form of

$$L(E_{eh} - E) = \frac{1}{\pi} \frac{\hbar/\tau_{in}}{(E_{eh} - E)^2 + (\hbar/\tau_{in})^2}, \quad (8.91)$$

where E_{eh} is the optical transition energy. The linewidth broadening expression can be generalized to take into account any scattering mechanism that causes broadening, in which case we replace $\hbar/\tau_{in}$, with the general broadening factor $\Gamma(E_{eh} - E)$. This factor is energy dependent and must take into account processes in both the conduction and valence bands. Alternative depictions of the broadening parameter are

$$\frac{1}{\pi} \frac{\gamma}{(E_{21} - E)^2 + \gamma^2}, \frac{1}{\pi} \frac{\hbar/\tau_{in}}{(E_{cv} - \hbar\omega)^2 + (\hbar/\tau_{in})^2}, \frac{1}{\pi} \frac{\gamma}{(E_{cv} - \hbar\omega)^2 + (\gamma)^2},$$

where γ is the half linewidth of the Lorentzian broadening lineshape. Taking this broadening into consideration, the absorption or the gain coefficient of Equation 8.81 can be written for the broadened case:

$$g_B(E) = \int g(E) L(E_{eh} - E) dE_{eh}$$

$$= \frac{\pi q^2 \hbar}{m_0^2 \varepsilon_0 n_r c E} \int_{-\infty}^{\infty} \rho_{cv}(E) \frac{1}{\pi} \frac{\hbar/\tau_{in}}{(E_{eh} - E)^2 + (\hbar/\tau_{in})^2} |M(E)|^2 [f_2(E) - f_1(E)] dE_{eh}.$$

$$(8.92)$$

In general, however, the single-particle model with k conservation cannot be used. This is particularly true in semiconductors with significant band mixing, which is the case in the valence band of GaN and related materials. In such a case, the gain expression, Equation 8.81, must be modified as follows (the subscript is dropped from the broadened gain, i.e., we will use $g(E)$ instead of $g_B(E)$ from here on):

$$g(E) = \frac{\pi \hbar q^2}{n_r \varepsilon_0 m_0^2 c E} \int \rho_{cv}(k_{xyz}) |M^{cv}|^2 \frac{1}{\pi} \frac{\hbar/\tau_{in}}{(E_{eh} - E)^2 + (\hbar/\tau_{in})^2}$$

$$\times [f_c(k_{xyz}) + f_v(k_{xyz}) - 1] dk_{xyz}$$

$$= \frac{\pi \hbar q^2}{n_r \varepsilon_0 m_0^2 c E} \int \frac{k_{xyz}^2}{\pi^2} |M^{cv}|^2 \frac{1}{\pi} \frac{\hbar/\tau_{in}}{(E_{eh} - E)^2 + (\hbar/\tau_{in})^2}$$

$$\times [f_c(k_{xyz}) + f_v(k_{xyz}) - 1] dk_{xyz}$$

$$= \frac{\pi \hbar q^2}{n_r \varepsilon_0 m_0^2 c E} \int_{k_z=-\infty}^{\infty} \int_{k_\perp=0}^{\infty} \frac{k_{xyz}^2}{\pi^2} |M^{cv}|^2 \frac{1}{\pi} \frac{\hbar/\tau_{in}}{(E_{eh} - E)^2 + (\hbar/\tau_{in})^2} \quad (8.93)$$

$$\times [f_c(k_\perp, k_z) + f_v(k_\perp, k_z) - 1] \frac{2\pi k_\perp \, dk_\perp \, dk_z}{4\pi k_{xyz}^2}$$

$$= \frac{\pi \hbar q^2}{n_r \varepsilon_0 m_0^2 c E} \int_{k_z=-\infty}^{\infty} \int_{k_\perp=0}^{\infty} \frac{k_\perp}{2\pi^2} |M^{cv}|^2 \frac{1}{\pi} \frac{\hbar/\tau_{in}}{(E_{eh} - E)^2 + (\hbar/\tau_{in})^2}$$

$$\times [f_c(k_\perp, k_z) + f_v(k_\perp, k_z) - 1] dk_\perp \, dk_z,$$

where $f_c(k_\perp, k_z)$ and $f_v(k_\perp, k_z)$ depict the electron occupancy and the hole occupancy in the conduction and valence bands, respectively $-[1 - f_v(k_\perp, k_z)]$ depicts the electron occupancy in the valence band. Also, E_{eh} in the broadening term is a function of $k_\perp, k_z$. While the conduction band in nitrides can be assumed parabolic, the valence band is much more complex with heavy and light hole band mixing. In this case, a numerical integral over the k-space is warranted. Furthermore, the valence band in nonpolar and semipolar orientations is anisotropic, necessitating the knowledge of the band structure not only along two directions but also for the full k'_x–k''_y plane or $k_\perp$ plane.

Because the electron's effective mass is much smaller than the hole mass, JDOS is dominated by the electron mass in both the Wz and zinc blende (ZB) structures, whereas the conduction band is s-like and isotropic. A reduced JDOS implies lower gain. But the same also requires a lower transparency current and thus a lower threshold current. In the ZB case, the valence band maximum (VBM) is degenerate, but lifted in QW structures. This in turn leads to a reduced DOS and, therefore, lowers transparency current. On the other hand, VBM in the Wz structure is separated into three twofold degenerate bands, one Γ_9 and two Γ_7 bands. These bands are not affected much by quantum wells grown in the [0001] direction.

The factor M^{cv} that is the squared *optical transition matrix* (OTM) element is given by

$$M^{cv} = \langle \Phi_c | \vec{p} | \Phi_v \rangle, \tag{8.94}$$

where $\vec{p}$ is the momentum operator and Φ_c and Φ_v are the actual electron and hole wavefunctions, respectively. For TE mode, the E field is in the plane of the junction and so is the matrix element M^{cv}. For the TM mode, the E field is normal to the plane of the junction and so is the matrix element. The dipole momentum matrix element for a Wz material in the c-direction is

$$\left| M^{cv}_{//} \right|^2 = \frac{\hbar^2}{2m_0} \left(\frac{m_0}{m_e^{//}} - \frac{m_0}{m_c^{//}} \right) \frac{(E_g + 2\Delta_2)(E_g + \Delta_1 + \Delta_2) - 2(\Delta_3)^2}{(E_g + 2\Delta_2)}, \tag{8.95}$$

where $m_e^{//}$ and $m_c^{//}$ represent the c-direction electron effective mass and the influence of higher energy bands on the effective mass, both along the k_z direction, respectively. The Δ_1, Δ_2, and Δ_3 parameters represent the diagonal and off-diagonal terms of the 6×6 Hamiltonian that describe the three valence bands. The momentum matrix element in the direction perpendicular to the c-axis is about the same as that along the c-direction, but with effective masses $m_e^\perp$ and $m_c^\perp$ in the $k_{x,y}$ plane. With the quasicubic approximation, that is $\Delta_{cr} = \Delta_1$, $\Delta_{so} = 3\Delta_2 = 3\Delta_3$, the matrix element reduces to

$$\left| M^{cv}_{//} \right|^2 \approx \frac{\hbar^2}{2m_0} \left(\frac{m_0}{m_e^{//}} - 1 \right) \frac{E_g(E_g + \Delta_{cr} + \Delta_{so}) + (2/3)\Delta_{cr}\Delta_{so}}{(E_g + (2/3)\Delta_{so})}, \tag{8.96}$$

where $E_g = 0.25$ Ry, $\Delta_{cr} = 72.9$ meV $= 5.36$ mRy, and $\Delta_{so} = 5.17$ meV $= 0.38$ mRy. The dimensionless Rydberg (Ry) denotes energy in terms of the lowest H energy level (13.6 eV), that is, 1 mRy $= 13.6$ meV.

In the direction perpendicular to the *c*-axis (in-plane), the momentum matrix element is approximately expressed:

$$\left|M_\perp^{cv}\right|^2 \approx \frac{\hbar^2}{2m_0}\left(\frac{m_0}{m_e^\perp}-1\right)\frac{E_g(E_g+\Delta_{cr}+\Delta_{so})+(2/3)\Delta_{cr}\Delta_{so}}{E_g+\Delta_{cr}+(2/3)\Delta_{so}+[(\Delta_{cr}\Delta_{so})/3E_g]}. \tag{8.97}$$

The third term in the integrals of Equations 8.92 and 8.93 is the occupation factor, which describes the carrier density distribution. Again, the stated form represents the electron occupation $f_c(E)$ in the conduction band and the hole occupation $f_v(E)$ in the valence band, as defined in Equation 8.90. As in the case of the matrix elements, the Fermi level too is dependent on the DOS in the valence band where the degeneracy is governed by the crystal symmetry as well.

The JDOS calls for an increase in the gain with increasing energy. On the other hand, increasing energy reduces the occupation factor. These two competing processes result in a maximum in the gain, the *maximum gain* or the *peak gain*, with respect to energy. The peak energy position necessarily increases with increasing injection along with an increase in the absolute value of the gain. In addition, the gain increases with the injected carrier density and thus the injection current. This is schematically depicted in Figure 8.14.

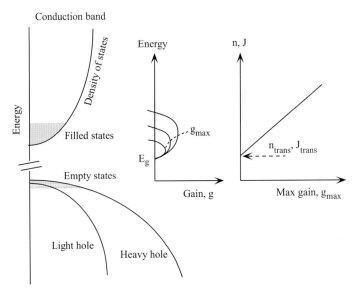

Figure 8.14 Schematic representation of conduction, heavy hole, and light hole band densities of states, gain versus energy for increasing injection, and the peak gain as a function of injection charge and/or current in WZ GaN. Note the transparency injection charge or current.

8.5.3
Gain in Quantum Wells

Due to the quantization in the z-direction (the growth direction), the crystal momentum in this direction is quantized and increases the effective bandgap. In the plane of growth, however, the band remains as in the bulk unless the *in-plane* effective masses change such as in the case of ZB strained wells. Another pertinent change is a staircase DOS function. Consequently, in writing the gain expression for a quantum well, the integral over E in Equation 8.81, and therefore in Equation 8.92, is replaced by a summation over all transitions between the conduction band energy levels and the valence band energy levels. Let us now go through the important steps involved in determining the gain coefficient in quantum well lasers.

The electron concentration in all the available, in aggregate, states in the quantum well for a parabolic band can be obtained by integrating the product of the Fermi–Dirac distribution function and 2D DOS over possible energies:

$$N = \int_{E_c}^{\infty} f(E)\rho_c(E) dE. \tag{8.98}$$

With the help of Equations 8.98 and 8.90, we can write for the electron concentration,

$$N = \frac{m_n^* kT}{\pi \hbar^2} \sum_i \ln\{1 + \exp[-(E_c^i - E_{fc})/kT]\}. \tag{8.99}$$

If we convert the electron concentration to an equivalent 3D form by dividing the above expression with L_z (this would allow the conventional gain equations to be used for quantum wells), we have

$$N = \frac{m_n kT}{\pi \hbar^2 L_z} \sum_i \ln\{1 + \exp[-(E_c^i - E_{fc})/kT]\}, \tag{8.100}$$

where L_z is the quantum well width, E_{fc} is the electron quasi-Fermi level, which we have also depicted by F_c or F_n, and E_c^i is the quantized conduction band energy levels in the quantum well. The summation is over all the quantized levels in the conduction band that are occupied.

Because the valence band is not parabolic, the hole concentration must be found through a numerical integration taking the dispersion into consideration:

$$P = \sum_j \int \rho^{2D}(k_{xy}) f_v[E_v^j(k_x, k_y)] dk_{xy} = \sum_j \int\int \frac{k_{xy}}{\pi} \frac{1}{L_z} f_v[E_v^j(k_x, k_y)] \frac{dk_x dk_y}{2\pi k_{xy}}$$

$$= \sum_j \int\int \frac{1}{4\pi^2 L_z} f_v[E_v^j(k_x, k_y)] dk_x dk_y, \tag{8.101}$$

where E_v^j is the hole energy (not for electrons) in the valence subbands. As in the case of the conduction band, the summation is over all the valence band quantized states that are occupied. Note that the hole concentration has also normalized to a three-dimensional form with the L_z term in the denominator. Similar to Equation 8.90, the Fermi–Dirac distribution function in the conduction and valence bands of the quantum well structure are given by

$$f_c = \frac{1}{1 + \exp[(E_c^i - F_n)/k_B T]} \quad \text{and} \quad f_v = \frac{1}{1 + \exp[(E_v^j - F_p)/k_B T]} \quad (8.102a)$$

and that for holes (not for electrons) in the valence band:

$$f_v = \frac{1}{1 + \exp[(E_v^j - F_p)/k_B T]}, \quad (8.102b)$$

where F_n and F_p are the quasi-Fermi levels for electrons and holes, respectively. We must state that the occupation probability of electrons in the valence is what enters into the gain expression. Because of that we must use $f_1 = 1 - f_v$ that makes the $f_2 - f_1$ term to be $(f_c + f_v - 1)$ for semiconductors, as in Equation 8.93.

As shown schematically in Figure 8.15, the dependence of the gain on energy in quantum well lasers differs from that of the bulk (Figure 8.14) due to the unique staircase-like DOS in which the gain maximum occurs at the confinement energy. This is because the DOS remains constant, while the occupation terms decrease exponentially.

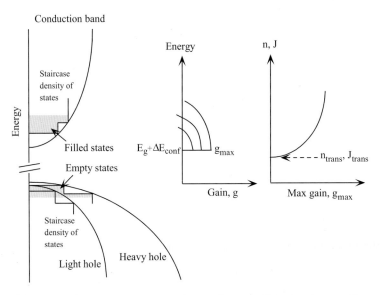

Figure 8.15 Schematic representation of conduction, heavy hole, and light hole band densities of states, gain versus energy for increasing injection, and the peak gain as a function of injection charge and/or current in a Wz GaN quantum well structure. Note the transparency injection charge or current.

The gain coefficient when modified takes the following form:

$$g(E) = \frac{\pi q^2 \hbar}{m_0^2 \varepsilon_0 n_r c E} \sum_{i,j} \int\int \frac{1}{4\pi^2 L_z} |M^{cv}|^2 \frac{1}{\pi} \frac{\hbar/\tau}{(E_{eh} - E)^2 + (\hbar/\tau)^2} [f_c(k) + f_v(k) - 1] dk_x dk_y. \tag{8.103}$$

Recall that the DOS in a 2D system is constant for a given quantum state. This equation must be modified to include carrier scattering, bandgap renormalization, and many-body effects. Assuming that the conduction and valence bands are parabolic, neglecting band mixing of the heavy and light holes in the valence band, and assuming the occupation of only the heavy hole band, the JDOS is

$$\rho_{cv}^{ij}(E) = \frac{1}{\pi \hbar^2 L_z} m_r u(E - E_c^i - E_v^j), \tag{8.104}$$

where u denotes the unity function. The reduced mass here represents the reduced effective DOS mass. In general, the valence band mass includes the effect of band mixing of the heavy and light hole bands, which is also applicable to nitrides. However, if the disparity between the heavy hole band mass and light hole band mass is very large and the momentum-conserving transitions are the only processes taking place, the DOS hole mass can be equated to the heavy hole mass.

The $|M^{cv}|^2$ term in Equation 8.103 is the squared OTM element. As in the bulk case, Equation 8.94, the matrix elements M_i^{cv} are calculated as

$$M_i^{cv} = \left\langle \Psi_{n_v,k}^{MQW} \middle| \hat{p}_i \middle| \Psi_{n_c,k}^{MQW} \right\rangle, \quad \text{with} \quad i = x, y, z, \tag{8.105}$$

where $\hat{p}_i$ is the momentum operator. For TE mode, $M^{cv} = M_x$ or $M^{cv} = M_y$ and for TM mode, $M^{cv} = M_z$.

In an ideal system with no perturbation and at the $k=0$ (Γ) point, the optical transitions from conduction subband states to the valence subband states obey the selection rule $\Delta n = 0$. The squared OTM for the TE mode represents mainly the contribution from electrons to heavy hole transitions, while for the TM mode the contribution comes from electrons to light hole and split-off hole band transitions. This is because M does not include the heavy hole wavefunction, while M_x (or M_y) includes it as well as the light hole and split-off-hole wavefunctions. However, a semiconductor laser with high injection level, as in nitride-based ones, does not represent this case, and symmetry-breaking transitions would be allowed. This necessitates numerical approaches, especially for gain calculations, which include the entire band structure over the momentum and energy spaces. The same is also true for the calculation of the OTM elements.

The optical transition energy in the case of quantum wells is given by

$$E_{eh} = E_{en_c} + E_{hn_v} + E_g^s, \tag{8.106}$$

where E_g^s is the bandgap energy in strained GaN. The GaN quantum well would be placed between the AlGaN waveguide layers and, therefore, the effect of strain on energy must be considered. Note that GaN is being used as a model quantum well material. In reality, the bulk of the laser work is based on InGaN. However, the concepts being discussed here would apply. The energy of the strained GaN is defined as

$$E_g^s = E_g + 2D_d^{cv}\left(1 - \frac{c_{12}}{c_{11}}\right)e_{xx}, \qquad (8.107)$$

where D_d^{cv} is the deformation potential, c_{11} and c_{12} are the elastic constants, and $e_{xx} = (a - a_0)/a_0$ is the in-plane strain in terms of the lattice constant of strained GaN, a, and relaxed GaN, a_0.

The spontaneous emission rate is given by

$$R_{sp}(E) = \frac{nq^2 E}{\pi \hbar^2 \varepsilon_0 m_0^2 c^3} \sum_{n_c, n_v} \iint \frac{|M^{cv}|^2}{4\pi^2 L_z} f_c(k_x, k_y) f_v(k_x, k_y) \frac{1}{\pi} \frac{\hbar/\tau}{(E_{eh} - E)^2 + (\hbar/\tau)^2} dk_x\, dk_y$$

or since the energy E_{eh} in the broadening term is a function of $k_\perp, k_z$,

$$R_{sp}(E) = \frac{nq^2 E}{\pi \hbar^2 \varepsilon_0 m_0^2 c^3} \sum_{n_c, n_v} \iint \frac{|M^{cv}|^2}{4\pi^2 L_z} f_c(E) f_v(E) \frac{1}{\pi} \frac{\hbar/\tau}{(E_{eh} - E)^2 + (\hbar/\tau)^2} dk_x\, dk_y. \qquad (8.108)$$

The radiative current density can be calculated from the spontaneous emission rate as

$$J_{rad} = qL_z \int R_{sp}(E) dE. \qquad (8.109)$$

8.5.3.1 Optical Gain

Due to the complexity of band structure, particularly the valence band, numerical calculations are employed to determine the spectral distribution of the material gain and its dependence on injected carrier density. Pertinent to the discussion in hand, in relation to the polar c-plane orientation, is that the hole masses of the uppermost two valence bands along the growth direction are very large. The mass of the CH band is small due to a strong coupling with CBM through the $k_z p_z$ perturbation. In the x-direction, in the plane of the layers, CBM is coupled to the mixed state of the uppermost two valence bands and, thus, the hole mass of the mixed band is small. The lowest valence band, CH, supports the optical gain of the TM mode. For this mode to become important, an extremely large carrier injection is required. This paves the way for the dominance of the optical gain by the TE mode under normal conditions. The above discussion would need to be modified when and if semipolar or nonpolar orientations are used as the valence band and its dispersion within the plane is affected.

To give a flavor, the calculated band structures for the topmost valence bands in a 3 nm $In_{0.2}Ga_{0.8}N$ quantum well on c-plane, semipolar ($11\bar{2}2$), and nonpolar ($11\bar{2}0$) crystal orientations are shown in Figure 8.16. Note that the in-plane energy dispersion relation is anisotropic for the non-c-plane crystal orientations. To calculate the gain, it is thus necessary to know the band structure not only along two

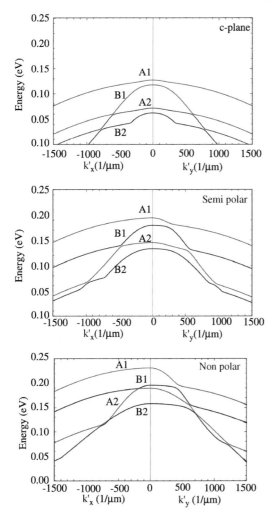

Figure 8.16 Topmost valence bands in a 3 nm In$_{0.2}$Ga$_{0.8}$N quantum well on c-plane (0001), semipolar (11$\bar{2}$2), and nonpolar (11$\bar{2}$0) crystal orientations for a sheet carrier density of 7×10^{12} cm^{-2}. To draw the contrast, the subbands corresponding to bulk valence bands of the same crystal orientations are labeled A and B for two different optical polarizations (subscripts 1 and 2 refer to first and second confined states). Terms k_x and k_y represent the in-plane momentum vector components. Note the in-plane symmetry in the c-plane case, but not the others. Courtesy of U. Schwarz and W. Scheibenzuber.

directions but also for the full k'_x–k'_y plane or $k_\perp$ plane to be consistent with previous nomenclature.

Many methods such as an adaptation of the *full-potential linearized augmented plane wave* (FPLAPW) and the 6×6 k·p Hamiltonian along with self-consistent Schrödinger–Poisson solver have been employed to calculate gain in nitride quantum wells. In the Hamiltonian, strain effects are included. In addition, full accounting of the field due to spontaneous and strain-induced polarization, when applicable, is taken.

Quantum wells at the very least serve to reduce the volume of the semiconductor that must be pumped, leading to very low-threshold currents. More importantly, by virtue of thin layers employed in quantum wells, coherently strained systems can be obtained. Strain-induced reduction of the *in-plane* hole mass in ZB structures leads to a much reduced transparency density and thus a reduced threshold. Consequently, the degrees of freedom in laser design are enhanced immensely with the advent of quantum wells. Unfortunately, the favorable effects of strain germane to ZB systems in relation to the valence band structure do not occur in Wz structures. In what follows, the optical gain in quantum wells, first without strain and later with strain, will be discussed.

Calculated material gain for TE mode (TM mode is either too small or negative, meaning material is absorptive) for a 3 nm $In_{0.2}Ga_{0.8}N$ quantum wells grown on c-plane, $(11\bar{2}2)$ semipolar, and $(11\bar{2}0)$ nonpolar ridge orientation parallel to the projection of the c-axis on the QW plane is shown in Figure 8.17 for the indicated injection levels. The calculations made use of an *ad hoc* bandgap renormalization and the screening of the quantum-confined Stark effect within the framework of a 6×6 *kp* Hamiltonian and a self-consistent Schrödinger–Poisson solver. Polarization due to compositional variation and strain, when applicable, that is, in polar and semipolar surfaces, as well as an inhomogeneous broadening factor of 60 meV corresponding to InN molar fraction variation of $\Delta x = 0.016$ has been considered.

8.5.3.2 Measurement of Gain in Nitride Lasers

Calculation of the gain is rather convoluted in that it requires an accurate knowledge of key parameters as well as of the critical mechanisms involved. For nitrides, the respective mechanisms and many of the key parameters have yet not been determined experimentally. They must therefore be calculated. It is thus imperative that gain measurements be made to provide the basis for obtaining an insight into the operation of nitride-based injection lasers even before they become widely available and to provide the needed calibration for the calculations. At the onset, it should be stressed that high-quality laser material is required to measure the pertinent parameters with the needed degree of confidence. Ironically, as is always the case, good material is not available in the early phases of development when it is needed the most. Because carrier injection is a major impediment, particularly during the early stage of development, due to the lack of a good p–n junction and the many leakage pathways, optical excitation/pumping is generally employed to glean an insight into the processes in new semiconductors, such as nitrides, under high excitation densities.

8.5.4
Gain Measurement via Optical Pumping

The optical pumping experiment provides an excellent environment to investigate the genesis of stimulated emission in semiconductors, particularly if pump and

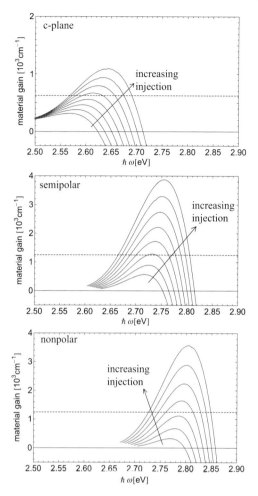

Figure 8.17 Material gain for TE mode 3 nm $In_{0.28}Ga_{0.72}N$ quantum wells on c-plane, $(11\bar{2}2)$ semipolar, and $(11\bar{2}0)$ nonpolar ridge orientation parallel to the projection of the c-axis on the QW plane. The dashed line depicts the estimated threshold gain. The sheet carrier densities are (from bottom to top) 5×10^{12}–8.5×10^{12} cm^{-2} for c-plane, 4×10^{12}–7.5×10^{12} cm^{-2} for semipolar, and 3×10^{12}–6×10^{12} cm^{-2} for nonpolar orientations in increments of 0.5×10^{12} cm^{-2}. Note that the TM modes are not really supported except for some gain in the semipolar case, which is below the threshold gain needed and thus not viable. Courtesy of W. Scheibenzuber and U. Schwarz.

probe experiments are carried out. Lack of metallization, ease of accessibility of the active layer with optical pump and probe beams, absence of the absorbing contacts, and ease with which the sample lattice temperature can be changed are among the reasons for the attractiveness of the method. Moreover, optical pumping experiments do not require the formation of good p–n junctions and complicated fabrication procedures, and as such are precursory to the current injection laser development. One of the methods used for studying the optical gain is the stripe excitation technique, which means that the light excitation source is focused on a stripe region. If the sample is excited sufficiently, spontaneously emitted light traveling along the excited stripe gets amplified by the stimulated emission. The intensity of light in such a case is then given as

$$\frac{dI}{dz} = \Gamma g(h\nu)I + \beta' \Gamma r_{sp}(h\nu), \qquad (8.110)$$

where $r_{sp}(h\nu)$ and β' are the spontaneous emission rate and spontaneous emission factor, respectively. The other terms have their usual meanings. When solved for a stripe length of l with the boundary condition $I(0) = 0$, we obtain

$$I(l) = \frac{\beta' r_{sp}(h\nu)}{g(h\nu)} [\exp(\Gamma g l) - 1]. \qquad (8.111)$$

The term $\beta' r_{sp}(h\nu)$ represents the spontaneous emission intensity I_0 and $I(l)$ is the intensity of the amplified spontaneous emission. A word caution, Equation 8.111 does not hold for long stripe lengths due to saturation effects. To account for this, a saturation term can be added in which case Equation 8.111 is modified as

$$g(h\nu)l = \alpha' I + \ln\left[\frac{gI(l)}{I_0} + 1\right], \qquad (8.112)$$

where α' is the saturation parameter.

Experimentally, the spectrum of the amplified spontaneous emission is measured for various values of l and used in conjunction with Equations 8.111 or 8.112 (depending on the lengths of the excited stripe) to determine the gain. This can be repeated for various wavelengths and pump intensities to obtain the gain spectra as well as the intensity dependence of the gain.

8.6
Threshold Current

The threshold gain can be related to the threshold carrier density through Equation 8.46 and the threshold current can be related to the threshold density through Equation 8.42 if the carrier lifetime is known. The gain (negative of the absorption coefficient) can be related to the spontaneous emission rate as in Equation 8.69. Once the spontaneous emission rate is calculated versus energy, the radiative current supporting the spontaneous emission rate, for example, Equations 8.69 and 8.108 can be expressed by Equation 8.109, is repeated here for convenience:

$$J_{rad} = q \int_{E_{cv}}^{\infty} R_{sp}(E) dE [A/cm^3]. \qquad (8.113)$$

In the case of a two-dimensional system, this expression should be replaced by

$$J_{\text{rad}} = qL_z \int_{E_{cv}^{i,j}}^{\infty} R_{\text{sp}}(E) dE \ \text{A/cm}^2 \tag{8.114}$$

where L_z is the total quantum well thickness.

One approach for finding the gain and current in an injection laser is to start with a certain concentration of injected carriers and calculate the quasi-Fermi levels. Knowledge of the quasi-Fermi levels paves the way for determining the occupation probabilities. One can then calculate the gain as a function of energy using Equation 8.93 for bulk and Equation 8.103 for quantum wells, which relies on momentum selection rule but includes band mixing. The prerequisite, of course, is that one needs to have an accurate knowledge of the lasing mechanism (i.e., single particle model in the framework of electron–hole plasma or not and excitonic while the derivations so far in this chapter are based on the former), the band structure, and the relevant matrix elements associated with the active layer. From the gain expression, the spontaneous emission rate and thus the radiative current can be calculated for that particular gain. The current should be consistent with the injected carrier concentration assumed at the beginning of this procedure. The radiative current density must be divided by the internal quantum efficiency to get the current density.

8.7
Analysis of Injection Lasers with Simplifying Assumptions

The above-mentioned calculations that were performed for GaAs lasers, together with experimental data, culminated in the conclusion that the gain is linearly proportional to the injected carrier concentration for a range of carrier densities and thus current, which is called the linear region. This holds reasonably well for bulk lasers where the DOS, and thus the gain, does not saturate. In quantum well lasers, the absorption coefficient is a staircase function of energy. Nevertheless, the linear approximation has been shown to explain the observed results away from the extremes even in quantum wells. It is therefore very useful to the understanding and the diagnosis of semiconductor lasers.

The gain of a semiconductor laser in the linear region can be expressed as

$$g = A_0\left(n - n_{\text{transp}}\right) = \beta\left(J - J_{\text{transp}}\right), \tag{8.115}$$

where n is the injected carrier concentration, n_{transp} is the injected carrier concentration to render the semiconductor transparent, J is the current density, and J_{transp} is the current density that causes the semiconductor to be transparent. A_0 and β are the differential gain and the gain coefficient, respectively.

At threshold, the net modal gain can be expressed as

$$g_{th} = A_0(n_{th} - n_{transp}) = \beta(J_{th} - J_{transp}). \tag{8.116}$$

The injected carrier concentration can be written as

$$n = \frac{g}{A_0} + n_{transp} \tag{8.117}$$

and at threshold $n_{th} = (g_{th}/A_0) + n_{transp}$

The injected carrier concentration needed to reach threshold can be calculated by

$$n = \frac{1}{A_0}\left[\frac{\alpha_i}{\Gamma} + \frac{1}{2\Gamma L}\ln\left(\frac{1}{R_1 R_2}\right)\right] + n_{transp}. \tag{8.118}$$

The current density is given by

$$J = \frac{qd}{\tau_s} n = \frac{qd}{\tau_s}\left\{\frac{g}{A_0} + n_{transp}\right\}, \tag{8.119}$$

where d represents the thickness of the pumped region and τ_s is the carrier recombination lifetime, which is related to radiative and nonradiative recombination lifetimes through

$$\tau_s^{-1} = \tau_{rad}^{-1} + \tau_{nonrad}^{-1}. \tag{8.120}$$

At threshold, the current density reduces to

$$J_{th} = \frac{qd}{\tau_s} n_{th} = \frac{qd}{\tau_s}\left\{\frac{1}{A_0}\left[\frac{\alpha_i}{\Gamma} + \frac{1}{2\Gamma L}\ln\left(\frac{1}{R_1 R_2}\right)\right] + n_{transp}\right\}. \tag{8.121}$$

When the quantum efficiency is less than one, that is, $\eta = \tau_s/\tau_r$, the threshold current density can alternatively be written as

$$J_{th} = \frac{qd}{\eta \tau_r} n_{th} = \frac{qd}{\eta \tau_r}\left\{\frac{1}{A_0}\left[\frac{\alpha_i}{\Gamma} + \frac{1}{2\Gamma L}\ln\left(\frac{1}{R_1 R_2}\right)\right] + n_{transp}\right\} \tag{8.122}$$

or

$$J_{th} = \frac{1}{\beta}\left[\frac{\alpha_i}{\Gamma} + \frac{1}{2\Gamma L}\ln\left(\frac{1}{R_1 R_2}\right)\right] + J_{transp}. \tag{8.123}$$

Here, the transparency current is that calculated with the injected carrier concentration and the radiative lifetime. In reality, this figure would increase by the inverse of the efficiency or the factor of radiative lifetime over the carrier lifetime.

From Equation 8.118, one can relate the gain to measurable quantities such as the current density and the differential gain by

$$g = \frac{\tau_s A_0}{qd} + A_0 n_{transp}. \tag{8.124}$$

The threshold gain can be obtained by replacing the current density with that at threshold.

8.7.1
Recombination Lifetime

The carrier recombination lifetime is an important parameter in that, if measured accurately, it gives one a qualitative assessment of recombination processes vis-à-vis radiative and nonradiative varieties. Moreover, as indicated in the preceding section, the lifetime is needed in relating the current to important parameters such as the transparency carrier density. The lifetime τ_s can be measured from the delay τ_d of the optical pulse emanating from the laser in response to a current pulse. This is determined from the rate equation, which states that the time rate of change of carriers is equal to the rate of carrier injection minus the rate of recombination. Mathematically, we have

$$\frac{dn}{dt} = \frac{kJ}{qd} - R(n) = \frac{kJ}{qd} - \frac{\partial R(n)}{\partial n} n = \frac{kJ}{qd} - \frac{n}{\tau_s(J)}, \quad (8.125)$$

where J, d, n, and τ_s represent the current density (assumed to be uniform across the active layer), the active layer thickness, the injected carrier concentration, and the carrier lifetime that is dependent on current, respectively. The term $R(n)$ is the recombination rate and k is the injection efficiency that ranges between 0 and 1 and accounts for the leakage current and the carrier overflow. The value of kJ relates to the part of the current that participates in the radiative recombination.

Assuming a current-independent recombination time, the solution of Equation 8.125 with the boundary condition $n(0)=0$, meaning that the laser is not prebiased, can be written as

$$n(t) = \frac{kJ\tau_s}{qd}\left[1 - \exp\left(\frac{-t}{\tau_s}\right)\right]. \quad (8.126)$$

At the onset of laser oscillations, the carrier concentration reaches the threshold value:

$$n_{th} = \frac{kJ_{th}\tau_s}{qd}. \quad (8.127)$$

Referring to the time needed to reach threshold and provided that k does not change with the injection current, one obtains

$$\tau_d = \tau_s \ln \frac{J}{J - J_{th}}. \quad (8.128)$$

In the case when the laser is prebiased with the current density J_0, which is done to reduce τ_d that is the delay between the electrical pulse and the onset of lasing, the delay time becomes

$$\tau_d = \tau_s \ln \frac{J - J_0}{J - J_{th}}. \quad (8.129)$$

Plotting the delay time versus the logarithmic term gives a straight line from which τ_s can be extracted. Once τ_s is determined and the thickness of the pumped region is known, the injected carrier concentration can be calculated for a given current density above threshold from Equation 8.119. The lifetime here is assumed to be constant in the range of injection levels, which are above threshold, very high.

In general, the recombination lifetime is not independent of current. The current-dependent lifetime can be determined from the 3 dB point of the modulation bandwidth through

$$F(\omega) = 10 \log \left[\frac{F_0}{|1 + j\omega\tau|} \right] \tag{8.130}$$

for a series of injection currents. Here, $F(\omega)$ and F_0 depict the frequency response and DC response of the laser power, respectively. The differential lifetime decreases with injection current as the excess carrier concentration increases, especially between a low injection level and the level of transparency.

Similarly, the injected carrier concentration can also be deduced from this method. When Auger recombination can be neglected, the lifetime can be expressed as

$$\frac{1}{\tau^2} = a^2 + \frac{4bkJ}{qd}. \tag{8.131}$$

From a plot of τ^{-2} versus J, one obtains a and bk/d. Again, from the dependence of the lifetime versus injection current, the excess electron concentration can be calculated from

$$n(J) = \frac{k}{qd} \int_0^J \tau(J') dJ'. \tag{8.132}$$

The recombination lifetime can be obtained using the delay between the optical output and electrical input in response to an electrical pulse and Equation 8.128. Figure 8.18 depicts the delay time versus the natural log of the injection current density divided by the injection current density above the threshold current density, as described by Equation 8.128. For this particular device, the threshold current density was 7 kA/cm^2 and the total active layer thickness was 150 Å. A fit to the data of Figure 8.18 results in a carrier lifetime of 5 ns. It should be mentioned that this figure is unreasonably long and the underlying assumption that the recombination lifetime is current independent in deriving Equation 8.128 may not be applicable. From the knowledge of the threshold current, the total thickness of the entire pumped region, which is 150 Å, and the lifetime just determined, one can calculate the carrier injection level at threshold, which comes out to be about 1×10^{20} cm^{-3}; this is extremely high. There are only three data points in Figure 8.18 that do not allow one to determine if, indeed, the delay at the onset of the lasing oscillation varies linearly with the parameter on the abscissa. Because the carrier lifetime decreases as the injection level is increased, a method not requiring a constant lifetime such as in the modulation bandwidth measurements described by Equations 8.130–8.132 is warranted.

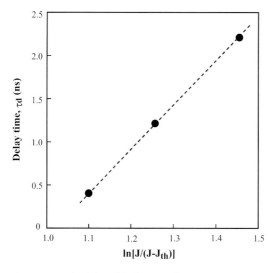

Figure 8.18 The delay of the laser oscillation onset with respect to the current pulse as a function of the current drive level in an InGaN MQW Nichia laser. Courtesy of S. Nakamura, then Nichia Chemical Ltd.

8.7.2
Quantum Efficiency

The external differential efficiency is a very useful parameter from which the *internal quantum efficiency* and, in turn, the internal loss can be determined. The differential quantum efficiency is simply the rate of change of the optical power with respect to the injection current. If the power is measured from one facet, the differential figure is doubled to account for two facets provided that the reflectivities of both mirrors are the same. The external efficiency of a laser is simply related to the internal efficiency through

$$\eta_{ext} = \eta_{int}\frac{\text{end losses}}{\text{total loss}} = \eta_{int}\frac{(1/2L)\ln[1/(R_1 R_2)]}{\alpha_i + (1/2L)\ln[1/(R_1 R_2)]}, \tag{8.133}$$

which reduces to

$$\eta_{ext}^{-1} = \eta_{int}^{-1} + \eta_{int}^{-1}\frac{\alpha_i}{(1/2L)\ln[1/(R_1 R_2)]}. \tag{8.134}$$

In other words, the *external quantum efficiency* beyond threshold is measured as the rate of change of the optical power converted to current versus the injection current. One method to obtain the *internal quantum efficiency* is to determine the external quantum efficiency of a series of lasers with varying cavity lengths. Plotting the inverse of the measured external quantum efficiency as a function of the cavity length leads to a straight line. In the limit of an infinitely long cavity, the end loss is zero at which point the internal quantum efficiency is the same as the external efficiency. From the slope of the straight line, one can obtain the internal loss. The

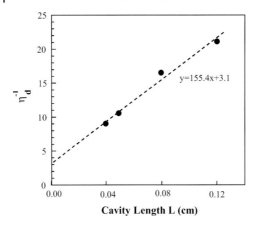

Figure 8.19 Plot of the inverse of the external quantum efficiency against the cavity length for a series of lasers with all the other parameters remaining the same. Courtesy of S. Nakamura, then Nichia Chemical Ltd.

threshold gain is then equal to the sum of the internal loss just deduced and the end loss $[(2L)^{-1}\ln(1/R_1R_2)]$, which can be calculated knowing the facet reflectivities and the cavity length. Figure 8.19 depicts the results of such an exercise with a plot of the inverse of the external quantum efficiency versus the cavity length. The data fit to the straight line expression $155.4x + 3.1$. The inverse of 3.1 is the internal quantum efficiency, which is about 33%. Using a reflectivity of 50% for the facets (coated), a confinement factor of 2.5%, and with the aid of Equation 8.26, one finds the gain at threshold to be $3200\,\text{cm}^{-1}$.

8.8
GaN-Based LD Design and Performance

The first InGaN QW diode laser heterostructures were grown on thick GaN buffer layers on the (0001) c-face of sapphire with large densities of extended defects and presumably point defects. Most commonly reported edge-emitting nitride laser diode is SCH/MQW InGaN/AlGaN heterostructure with index-guided mesa structure. Because nitrides are nearly impervious to wet chemical etches, dry etching techniques are employed to fabricate the structure, among which are the electron–cyclotron resonance plasma (ECR), chemically assisted ion beam etching (CAIBE), and standard reactive ion etching varieties.

Initially reported was CW InGaN MQW diode lasers operating for several hundred hours with a threshold current density of $3.6\,\text{kA/cm}^2$. For operational longevity, the dislocation density had to be lowered through the defect reduction method of epitaxial lateral overgrowth (ELO), see Chapter 3 of Morkoç (2008). This allowed CW operation lifetimes in excess of 1000 h under high-power ($\sim 50\,\text{mW}$) CW operation. Spatially resolved photoluminescence (PL) confirmed enhanced efficiency in the coalesced wing regions. As for the laser fabrication, the sapphire substrate was thinned down to about 100 μm to facilitate laser diode facet formation by cleaving, which resulted in roughness of about 1 nm compared to 10 nm or so achieved in non-ELO material.

In addition to the ELO process alone, very thick HVPE-grown layers can be incorporated into the process to lower the defect density and thus increase the longevity of laser diodes. Another approach involves the use of freestanding GaN in conjunction with ELO followed by laser structure. Because this is GaN system in totality, cleaving techniques can be used, as opposed to dry etching, to form the facets along the $\{12\bar{2}0\}$ direction. The lifetimes at 30 mW average power increased from the reported 700 h in simple ELO to 15 000 h in two ELO together with substrate removal. Figure 8.20a sketches the compositional variation of an early variety of Nichia laser structure with InGaN MQW. A pair of modulation doped superlattices for both n- and p-sides of the junction and the active MQW region sandwiched between AlGaN waveguide layers are featured.

When the AlGaN cladding layers exceed certain values for a given mole fraction, cracks occur that must be mitigated by modification of the growth conditions and insertion of strain reducing layer composites. Use of GaN/AlGaN short-period superlattices (SLS) that emulate bulk AlGaN can ameliorate this compromise in designer's favor because thicker layers and/or higher mole fractions can be employed.

A different version of ELO wherein troughs are etched all the way down into the sapphire substrate after GaN growth has also been explored. In the subsequent growth, the standing GaN stripes are used as the seed layer to grow the ELO template on which the rest of the laser structure was grown, an implementation of which by a group at Sony is shown in Figure 8.21. The thickness of the ELO-GaN layer was kept below 5 μm to suppress wafer bowing and possible cracking of the layers. A 1.5 μm-wide ridge stripe was placed on the wing region and coated with a Si on SiO_2 stack to obtain kink-free high-output power. The ridge depth was adjusted to yield $\theta_{||}$ of 9° on top of the aforementioned divergence perpendicular to the junction, $\theta_{\perp} = 22°$.

The (Al,Ga)N waveguide layers, more typically made of just GaN, feature an additional thin AlGaN layer, shown in Figure 8.20a, for reducing the electron overflow elaborated in Chapter 7. Variants of this structure have also been explored where an additional InGaN spacer layer, interlayer, is placed between the quantum well region and the p-AlGaN current blocking layer, as shown in Figure 8.20b. This low mole fraction InGaN serves to make the transition from the active layer to the GaN, which is grown some 300 °C above that used for the active layer.

As an example of the device performance of the InGaN MQW violet laser, operating in the 405–415 nm wavelength range, Figure 8.22 plots the light and voltage versus current characteristics of a Sony laser taken between 20 and 80 °C case temperatures. This led to a characteristic temperature:

$$J_{th} = J_0 \exp(T/T_0), \tag{8.135}$$

T_0, of 235 K. Traditionally, quantum well devices fared better on this front, everything else being equal. These devices operated more than 1000 h under CW operation at a power level of 30 mW and ambient temperature of 60 °C. The

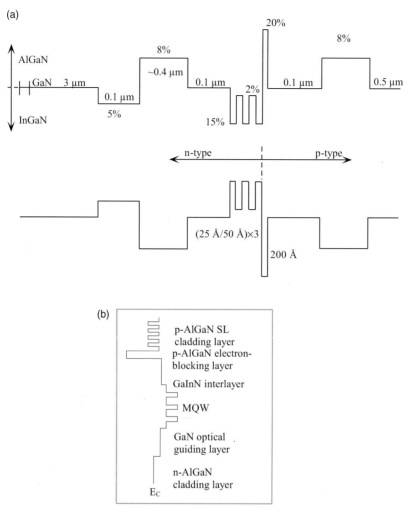

Figure 8.20 (a) Schematic diagram of the conduction band edge structure of an InGaN MQW active layer Nichia laser structure. (b) Schematic of the conduction band edge of the Sony laser with an InGaN interlayer leading to a characteristic temperature of 146 K. The AlGaN interlayer variety led to higher characteristic temperature of 235 K. Courtesy of S. Nakamura, then Nichia Chemical Ltd. Courtesy of Sony Corp.

temperature of operation and power levels are more or less what is required for CD writing applications. As mentioned, insertion of an InGaN and an AlGaN interlayer on top of the quantum well region led to characteristic temperatures of 146 and 235 K, respectively.

For just about any applications, one would aim for a low-threshold current, high slope efficiency, and low forward voltage. These requirements are strictly required for mobile devices due to limited battery power. In this frame of mind, Figure 8.23 shows the output characteristics of the above-mentioned green, blue, and violet laser

8.8 GaN-Based LD Design and Performance | 315

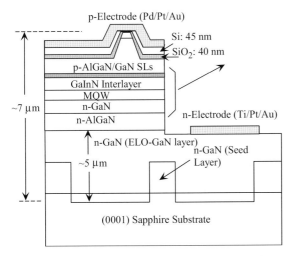

Figure 8.21 Cross-sectional view of a Sony InGaN/GaN/AlGaN DH laser structure that takes advantage of epitaxial lateral overgrowth (ELO), not shown, as well as growth on patterned sapphire.

diodes prepared on GaN templates. The threshold currents and forward voltages corresponding to the threshold are 36, 24, and 104 mA and 4.9, 4.2, and 6.4 V, respectively. As expected, the slope efficiency decreases with increasing wavelength (InN molar fraction) with values of 1.30 W/A for violet, 1.06 W/A for blue, and 0.36 W/A for green laser diodes. The same trend is reflected in the threshold current as well in that the green laser diode exhibits a factor of three to four times larger threshold current compared to the violet and blue varieties. As will shortly been seen, stating the obvious, the high-threshold current of the green laser diode is a

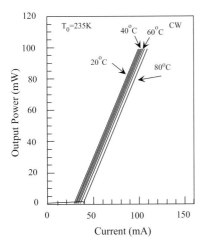

Figure 8.22 Light versus current characteristics of a Sony ridge laser, built on ELO structures, between 20 and 80 °C, featuring an AlGaN spacer layer on top of the MQW region.

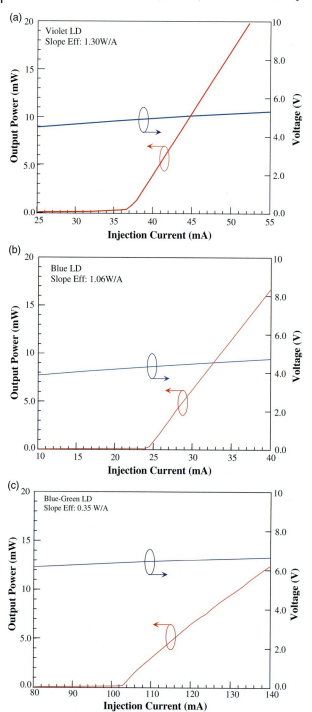

Figure 8.23 Light–current–voltage, L–I–V, curves for the violet, blue, and green LDs: violet (a), blue (b), and green (c) laser diodes studied. Courtesy of W. Scheibenzuber and U. Schwarz.

direct result of substantially reduced optical gain for a given current, to be discussed shortly, compared to the blue and violet counterparts. Reduced quality with increasing InN molar fraction necessitates high currents, which in turn exacerbates the situation by the detrimental self-heating effects such as thermal rollover of the output power curve. At 10 mW optical power, the overall efficiency ("wall-plug" efficiency), which is the quotient of output and input power, is 0.01, 0.08, and 0.05 for the green, blue, and violet LDs, respectively.

Changing our focus now to the quality of radiation as viewed from afar, owing to the small ridge width, a rather stable fundamental transverse mode operation is possible. However, in order to avoid kinks, characterized as the dependence of the far-field pattern on the injection level, the ridge waveguide design requires high accuracy of the refractive index data as well as high precision in the ridge etch depth. By tweaking the waveguide and active layer parameters, far-field FWHM angles of 8° and 27° parallel and perpendicular to the junction plane, respectively, have been reported with an aspect ratio of around 3.4. Eventually, FWHM angles parallel and perpendicular to the junction planes of 8.8° and 20.8°, respectively, with an aspect ratio of 2.36 were obtained, as shown in Figure 8.24.

8.8.1
Gain Spectra of InGaN Injection Lasers

The gain spectra can be obtained employing, for example, the method developed by Hakki and Paoli (1973, 1975). In this method, the device is pumped to just below threshold, and the gain is calculated from the interference fringes. This can be done as a function of injection current and wavelength, that is,

$$R_1 R_2 \exp(-2\alpha L) = \frac{\text{power at peak}}{\text{power at valley}}, \tag{8.136}$$

where α is the *negative loss* that is equal to the modal gain minus the internal loss. In other words,

$$-\alpha = \Gamma g - \alpha_{tot} = \frac{1}{2L}\left[\ln\left(\frac{1}{R_1 R_2}\right) + \ln\frac{\text{power at peak}}{\text{power at valley}}\right]. \tag{8.137}$$

Having already found the internal loss, the gain can be calculated from Equation 8.137 as a function of wavelength for a given injection current. The peak power and the valley power can be determined at any wavelength. It is customary to average the intensity of the two adjacent peaks for the peak power (Figure 8.25). The minimum (V_A) straddled by the two adjacent maxima (P_A and P_B) is taken as the power-at-valley. The output power of the ridge laser versus the wavelength above threshold for current levels of 50 and 53 mA is shown in Figure 8.26. The modal gain versus the injection current for an InGaN laser at the wavelength of 400.2 nm is depicted in Figure 8.27. For the InGaN laser employed in this experiment, the threshold current density is 8.8 kA/cm^2, the internal loss is 46 cm^{-1}, and the recombination lifetime is 3.5 ns. The modal gain can be represented by a straight line expressed by $-180 + 0.03J$ (cm^{-1}). It should be noted that the first term is

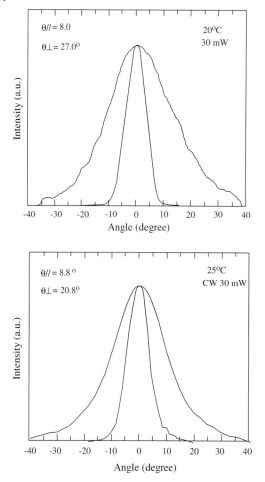

Figure 8.24 Far-field emission pattern from the transverse mode-stabilized device at 30 mW output power at 25 °C of a Sony laser.

constant, whereas the second term is a function of the injection current. This is in agreement with the form of Equation 8.116, which applies to the linear operating regime. The constant term and the slope can be utilized to determine the differential gain coefficient and the transparency current, which for the device under test are 5.8×10^{-17} cm^2 and 9.3×10^{19} cm^{-3}, respectively. If the carrier lifetime were about 0.5 ns, the transparency density would reduce to about 1×10^{19} cm^{-3}. The single carrier model indicates the transparency carrier density to be around 5×10^{18} cm^{-3}.

The measured gain spectra just below the threshold over a range of indium compositions in QW InGaN lasers, from blue–green to violet region of the spectrum, are shown in Figure 8.28. The particular prototype laser diodes studied by Scheibenzuber and Schwarz are grown on freestanding GaN substrates at OSRAM Opto Semiconductors. The structures feature a MQW-active region with

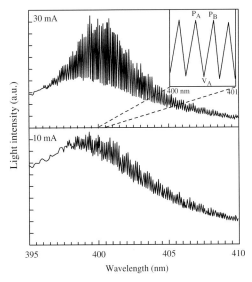

Figure 8.25 Light intensity of a ridge laser versus wavelength below the threshold current (45 mA) for current levels of 10 and 30 mA indicating the modes used in calculating the modal gain through the method of Hakki and Paoli. Courtesy of S. Nakamura, then Nichia Chemical Ltd.

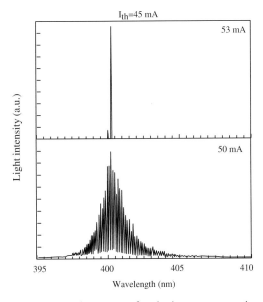

Figure 8.26 Light intensity of a ridge laser versus wavelength above the threshold for current levels of 50 and 53 mA. Courtesy of S. Nakamura, then Nichia Chemical Ltd.

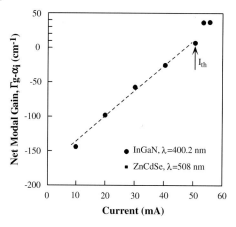

Figure 8.27 Net modal gain versus the injection current for an InGaN laser at a wavelength of 400.2 nm that can be expressed as 0.03J−180 (cm^{-1}). Courtesy of S. Nakamura, then Nichia Chemical Ltd.

thin InGaN quantum wells (QW < 4 nm). The emission wavelengths are 409 nm (violet), 457 nm (blue), and 511 nm (green).

Note the need for increased injection current for a given gain as the wavelength increases. Normally, increasing current needed as the wavelength is increased through larger InN molar fraction increases the junction temperature, which is in part responsible for the increased current needed for longer wavelengths to note the modes. In order to extricate the effect of junction temperature, the heat sink temperature for the three sets of devices was adjusted in such a manner as to

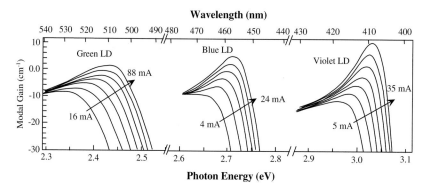

Figure 8.28 Measured gain spectra just below threshold over a range of indium compositions in QW InGaN lasers, from blue–green to violet region of the spectrum. Note the need for increased injection current for a given gain as the wavelength increases. The injection current increments are in steps of 4 mA for the blue–green and blue LDs, and 5 mA for the violet LD. Note that absolute value of the peak gain increases with injection and the wavelength at which it occurs blueshifts. The lasing wavelength depends on the energy position of the peak gain as well as the modal properties of the cavity. Courtesy of W. Scheibenzuber and U. Schwarz.

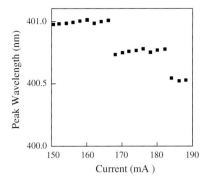

Figure 8.29 Mode hopping behavior of an InGaN laser with an injection current in the range 150–190 mA. Courtesy of S. Nakamura, then Nichia Chemical Ltd.

maintain the same junction temperature for the three LDs. The spectra shown in Figure 8.28 are also vertically shifted depending on the mirror losses to remove this variable as well from the comparison.

8.8.2
Mode Hopping

Another issue needing attention is that as the spectra displayed in Figure 8.26 indicate, many modes are supported by a gain versus energy (wavelength)–cavity combination. Even though only one mode appears when the injection current is at 53 mA, single-mode operation is not maintained, as the current is changed and/or the temperature is controlled. Figure 8.29 illustrates the mode hopping behavior with current. It appears that there is a slight blueshift at each mode as the injection current is increased, which appears to be indicative of filling effects. On the other hand, Figure 8.30 exhibits the mode hopping behavior as the device temperature is

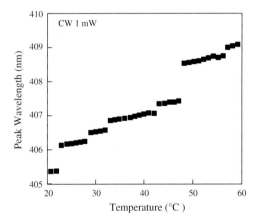

Figure 8.30 Mode hopping as the case temperature is increased from 20 to 60 °C in an InGaN laser, with the redshift being clearly observed with increasing temperature. Courtesy of S. Nakamura, then Nichia Chemical Ltd.

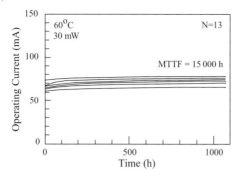

Figure 8.31 Evolution of the current through 13 CW Sony InGaN lasers with initial threshold currents in the range 60–70 mA. The injection current was adjusted to maintain the light output at 30 mW at 60 °C. The mean time to failure (MTTF) is estimated at 15 000 h. Courtesy of Sony Corp.

increased from 20 to 60 °C. As expected, a redshift is clearly observed with increasing temperature.

During the evolution of GaN laser longevity, many types of devices exhibited varying lifetimes. For example, lasers grown without the epitaxial lateral overgrowth process, discussed in great detail in Chapter 3 of Morkoç (2008), exhibited operating lifetimes in the tens of hours under CW testing at room temperature. With the incorporation of ELO in the structures, lifetimes increased steadily and dramatically. For example, early versions of ELO-based lasers produced by Nichia operated at room temperature for up to about 3000 h under CW testing. The life testing improved devices where the current drive is automatically adjusted to maintain a CW per facet power of 2 mW led to lifetimes of up to about 1000 h without any marked degradation. The Sony devices, again utilizing the ELO process, improved on this figure tremendously with extrapolated room temperature lifetimes of about 15 000 h. Figure 8.31 displays the evolution of the current through 13 InGaN MQW lasers over 1000 h when the per facet output power is kept at 30 mW at 60 °C. The power levels and temperature are acceptable for commercial write and definitely read applications.

8.9
Thermal Resistance

Deleterious effect rising junction temperature manifests itself as increased threshold current, redshift of the emission, and unstable operation over time. The goal is, therefore, to remove the heat emanating from I^2R, nonradiative recombination, and other losses and to minimize the thermal resistance as well as equip the device with an efficient heat sink. Increased junction temperature flattens the dependence of the quasi-Fermi levels on injection and thus reduces the optical gain and consequently increases the injection current needed for transparency as well as redshifting the gain spectrum. In terms of the CW output power, its dependence on injection current begins to roll over as heating effects become so severe that the output power no more increases with injection but decreases. The characteristic temperature T_0, see Equation 8.135, is a measure of the sensitivity of the threshold current with respect to temperature with higher T_0 leading to lesser sensitivity. The only solace specific to GaN is that increased junction temperature leads to higher hole concentration due to the deep Mg acceptor and reduces the forward voltage and thus the I^2R loss.

The thermal resistance can be measured conveniently as above threshold the carrier concentration can be assumed constant due to Fermi level pinning, and the change in the wavelength due to increasing injection current can be ascribed to the refractive index change only. The longitudinal mode spectrum shifts due to the temperature dependence of the effective refractive index of the laser resonator. To determine the thermal resistance, a small part of the longitudinal mode spectrum is recorded at threshold current and beyond. Then, the heat sink temperature is decreased until the initial mode position is restored. Assuming temperature-independent thermal resistance, change in the heat sink temperature would equal the change in junction temperature. Repeating this procedure for several injection currents would allow a linear fit of the heat sink temperature as set following the above and the dissipated power that would lead to the thermal resistance through the heat flux expression:

$$\phi_{th}(W/cm^2) = \Delta T(K)/R_{th}(cm^2\, K/W) = (P_{in} - P_{out})(W)/A(cm^2), \quad (8.138)$$

where ϕ_{th} is the heat flux out of the device, P_{in} and P_{out} represent the input electrical power and optical power out, respectively, and A is the total area of the chip participating in the thermal conductance assuming uniform lateral temperature. The units are given in parentheses for further clarity, as it is often the case that the area is dropped from both the thermal resistance term and the power term in Equation 8.138. This then leads to thermal resistance in terms of K/W. The thermal resistance here represents the entire stack between the junction and the heat sink and thus values larger than that corresponding to the thermal conductivity of GaN should be expected.

8.10
Nonpolar and Semipolar Orientations

To overcome charge-polarization-induced problems, real or perceived, and also perhaps to enhance laser performance at longer wavelengths, particularly green, nonpolar, and semipolar orientations of GaN have been explored. To restate deteriorating materials quality as the InN molar fraction is increased, which manifests itself as broadening of the gain spectrum as well as reduction in its absolute value coupled with piezoelectric and composition-induced polarization, exacerbates attempts to obtain high-performance lasers at relatively longer wavelengths. Setting the materials quality aside, the polarization-charge-induced

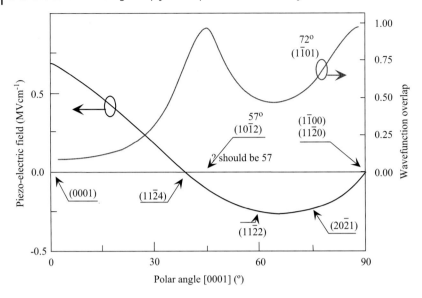

Figure 8.32 Longitudinal piezoelectric field, which is a measure of polarization, in GaN and wavefunction overlap in a 3 nm $In_{0.25}Ga_{0.75}N$ quantum well straddled by GaN as a function of angle of deviation from the c-direction. Nonpolar $\{11\bar{2}0\}$ (a-plane) and $\{1\bar{1}00\}$ (m-plane), which are perpendicular to the [0001] direction, as well as semipolar $\{1\bar{1}01\}$, $\{11\bar{2}2\}$, and $\{20\bar{2}1\}$ planes that are inclined with respect to [0001] direction are specifically indicated. Adapted from Takeuchi et al. and W. Scheibenzuber and U. Schwarz.

separation of the electron and hole wavefunctions, thus reduction in the overlaps that is ameliorated with increasing injection current, can be mitigated by exploring nonpolar and if not at least semipolar orientations (see Figure 8.32) provided that InN can be incorporated and the quality can be maintained at the standards set by c-plane examples if not improved upon. The experimental data available at this time of writing indicate a lower quality than that available on c-plane. Nevertheless, a short discussion is provided for general knowledge. On the nonpolar arena, the choices are a-plane or the m-plane, with the latter showing better materials quality than the former, the main nemesis being the large line density of stacking faults.

Experimental investigation in general focused on low-index nonpolar planes of wurtzite GaN, $\{11\bar{2}0\}$ (a-plane), and $\{1\bar{1}00\}$ (m-plane), which are parallel to the [0001] direction, as well as semipolar $\{1\bar{1}01\}$, $\{11\bar{2}2\}$, and $\{20\bar{2}1\}$ planes that are inclined with respect to [0001] direction, as shown in Figure 8.32. Clearly, where the strain vanishes, the wavefunction overlap is unity. Considering some of the technological impediments involved with nonpolar orientations, the semipolar orientations are shown to improve the wavefunction overlap, the degree of which declines with increasing injection level.

In considering the non-c-plane orientations, one must consider the optical polarization as well as any birefringence. For example, light emanating from an m-plane-oriented emitter is polarized with its field perpendicular to the c-direction. Experimentally observed polarization is less than unity due to dispersion, lack of high quality and purity of crystalline orientation, and so on. Wurtzite nitrides show uniaxial birefringence through the dispersion between the ordinary and extraordinary refractive indices, which amounts to about 2% of the refractive index. This birefringence must be taken into consideration when calculating the waveguide modes. The extraordinary direction is orthogonal to the *c*-direction and the associated modes are termed as the "transverse electric TE (with its optical mode normal to the *c*-axis) and transverse magnetic TM (with its optical mode parallel to the *c*-axis) modes." Therefore, when using orientations that are inclined with respect to the *c*-axis unless the ridge orientation is carefully chosen, there may be a competition between the transverse index profiles paving the way for polarization parallel or normal to the growth plane. Also, optical anisotropy forces the polarization of the modes to be aligned with ordinary and extraordinary direction that are innate to the crystal. Possible orientations of the ridge are parallel or normal to the projection of the *c*-axis. The parallel orientation allows matching of the TE and TM polarization directions, but one is forced to form the facets on high-index planes, which is difficult. The normal orientation allows the facets to be formed on low-index planes, but the extraordinary direction is rotated in a plane normal to the direction of propagation that causes optical modes different from TE and TM modes. In addition, the optical polarization must be considered in gain calculations.

8.11
Vertical Cavity Surface-Emitting Lasers (VCSELs)

The most salient feature of vertical microcavity-based laser is the elimination of the cleaved or etched facets needed for edge emitters. This makes vertical microcavity laser amenable for integration as two-dimensional arrays on the wafer level for high-density optical storage with significantly reduced readout time and high-speed/high-resolution laser printing/scanning technology. They are also well suited for parallel optical communication and potential optical computing. VCSELs exhibit completely circular field patterns as opposed to the elliptical beam profile with an aspect ratio of about 4 between the vertical and horizontal modes in the edge-emitting lasers, which makes light coupling relatively easier in the former. The critical problems facing the microcavity lasers are the need for very high reflectivities demanded of the cavity reflectors (due to short cavity lengths and thus low gain), lateral current confinement, and ways for current conduction without large Joule's heating. Short cavity lengths mean only a few, even one, cavity mode to be present in which case the active layer must be precisely positioned at the antinode of the cavity field. When extreme control of the thicknesses is lacking, longer waveguides with multiple active layers are used as to increase chances of one of the cavity antinodes to be spatially aligned with at least one of the active layers.

The reflectors are made of either heterostructures composed of materials with similar lattice structure but dissimilar refractive indices (the more the dissimilar, the less the number of pairs needed for high reflectivity and wide stopband) or dielectric stacks. Two of these stacks are required, one for the bottom and the other for the top. Due to epitaxial growth requirements often times, the bottom reflector stack is made of the semiconductor variety, while the top can be either, but injection current must be allowed.

Another quality factor, which embodies the entire VCSEL structure, is the spontaneous emission factor β that represents the spontaneous emission coupled to the cavity mode of interest. This entails the ratio of the rate of spontaneous emission contributed to the lasing mode at the emission wavelength of λ to the total spontaneous emission rate. Assuming the rate of spontaneous emission contributing to the cavity mode at the wavelength of λ to be represented by $R_{sp}(\lambda)$, the spontaneous emission factor β at λ is defined as

$$\beta(\lambda) = \frac{R_{sp}(\lambda)}{R_{sp}}, \tag{8.139}$$

where $R_{sp}(\lambda)$ is the total spontaneous emission rate. Spontaneous emission factors near $\beta \approx 2 \times 10^{-3}$ have been reported for GaN-based VCSELs, which is comparable to that for conventional semiconductor-based varieties and compares with about 10^{-4} in edge emitters based on conventional semiconductors.

Another microcavity-based laser is the *polariton laser*, discussed in some detail below, which is very attractive owing to its nearly thresholdless operation. Cavity polaritons whose genesis lies in the interaction between photons and excitons are the basis for its operation. Operated in the strong coupling regime, polariton lasers can produce ultracoherent light. As one can construe, polariton lasers are technologically more demanding. Polariton lasing at high temperatures, such as room temperature and beyond, requires material systems in which excitons actually survive at the temperature of operation. Wide-bandgap semiconductors such as GaN and ZnO are candidates for achieving low-threshold, ideally thresholdless, polariton lasing at room temperature, owing to large exciton binding energies in these materials systems.

Although every aspect of a microcavity-based laser is very crucial, the cavity itself must be a very good resonator with very high Q factor. This then necessarily means the power out cannot compete with edge emitters that can tolerate low reflectivities, such as 0.3, owing to long cavity lengths and thus high gain, but applications of microcavity lasers are different. In lattice-mismatched system and in the context of semiconductor bottom reflector stack, one must consider strain, particularly the sign of it. The AlGaN/GaN system is problematic in that the resultant strain is tensile and cracks ensue. Even without cracks, strain-induced bowing reduces the Q factor as well as making it cavity footprint size dependent, with smaller sizes leading to high Q factors. To combat cracking, the AlGaN layer in the stack is periodically replaced, for example, every fifth, with an equivalent $\lambda/4$ optical length of AlN/GaN short-period stack. Q factors approaching 1000 has been reported by several groups in the fully fabricated electrically pumped microcavity lasers. The crack impediment can be assuaged to some extent by patterning the wafer, leaving periodical troughs in the form of a grid as shown in Figure 8.33.

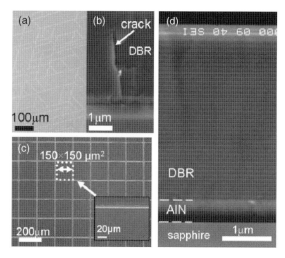

Figure 8.33 (a) Optical microscope image and (b) cross-sectional SEM image of a 30.5 pair planar $Al_{0.46}Ga_{0.54}N/GaN$ DBR. (c) Plan view and (d) cross-sectional SEM image of a selectively grown 40.5 pair $Al_{0.46}Ga_{0.54}N/GaN$ DBR.

Fortuitously for GaN, a lattice matched, one with a large refractive index contrast, variety is available, which can be found in InAlN/GaN. This system can be made crack-free and if particular attention is given to make certain that the vertical thickness is well controlled, very high Q factors can be obtained. Despite the residual strain present in the GaN system owing to a large extent to sapphire substrates, high Q factors are reported in the context of micro PL experiments wherein the area sampled can be made very small, on the order of $\sim 1\,\mu m$. The Q factor values reported steadily increased from near 3000 to 6400, albeit in micro PL experiments. A typical construction features a $5\lambda/2$GaN cavity containing three $In_{0.15}Ga_{0.85}N/GaN$ QWs placed between the bottom 50 pairs $Al_{0.83}In_{0.17}N/GaN$ distributed Bragg reflectors (DBRs) and atop 20 pairs SiO_2/Si_3N_4 DBRs operative at a wavelength of ~ 421 nm.

There are many choices for the top reflector stack, it can be all semiconductor or dielectric based but with the caveat that injection current must be allowed. The advantage of dielectric stacks is that the number of periods needed for >98% reflection totals around 10 where three times as many are needed if all semiconductor stacks are used. For the reflectors to work well, the interfaces must be very smooth over a long range and the thicknesses must be accurate. Among the dielectric stacks, TiO_2/SiO_2, HfO_2/SiO_2, SiO_2/ZrO_2, and so on have been explored for the GaN-based microcavities. Room-temperature current injection microcavity lasers based on lift-off technology have also been reported.

All dielectric reflector-based microcavity structures have also been explored. In one approach, the structure is completed but without the bottom dielectric stack, which is applied on the backside after the structure is separated from the substrate. The other approach uses lateral overgrowth of the waveguide, including the active

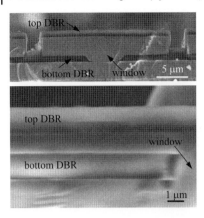

Figure 8.34 Cross-sectional SEM images of full cavity structures consisting of ELO GaN between 13-pair SiN$_x$/SiO$_2$ bottom and top DBRs. The cavity shown in the top panel is ~3 μm thick GaN, whereas that in the bottom panel is thinner, ~1.2 μm, and incorporates InGaN QWs emitting at ~400 nm. The individual layers of the bottom dielectric stack can be identified in the bottom panel image.

region, through a small opening in the bottom dielectric. The cavity in this case would be formed on the wing regions, as shown in scanning electron microscope cross-sectional image of Figure 8.34. In concert with the above all dielectric, 10-pair SiO$_2$/HfO$_2$ stacks and laser lift-off (LLO) technology, microcavities with a Q factor of about 600 have been reported.

8.11.1
Microcavity Fundamentals

A typical MC structure is formed by two DBRs, which consist of $\lambda/4$ wave ($\lambda = \lambda_{\text{air}}/n$ is the design wavelength, with n being the refractive index at the wavelength of particular interest) alternating dielectric stacks with low and high refractive index layers. This gives rise to a broad stopband in the high-reflectivity region centered at λ_{air}, with oscillating side lobes on both sides. In the stopband, the mirror reflectivity is given by

$$R = \left(\frac{1 - (n_{\text{ext}}/n_{\text{c}})(n_{\text{L}}/n_{\text{H}})^{2N}}{1 + (n_{\text{ext}}/n_{\text{c}})(n_{\text{L}}/n_{\text{H}})^{2N}} \right)^2, \tag{8.140}$$

where n_{L}, n_{H}, n_{c}, and n_{ext} are the refractive indices of the low- and high-index layers, the cavity material, and the external medium, respectively, and N is the number of stack pairs. If the refractive index contrast between the layers of the dielectric stack is relatively small, a large number of pairs (~10 for all dielectric and ~50 for all semiconductors) are needed for high reflectivity. The required cavity thickness L_{c} is an integer multiple m of $\lambda/2$ in the medium.

The semiconductor MC is very similar to a simple Fabry–Pérot resonator with planar mirrors. However, due to the penetration of the cavity field into DBRs, the

Fabry–Pérot cavity length must be replaced by a larger effective length, $L_{\text{eff}} = L_c + L_{\text{DBR}}$, where L_{DBR} represents the penetration into the DBRs and is given by

$$L_{\text{DBR}} = \frac{\lambda}{2n_c} \frac{n_L n_H}{n_H - n_L}. \tag{8.141}$$

The cavity mode frequency is given by $\omega_m = (L_c \omega_c + L_{\text{DBR}} \omega_s)/L_{\text{eff}}$, where ω_c is the Fabry–Pérot frequency defined by the length of the cavity and ω_s is the center frequency of the DBR stopband. For $\omega_c \neq \omega_s$, which may arise from inadequately controlled quarter-wave stack thicknesses, the observed ω_m is no longer equal to ω_c and is in fact more sensitive to ω_s than to ω_c since L_{DBR} is greater than L_c. Because the mirrors have a finite transmission probability, the cavity mode has a finite width (FWHM) for $R \to 1$:

$$\hbar \Delta_c = \frac{\hbar c (1 - R)}{n_c L_{\text{eff}}}. \tag{8.142}$$

This width can be considered as homogeneous or lifetime broadening of the confined cavity mode. A typical width of ~ 1 meV corresponds to a cavity mode lifetime of ~ 4 ps.

A planar cavity provides no in-plane confinement perpendicular to the growth axis, just as for electronic states in a QW, and therefore, photon has only in-plane dispersion. Recall that photons are quantized along the cavity as the mirrors force the axial wave vector k_z in the medium to be $2\pi/L_c$. Hence, the cavity photon energy is approximately

$$E = \frac{\hbar c}{n_c} k = \frac{\hbar c}{n_c} \left[\left(\frac{2\pi}{L_c} \right)^2 + k_{//}^2 \right]^{1/2}. \tag{8.143}$$

For small $k_{//}$, the in-plane dispersion is parabolic (as depicted in the above equation), and therefore, it can be described by a cavity photon effective mass $M = 2\pi^2 \hbar n_c / c L_c$. This effective mass is very small, $\sim 10^{-5} m_e$, and the dispersion can be measured directly in angle-resolved experiments allowed by the introduction of an in-plane component to the photon wave vector. At the cavity mode wavelength, the electric field peaks at the center of the cavity. Experiments involving off-normal incidence can also be modeled by including an appropriate *in-plane* wave vector for the field.

Because the gain region is short, and thus the gain is low, in vertical cavity devices, the reflectivity of the top and bottom reflectors must be high, in the high 90% range, in order to overcome optical losses for lasing. In fact, the bottom one should be as close to 100% as possible, while the top one is made slightly less depending on the optical power desired for extraction. The thickness control required for high-reflectivity DBRs scale with the wavelength exacerbating the situation for short wavelengths, the realm of nitride-based lasers. Even small errors in the thickness of each layer in the stack can easily lead to large deviations from the desired central wavelength of the reflection band. It is also imperative that the reflector region be made of materials that have no loss at the operation wavelengths.

The maximum reflectivity of a DBR mirror consisting of alternating $m + 1/2$ pairs of quarter wavelength layers of refractive indices n_1 and n_2 is given by

$$R = \tanh^2\left[\frac{1}{2}\ln\left(\frac{n_0}{n_s}\right) + m \cdot \ln\left(\frac{n_1}{n_2}\right)\right], \tag{8.144}$$

where n_0 and n_s are the refractive indices of air and substrate, respectively.

Figure 8.35a shows the reflectivity of a 10-pair SiO_2/Si_3N_4 quarter-wave reflector stack calculated using the transfer matrix method for a central design wavelength of 380 nm. For comparison, Figure 8.35b shows the calculated (solid line) and measured (dotted line) reflectivity of a 30-pair $Al_{0.50}Ga_{0.50}N/GaN$ quarter-wave stack designed for a central wavelength of 380 nm. Note that the better contrast in

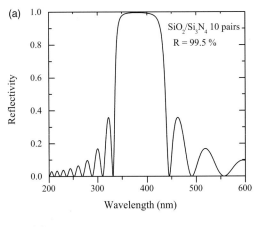

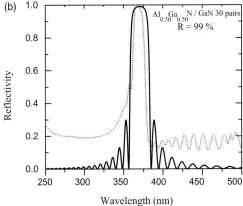

Figure 8.35 Calculated reflection spectra of (a) a 10-pair SiO_2/Si_3N_4 quarter-wave reflector and (b) 36-pair $Al_{0.50}Ga_{0.50}N/GaN$ quarter-wave stack designed for a central wavelength of 380 nm. The dotted line in (b) shows the measured reflection spectrum. Note that the better contrast in refractive indices provided by the SiO_2/Si_3N_4 pair even with fewer pairs produces better reflectivity and wider stopband compared to the $Al_{0.38}Ga_{0.62}N/GaN$ quarter-wave stack. Courtesy of R. Shimada and J. Xie.

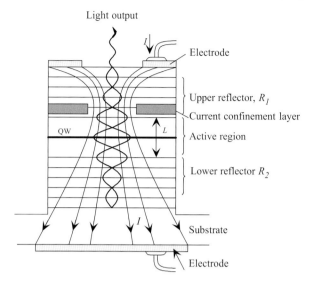

Figure 8.36 Schematic diagram of a vertical cavity surface-emitting laser structure along with equicurrent lines and wave pattern.

refractive indices provided by the SiO_2/Si_3N_4 pair even with fewer pairs produces better reflectivity and wider stopband compared to the $Al_{0.50}Ga_{0.50}N/GaN$ quarter-wave stack. Higher refractive index contrast can be obtained when SiO_2 is used in combination with, for example, TiO_2 or ZrO_2. Figure 8.36 presents a schematic diagram of a vertical cavity surface-emitting laser structure with imbedded equicurrent lines and wave pattern.

The threshold gain g_{th} is given by Equation 8.26. Assuming an optical gain of $10^3\,cm^{-1}$ for the active region, a confinement factor $\Gamma = 1.0$, and $R_1 = R_2 = R$, the minimum reflectivity R is 0.61, which can be easily provided by thin metallic or multiple layer Bragg reflectors. In order to place the maximum of the E field where the gain is generated, the active layer is placed at the antinode of the optical standing wave in a laser cavity.

Lateral current confinement structure is necessary to achieve a carrier population inversion state in the active region with low operation current. This is to avoid fruitlessly pumping the regions that do not contribute to the gain of interest. Because the typical size of a VCSEL is only 10 μm in diameter, it is very difficult to use regrowth methods to form current confinement layers, which are widely employed in edge-emitting laser diodes. There are two typical methods to form current confinement structure in VCSELs. One is the ion implantation technique in which implanted proton and/or other ions such as oxygen have deep levels in the bandgap and create highly resistive regions through carrier trapping, thereby guiding the current away from the implanted regions because the current flows through the path of the least resistance. The other is the lateral oxidization technique made possible by high mole fraction Al-containing layers, such as AlAs, AlInAs, AlGaAs, or AlGaInP in the conventional emitters and AlInN in the nitride family, which are formed near the active region. After etching a mesa structure, the Al-containing layers are selectively oxidized by wet oxidization methods. The current is then funneled from electrodes through the small unimplanted or unoxidized regions to achieve a high injection current density in the active region, which is what is needed for population inversion and thus gain.

Both optically pumped and electrically pumped VCSELs in GaN have been achieved, the former typically leading the latter in time. An example of current injection VCSEL operating CW at 77 K, Figure 8.37a shows the light output power versus CW injection current and current–voltage characteristics. The turn-on voltage is ~4.1 V, which indicates good electrical contacts and efficient intracavity

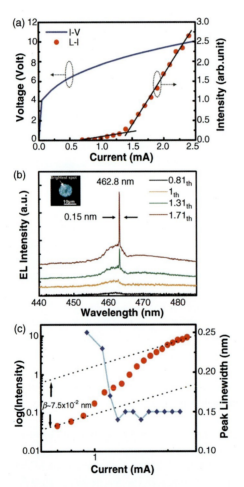

Figure 8.37 (a) The light output intensity versus injection current and current–voltage characteristics under the CW condition at 77 K. (b) The laser emission spectra at different injection current levels at 77 K. Inset shows CCD image of the emission from the aperture. (c) The logarithm light output intensity and the laser emission linewidth versus injection current. Courtesy of H.C. Kuo.

current injection. Lasing action at 462.8 nm occurred at a threshold current of ~1.4 mA. The emission spectra versus injection current levels are shown in Figure 8.37b where the laser emission line at 462.8 nm has a linewidth of ~0.15 nm. From Figure 8.37c, the spontaneous emission coupling factor was estimated at ~7.5×10^{-2}, which is nearly four orders of magnitude higher than that of a typical edge-emitting laser. The laser beam divergence angle was ~11.7° and the angle of polarization was ~80%. Soon after, CW lasing at room temperature by electrical injection in a GaN-based VCSEL structure was reported with all dielectric reflector stack made possible by laser lift-off followed by what would be the bottom dielectric stack formation.

8.11.2
Polariton Lasers

Interaction between photons and excitons can be manipulated in microcavities (MC) to pave the way for coherent optical sources such as polariton lasers, which are based on Bose–Einstein condensation (BEC). In contrast with the bulk polariton, cavity polariton has a quasi-two-dimensional nature with a finite energy at zero wave vector, $k=0$, and is characterized by a very small in-plane effective mass. These characteristics lead to bosonic effects in MCs that cannot be achieved in bulk material. In particular, the large occupation number and BEC, or, more strictly, nonequilibrium polariton population, at the lower polariton branch can be accessible at densities well below the onset of exciton bleaching. This can potentially lead to ultralow-threshold polariton lasers, the operation of which is markedly different from the conventional lasers wherein population inversion is needed. Planar microcavities, also used in VCSELs, whose optical length is a half-integer multiple of the quantum well exciton transition wavelength (λ), are highly suited for the manipulation of cavity polaritons. The requisite high reflectivity mirrors are formed by DBRs composed of alternating $\lambda/4$ stacks of semiconductor and/or dielectric materials possessing high refractive index contrast as in the case of VCSELs.

The vertical cavity causes photon quantization in the growth (vertical) direction. However, the in-plane photon states are unaffected by the cavity confinement. As a result, the dispersion of cavity photons is strongly modified relative to that of photons in free space. The exciton states are also quantized parallel to the growth direction with a continuum of the in-plane excitonic wave vector states. Neglecting the effects of disorder, coupling between exciton and photon states can occur only for the same in-plane wave vector, and the coupled mode eigenstates are termed "cavity polaritons." The energy splitting between the two coupled modes, called Vacuum Rabi splitting (Ω_i) by analogy to the atom–cavity coupling in atomic physics, is typically on the order of 5 meV. The dispersion of cavity modes and that of the exciton–polariton that otherwise has no dispersion interact with each other as shown in Figure 8.38 to form the Rabi splitting.

In order to obtain a significant interaction between the cavity mode and excitonic states, the exciton energy is chosen to be at or close to the resonance frequency of the cavity mode. The coupling is determined by the exciton oscillator strength and the

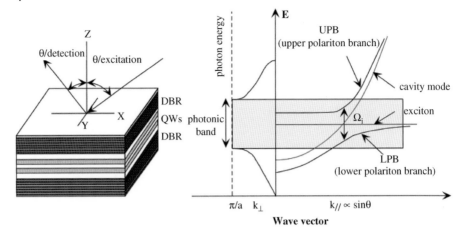

Figure 8.38 Schematic showing the interaction of cavity mode dispersion and exciton–polariton dispersion leading to Rabi splitting.

amplitude of the cavity field at the QW position. It is characterized by an energy equal to the vacuum Rabi splitting, which for QWs is close to the electric field antinodes and is given by

$$\Omega_i \approx 2\hbar \left(\frac{2\Gamma_0 c N_{qw}}{n_c L_{eff}} \right)^2, \qquad (8.145)$$

where N_{qw} is the number of quantum wells in the cavity. The term $\hbar\Gamma_0$ is the radiative linewidth of a free exciton, which can be expressed in terms of the exciton oscillator strength per unit area, f_{ex}:

$$\hbar\Gamma_0 = \frac{\pi}{n_c} \frac{e^2}{4\pi\varepsilon_0} \frac{\hbar}{m_e c} f_{ex}. \qquad (8.146)$$

In the strong coupling regime, where the vacuum Rabi splitting is greater than the linewidths of the cavity and exciton modes, the splitting can be measured in the optical spectrum where the two modes anticross. At resonance, the two polariton modes arise from symmetric and antisymmetric combinations of the exciton and cavity modes. Only the cavity mode couples directly to the external photons away from resonance, while the excitonic mode becomes weak. Equation 8.145 is accurate only for the vacuum Rabi splitting much greater than the linewidths of the exciton and cavity modes.

The basic principle of a polariton laser is depicted in Figure 8.39, which is based on the dispersion curve of the lower branch exciton–polaritons. The strong coupling regime creates a trap containing a small number of polariton states at energies below all the other states in the semiconductor. This polariton trap is sharp with a depth equal to nearly half the splitting Ω_i between the two polariton modes. Polaritons in the trap have properties suitable for BEC of exciton–polaritons. Recombination from this state in the BEC regime is coherent, monochromatic, and sharply directed,

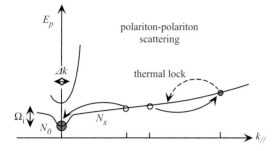

Figure 8.39 Dispersion relations of upper and lower polaritons, showing the dominant pair scattering of polaritons feeding energy into the trap at $k = 0$. Courtesy of J.J. Baumberg.

representing characteristics of laser emission. The relaxation of polaritons into the $k = 0$ state is stimulated if the population of the final state is >1. This lasing process is fundamentally different from that for conventional lasing. The lasing threshold in conventional lasers depends on population inversion, which balances absorption (loss) by stimulated emission. In polariton lasers, however, the threshold is dependent only on the lifetime of lower polariton ground state. When the relaxation to the ground state of the trap is faster than the radiative recombination from this state, optical amplification occurs. The absorption and reabsorption of light are already taken into account within the polariton picture and inversion of population is not required for lasing.

When Ω_i is significantly greater than both the exciton and the cavity linewidths, the system is in strong coupling regime (Figure 8.38). This strong coupling regime is completely different from that for excitons in isolated QWs not embedded in an MC. In the case of QWs without the MC, excitons decay irreversibly into the continuum of photon states along the growth direction due to the discontinuity of translational symmetry along the growth direction. Thus, excitons in QWs would have a finite lifetime (~ 20 ps). The photon states in MCs, which are quantized along the growth direction (along the cavity), and each of the in-plane exciton states can couple to only one external photon state. Quasistationary eigenstates exist as the cavity polaritons, and irreversible exciton decay does not arise. In the time domain, excitations oscillate between the two modes on a subpicosecond timescale with a period h/Ω_i before the leakage of photons from the cavity.

Bulk and cavity polaritons differ drastically. In bulk polaritons, photon dispersion is linear ($\omega = kc/n$), where n is the refractive index of the material. The splitting between the upper and lower polariton branches at resonance is around 16 meV in GaAs. Although this is larger than that in MCs, it is difficult to probe the properties of exciton–polaritons in bulk materials due to the photon–polariton conversion process at the air–semiconductor interface. Unlike the bulk case, in MCs the polaritons are converted directly into external photons as a result of the finite lifetime of their photon component within the cavity and propagation to the surface is not required. Moreover, since the wave vector perpendicular to the sample surface

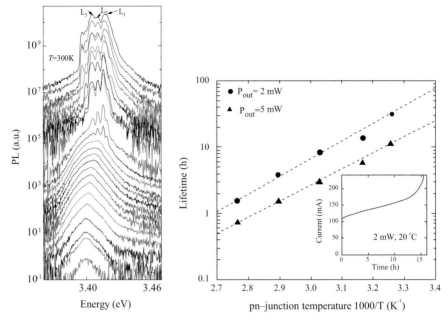

Figure 8.40 Laser diode lifetimes for constant power output levels of 2 and 5 mW, as defined when the current needed to maintain the constant output power increased by 50%, versus the junction temperature. The junction temperature was determined from the thermal resistance of the packaged device, which in turn was determined from the wavelength shift by applied electrical power to the device. Courtesy of M. Kneissl.

is quantized, the complications due to spatial dispersion along the growth direction are eliminated and there is a one-to-one correspondence between the internal polariton states and the external photons. This allows the polariton dispersion curves to be measured easily in reflectivity orPL experiments. A cavity polariton has a quasi-2D nature with a finite energy at zero wave vector ($k=0$) and is characterized by a very small in-plane effective mass (of order $10^{-5} m_e$) that gives rise to narrowing of the linewidth. This feature is due to the half-photon and half-exciton character of the polaritons, the photon fraction having very strong dispersion.

Polariton lasing at room temperature in bulk GaN-based microcavities in the strong coupling regime has been observed under nonresonant pulsed optical pumping. Figure 8.40 shows the emission spectra collected at normal incidence for aforementioned condition. A clear nonlinear behavior can be noted for the emission at $\lambda \approx 365$ nm, with an increase of over 10^3 at the critical threshold of around $I_{th} = 1.0$ mW. This corresponds to a density of $N_{3D} \approx 2.2 \times 10^{18}$ cm^{-3}, which is an order of magnitude below the Mott density $\approx 1-2 \times 10^{19}$ cm^{-3} in GaN at 300 K. The observation of a low-threshold coherent emission, one order of magnitude smaller than that in previously reported nitride-based VCSELs, and the emission line blueshift due to polariton–polariton interaction demonstrate the first room-temperature polariton lasing.

8.12
Degradation

A mention was made that with combination of ELO-, HVPE-, and OMVPE-grown layer structures, the projected laser operation lifetimes are in excess of 15 000 h, albeit at a low power of 30 mW at 60 °C. At higher powers such as 60 mW and higher case temperatures, which are imperative for many applications, however, the operating lifetime of the lasers grown on only ELO or combinations of ELO material is not good enough. For the reasons, as should have been expected, the laser community moved aggressively toward the use of what we can term as pseudo-GaN substrates. These are the freestanding GaN wafers grown by a fast growth technique such as HVPE. In time, however, other bulk growth methods such as ammonothermal and growth from liquid may take over. In this section, we will discuss the degradation of lasers prepared in the pre-pseudo-GaN substrate era first followed by the degradation analysis of lasers grown on pseudo-GaN wafers.

The degradation characteristics of CW InGaN multiple quantum well laser diodes on sapphire substrates have been reported in structures containing a TEM-determined defect density of $5 \times 10^7 \, \text{cm}^{-2}$ with longevities nowhere near those on GaN substrates. The intent here is to illustrate the degradation mechanism(s) in defective structures. From the temperature dependence of the laser diode lifetimes, activation energy of 0.50 ± 0.05 eV was determined. Similar investigations on SiC substrates have also been conducted. Unlike the sapphire substrates, use of SiC substrates enables a vertical current path, cleaved facets, and excellent heat spreading. The temperature rise during CW operation is measured for different mountings in an effort to determine the thermal resistance. A p-side up mounted diode exhibited a thermal resistance of 18 K/W and a 143 h under CW lasing at 1 mW optical power ($T = 25$ °C). DC and pulsed aging show the current flow through the device as being the main degradation source compared to heat for InGaN LDs on SiC substrates. The same may not be said on sapphire due to difficulties associated with heat spreading.

For the $2 \, \mu\text{m} \times 800 \, \mu\text{m}$ ridge waveguide laser diode on sapphire, the increase in threshold current corresponds to an increase in p–n junction temperature of about 18 °C when operated under CW conditions. This is in good agreement with the temperature increase determined from the shift in emission wavelength. For the total power dissipated under CW operation condition, this change in device temperature corresponds to a thermal resistance of 23 K/W, which is rather remarkable compared to the above-mentioned SiC substrate case and speaks of the quality of packaging for heat dissipation.

For the lifetime testing, the output optical power was kept at 2 or 5 mW, and as much current as necessary was allowed to flow to maintain a constant optical power. For the first 13 h, the operating current increased almost linearly and then started to degrade at a faster rate, while the device continued to lasing action after more than 15 h of operation. The lifetime in the measurement is operationally defined as the elapsed time until the laser operating current for the constant output power has increased by 50% from its initial value. The measured lifetimes for two output powers of 2 and 5 mW versus the p–n junction temperature are plotted in Figure 8.40. Joule heating of the device was taken into account and the actual p–n junction temperature was calculated from the thermal resistance, the respective operating current and voltage of the device averaged over the time of operation, and the heat sink temperature. The degradation rate can be fit exponentially on the reciprocal temperature:

$$L(T) = e^{E_A/kT}, \tag{8.147}$$

where E_A is the activation energy, T is the device temperature, and k is the Boltzmann constant. As can be seen from Figure 8.40, this empirical formula describes the measured temperature dependence quite well, and through a least-squares fitting, activation energies of $E_A = 0.50 \pm 0.05$ eV and 0.46 ± 0.04 eV can be derived for 2 and 5 mW output powers, respectively. These activation energies are similar to those exhibited by the Nichia Chemical laser exhibiting lifetimes of several 1000 h. It has been argued that although the activation energy did not seem to be strongly dependent on the light output power of the laser diodes, the overall laser diode lifetime decreased significantly with increasing light output. This reduced lifetime when operated at higher output power levels, however, cannot be solely attributed to a temperature increase. The increase in electric power dissipation, when the laser light output is changed from 2 to 5 mW, results only in a p–n junction temperature increase of ~ 1 K. This temperature increase is much too small to explain the drastic drop in lifetime for the higher output power. Therefore, the degradation mechanism is not only thermally induced but also photon assisted.

The reliability of lasers can be divided into two categories, facet degradation and that due to material properties. In terms of the latter, the formation of In–In bonds at the interface between the InGaN QW and the barrier layers, which result in inversion domains, must be suppressed using either AlGaN alloy or GaN as barrier layers instead of InGaN. There remain compositional inhomogeneities and the notorious V-shape defects, which are linked to dislocations extending to the quantum well region. These V-shaped defects have been reported to take the form of an open hexagonal, inverted pyramid with {1011} sidewalls. Other defects not emanating from the dislocations and running upward toward epilayer surface have been observed in conjunction with quantum well laser structures. These defects have combined the planar defect and the dislocation nature, and so termed as multiple defects. High-resolution lattice imaging indicates that these planar defects are formed at the interface between the quantum well region and the barrier layers. Conversion beam electron diffraction studies indicated the polarity to be opposite of the surrounding matrix, which means that they represent inversion domains (IDs). These ID defects are suspected to form due to excess In–In bonds at the interface between the quantum wells and the barrier layers. The need to suppress the formation of metal–metal bonds, particularly at the interface between the quantum well region and the barrier, is obvious.

As a result of the measures such as using GaN substrates and improved growth conditions, the defect levels are reduced and the laser degradation is now taking a route similar to other semiconductors before GaN and focusing on intrinsic degradation of the material such as production of nonradiative recombination centers and dark line defects and laser facet damage. Pertinent to power extraction from edge-emitting lasers, one facet that is called the back facet is coated with high-reflection coating to prevent light emission that is wasted. The other facet, the front facet, might be coated with an antireflection coating to the extent desired for power extraction. To increase power extraction, which is termed as the edge loss, the laser cavity length can also be made shorter, which requires higher levels of pumping as in the case of partial antireflection coating.

The low dislocation mobility in GaN argues against degradation related to the dislocation multiplication and glide, dubbed in lasers as dark line defects (DLD). However, dislocations introduce inhomogeneous local strain, which could in turn, under severe device operation, pave the way for generation and then multiplication of nonradiative point defects. Any Mg dopant used for p-type AlGaN and GaN diffusion could also be responsible for nonradiative recombination centers. There is evidence that degradation in well-developed lasers is a thermally activated process with a characteristic activation energy of 0.32–0.81 eV.

Occurrence of sudden failure with reduction of cavity length in high-power GaN-based laser diodes is a serious reliability concern, as shown in Figure 8.41a. These failures are in part caused by excess heating of the front facet that has its origin in increasing capture cross section of nonradiative recombination centers by the diffusion of point defects/impurities (such as Mg dopant) during laser operation. Employing a current injection-free region near the facet, termed as the noninjected facet (NIF), by as long as 45 μm, formed simply by leaving out of the p-electrode metal near the laser facet, significantly reduces the catastrophic optical damage and increases the lifetime, as shown in Figure 8.41b, which is estimated to be more than 10 000 h under 0.75 W (reduced to 0.65 W after 700 h under 6.2 kA/cm^2 drive current) CW operation at room temperature.

The premature degradation has also been attributed to the photon-mitigated carbon deposition on the dielectric stack on the front facet. When the residual carbon contamination is successfully removed by plasma cleaning just before cap welding, the lifetime of the laser diode packed with argon gas has been reported to

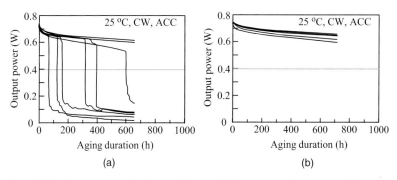

Figure 8.41 Aging test results of laser diodes with (a) conventional structure and (b) with current injection-free structure. Courtesy of S. Tomiya, Sony Corp.

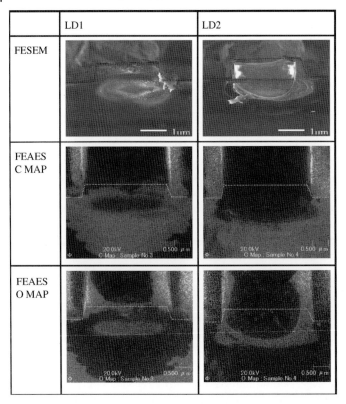

Figure 8.42 Field emission SEM (FESEM) and field emission Auger electron spectroscopy (FEAES) images based on carbon and oxygen mapping for two LD samples. Courtesy of C.C. Kim, L.G. Electronics.

exceed 2000 h under 90 mW CW operation at 60 °C. This particular investigation has been facilitated by field emission SEM (FESEM) and field emission Auger electron spectroscopy (FEAES), the results of which are shown in Figure 8.42 for two tested LDs, where the dotted lines outline the laser ridge for guide to the eye. During the aging tests, the CW output power of two LDs, LD1 and LD2, were kept at 40 mW and the current increase rate was observed to be much larger in LD1, where the damage was more drastic. The FESEM image of LD1 shows extensive damage to the antireflection (AR) facet, the emitting surface (see top row in Figure 8.42). The oval-shaped damage developed during aging resembles the near-field emission pattern whose center is just below the ridge bottom.

The damage so studied has been linked to the presence of C, which covered the dielectric facet coating whose shape is similar to that of the near-field pattern, as shown in Figure 8.42 (middle row). Interestingly, C deposited less in the core area of the oval-shaped damage where the optical density of the emitted light is higher than that in the areas near the circumference. In addition, the FESEM image of LD2 in Figure 8.42 showed layer peeling observed by others as well. The results lead to the conclusion that the premature degradation during the early stages of life tests can also be attributed to the photon-induced carbon build up on the light-emitting facet.

Attempts have been made to delineate degradation caused by facet damage and nonradiative processes. The threshold current and voltage, differential efficiency, and its characteristic temperature T_0 were monitored with the conclusion that facet and near-facet degradation, bulk degradation, and, quite rarely, contact degradation (attributed to processing flaws) occurred. In a majority of devices, high temperature near the facets was noted accompanied by the SiO_2/TiO_2 coating delamination or large catastrophic degradation within the stripe area adjacent to the facet.

Returning to the NIF, the unpumped region of the waveguide should, in principle, increase the optical loss and thus the threshold current, which was not observed. This is in part because of the compensating effect of decreasing pumped area and pertinent to reliability possibly reduced heating and resultant minimization of defect generation. Inclusion of an unpumped region certainly reduces the temperature of the near-facet region. In most cases, degradation was found to manifest itself mainly by an increase in the threshold current through a square root dependence on the aging time, suggestive of the diffusion process being responsible for damage. The diffusion-mitigated failure process may be driven by the current itself, current-induced increase of the junction temperature, and optical field or in aggregate. Aging test below and above lasing threshold can help to delineate the effect of optical field from other processes. Doing so led to the conclusion that the optical field does not influence the degradation rate. Optical field being out of the way, time derivative of the evolution of the threshold current density can be used as a measure of the degradation rate. A nearly exponential dependence of the so-defined degradation speed on the operation current has been observed. The results can be interpreted by assuming a normal Joule heating that can be described as

$$D = C \exp\left[\frac{-E_A}{l(T_{RT} + \alpha_1 I)}\right] \quad \text{or} \quad D = C \exp\left[\frac{-E_A}{l(T_{RT} + \alpha_2 P)}\right], \tag{8.148}$$

where D is the degradation rate, C is the preexponential constant, E_A is the characteristic activation energy of the degradation process, T_{RT} is the room temperature, P is the power dissipation, I is the current, and α_2 is a constant connected with thermal resistance, which is on the order of 12 K/W (α_1 is similar to α_2 in that it is tied to current as opposed to the power). Fitting Equation 8.148 to the experimental data, an α_1 coefficient of 200 K/A and an activation energy E_A of 0.42 eV have been deduced. The good fitting lends support that the current-induced increase of the junction temperature is indeed one of the driving forces for the degradation.

Let us now turn our attention to aging-related increase of the nonradiative recombination in the active area and the increase of the leakage current, which represents the current due to carriers traversing through the active region and recombination somewhere else. To delineate which process plays a more important role, a series of experiments, namely, cathodoluminescence, have been performed to monitor nonradiative recombination in the active region and the characteristic temperature T_0 in Equation 8.135.

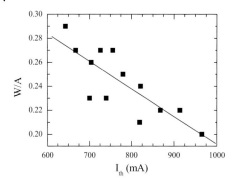

Figure 8.43 Correlation between slope efficiency (W/A) and threshold current (I_{th}) after aging. Courtesy of P. Perlin.

The CL signal of aged devices showed a decrease of 5–30% compared to virgin control devices. Nevertheless, the consistent decrease of CL intensity statistically confirms the appearance of nonradiative centers in the active area of these devices. The contrast changes in the CL images are typically uniformly distributed over the entire stripe area that represents a notable deviation from that observed in the GaAs/AlGaAs lasers in which the degradation is manifested through the appearance of black dots or lines (dark line defects) in the CL microphotographs.

Correlation of the slope efficiency (the slope of the I–L curve) with the threshold current is also illuminating (Figure 8.43) in that it would indicate the increase of nonradiative recombination as the main cause of degradation. A strong correlation has been observed in some cases, but in other cases no such correlation has been noted. This perplexing behavior can be explained by a current leakage path. Due to the formation of a parallel path with the aging time, an increasingly larger portion of the current flows without participating in the radiative process. The fact that the slope efficiency in some cases remains constant during aging and in some cases is nearly proportional to the change of the threshold current indicates that both mechanisms (leakage and nonradiative recombination) are indeed contributors to varying degrees to this process.

Leakage current is typically reflected in the magnitude of the characteristic temperature T_0 in Equation 8.135. A larger leakage current would imply a lower value of T_0. There have been reports of anomalous values of T_0, including negative values. The negative value has been attributed either to the anomalous temperature dependence of the carrier capture rates or to the increasing temperature facilitating a relatively more homogeneous hole distribution through the MQWs. The characteristic temperature measured before, during, and after the aging procedure is shown Figure 8.44. The initial value of $T_0 = 274$ K increased by a factor of 4 after 45 h aging. For room-temperature aging, the threshold current changed significantly, but for higher temperature aging (~80 °C), the increase was found to be relatively small. The higher temperature behavior can be explained by the formation or increase of a thermal barrier for holes.

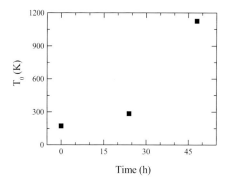

Figure 8.44 Characteristic temperature of laser diodes (T_0) as a function of aging time. Courtesy of P. Perlin.

To summarize, the laser degradation in terms of the materials issues appears to be related to multiple defects formed at the interface between the quantum wells and barrier and to the generation of nonradiative recombination centers within the quantum wells. Furthermore, the facet damage that occurs can be mitigated to some extent by introducing a noninjecting stripe near the front facet. The material/device designs mitigating carrier flyover, as in the case of LEDs, and processing details to either prevent or remove C deposits on the facets would help reduce tendency toward degradation.

Appendix 8.A: Determination of the Photon Density and Photon Energy Density in a Cavity

First, we can describe the cavity as shown in the figure below with the dimensions indicated:

To determine the photon density, we begin with the Maxwell's equations. Assuming a lossless cavity, which is source-free, linear, isotropic, and homogeneous, the Maxwell equations can be written as

$$\nabla \times \vec{E} = -\mu \frac{\partial \vec{H}}{\partial t} \text{ Faraday's law of induction } \vec{B} = \mu \vec{H}, \quad (8.A.1)$$

$$\nabla \times \vec{H} = \varepsilon \frac{\partial \vec{E}}{\partial t} \text{ Ampere's circuit law } \vec{D} = \varepsilon \vec{E}. \quad (8.A.2)$$

$\nabla \cdot \vec{E} = \rho/\varepsilon$ is the Gauss' law, but $\rho = 0$ here because there is no source and no charge and the propagation velocity is $u = 1/\sqrt{\varepsilon\mu}$.

$$\nabla \cdot \vec{E} = \nabla \cdot \vec{E} = 0. \tag{8.A.3}$$

Since the cavity is homogeneous and lossless, the permittivity ε and permeability μ of the cavity are real and constant. Therefore, the Maxwell equations above can be modified as

$$\nabla \times \left(\nabla \times \vec{E}\right) = \nabla \times \left(-\mu \frac{\partial \vec{H}}{\partial t}\right) = \nabla(\nabla \cdot \vec{E}) - \nabla^2 \vec{E}. \tag{8.A.4}$$

$$\nabla \times \nabla \times \vec{E} = -\mu \frac{\partial(\nabla \times \vec{H})}{\partial t}. \tag{8.A.5}$$

$$\nabla \times \nabla \times \vec{E} = -\mu \frac{\partial\left(\varepsilon(\partial \vec{E}/\partial t)\right)}{\partial t}. \tag{8.A.6}$$

$$\nabla\left(\nabla \cdot \vec{E}\right) - \nabla^2 \vec{E} = -\mu\varepsilon \frac{\partial^2 \vec{E}}{\partial t^2}. \tag{8.A.7}$$

From Equation 8.A.3, $\nabla \cdot \vec{E} = 0$. Therefore, Equation 8.A.7 can be written as

$$\nabla^2 \vec{E} = \mu\varepsilon \frac{\partial^2 \vec{E}}{\partial t^2} = \frac{1}{u^2} \frac{\partial^2 \vec{E}}{\partial t^2} \quad \text{or} \quad \frac{\partial^2 \vec{E}}{\partial x^2} + \frac{\partial^2 \vec{E}}{\partial y^2} + \frac{\partial^2 \vec{E}}{\partial z^2} = \frac{1}{(c/n_r)^2} \frac{\partial^2 \vec{E}}{\partial t^2}. \tag{8.A.8}$$

Here, $\vec{E}$ is the electric field of the electromagnetic wave, c is the speed of light in free space, and n_r is the refractive index of the medium. If we assume the lightwave to be a plane wave, the electric field can be written as

$$\vec{E} = \vec{E}_0 \exp\left[j\omega t - j(k_x x + k_y y + k_z z)\right]. \tag{8.A.9}$$

In this equation, k_x, k_y, and k_z are the wave vectors along the x-, y-, and z-axes, and $\hat{x}k_x + \hat{y}k_y + \hat{z}k_z = \vec{k}$.

Substituting Equation 8.A.9 into Equation 8.A.8, we have

$$\left(k_x^2 + k_y^2 + k_z^2\right) = \omega^2 \mu\varepsilon. \tag{8.A.10}$$

We know

$$\left|\vec{k}\right| = \omega\sqrt{\mu\varepsilon} = \frac{2\pi}{\lambda}.$$

Therefore, Equation 8.A.10 can be modified as

$$\left(k_x^2 + k_y^2 + k_z^2\right) = \left|\vec{k}\right|^2 = \frac{(2\pi)^2}{\lambda^2} = \frac{(2\pi\nu)^2}{(c/n_r)^2} = \frac{(2\pi\nu n_r)^2}{c^2}. \tag{8.A.11}$$

Here ν is the wave frequency.

Now we consider the selected modes of this cavity by using the equation above. The boundary conditions for an oscillating electromagnetic wave in cavity are that the electromagnetic waves travel back and forth in the cavity and form standing waves.

Appendix 8.A: Determination of the Photon Density and Photon Energy Density in a Cavity

The boundary conditions for a cavity with dimensions of a, b, and c require the electric field strength at the walls of the cavity is zero:

$$\vec{E}(x=a) = E_0 e^{-j(m_x \pi)} = 0, \quad \vec{E}(y=b) = E_0 e^{-j(m_y \pi)} = 0, \quad \text{and} \quad \vec{E}(z=c)$$
$$= E_0 e^{-j(m_z \pi)} = 0,$$

(8.A.12)

which leads to discretization of the wavenumbers in x-, y-, and z-directions as

$$k_x = \frac{m_x \pi}{a} \quad (m_x = 1, 2, \ldots), \quad k_y = \frac{m_y \pi}{b} \quad (m_y = 1, 2, \ldots), \quad \text{and} \quad k_z$$
$$= \frac{m_z \pi}{c} \quad (m_z = 1, 2, \ldots),$$

(8.A.13)

where a, b, and c represent the dimensions of the cavity. After assuming $a = b = c$ for simplicity,

$$\left(k_x^2 + k_y^2 + k_z^2\right) = \left(\frac{m_x \pi}{a}\right)^2 + \left(\frac{m_y \pi}{a}\right)^2 + \left(\frac{m_z \pi}{a}\right)^2 = \left(\frac{2\pi v n_r}{c}\right)^2. \quad (8.A.14)$$

Solving the equation above, we can get

$$v = \left(\frac{c}{2n_r a}\right) \sqrt{m_x^2 + m_y^2 + m_z^2} \quad (8.A.15)$$

and

$$m_x^2 + m_y^2 + m_z^2 = \left(\frac{2n_r a}{c}\right)^2 = R^2. \quad (8.A.16)$$

According to the relation of energy and photon wave frequency, we know that

$$E = \hbar \omega = \frac{\hbar c}{n_r} |\vec{k}|. \quad (8.A.17)$$

That means that if we set all of the k values as positive, a single k value corresponds to a particular photon energy and a photon frequency. According to Equation 8.A.16, the number of modes between $v = 0$ and v can be found by calculating the number of k vectors in the sphere. Since only the positive integers are taken for the k values, we can limit the calculation to only one-eighth of the sphere with the radius of $R = (2n_r a v)/c$. The volume of one-eighth of the sphere is

$$V = \frac{1}{8}\left(\frac{4\pi R^3}{3}\right). \quad (8.A.18)$$

Since there are two separate modes, TE and TM, for the electromagnetic fields, the total number of modes can be expressed as

$$N = 2xV = 2 \times \frac{1}{8}\left(\frac{4\pi R^3}{3}\right) = \frac{8\pi n_r^3 v^3}{3c^3} a^3. \quad (8.A.19)$$

The mode density between the frequencies of 0 and v is

$$\frac{N}{a^3} = \frac{8}{3}\pi \left(\frac{vn_r}{c}\right)^3 \tag{8.A.20}$$

and its gradient that is the mode density in a frequency interval between v and $v + dv$ (neglecting the dispersion of the refractive index that means n_r is a constant):

$$\frac{d}{dv}\left[\frac{N(v)}{a^3}\right] = \frac{8\pi n_r^3}{c^3} v^2. \tag{8.A.21}$$

If the dispersion of the refractive index is taken into account, the mode density in a frequency interval between v and $v + dv$ is

$$\frac{d}{dv}\left[\frac{N(v)}{a^3}\right] = \frac{d}{dv}\left[\frac{8\pi}{3}\left(\frac{vn_r}{c}\right)^3\right] = \frac{8\pi}{3c^3}\left[n_r^3 \frac{d(v^3)}{dv} + v^3 \frac{d(n_r^3)}{dn_r}\frac{dn_r}{dv}\right]$$

$$= \frac{8\pi}{c^3}\left[n_r^3 v^2 + v^3 n_r^2 \frac{dn_r}{dv}\right] = \frac{8\pi n_r^2 v^2}{c^3}\left[n_r + v\frac{dn_r}{dv}\right] = \frac{8\pi n_r^2 \bar{n}_g v^2}{c^3},$$

where $\bar{n}_g = n_r + v\dfrac{dn_r}{dv}$. \hfill (8.A.22)

In a semiconductor, the DOS should be considered to get the photon density and photon energy density. Considering Boltzmann distribution, the average number of photons for each mode is $\langle \rho \rangle = 1/(e^{hv/kT} - 1) = 1/(e^{E/kT} - 1)$ and the average energy of each mode is $\langle E \rangle = hv/(e^{hv/kT} - 1) = E/(e^{E/kT} - 1)$.

Therefore, the mode density in the frequency interval between v and dv for a unit volume can be expressed as (in units of $1/m^3$)

$$\rho(v)dv = \frac{1}{V}\frac{dN}{dv}dv = \frac{8\pi n_r^3 v^2}{c^3}dv. \tag{8.A.23}$$

Now we can consider the mode density in the k space. As mentioned above, $\hat{x}k_x + \hat{y}k_y + \hat{z}k_z = \vec{k}$.

$$dN(k) = 2 \times \frac{1}{8}(4\pi k^2 \cdot \Delta k) \Big/ \left[\left(\frac{\pi}{a}\right)^3 \cdot a^3\right] = \left(\frac{k}{\pi}\right)^2 \cdot \Delta k. \tag{8.A.24}$$

As $k = 2\pi/\lambda = (2\pi v n_r)/c$, the Equation 8.A.24 can be written as

$$dN(v) = \frac{8\pi n_r^3 v^2}{c^3}\left[1 + \left(\frac{v}{n_r}\right)\left(\frac{dn_r}{dv}\right)\right]dv. \tag{8.A.25}$$

As we know,

$$k = \frac{2\pi n_r}{c} v. \tag{8.A.26}$$

Therefore,

$$dk = \frac{2\pi n_r}{c}\left[1 + \frac{v}{n_r}\cdot\frac{dn_r}{dv}\right]\cdot dv \tag{8.A.27}$$

Appendix 8.A: Determination of the Photon Density and Photon Energy Density in a Cavity | 347

and the Equation 8.A.27 can be rewritten as

$$n_g = n_r + v \cdot \frac{dn_r}{dv} \tag{8.A.28}$$

$$dN(v) = \frac{8\pi n_r^2 n_g v^2}{c^3} dv. \tag{8.A.29}$$

If the refractive index dispersion is neglected, $n_g = n_r$, the density of modes can be expressed as

$$dN(v) = \frac{8\pi n_r^3 v^2}{c^3} dv. \tag{8.A.30}$$

The number of modes in a volume V in a frequency interval of around a central frequency is obtained by multiplying Equation 8.A.30 with volume V:

$$dN'(v) = \frac{8\pi n_r^3 v^2 V}{c^3} dv. \tag{8.A.31}$$

Now we can convert the number of modes in terms of frequencies into that in terms of photon energy by replacing v with E/h and dv with dE/h:

$$dN'(E) = \frac{8\pi n_r^3 E^2 V}{h^3 c^3} dE. \tag{8.A.32}$$

The number of modes per unit volume and per unit frequency can be written as

$$\rho(E) = \frac{1}{V} \frac{dN'(E)}{dE} = \frac{8\pi n_r^3 E^2}{h^3 c^3}, \tag{8.A.33}$$

$$\rho(v) = \frac{1}{V} \frac{dN'(v)}{dv} = \frac{8\pi n_r^3 v^2}{c^3}. \tag{8.A.34}$$

As we know, at thermal equilibrium, the possibility of each state occupied by photon follows the Bose–Einstein distribution. The average energy per mode can be expressed as

$$\langle E \rangle = \frac{hv}{e^{hv/kT} - 1} = \frac{E}{e^{E/kT} - 1}. \tag{8.A.35}$$

Therefore, the average number of photons for each mode can be written as

$$\langle \rho \rangle = \frac{hv}{e^{hv/kT} - 1} = \frac{E}{e^{E/kT} - 1}. \tag{8.A.36}$$

The photon mode density can be expressed as

$$\rho(v) = \frac{8\pi v^2 n_r^3}{c^3} \frac{hv}{e^{hv/kT} - 1} \tag{8.A.37}$$

and

$$\rho(E) = \frac{8\pi E^2 n_r^3}{h^3 c^3} \frac{1}{e^{E/kT} - 1}. \tag{8.A.38}$$

Finally, including the dispersion of the refractive index, the above equations can be written as

$$\rho(\nu) = \frac{8\pi \nu^2 n_r^3}{c^3} \frac{1 + [(\nu/n_r)(dn_r/d\nu)]}{e^{h\nu/kT} - 1}, \qquad (8.A.39)$$

$$\rho(E) = \frac{8\pi E^2 n_r^3}{h^3 c^3} \frac{1 + [(E/n_r)(dn_r/dE)]}{e^{h\nu/kT} - 1}. \qquad (8.A.40)$$

Further Reading

Bernard, M.G.A. and Duraffourg, G. (1961) Phys. Status Solidi B, **1**, 699.

Casey, H.C., Jr. and Panish, M.B. (1978) Heterostructure Lasers Part A: Fundamental Principles, Academic Press, New York.

Hakki, B.W. and Paoli, T.L. (1973) J. Appl. Phys., **44**, 4113.

Hakki, B.W. and Paoli, T.L. (1975) J. Appl. Phys., **46**, 1299.

Kogelnik, H. (1990) in Guided-Wave Optoelectronics, 2nd edn (ed. T. Tamir), Springer Series in Electronics and Photonics, vol. 26, Springer, Berlin.

Kressel, H. (ed.) (1982) Semiconductor Devices for Optical Communication, 2nd edn, Topics in Applied Physics, vol. 39, Springer, Berlin.

Kressel, H. and Butler, J. (1979) Heterojunction laser diode, in Semiconductors and Semimetals, vol. 14, Academic Press, New York, pp. 65–194.

Lasher, G. and Stern, F. (1964) Phys. Rev. A Gen. Phys., **133**, 553–563.

Morkoç, H. (2008) Handbook on Nitride Semiconductors and Devices, vol. 1, Wiley-VCH Verlag GmbH, Weinheim.

Morkoç, H. (2009) Handbook on Nitride Semiconductors and Devices, vol. 3, Wiley-VCH Verlag GmbH, Weinheim.

Pankove, J.I. (1971) Optical Processes in Semiconductors, Prentice Hall, Englewood Cliffs, NJ.

Weisbuch, C. and Vinter, B. (1991) Quantum Semiconductor Structures, Academic Press, New York.

9
Field Effect Transistors

9.1
Introduction

Excluding the applications dominated by the ubiquitous Si-MOSFET, heterojunction field effect transistors (HFETs) are used to meet low noise, high power, and high switching power needs. In a well-designed field effect transistor (FET), the speed depends on the carrier transit time under the gate from the source side to the drain side and the delays inherent in the devices such as those caused by parasitic capacitances and resistances. The semiconductor form that was first used for HFETs was GaAs based. While exploring the properties of quantum wells and superlattices, it was discovered that when the large-bandgap AlGaAs is doped with a donor impurity near the junction it forms with the adjacent small-bandgap GaAs, the electrons donated by donors in the wider bandgap material diffuse to the lower energy conduction band of GaAs where they are confined due to the heterointerfacial potential barrier. The electron gas formed in the process does not get affected, except remotely, by the donors and possesses nearly impurity scattering-free transport.

A large bandgap, large dielectric breakdown field, good electron transport properties (electron mobility possibly in excess of $2000\,cm^2/(V\,s)$ and a predicted peak velocity approaching $3 \times 10^7\,cm/s$ at room temperature, although the velocity deduced from device current or speed is lower than expected and a topic of raging debate), and good thermal conductivity are trademarks of GaN and high-power/temperature electronic devices. These attributes have helped catapult GaN-based HFETs to the realm of cutoff frequencies in the vicinity of $300\,GHz$ and communication band power levels approaching $1\,kW$. Applications of high-power GaN-based HFETs include amplifiers operative at high power levels, at high temperatures, and in unfriendly environments. Examples include radar, missiles, and satellites as well as low-cost compact amplifiers for wireless base stations. A great deal of these applications have been met by pseudomorphic HFETs developed earlier, employing the GaAs system.

9.2
Operation Principles of Heterojunction Field Effect Transistors

With its reduced impurity scattering and unique gate capacitance–voltage characteristics, the HFET has become the dominant high-frequency device. Among the HFET's most attractive attributes are the close proximity of the mobile charge to the gate electrode and high drain efficiency. As was the case for emitters in terms of optical efficiencies and brightnesses, the GaN-based HFETs have quickly demonstrated record power levels at high frequencies with very respectable noise performance and large drain breakdown voltages.

In HFETs, the carriers that form the channel in the smaller bandgap material are donated by the larger bandgap material, ohmic contacts, or both. Because the mobile carriers and their parent donors are spatially separated, short-range ion scattering is nearly eliminated, which leads to mobilities that are characteristic of nearly pure semiconductors. A Schottky barrier is then used to modulate the mobile charge that in turn causes a change in the drain current. Because of this heterolayer construction, the gate can be placed very close to the conducting channel, resulting in large transconductances. Figure 9.1 illustrates a schematic of a GaN/AlGaN HFET heterostructure in which the donors in the wider bandgap AlGaN provide the carriers. In an HFET device under bias, the carriers can also be provided by the source contact. Owing to the polar nature of GaN, HFETs based on GaN, free electrons in the system tend to accumulate in the well and can originate from intentional or unintentional donor in the barrier or additionally from surface states.

9.2.1
Heterointerface Charge

A succinct mention of the polarization effect is made here for continuity while leaving the in-depth discussion to Chapter 2. Polarization induces a field, which in turn causes mobile carriers to move toward the fixed polarization charge with opposite polarity. Because nitrides are large-bandgap materials, they tend to be n-type and the hole concentration is extremely low. Consequently, the mobile carriers are normally electrons donated by intentional donors and surfaces, donor-like defects, or contacts. As the free electrons build up, they occupy and screen the potential minima (which arose due to the polarization charge) until equilibrium is established for the entire heterostructure.

The Schrödinger equation and the Poisson equation can be used self-consistently in order to study the channel formation and current flow mechanisms in a GaN-based HFET. Several approaches have been used to define the system Hamiltonian used in the Schrödinger equation, namely, effective masses, $k \cdot p$ expansion, and tight-binding expansion. The use of sophisticated models such as $k \cdot p$ or tight binding is justified, perhaps even made necessary, by the complex wurtzite band structure, particularly for determining the valence band states. When only the conduction band processes are of interest, effective mass (EM) approximation is a

9.2 Operation Principles of Heterojunction Field Effect Transistors | 351

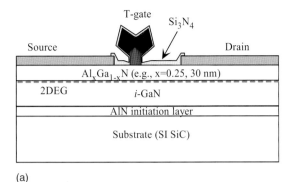

(a)

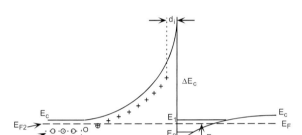

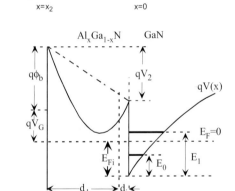

(b)

Figure 9.1 (a) Schematic representation of an AlGaN/GaN modulation-doped field effect transistor with a T-gate HFET. (b) Schematic band structure of an AlGaN/GaN modulation-doped heterostructure in which the free carriers are provided to the GaN layer by the dopant impurities placed in the larger bandgap AlGaN barrier layer. The band bending in the barrier for the doped barrier case is represented by a solid line whereas that for the literally undoped barrier is shown with a broken line. In this case, the free carriers would be provided by a surface, contact, or unintentional impurities in the barrier or bulk.

reasonable approach. Within the effective mass theory, the Schrödinger equation takes the form

$$-\frac{\hbar^2}{2}\frac{d}{dz}\left(\frac{1}{m^*(z)}\frac{d}{dz}\right)\psi_m(z) + V(z)\psi_m(z) = E_m\psi_m(z), \quad (9.1)$$

where $m^*(z)$ is the (position-dependent) effective mass, E_m is the eigenenergy for the mth subband, ψ_m is the wavefunction corresponding to this eigenenergy, and $V(z)$ is the potential energy associated with the conduction band discontinuity and static potential φ by

$$V(z) = -q\varphi(z) + \Delta E_C, \quad (9.2)$$

where ΔE_C is the conduction band discontinuity at the interface.

There are quite a few numerical approaches that can be used to solve the one-dimensional Schrödinger equation in a quasitriangular well (see Figure 9.2) among which are finite difference methods, variational methods, the Rayleigh–Ritz method, and the moment method. The Rayleigh–Ritz method gives only finite numbers of the eigen wavefunctions and eigenenergy levels. Letting z be in the $-c$ direction with $z=0$ marking the interface between the channel and the barrier, the mth wavefunction's Fourier expansion over the range from the top surface ($z=-d$) to the bulk area of GaN ($z=a-d$, $a \gg d$) can be expressed as

$$\psi_m(z) = \sum_{n=1}^{N} a_{nm} \sin\left(\frac{n\pi z}{a}\right). \quad (9.3)$$

In order to determine coefficients a_{nm}, one substitutes ψ_m into Equation 9.1 and takes the integral after multiplying by each orthogonal basis function. Doing so leads to

$$\sum_{n=1}^{N} A_{nm} a_{nm} = E_m a_{nm}. \quad (9.4)$$

The matrix A is given by

$$A_{nm} = \frac{\hbar^2}{2}\left(\frac{n\pi}{a}\right)^2 \frac{2}{a} \int_{-d}^{a-d} \frac{1}{m^*(z)} \sin\left(\frac{n\pi z}{a}\right) \sin\left(\frac{m\pi z}{a}\right) dz$$

$$+ \frac{2}{a} \int_{-d}^{a-d} V(z) \sin\left(\frac{n\pi z}{a}\right) \sin\left(\frac{m\pi z}{a}\right) dz. \quad (9.5)$$

If the effective mass is constant over the entire range, the first term in A_{nm} will reduce to a δ_{nm} function. In a two-dimensional system, the density of states of each subband is constant, and given as

$$\rho_{2D} = \frac{m^*}{\pi\hbar^2}. \quad (9.6)$$

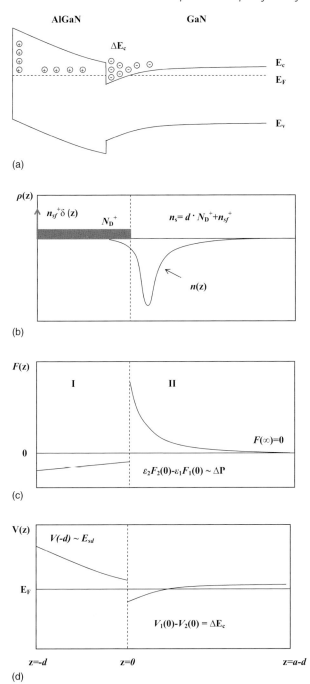

Figure 9.2 (a) Schematic band structure of a GaN/AlGaN heterostructure in which the AlGaN may or may not be doped (doped version is shown) with two-dimensional electron gas in the quasitriangular well in GaN. (b) The same showing the surface δ charge, AlGaN bulk charge, and interface electron charge in GaN. (c, d) Illustration of field ($F(z)$) and potential ($V(z)$) along with the boundary conditions.

Once the wavefunction is calculated, the free electron distribution n is then obtained from

$$n(z) = \sum_m \int_{E_m}^{\infty} \rho_{2D} f(E) |\psi_m(z)|^2 \, dE. \tag{9.7}$$

If one takes f as the Fermi–Dirac distribution function, Equation 9.7 will become

$$n(z) = \frac{m^*}{\pi \hbar^2} \sum_m \int_{E_m}^{\infty} \frac{|\psi_m(z)|^2}{1 + \exp[(E - E_F)/k_B T]} \, dE$$

$$= \frac{m^* k_B T}{\pi \hbar^2} \sum_m \ln\left[1 + \exp\left(\frac{E_F - E_m}{k_B T}\right)\right] \cdot |\psi_m(z)|^2 \tag{9.8}$$

and the total sheet electron density becomes

$$n_s = \int_0^{\infty} n(z) \, dz = \frac{m^* k_B T}{\pi \hbar^2} \sum_m \ln\left[1 + \exp\left(\frac{E_F - E_m}{k_B T}\right)\right]. \tag{9.9}$$

Considering Equation 9.2, the static potential $\varphi(z)$, electrical field $F(z)$, and charges can be correlated by the Poisson equation. If one considers both shallow and deep donors/acceptors in the space charge term, then within each layer the one-dimensional Poisson equation is given by

$$\frac{d[D(z)]}{dz} = \varepsilon \frac{d[F(z)]}{dz} = -\varepsilon \frac{d^2 \varphi(z)}{dz^2} = q\left[p(z) - n(z) + \sum_i N_{Di}^+ - \sum_i N_{Ai}^-\right], \tag{9.10}$$

where ε is the dielectric constant, $n(z)$ and $p(z)$ are the electron and hole distributions, respectively, and N_{Di}^+ and N_{Ai}^-, respectively are the ionized donor and acceptor densities but with different ionization energies. Because we are dealing with polarized materials, spontaneous and piezoelectric polarization effects must be taken into account in the balance of the boundary condition at the interface as well as a sheet charge of donor states present at the surface. In this respect, the schematic band structure, charge distribution, and field and potential distributions are shown in Figure 9.2.

The conservation of the normal component of the electrical displacement leads to

$$\varepsilon_2 \cdot F_2(z=0) - \varepsilon_1 \cdot F_1(z=0) = P_1 - P_2 = \Delta P. \tag{9.11}$$

In region I characterized by the AlGaN layer, the Poisson equation can be written as

$$\varepsilon_1 \frac{dF_1}{dz} = q n_{sf}^+ \delta(z+d) + q N_D^+, \tag{9.12}$$

where n_{sf}^+ is the density of ionized surface donor states, an integration of which leads to the electric field in the same region as

$$F_1 = \frac{q}{\varepsilon_1} n_{sf}^+ + \frac{q}{\varepsilon_1} N_D^+(z+d) + C, \tag{9.13}$$

with C being the integration constant. Doing the same for region II leads to

$$\varepsilon_2 \frac{dF_2}{dz} = -qn(z) \quad \text{and} \quad F_2 = -\frac{q}{\varepsilon_2} \int n(z)dz + C'$$

$$= -\frac{m_C k_B T q}{\pi \varepsilon_2} \sum_l \int_0^z |\psi_l(x)|^2 dx \ln\left[1 + \exp\left(\frac{E_F - E_l}{k_B T}\right)\right] + C'. \quad (9.14)$$

The charge neutrality condition requires that

$$n_{sf}^+ + N_D^+ d = n_s \quad \text{with} \quad n_s = \int_{z=0}^{\infty} n(z)dz. \quad (9.15)$$

F_1 at the interface will be

$$F_1(z=0) = \frac{q}{\varepsilon_1} n_{sf}^+ + \frac{q}{\varepsilon_1} N_D^+ d + C = \frac{q}{\varepsilon_1} n_s + C. \quad (9.16)$$

Noting that $F_2(z=0) = C'$, from Equation 9.11 we can determine C' by

$$C' = \frac{\Delta P + C\varepsilon_1 + qn_s}{\varepsilon_2}. \quad (9.17)$$

Note that the electron concentration in the barrier is very small, so for large values of z, $\int_0^z |\psi_l(x)|^2 dx = 1$. The constant C can be determined noting that the electric field in the bulk of GaN vanishes, that is, $F_2(z \to \infty) = 0$, as

$$C = -\Delta P / \varepsilon_1. \quad (9.18)$$

The electric fields in regions I and II can then be expressed as

$$F_1 = \frac{q}{\varepsilon_1} n_{sf}^+ + \frac{q}{\varepsilon_1} N_D^+(z+d) - \frac{\Delta P}{\varepsilon_1} = \frac{q(N_D^+ z + n_s) - \Delta P}{\varepsilon_1},$$

$$F_2 = -\frac{q}{\varepsilon_2} \int n(z)dz + \frac{qn_s}{\varepsilon_2}. \quad (9.19)$$

For the AlGaN/GaN heterojunction that has an AlN interfacial layer, there are similar relationships among the electrical fields in each of the three regions that are described in the appendix. When there is a Schottky metal on the surface, the surface states will be depleted and the potential $V(z=-d)$ at the surface will be set together by the barrier height and applied voltage. Without the Schottky metal, the prevailing viewpoint is that the surface donor states are the main source of electrons.

Including surface state distribution and the associated statistics will add extra complexity to the self-consistent method. One can instead neglect the hole and acceptor concentrations and consider only the surface donor states, in which case the charge neutrality leads to

$$n_{sf}^+ + \sum_i \int N_{Di}^+ dz = n_s = \int n(z)dz, \quad (9.20)$$

where n_{sf}^+ is the density of ionized surface donor states, which follows the Fermi–Dirac distribution given that only a single energy level E_{sd} is present:

$$n_{sf}^+ = \frac{n_{sf}}{1 + g \exp[(E_F - E_{sd})/k_B T]}. \quad (9.21)$$

The degeneracy factor g is 2 and surface donor state's energy level is approximately 1.2 eV below the conduction band according to scanning Kelvin probe microscopy measurements. The density of total surface donor states N_{surf} can be assumed to be up to 10^{15} cm^{-2}, sufficient to pin E_F at the surface donor energy level. The boundary condition for $V(z=-d)$ (E_F at the surface) must be solved by combining Equations 9.20 and 9.21.

Figure 9.3 shows a self-consistently calculated potential energy (conduction band diagram) and free electron distribution for an $Al_{0.3}Ga_{0.7}N/GaN$ heterojunction.

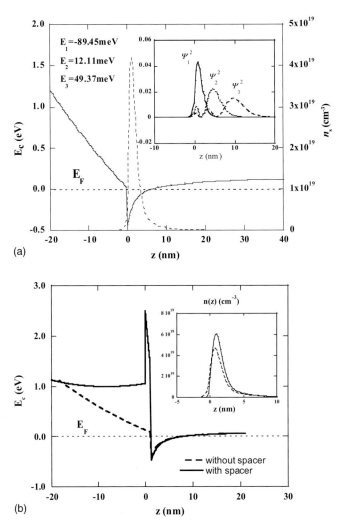

Figure 9.3 (a) Conduction band profile, electron distribution, and eigenenergy/wavefunction for the first three subbands in an $Al_{0.3}Ga_{0.7}N/GaN$ heterojunction. (b) Band structure for AlGaN/GaN HFET structure with (solid line) and without (dashed line) an AlN spacer; electron density distributions for the same are shown in the inset.

During the simulation, the surface state density was assumed to be extremely large (10^{15} cm^{-2}), with a single-donor energy level located at 1.2 eV to make sure a sufficient number of electrons are available from this source alone for the system to reach equilibrium. Additionally assumed shallow and deep donors in the barrier layer have concentrations of 8×10^{17} and 5×10^{17} cm^{-3}, respectively. The detailed spontaneous and piezoelectric polarization calculations for AlGaN and GaN could be found in Chapter 2. The room-temperature total sheet electron density is approximately 1×10^{13} cm^{-2}, depending on the particulars of the structure, and the 2DEG is confined within a 5 nm channel under the interface. The eigenenergy and wavefunction of first three subbands are also calculated. Experiments also revealed a similar range for the ground energy (~100 meV) of the 2DEG by Shubnikov–de Haas (SdH) measurements. The sheet density depends on the barrier thickness and also on the AlN spacer layer thickness.

The potential energy $V(z)$ at the heterointerface/channel area can be approximated by a triangular potential well, so that an analytical solution to the Schrödinger equation can be formulated. In terms of the Hamiltonian, the Schrödinger equation can be expressed as

$$H = H_0 + eF_z z = -\frac{\hbar^2}{2m_z^*}\frac{\partial^2}{\partial z^2} + eF_z z. \qquad (9.22)$$

Nonparabolicity may induce deviations from the simple parabolic band model. However, this will not substantially change the results that are presented here. For a triangular potential barrier that can approximately represent the potential distribution at the heterointerface with a 2D charge being present, we can use the following boundary conditions:

$$V(z) = \begin{cases} qF_z z, & z > 0, \\ \infty, & z \leq 0, \end{cases} \quad \text{with} \quad F_z = \frac{\Delta P}{\varepsilon}, \qquad (9.23)$$

where F_z is the electric field and z is the distance along the growth direction, normal to the interface, being 0 at the interface.

The solution of the wavefunction is given by

$$\psi(z) = \text{Ai}\left[\frac{2m_z^* eF_s}{\hbar^2}\left(z - \frac{E_i}{eF_z}\right)\right], \qquad (9.24)$$

with $\text{Ai}(\cdot)$ being the Airy function given by

$$\text{Ai}(u) = \frac{1}{\pi}\int_0^\infty \cos\left(\frac{1}{3}t^3 + ut\right) dt. \qquad (9.25)$$

The eigenvalues for energy are

$$E_i \approx \left(\frac{\hbar^2}{2m_z^*}\right)^{1/3}\left[\frac{3\pi eF_z}{2}\left(i + \frac{3}{4}\right)\right]^{2/3}, \qquad (9.26)$$

with the $(i+3/4)$ replaced with 0.7587, 1.7540, and 2.7575 for the first three lowest subbands, respectively, for the exact eigenvalues. The parameter i takes values of

$i = 0, 1, 2$, and 3, with 0 representing the ground state and the rest the excited states. The average value for z, where the 2DEG can be approximated to be, for one subband occupation can be found from

$$\langle z \rangle = \int \psi^* z \psi \, dz \approx \frac{2E_i}{3eF_z}. \tag{9.27}$$

A variational method introduced by Fang and Howard can be used as well in which case $\psi(z) = Az\exp(-az)$, where A is a normalization constant and a is a variational parameter, is a wavefunction of the ground state. From the normalization condition $\int \psi^*(z)\psi(z)dz = 1$, the normalization constant A can be found as $A = 2a^{3/2}$.

The expectation value for the total energy is given by

$$\langle E \rangle = \int \psi^*(z) H \psi(z) dz = \int \psi^*(z) \left(-\frac{\hbar^2}{2m_z^*} \frac{\partial^2}{\partial z^2} + eF_z z \right) \psi(z) dz$$
$$= \frac{\hbar^2 a^2}{2m_z^*} + \frac{3}{2} \frac{eF_z}{a}. \tag{9.28}$$

The variational parameter can be found from $d\langle E \rangle / da = 0$, that is, $2\hbar^2 a^3 - 3m_z^* eF_z = 0$.

Then wavefunction of the ground state is

$$\psi(z) = 2(a)^{3/2} z \exp[-az] \quad \text{with} \quad a = \left(\frac{3}{2} \frac{m_z^* eF_z}{\hbar^2} \right)^{1/3}. \tag{9.29}$$

The average position of electrons in the ground state is (see Section 4.9 for a more in-depth treatment)

$$\langle z \rangle = \int \psi^*(z) z \psi(z) dz = \frac{3}{2a} = \frac{3}{2[(3/2)(m_z^* eF_z/\hbar^2)]^{1/3}} = \frac{3\hbar^{2/3}}{(12 m_z^* eF_z)^{1/3}}. \tag{9.30}$$

Assuming that only the ground state is occupied at an $Al_{0.2}Ga_{0.6}N/In_{0.1}Ga_{0.9}N$ interface with $-0.0055 \, C/m^2$ interfacial charge, the electric field is $F_z = -5.7068 \times 10^7 \, V/m$, the variational parameter $a = 5.9819 \times 10^8 \, m^{-1}$, and the average position of electrons is $\langle z \rangle = 2.5 \, nm$ from the interface. Here effective masses used for GaN and InN are 0.2 and 0.11, respectively, and a linear interpolation is used to find the effective mass of $In_{0.1}Ga_{0.9}N$.

9.2.2
Analytical Description of HFETs

In order to qualitatively demonstrate the effect of charge stored at the heterointerface on mobility and carrier velocity, we present in the following an analytical description of the operation of HFET, albeit approximate and compact. The model developed a priori for the GaAs/AlGaAs system can be modified to include polarization charge. It is based on the concept that the amount of charge that is depleted from the barrier donor layer (applicable to conventional heterojunctions) or provided by the surface (applicable to the semiconductor nitride family) is accumulated at the interface, while the Fermi level is kept constant across the heterointerface. The polarization charge will later be added to the model. The electron sheet charge with no external gate bias (or hole charge in the case of p-channel HFET, which is very unlikely in the case of GaN) provided by the donors in the barrier layer may be given by

$$n_{s0} = \left[\frac{2\varepsilon_2 N_d}{q}(\Delta E_c - E_{F2} - E_{Fi} + N_d^2 d_i^2)\right]^{1/2} - N_d d_i, \quad (9.31)$$

where E_{F2} is the separation between the conduction band in the barrier layer and the Fermi level, N_d is the donor concentration in the barrier layer, ε_2 is the dielectric constant of the barrier layer, q is the electronic charge, ΔE_c is the conduction band discontinuity, E_{Fi} is the Fermi level with respect to the conduction band edge in the channel layer, and d_i is the thickness of the undoped layer in the barrier layer at the heterointerface. A graphical description of the aforementioned parameters as well as the band edge profile for an AlGaN/GaN heterostructure with gate bias on the surface is shown in Figure 9.1.

Assuming only the ground- and first excited-state occupation, the electron charge at the heterointerface is given by

$$n_s = \frac{\rho}{\beta}\ln\left[\left\{1 + e^{\beta(E_{Fi} - E_0)}\right\}\left\{1 + e^{\beta(E_{Fi} - E_1)}\right\}\right], \quad (9.32)$$

where $\beta = q/kT$, and $E_0 = \gamma_0 n_s^{2/3}$ and $E_1 = \gamma_1 n_s^{2/3}$ are the positions of the first and the second quantum (potential) states at the interface, respectively. These states correspond to a triangular well formed by the interfacial stored charge. The energy reference is the bottom of the conduction band edge in GaAs. It is assumed here that these lowest energy states are the only ones that are either filled or partially filled. The constants γ_0 and γ_1, which are dependent on the effective mass of the channel material used, and the density of states ρ ($\rho = qm^*/\pi h^2$, where m^* is the effective mass of electrons and h is the Planck constant), are derived on the basis that the quantum well may be reasonably triangular in shape. Depending on the value of the applied voltage, the gate on the surface of the barrier layer depletes some or all of the stored charge at the interface. Thus, only a simultaneous solution of Equations 9.31 and 9.32 can result in the determination of the Fermi level provided the interface sheet charge concentration is known. The determination of the sheet charge concentration can similarly be carried out from the same equations if the Fermi level is known. With a gate voltage present, Equation 9.31, which depicts the equilibrium situation, must be replaced by

$$n_s = \frac{\varepsilon_2}{qd}\left[V_g - \left(\phi_b - V_{p2} + \frac{1}{q}E_{Fi} - \frac{1}{q}\Delta E_c\right)\right], \quad (9.33)$$

where ϕ_b is the Schottky barrier height of the gate metal deposited on the barrier layer, V_g is the applied gate to source bias voltage, d_d is the thickness of the doped barrier layer, $d = d_d + d_i$, $V_{p2} = qN_d d_d^2/2\varepsilon_2$, and ε_2 represents the dielectric constant in the barrier (AlGaN).

The interface charge concentration in the presence of a gate bias may be expressed by

$$n_s = \frac{\varepsilon_2}{q(d + \Delta d)}(V_g - V_{off}), \tag{9.34}$$

where $\Delta d = \varepsilon_2 a_F/q$ and for the case without polarization-induced interfacial charge

$$V_{off} = \varphi_b - \frac{1}{q}\Delta E_c - V_{p2} + \frac{1}{q}\Delta E_{F0}. \tag{9.35}$$

Δd is typically about 2–4 nm for GaN and 8 nm for GaAs, and represents in pictorial terms the location of the peak density of the 2DEG from the heterointerface. In addition, the terms ΔE_{F0} and a_F are determined from the extrapolations. For example, ΔE_{F0}, which is a temperature-dependent quantity, is the residual value of the Fermi level for zero interface sheet density obtained by extrapolating linearly from the linear region of the curve.

Under ideal conditions, V_{off} would represent the threshold voltage. However, experimentally the current and through it the free sheet charge in the channel is plotted against the gate voltage and the extrapolation of the linear region to zero charge would result in the threshold voltage. Due to substantial subthreshold voltage, the gate voltage leading to nearly zero current flow in the channel would be smaller (meaning negative, in absolute value it is larger for an n-channel FET).

The parameter a_F represents the slope of the curve, which is reasonably linear for a wide range of sheet charge except near the vanishing values, relating the Fermi level to the sheet charge. Utilization of this displaced linear approximation leads to the interface Fermi level to be expressed as

$$E_{Fi} = \Delta E_{F0}(T) + a_F n_s. \tag{9.36}$$

For example, for the GaAs/AlGaAs system, $a_F \approx 0.125 \times 10^{-16}$ V/m^2 and $\Delta E_{F0} \approx 0$ at 300 K and 0.025 meV at $T \leq 77$ K. Similar figures can be obtained for the GaN/AlGaN system neglecting, for the time being, the effect of the polarization charge at the interface.

In a field effect transistor, the drain bias produces a lateral field. For long-channel devices and/or for very small drain biases, it is generally assumed that the channel voltage, which varies along the channel between the source and the drain and finally reaches a value equal to the drain voltage, is added to the gate potential. When it is done, Equation 9.34 becomes a function of the distance, x, along the channel:

$$n_s = \frac{\varepsilon_2}{q(d + \Delta d)}[V_g - V_{off} - V(x)], \tag{9.37}$$

with $V(x)$ as the channel potential.

For a GaN-based modulation-doped FET including the variety dependent only on the polarization-induced charge, the terms Δd and ΔE_{F0} will be neglected in which case one can relate the sheet carrier concentration to the gate voltage above threshold as

$$n_s = \frac{\varepsilon_2}{qd}(V_g - V_{off}) \quad \text{or the interface charge } Q_s = \frac{\varepsilon_2}{d}(V_g - V_{off}) \qquad (9.38)$$

and with interfacial polarization charge of n_p and again neglecting the effect of the Δd and ΔE_{F0} on the threshold voltage, but including the charge due to bulk doping $N_B W_d$ (donor level and depletion layer thickness in the channel layer), one gets

$$V_{off} = \phi_b - \frac{1}{q}\Delta E_c - V_{p2} - \frac{qd}{\varepsilon_2}(n_p + N_B W_d). \qquad (9.39)$$

It is assumed that the doped layer thickness is very small compared to the AlGaN barrier because Equation 9.39 lumps the bulk charge as interfacial charge of equal amount without considering its distributed nature and assumes its distance to the gate metal to be the same as the thickness of the barrier layer. The polarization-induced charge can be determined using the methodology discussed in Chapter 1, but suffice it to say that it is about $10^{13}\,\text{cm}^{-2}$ under equilibrium conditions.

For long-channel HFETs, the following expressions are obtained. The current–voltage relationships can be found in many elementary device texts. Suffice it to say that the drain current in an n-type long-channel device before saturation, where constant mobility can be assumed, is given by

$$I_D = G_0 V_P \left[\frac{V_D}{V_P} + \frac{2}{3}\left(\frac{-V_G}{V_P}\right)^{3/2} - \frac{2}{3}\left(\frac{V_D - V_G}{V_P}\right)^{3/2} \right], \qquad (9.40)$$

where G_0 and V_P represent the full channel conductance (assuming no depletion at all) and total gate voltage inclusive of the gate built-in voltage, required to pinch off the channel, respectively. They are given by

$$G_0 = \frac{qN_D\mu aZ}{L} \quad \text{and} \quad V_P^2 = \frac{qN_D}{2\varepsilon}a^2, \qquad (9.41)$$

where N_D is the channel doping level assumed to be equal to the electron concentration (should be the electron concentration when these two parameters deviate substantially), Z is the width of the gate, L is the gate length, a is the total channel thickness, and ε is the dielectric constant of the channel.

For small values of $V(z)$, it may be assumed that the constant-mobility regime is inapplicable and that

$$I_d = qn_s\mu Z\frac{dV(x)}{dx} = \frac{\mu Z\varepsilon_2}{d}[V_G - V_{off} - V(x)]\frac{dV(x)}{dx}, \qquad (9.42)$$

where μ is the charge carrier mobility and Z is the width of the gate. By integrating Equation 9.42 from the source to the drain while keeping in mind that the drain current remains constant throughout the channel and $V(x=0)=0$ and $V(x=L)=V_{DS}$, one obtains

$$I_{DS} = \frac{\mu Z\varepsilon_2}{L\,d}\left[(V_G - V_{off})V_{DS} - \frac{V_{DS}^2}{2}\right] = \beta_d\left[V_{Geff}V_{DS} - V_{DS}^2/2\right], \qquad (9.43)$$

where $V_{Geff} = V_G - V_{off}$, V_{DS} is the drain–source voltage, $\beta_d = \mu Z\varepsilon_2/dL$, and L is the intrinsic channel length or in practical terms the gate length. The current reaches saturation when the drain voltage is increased to the point where the field in the channel exceeds its critical value, thereby causing the velocity to saturate.

Under these circumstances and utilizing Equation 9.37, the drain current may be calculated as follows:

$$I_{DS} = qZv_{sat}n_s = \frac{\varepsilon_2 Zv_{sat}}{d}(V_G - V_{off} - V_{DSS}) = \beta_d V_0(V_{Geff} - V_{DSS}), \quad (9.44)$$

where V_{DSS} is the saturation drain voltage, I_{DS} is the saturation current, $V_0 = v_{sat}L/\mu$, and v_{sat} is the saturation velocity. For GaN devices, Δd can be neglected. The maximum of the 2DEG is only about 4 nm from the interface, which is smaller than the thickness of the barrier layer that is typically greater than 20 nm.

Equating Equations 9.43 and 9.44 at the saturation point, one can solve for drain saturation voltage as

$$V_{DSS} = V_G + V_0 - (V_0^2 + V_{Geff}^2)^{1/2}, \quad (9.45)$$

which when substituted back into Equation 9.44 yields

$$I_{DS} = \beta_d V_0^2 \left\{ \left[1 + \left(\frac{V_{Geff}}{V_0}\right)^2\right]^{1/2} - 1 \right\}. \quad (9.46)$$

The treatment presented above is known as the two-piece model, implying that an abrupt transition takes place from the constant-mobility regime to the constant-velocity regime. A more accurate picture is one in which this transition is smoother allowing the use of a phenomenological velocity–field relationship for a more accurate description of the HFET operation. The simplest of all these pictures is one that neglects the peak in the velocity–field curve and assumes Si-like velocity–field characteristics. One such characteristic may be expressed by

$$v = \frac{\mu F(x)}{1 + \mu F(x)/v_{sat}} = \frac{\mu_0 F(x)}{1 + F(x)/F_C}, \quad (9.47)$$

where $F(x)$ represents the electric field in the channel, and is equal to $F(x) = dV(x)/dx$, μ is low-field mobility, and $F_C = v_{sat}/\mu$ is the electric field at the saturation point. It may be noted that this field is not constant throughout the channel. Recognizing that

$$I_D = Q_{total}(x)Zv(x), \quad \text{where} \quad Q_{total} = \frac{\varepsilon_2}{d}[V_G - V_{off} - V(x)], \quad (9.48)$$

we get for the drain current I_D:

$$I_D = \frac{Z\varepsilon_2 v(x)}{d}[V_G - V_{off} - V(x)]. \quad (9.49)$$

Using the expression for $v(x)$ given in Equation 9.47, Equation 9.49 may be simplified as

$$I_D = \frac{Z\varepsilon_2 \mu v_{sat}}{d} \left\{ \frac{dV(x)/dx}{v_{sat} + \mu dV(x)/dx}[V_G - V_{off} - V(x)] \right\}, \quad (9.50)$$

where $v_{sat} = \mu F_C$, F_C being the field where the velocity assumes its saturation value. By integrating Equation 9.50 from the source end ($x = 0$) of the channel to the drain end of the channel ($x = L$), keeping in mind that the drain current must be constant throughout the channel, one may obtain an expression for the drain current in the framework of a procedure elucidated by Lehovec and Zuleeg. One thus obtains

$$I_D = \frac{Z\varepsilon_2 \mu v_{sat}}{d} \left[\frac{V_{Geff} V_D - V_D^2/2}{v_{sat} L + \mu V_D} \right]$$
$$= \frac{1}{1 + \mu V_D / v_{sat} L} \left\{ \frac{\mu Z \varepsilon_2}{L\, d} \left[(V_G - V_{off}) V_D - \frac{V_D^2}{2} \right] \right\}. \tag{9.51}$$

Note that when the saturation velocity v_{sat} approaches infinity, Equation 9.51 reduces to Equation 9.43, which is valid for the constant-mobility case, and corroborates with the gradual channel approximation that is valid for long-channel HFETs. Following the procedure of Lehovec and Zuleeg using Equations 9.50 and 9.51, and assuming velocity saturation, the drain saturation current I_{DSS} may be determined as

$$I_{DSS} = \frac{2(V_G - V_{off})^2 Z\varepsilon \mu}{L(d + \Delta d)} [1 + (1 + \xi_d)^{1/2}]^{-2}, \tag{9.52}$$

where

$$\xi_d = \frac{2\mu(V_g - V_{off})}{v_{sat} L}. \tag{9.53}$$

Note that when $V_0 \gg (V_g - V_{off})$ drain current saturation occurs due to pinch-off, but not due to velocity saturation as is expected for the long-channel devices.

The transconductance is an important parameter in HFETs and is defined by

$$g_m = \left(\frac{\partial I_d}{\partial V_g} \right)_{V_d = const}. \tag{9.54}$$

For the saturation regime, the transconductance may be expressed by

$$g_m^{sat} = \left(\frac{\partial I_{dss}}{\partial V_g} \right)_{V_d = const} = \frac{2Z\varepsilon\mu(V_g - V_{off})}{(d + \Delta d) L} [(1 + \xi_d) + (1 + \xi_d)^{1/2}]^{-1}. \tag{9.55}$$

The maximum transconductance is obtained when the sheet charge density is fully undepleted under the gate, which leads to

$$g_m^{max} = \frac{Zq\mu n_s}{L} \left\{ 1 + \left[\frac{q\mu n_s (d + \Delta d)}{\varepsilon v_{sat} L} \right]^2 \right\}^{-1/2}. \tag{9.56}$$

For very short gate lengths, which occur essentially in all modern HFETs, the second term in the bracket dominates, and Equation 9.56 reduces to

$$g_m^{max} \approx \frac{Z\varepsilon v_{sat}}{d + \Delta d}. \tag{9.57}$$

The measured transconductance is actually smaller than that given by Equation 9.57 in that the source resistance, which will be defined shortly, acts as a negative feedback. Taking the circuit effects into account, the measured extrinsic transconductance may be given by

$$(g_m^{max})_{ext} = \frac{g_m^{max}}{1 + R_s g_m^{max}}. \quad (9.58)$$

9.3
GaN and InGaN Channel HFETs

Let us now consider cases with GaN and InGaN channels having either AlGaN or InAlN (can be lattice matched to GaN) barrier layers. If the GaN channel is grown on a thick GaN buffer layer, it can be assumed to be relaxed. Therefore, no piezoelectric charge would be induced by GaN. If the InGaN channel is made thin enough for it to be coherently strained, it would be under compressive strain and there would be a piezoelectric polarization. The AlGaN layer on the GaN channel layer would be under tensile strain leading to piezoelectric polarization charge. If an InAlN barrier is used, it can be lattice matched to GaN in which case the piezoelectric component of the polarization would not exist. However, it can be grown with smaller In concentration than the roughly 18% matching composition for it to be under tensile in-plane strain or with larger In concentration for it to be under compressive in-plane strain. For simplicity, let us assume that in the case of InGaN channel, both AlGaN and InGaN layers assume the in-plane lattice constant of GaN (Table 9.1). This example is depicted in Figure 9.4, which also shows the compositional gradient-induced spontaneous polarization.

Briefly, tensile-strained AlGaN and compressively strained InGaN would have piezoelectric polarization as indicated in Figure 9.4. In addition, all three layers would have spontaneous polarization at each compositional gradient, again as shown in Figure 9.4, where the length of the arrows represents the relative values. We should point out that the spontaneous polarization is negatively larger in InGaN than GaN, which means that InGaN on GaN interface would attract holes that are not assumed

Table 9.1 Piezoelectric constants, elastic constants, and spontaneous polarization in nitride-based binaries (see Chapter 2 for details).

	AlN	GaN	InN
e_{33}^* (C/m^2)	1.5	0.67	0.81
e_{31} (C/m^2)	−0.53	−0.34	−0.41
e_{31}^p, GGA	−0.62	−0.37	−0.45
C_{33} (GPa)	377	354	205
C_{31} (GPa)	94	68	70
P_0 (C/m^2)	−0.090	−0.034	−0.042
$e_{31} - (C_{31}/C_{33})e_{33}$	−0.86	−0.68	−0.90
a-lattice parameter (Å)	3.11	3.199	3.585

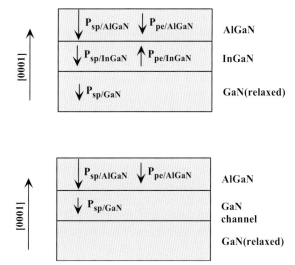

Figure 9.4 Schematic representation of an AlGaN/GaN and an AlGaN/InGaN channel structure indicating the piezoelectric and spontaneous polarization under the assumption that both InGaN and AlGaN assume the in-plane lattice constant of GaN in which case InGaN would be under compressive strain and AlGaN would be under tensile strain.

present in the system and therefore neglected. However, that at the AlGaN/GaN interface and also that at the AlGaN/InGaN interface would attract electrons.

Let us calculate the output I–V characteristics for an AlGaN/InGaN HFET and AlGaN/GaN HFET. The layer structure consists of a GaN buffer that is relatively thick and relaxed, the InGaN channel is 20 nm thick, and the top layer is AlGaN, part of which next to the interface is undoped. Let us neglect the effect of strain on the band structure while using the appropriate bowing parameters for bandgap calculations. Let us use the linear interpolation for determining the spontaneous and piezoelectric polarization charges; although not strictly correct, see Chapter 1 for an accurate treatment.

The model discussed in the previous section can be used to calculate the output characteristics of HFETs. The parameters discussed in Chapters 1 and 2 together with the FET-specific ones given below would suffice. Let us assume that the mole fraction in $Al_xGa_{1-x}N$ is 30% ($x = 0.3$), the doping level in $Al_xGa_{1-x}N$ is $10^{18}\,\text{cm}^{-3}$ (N_D) for generality although this can be assumed undoped, and the thickness of the doped $Al_xGa_{1-x}N$ is 20 nm. The thickness of the undoped $Al_xGa_{1-x}N$ is 2 nm. The gate length is 1 μm and centered in a 3 μm channel. The gate width is $Z = 100\,\mu\text{m}$ and the Schottky barrier built-in voltage is 1 V. Let us also assume a saturated velocity of $1 \times 10^7\,\text{cm/s}$. To make the picture simpler, let us consider the access resistances to be negligible or zero for the sake of the model. Doing so for an $Al_{0.3}Ga_{0.7}N/GaN$ HFET having a 1 nm AlN spacer with a gate length $L = 1\,\mu\text{m}$, low-field electron mobility $\mu = 1000\,\text{cm}^2/(\text{V s})$, and saturation velocity $v_s = 10^7\,\text{cm/s}$ leads to output characteristics shown in Figure 9.5.

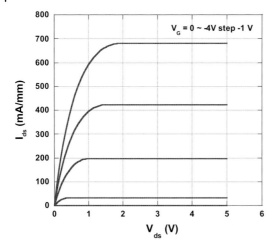

Figure 9.5 Calculated output characteristic for the Al$_{0.3}$Ga$_{0.7}$N/GaN HFET with a 1 nm thick AlN spacer layer and parameters described in the text. Courtesy of Q. Fan.

The effect of electromechanical coupling is sufficiently small to be neglected. A step-by-step approach can be found in Morkoç (2009).

9.4
Equivalent Circuit Models: De-embedding and Cutoff Frequency

Modeling of FETs is essential from an engineering point of view, not only because it can provide an accurate prediction of circuit performance under the operating conditions, but also because it facilitates the design and integration routines in conjunction with microwave applications such as amplifiers, mixers, oscillators, or filters that are widely used in burgeoning communication systems. In general, the modeling of FETs falls into two categories: physical and empirical models. The former constructs the device model based on the electromagnetic and carrier transport theory. Monte Carlo calculations are often adopted based on the real device geometry, with some parameters required to be adjusted according to the experiments. The merits of physical device modeling lie in that the theoretical explanations are attempted to undercover the basic physics behind the device operation, which indeed could help us to better understand device design. Unfortunately, a great deal of simplifications and approximations are required in order to reduce the complexity of numerical calculations and bring them into the feasible category, albeit at the expense of some accuracy. Even then, the computational burden is quite heavy. Also, the lack of clarity for many mechanisms that cause degradation of devices needs to be addressed for a more precise prediction of device performance. Especially for GaN-based HFETs, the presence of more defects and the applications under high-power conditions bring about extra difficulty in fitting the experiments within the available models. In practice, physical device modeling can provide the preliminary guides for device optimization, but the routines used still have a high computational complexity that must be streamlined further before incorporation into the circuit design techniques.

Empirical or analytical device modeling in the context of parameter extraction is driven by circuit models consistent with the measurements in fabricated devices. It can also predict the device behavior in a range where the measured data are not available. Empirical models need to construct analytical equations for the description of the device; therefore, the computational efforts are much smaller than those in the physical modeling methods. The traditional empirical device modeling methods are often referred to as the equivalent circuit methods that apply discrete circuit elements to the physical attributes of the transistors. For example, a gate–source capacitor represents the capacitance of the depletion region under the gate, a drain–source resistor represents the output resistance of the channel, and a drain inductor represents the parasitic lead inductance of the metallization running to the device.

Relatively speaking, the empirical methods could be much more flexible, despite their somewhat black-box nature, especially for a better fitting in large-signal situations or interpolating some special effects such as self-heating, trap-assisted current collapse, or current lag, compared to physical models. For example, table-based modeling has been studied by many researchers with great interest. Instead of using linear circuit components, the equivalent circuit uses nonlinear transconductances and transcapacitances. These components are symbolized as lookup tables obtained from measurement data. Another hotly pursued approach utilizes neural networks for mapping the nonlinearities of an equivalent circuit and training the model through a certain algorithm from the measured data. In short, the modeling approaches that offer the flexibility to describe the device behavior in regions where the physical or predefined equation-based models fail to simulate are valuable. The various empirical methods may vary in data acquisition, extraction procedures, computational complexity, and accuracy, but normally their execution time is much shorter than that for physical modeling. Therefore, when properly defined and extracted, empirical device modeling is more practical for design purposes.

9.4.1
Small-Signal Equivalent Circuit Modeling

To address the above discussion, we will present the theories for empirical equivalent circuit models on III–V-based HFETs. The small-signal modeling is discussed first followed by the large-signal modeling. In terms of the small-signal configuration, the transistor is biased in the saturation region, and the small-signal nature of input signal will not affect the DC bias conditions of the device. Therefore, a small input voltage will produce approximately a linear response in the drain–source current. On the other hand, if the amplitude of the input signal is increased considerably, the extent of which depends on the device linearity, at a certain point the drain–source current no longer changes linearly. Also, the large

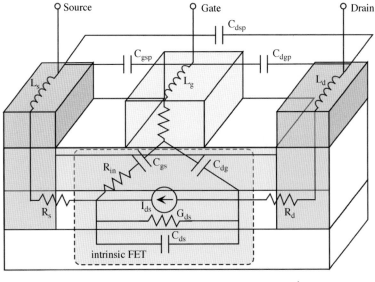

$$I_{ds}=V_i g_{mo} e^{-j\omega\tau}$$

Figure 9.6 High-frequency equivalent circuit of an FET superimposed on a 3D schematic of the device where the stray parasitic capacitances are also included. The equivalent circuit in the dashed box represents the intrinsic device without the contributions from the access regions and the stray capacitances.

signal will bring about a shift in the quiescent operating point of the transistor and strong nonlinearity sets in. In the frequency domain, no additional frequency components are created in the small-signal case; only phase and amplitude of the output signal are changed. However, in large-signal operation, the introduction of multiple harmonics will make the circuit analysis different since the Kirchhoff's law is not satisfied anymore.

A schematic of the equivalent circuit elements overlaid onto a cross-sectional FET diagram is shown in Figure 9.6. Here we consider only the lumped rather than the distributed nature of the elements. For small-signal analysis, all components are assumed to have frequency-independent values. It is customary to establish an equivalent circuit for FETs whose complexity is determined by the depth to which one wants to treat the device in a circuit environment or undertake diagnosis.

For an AlGaN/GaN HFET, a slightly more complicated equivalent circuit including the parasitic capacitances between the terminals and the ground can be developed as shown in Figure 9.7a. The model can be divided into two subparts: a nonlinear, bias-dependent intrinsic part corresponding to the inner device that excludes the contribution from the access regions as well as the stray capacitances, which is outlined with the dashed box, and a linear bias-independent extrinsic part corresponding to parasitic access elements. Regarding the

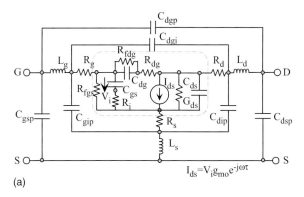

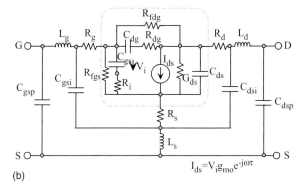

Figure 9.7 The equivalent circuit gleaned from Figure 9.6 (a), and a somewhat simplified version of the equivalent circuit (b).

extrinsic part, C_{gsp}, C_{dgp}, and C_{dsp} represent the parasitic capacitances introduced by the pad connection and probe contacts, and C_{gsi}, C_{dgi}, and C_{dsi} account for the interelectrode capacitances. If we neglect the extrinsic parasitic capacitances and change to intrinsic differential resistance of the gate–source and gate–drain diodes (R_{fgs} and R_{fgd}), the model can be further simplified to that shown in Figure 9.7b, which is more applicable for a GaAs MODFET. It is worth mentioning that charging resistance, R_i in the intrinsic part, represents the electron recombination with the depleted donors under the gate depletion region, and the delay time τ is the time it takes for the depletion region to respond to the change in the gate signal. By matching the two-terminal y parameters calculated from the s parameters and of the equivalent circuit of Figure 9.7, one can determine the values of the elements shown in the same. The crux of the problem, however, is to converge on a solution set that is unique and representative of the FET under consideration. To ensure that this is the case, many types of measurements, frequency and bias dependent, are made to attain as much confidence as possible on a given parameter or a set of parameters, which is a very cumbersome task relying on this matching only.

To facilitate equivalent circuit development, it is more convenient to work with y parameters in which case it is necessary to convert the measured s parameters to y parameters. This is done as follows:

$$y_{11} = \frac{(1-s_{11}^*)(1+s_{22}) + s_{12}^* s_{21}}{(1+s_{11})^*(1+s_{22}) - s_{12}^* s_{21}},$$

$$y_{12} = \frac{-2s_{12}}{(1+s_{11})^*(1+s_{22}) - s_{12}^* s_{21}},$$

$$y_{21} = \frac{-2s_{21}}{(1+s_{11})^*(1+s_{22}) - s_{12}^* s_{21}},$$

$$y_{22} = \frac{(1+s_{11}^*)(1-s_{22}) + s_{12}^* s_{21}}{(1+s_{11})^*(1+s_{22}) - s_{12}^* s_{21}}.$$

(9.59)

The network analyzer software is generally able to make the conversion from the y parameters to s parameters, which is very convenient. From the y parameters thus determined and augmented by a variety of supporting measurements such as those conducted in pinch-off and low drain bias conditions – representing the linear regime – an equivalent circuit model can be developed as shown in Figure 9.8, details of which can be found in many publications including Morkoç (2009), with references to prior work.

9.4.2
Cutoff Frequency

The maximum speed of operation of an FET is ultimately limited by the transit time of carriers under the gate together with charging times of the input and feedback capacitances. In addition, extrinsic resistive elements such as the output conductance contribute adversely to lowering the gain. It is, therefore, warranted to present a

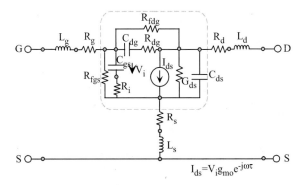

Figure 9.8 A modified equivalent circuit applicable for GaN HFET using the optimization-based extraction method.

succinct discussion of the speed of the device in terms of current gain and power gain cutoff frequencies.

The transit time under the gate of a submicron HFET is on the order of a few picoseconds. In view of this, the charging time of the input and the feedback capacitances (C_{gs} and C_{dg}) through the input resistance R_{in} in the equivalent circuit, shown in Figure 9.7, determines the speed of response. Generally, two parameters, the current gain cutoff frequency and maximum oscillation frequency, are figures of merit to gauge the expected high-frequency performance of an HFET. The *current gain cutoff frequency* is defined as the frequency at which the current gain goes to unity:

$$f_T = \frac{g_m}{2\pi(C_{gs} + C_{gd})} \approx \frac{v_{sat}}{2\pi L}, \qquad (9.60)$$

where C_{gs} and C_{gd} are the gate–source capacitance and gate–drain feedback capacitance, respectively. The feedback capacitance (C_{gd}) is smaller than the gate–source capacitance (C_{gs}) and is typically neglected, which when utilized leads to the current gain cutoff frequency in its simplest form:

$$f_T = \frac{g_m}{2\pi C_{gs}} = \frac{v_{sat}}{2\pi L} = \frac{1}{2\pi \tau_{int}}, \qquad (9.61)$$

where τ_{int} is the intrinsic delay time that represents the time required for electrons to transverse the physical length of the gate. When the delays due to the extension of the depletion region under the gate due to drain bias and the RC time constant are included, Equation 9.58 must be modified as

$$f_T = \frac{1}{2\pi \tau_{total}}, \qquad (9.62)$$

with τ_{total} representing the total delay time that is defined as $\tau_{total} = \tau_{int} + \tau_d + \tau_{RC}$, where τ_d represents added drain delay due to the extension of the gate depletion region due to drain bias and τ_{RC} represents the RC delay time constant. To elaborate further on τ_d and τ_{RC}, as the drain bias is increased further into the saturation regime, the depletion under the gate is no longer limited to the physical dimension of the gate. Rather, it extends toward the drain contact, effectively increasing the gate length, and the additional time required to traverse the added effective length is represented by the drain time constant, τ_d. As for the RC time constant, the modulated gate voltage changes the charge distribution, and to move charge in and out of a given element requires an additional RC time constant. The RC time constant would approach zero if the drain current approaches infinity. Therefore, the total delay time can be determined from Equation 9.62 with the measured current cutoff frequency in hand for each of the drain current values; one can plot the total delay time versus the inverse of the drain current (by changing the gate bias) for a constant drain voltage bias. Extrapolating the linear portion of the data to zero would allow the determination of τ_{RC} and $\tau_d + \tau_{int}$, as shown in Figure 9.9. In short, the delay time for zero inverse current represents the sum of the intrinsic and drain delay times. Isolation of the drain delay can be accomplished by plotting the total delay time versus the channel voltage defined as $V_{ch} = V_{DS} - I_D(R_s + R_d)$, as

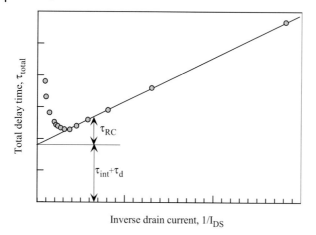

Figure 9.9 Charging delay plot demonstrating the extrapolation method for determining τ_{RC} and $\tau_d + \tau_{int}$. The total delay time is determined from the measured current cutoff frequency and Equation 9.62. The experimental total delay time data are patterned after GaAs-based MODFETs reported by N. Moll.

shown in Figure 9.10. Here V_{DS}, I_D, R_s, and R_d represent the applied drain–source voltage, drain current, source resistance, and drain resistance, respectively. Extrapolation from the linear region to vanishing channel voltage will lead to the sum of intrinsic and RC time constants. For a given channel voltage, the difference between the total delay and the sum of the RC and intrinsic delays would lead to the drain

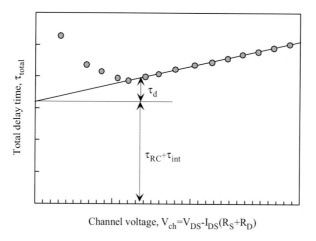

Figure 9.10 Drain delay plot demonstrating the extrapolation method for determining τ_d and $\tau_{RC} + \tau_{int}$. The total delay time is determined from the measured current cutoff frequency and Equation 9.62. The experimental total delay time data are patterned after GaAs-based MODFETs reported by N. Moll.

delay. Having measured the total delay from the cutoff frequency and isolated the drain and RC delays would allow one to determine the intrinsic delay time.

Experimentally, the current gain cutoff frequency (unity current gain frequency output under short-circuit conditions) is determined by extrapolating $|h_{21}|^2$ to 0 dB. The term h_{21} is defined in terms of the s parameters as

$$h_{21} = \frac{-2s_{21}}{(1-s_{11})(1+s_{22}) + s_{12}s_{21}}. \tag{9.63}$$

Similarly, the unity power gain frequency is determined by extrapolating the unilateral power gain, U, to 0 dB, which represents gain if one assumes the scattering parameter $s_{12} = 0$ or negligibly small.

The *maximum oscillation frequency*, defined as the frequency at which the power gain goes to unity, may be given by

$$f_{max} = \frac{f_T}{2(r_1 + f_T \tau_3)^{1/2}}. \tag{9.64}$$

If R_s is the series resistance, C_{dg} is the drain–gate capacitance, and G_d is the differential drain conductance, then the parameter r_1 is

$$r_1 = (R_g + R_i + R_s)G_d \tag{9.65}$$

and the feedback time constant τ_3 is

$$\tau_3 = 2\pi R_g C_{dg}. \tag{9.66}$$

If we consider yet more of the extrinsic elements, the f_{max} term can be expressed as

$$f_{max} = \frac{f_T}{2\{1 + (R_s + R_g)G_d + 2(C_{dg}/C_{gs})[(C_{dg}/C_{gs}) + g_m(R_s + G_d^{-1})]\}^{1/2}}. \tag{9.67}$$

The parameters have their usual meanings in that R_g and R_s represent the gate and source extrinsic resistances, respectively. The terms C_{dg} and C_{gs} represent the drain to gate feedback capacitance and gate to source capacitance, respectively. From Equation 9.62, it may be noted that the higher the saturation velocity and the smaller the gate length, the higher the value of f_T.

9.5
HFET Amplifier Classification and Efficiency

GaN-based FETs are used primarily for power in switching or RF applications. Consequently, traditional small-signal considerations have to be augmented by large-signal specific issues. As in small-signal modeling, the first step in power modeling is to establish the basic device geometrical factors needed to calculate the current–voltage characteristics. Once these are known, the output characteristics superimposed with the load line can be used to estimate the RF power level that can be obtained from the device provided that it is not limited by the input drive. Details such as the amplifier classification, determined by biasing point, must be settled first. To this end, a short description of the various types of amplifier configurations is given first.

Depending on the operating point and the load line in the DC characteristic of the transistor, amplifiers can be classified as class A, class B, class AB, and class C, where the transistor is used as a controlled current source in an amplifier, as schematically shown in Figure 9.11. For even higher efficiency, there are class D, E, and F amplifiers in which the transistor is used as a switch.

In a class A amplifier, the transistor is biased and driven by RF in a manner in which the complete swing of the output current is in the saturation region of the transistor and the transistor obviously is on for 360° phase angle of the RF swing. The linearity in the class A amplifier is the highest and the bandwidth is the widest, provided impedance matching is in place, in comparison with other classes of operation. In switched modes, high output resistances mitigate impedance matching and thus increase the bandwidth. The major drawback, however, is the low efficiency due to high power consumption. The upper drain efficiency of an ideal class A amplifier is 50% with the feasible value being in the range 30–40%.

The operating point of a class B amplifier is at the cutoff voltage of the transistor (Figure 9.11), with a conduction phase angle of 180°. With the push–pull configuration using two transistors in parallel, however, two half-waves can be amplified and

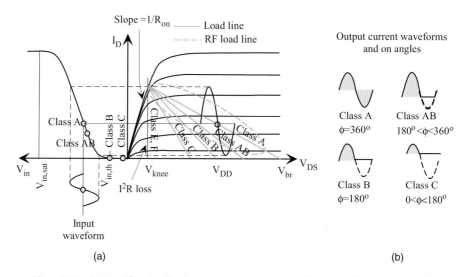

Figure 9.11 (a) Amplifier classification regarding their operating points and load lines. The high-frequency load line for class A operation is also shown as oval with broken lines. The divergence between the DC and high-frequency RF load lines is caused by inductive and capacitive elements of the transistor and matching elements with their associated phase dispersion and thus a nonlinear load line. (b) Conduction phase angles for class A, class AB, class B, and class C operations.

combined to a full sinusoidal output with a conduction phase angle of 360°. The maximum efficiency of a class B amplifier is therefore about 80%, while experimentally 40–50% is feasible but at a relatively lower degree of linearity and smaller bandwidth compared to class A.

The operating point of a class AB amplifier represents a compromise between the class A and the class B operating points as seen in Figure 9.11. Depending on whether the operation point is closer to the class A operation or the class B operation, the conduction phase angle can vary between 180° (for class B) and 360° (for class A). Therefore, a class AB amplifier is capable of providing an efficiency between the theoretical 50% and 80%. The bandwidth of class AB is larger than that of class B, but smaller than that of class A. Class AB amplifiers are suitable for mobile base stations that require both high efficiency and high linearity.

In a class C amplifier, the transistor is on for less than the half phase cycle; that is, the conduction phase angle is <180°. Therefore, relatively less power is dissipated and the efficiency is better than that in class B at the expense of only a fraction of the peak output power. The bandwidth of this class is inferior even to class B and it is very nonlinear, necessitating filtering to eliminate the unwanted harmonics.

For switching applications, a class E amplifier is employed. The operating point in class E is the same as in class B, but with the distinguishing feature that the transistor is loaded in such a way that either the output voltage or the output current is present at a given moment in time but not both. Theoretically, therefore, the power loss is zero (or very small in reality) during the operation leading to an efficiency near to 100%. The inevitable nonlinearity can be reduced by linearization techniques. This amplifier class is suitable for mobile base station applications where the high efficiency is of prime concern. The bandwidth of this amplifier class is very narrow as compared to the class A, B, and AB operations.

As mentioned above, class F also represents a switched amplifier, which is also referred to as a harmonic controlled amplifier owing to the fact that it boosts mainly the second and the third harmonics of the input signal. For this to happen, the load must provide the switching transistor with high termination impedances at both second and third harmonics. The edge of the nearly rectangular output voltage waveform is sharper than a sinusoid making the transition time small resulting in lower power loss. As in the case of class E, this class of amplifiers is suitable for base station power amplifiers owing to its high efficiency.

In class A operation (see Figure 9.11), the maximum power that can be expected from the drain circuit of a device is given by

$$P_{max} = \frac{I_{dson}(V_b - V_{knee})}{8}, \qquad (9.68)$$

where I_{dson} is the maximum drain current (this is the drain current with a small positive voltage on the gate electrode), V_b is the drain breakdown voltage, and V_{knee} is the knee voltage as shown in Figure 9.11a. The allowable positive gate voltage (≈ 1 V) will depend on the channel doping and the work function of the gate metal. The positive gate voltage is limited by the onset of forward Schottky diode current. The DC load line shown in Figure 9.11a would be used in a class A RF amplifier with the maximum drain voltage $V_d = V_b/2$. The slope of the load line is $1/R_L$, where R_L is the value of the load resistance at the output of the FET. It can be gleaned from Equation 9.68 that V_b and I_{dson} must be made as large as possible. The wide-bandgap semiconductors such as GaN are used in this juncture because their drain breakdown voltage is larger than that in conventional group III–V semiconductors.

In general, the drain can be swung up to voltages within 80% of the drain breakdown for a 20% margin of safety. It should be pointed out that the maximum drain current in nitride semiconductor-based HFETs is in the same ballpark as that of more conventional semiconductors. This implies that increased power handling capability is a direct result of large breakdown voltages and thermal conductivity and the fact that higher junction temperatures can be tolerated. The ability to increase drain bias increases the load resistance and makes it easier to match impedances, particularly in devices with large gate widths.

The *efficiency* of an amplifier is a value that quantifies the portion of the supplied DC power converted to RF power by the amplifier. The term drain efficiency (η_d) indicates simply the ratio between the output power of an amplifier and the supplied DC power:

$$\eta_d = \frac{P_{out}^{rf}}{P_{DC}}. \qquad (9.69)$$

Another efficiency denomination is the power added efficiency (PAE) that indicates the conversion efficiency of supplied DC power (P_{DC}) to the high-frequency power added to the input signal resulting in the output signal:

$$\text{PAE} = \frac{P_{out}^{rf} - P_{in}^{rf}}{P_{DC}} = \eta_d \left(1 - \frac{1}{G}\right), \qquad (9.70)$$

where G is the gain of the amplifier. If one assumes that the gain G is infinitely large, the PAE reduces to

$$\text{PAE} = \frac{V_b - V_{knee}}{2(V_b + V_{knee})}. \qquad (9.71)$$

Safe operation requires that the maximum drain–source voltage be smaller than V_b by say 20%. If $V_{knee} = 0$, representing the unattainable ideal case, an efficiency of 50% is obtained. Efficient conversion from DC to high-frequency signals leads to low power dissipation in the device and thus reduced heating. In this case, the cooling system of the power amplifier (PA) can be made simpler, which translates to small size and cost efficiency.

The output power of each amplifier stage is a function of the RF drain current and the supply voltage. For PAs, the output of the transistor is matched to an optimum load resistance to provide maximum available output power. In the DC characteristic of the transistor device used in the PA, the load can be depicted as a load line. At high frequencies, this straight line is replaced with that shown in dashed lines in Figure 9.12a due to inductive and capacitive components in the circuit. The RF voltage swing along this load line is limited by the knee voltage V_{knee} and the breakdown voltage V_{br}. The drain current through the load is

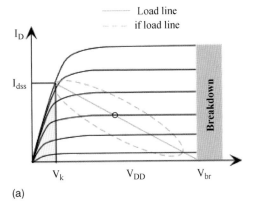

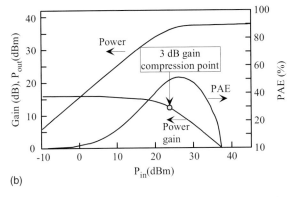

Figure 9.12 (a) Simplified schematic representation of I–V characteristics with a DC or low-frequency load line and for high-frequency load line, the latter shown as oval with broken lines. The divergence between the DC and high-frequency RF load lines is caused by inductive and capacitive elements of the transistor and matching elements with their associated phase dispersion and thus a nonlinear load line. (b) Output power and power gain of a power amplifier as a function of the input power. The 3 dB power gain compression point is shown with a solid circle on the gain curve.

maximum ($I_{ds} = I_{dss}$) at the bias point, where $V_{ds} = V_{knee}$. Voltage exceeding the breakdown voltage leads to a large drain current flow and breakdown of the gate–drain diode (see Figure 9.12a). The optimum DC load resistance for the maximum power in a class A amplifier is determined by

$$R_{opt} = \frac{V_{br} - V_{knee}}{I_{dss}}. \tag{9.72}$$

Figure 9.12b displays a plot of the output power, the power gain of a power amplifier, and the power added efficiency. The output power increases as the input power is raised until it reaches saturation where the power gain begins to drop. A PA is normally driven to the beginning of the saturation, where the gain drops. The point at which the gain drops by 1 dB is the so-called 1 dB compression point and the point at which the gain drops by 3 dB is the so-called 3 dB compression point. The efficiency is maximum around these points. A trade-off between the gain and output power generally would have to be made in determining the device size noting that the output power increases with increasing device size (total gate width in our case) due to the increased maximum drain current. On the other hand, the gain decreases with increasing device size due to various reasons among which are the increasing parasitics, such as gate to source capacitance, gate resistance, and source inductance. Moreover, the phase dispersion among the various gate fingers is not negligible in large devices, since the geometry of the active area is not insignificantly small compared to the wavelength of the signal propagating through it.

Furthermore, in a large device with multiple gate fingers, the thermal impedance is higher than that in a device with only one finger due to the mutual heating between adjacent channels. This can be mediated to some extent by placing the gate fingers farther apart from each other, which comes at the expense of the large chip size and phase dispersion. This is one of the reasons why citing only the normalized output power to 1 mm of gate width is not as meaningful as it has been made out to be. The real figure of merit should be the total output power per chip size along with the associated efficiency.

9.6
Drain Voltage and Drain Breakdown Mechanisms

A somewhat unrivaled feature of GaN in relation to power devices is its large breakdown field and relatively large thermal conductivity. Due to much less than unity power conversion efficiency, however, thermal limitations come into play seriously. Thermal issues depend very much on the pitch of the device, meaning the density of gates in the context of FETs, heat sinking, and most importantly heat conduction or power dissipation. The thermal conductivity is only a measure of the LA phonon velocity and does not take into account the roadblocks put in place by LO phonon disintegration to LA phonons, which will be discussed in some detail later on.

The output power that can be extracted from the device is determined by the drain output characteristics, specifically the maximum current and maximum voltage that can be attained. To increase the output power, either the maximum drain current or the maximum drain voltage or both must be increased. Increased current reduces the output resistance, whereas increased voltage increases the output resistance. The latter is easier to attain by using a semiconductor material with large breakdown field and appropriate device designs, and therefore more conducive to impedance matching. If the gain and the input drive are sufficient and there are no anomalies, using class B operations drain efficiencies in the range of 90% can be obtained as has been done in GaAs-based devices. In GaN, there are anomalies, which are discussed next, that prevent drain efficiency from reaching anywhere near these values.

9.6 Drain Voltage and Drain Breakdown Mechanisms

The impetus for large drain voltage is clear, which is one of the reasons why GaN is such an attractive material for power applications. Intrinsically, excluding the surface-related issues and bulk bound contributions to premature current rise, the avalanche breakdown of the revere-biased gate–drain junction determines the upper voltage limit that can be applied even momentarily as the one that occurs during a swing of the RF voltage. In avalanche breakdown, carriers introduced into a region of high field, such as the gate–drain region, multiply through avalanche multiplication causing a large current to flow, and thus the term breakdown. It is a reversible process if the Joule heating is limited.

In the bulk, the critical field at which avalanche breakdown occurs scales approximately with the square of the energy bandgap, or $E_{br} \propto E_g^2$. The breakdown voltage then scales with the fourth degree of the bandgap. At higher dopant concentrations, the avalanche breakdown voltage is reduced, so the known concentration dependence $V_B \propto N_D^{-0.75}$ can be used. The final result for the avalanche breakdown voltage for the asymmetric p^+–n junction, or Schottky contact–n junction, where n-type material has an energy bandgap E_g and doping concentration N_D, is

$$V_B = 23.6 E_g^4 \left(\frac{N_D \, (\text{cm}^{-3})}{10^{16}} \right)^{-0.75}. \tag{9.73}$$

Here the energy bandgap must be in eV.

FETs are planar devices and surfaces play a large role as the actual breakdown voltage would be different from that predicted by Equation 9.73. However, Equation 9.73 is very useful in judging the semiconductor and doping level dependence of the breakdown voltage. In addition, the issue of gate–drain junction breakdown is always marred with the gate leakage current whose origin is often times not well understood in an evolving technology such as GaN. Furthermore, a standard definition of breakdown voltage is also lacking. Already inconsistent temperature dependences of Schottky diode breakdown voltages have been reported, some indicating that in at least some devices impact ionization process (avalanching) does not dominate the breakdown mechanism. If so, the surface states would then be implicated. However, the field is yet in its infancy and not much is known about surface states in GaN except that passivation seems to have an effect.

Breakdown properties have been investigated with a full-band ensemble Monte Carlo simulation having a numerical formulation of the impact ionization transition rate. The breakdown voltage was defined as the drain–source voltage for which the drain current calculated with impact ionization is 3% higher than the drain current calculated without it. For RF breakdown, a large-signal RF bias was applied between the drain and source simulating on-state breakdown instead of the gate, which is common in experiments to avoid computational complexity. The RF signal is assumed sinusoidal oscillating between V_{hi} and V_{lo} with an angular frequency ω. In order to determine the RF breakdown, the drain current was calculated with and without impact ionization.

The calculated drain currents, I_d, versus drain–source voltage, V_{ds}, for gate–source voltage (inclusive of the Schottky barrier height), V_{gs}, varying from -0.1 to -5.1 V in

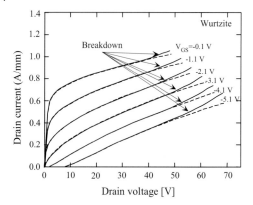

Figure 9.13 Output characteristics for a wurtzite polytype GaN MESFET. The gate–source voltage, V_{gs}, includes the Schottky barrier height. The drain currents have been calculated with (solid lines) and without (dashed lines) impact ionization. The ovals show the breakdown locations on the I–V curves. The gate length is 0.1 μm and the channel donor concentration is $3 \times 10^{17}\,\mathrm{cm}^{-3}$. Courtesy of P.P. Ruden.

1 V steps are shown in Figure 9.13 for wurtzite-phase metal semiconductor field effect transistors (MESFETs). The drain current for each bias was determined both with and without impact ionization. The solid and dashed lines in Figure 9.13 represent the calculated drain current in the presence and absence of impact ionization, respectively. The relatively high output conductance of the devices is a consequence of the small gate length. The DC breakdown voltage is significantly higher in the wurtzite-phase device than in the zincblende-phase device for all gate biases. As an example, the breakdown voltage for $V_{gs} = -0.1\,\mathrm{V}$ is 45 V for the wurtzite GaN.

The gate leakage current is attributed to mainly surface states based on the observation that Schottky diodes of varying diameter fabricated in FET layers indicated the leakage current to be proportional to the periphery of the diodes, as opposed to the area. In addition to scaling with area or periphery, the temperature gate leakage current under dynamic bias with grounded source and reverse-biased gate–drain terminals or static bias with floating source and reverse-biased gate–drain terminals can be used to gain insight into the mechanisms involved. If the gate–drain bias is larger than the channel pinch-off voltage (the gate voltage at which the channel is depleted for zero drain bias), the measurements are considered performed in a regime beyond pinch-off. The surface-inspired gate leakage is attributed to hopping along the surface states, which beyond pinch-off and with increasing drain bias is fueled by electrons tunneling from the drain edge of the gate, where the field is highest, into surface states. These electrons tend to screen the depletion charge in the region between the gate and the drain, and under equilibrium the electron transport rate is limited by their collection rate by the drain, which relates to electron mobility. Illumination of the surface with light or an increase in temperature enhances this conduction.

The drain breakdown voltage is somewhat elusive in that its determination is convoluted by leakage current. In fact, many practitioners define it as the voltage when the gate leakage reaches 1 mA/mm, which is determined by the amount of leakage that can be tolerated rather than with any physics related to a breakdown mechanism. It is also of interest that in some cases observations are consistent with negative temperature coefficient to a critical power density in the surface leakage current, beyond which surface heating exceeds cooling leading to a positive temperature feedback, thermal runaway, and subsequent breakdown. In other cases, a positive temperature coefficient of the breakdown voltage is observed, which naturally is attributed to impact ionization. The apparent conflict may be attributed to the status of the device and the surface between the gate and the drain. If the surface-bound issues are eliminated, the quality of the channel is improved and the impact ionization-induced breakdown would be uncovered. The case showing positive temperature coefficient is illustrated in Figure 9.14, where the breakdown voltage increases at a rate of 0.12 V/K with temperature.

The analytical expression for impact ionization coefficient $\alpha(E)$ provided by P.A. Wolff is as follows:

$$\alpha(E) = \frac{qE}{E_i} \exp\left(\frac{-3E_{ph}E_i}{(qE\lambda)^2}\right), \tag{9.74}$$

where E is the electric field, E_i is the threshold energy for electron–hole pair generation, which can be assumed slightly larger than the bandgap of the semiconductor, E_{ph} is the optical phonon energy, and λ is the mean free path for optical phonon scattering assuming that this is the only scattering mechanism. In order to determine the optical phonon energy, one must know the momentum conserving

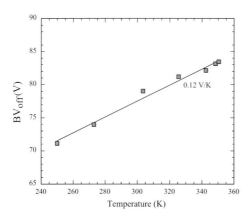

Figure 9.14 Temperature (case) dependence of the drain breakdown voltage BV_{off} in the off state, meaning that the device is in pinch-off state and electron supply is not from the source. A positive temperature coefficient of 0.12 V/K is obtained. Devices with varying gate length (L_g) from 1 to 10 μm are used; similarly, the gate–source and gate–drain spacings are 1.5 and 2 μm, respectively, and the gate width for all the devices is 20 μm. Courtesy of T. Nakao.

transitions accurately. Ascribing a length l within which the impact ionization occurs, further assuming that the field strength in that region is constant and given as $E_{max} = V_{DG}/l$, and I_{leak} is the primary electron current, the impact ionization current is given by

$$I_{ii} = \alpha E_{max} I_{leak} l. \tag{9.75}$$

Definition of the breakdown voltage, unless the breakdown is very sharp, requires some assumptions that are somewhat arbitrary and yet practiced widely. Generally, 1 mA for 1 mm of gate periphery is used although other criteria have been used. When the breakdown voltage is assumed to be drain–gate voltage V_{DG} at which the increase in drain current (ΔI_D) due to I_{ii} becomes a certain value (0.05 mA/mm in the present case),

$$I_{leak} = \frac{E_i \Delta I_D}{qV_{Boff}} \exp\left[3E_{ph} E_i \left(\frac{1}{q\lambda V_{Boff}}\right)^2\right]. \tag{9.76}$$

Equation 9.76 is plotted in Figure 9.15 with the following assumptions: $E_i = 3.7$ eV and $E_{ph} = 92$ meV. l/λ (=220) is a fitting parameter, where λ is 3 nm and l is 0.7 μm. The parameters are reasonable for a gate–drain spacing of 2 μm. Despite many fitting parameters in the form of assumptions that are not consistent across the board for GaN, the shape of the dependence of the breakdown voltage on the primary current indicates that at least in the particular set of devices, the impact ionization may be the primary breakdown mechanism, which implies that other surface- and/or quality-related processes are not dominant.

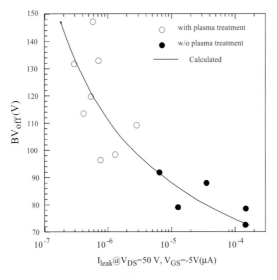

Figure 9.15 The values of BV_{off} for devices plotted as a function of I_{leak}. The solid line is given by Equation 9.76. Courtesy of T. Nakao.

9.7
Field Plate for Spreading Electric Field for Increasing Breakdown Voltage

In FETs, the electric field is nonuniform particularly in the gate edge on the drain side where the breakdown occurs. Spreading the field, with a field plate, would increase the breakdown field, but at the expense of increasing the gate capacitance as well as complexity of fabrication. Implementation can be with a p–n junction below the 2DEG channel as well as one or two field plates, one extending from the gate and the other from the drain. A schematic of the cross section of the device with a field plate is shown in Figure 9.16, where L_{sg} represents the source–gate distance, L_g represents the gate length, L_F represents the plate length, and L_{FD} represents the gap between the field plate and the drain. The term t indicates the thickness of the dielectric layer used under the field plate.

If the added capacitance due to the field plate is C_{FP} and the gate capacitance without the field plate is C_G, the field plate capacitance would be effectively the same as increasing the gate length by a factor of $L_g + C_{FP}/C_G$. Because the transconductance is inversely proportional to the gate length and the cutoff frequency is proportional to the transconductance and inversely proportional to the gate capacitance, one can construct an expression for the current gain cutoff frequency as

$$\frac{f_T|_{FP}}{f_T|_P} = (1 + C_{FP}/C_G)^{-2} = \left[1 + \frac{L_F/L_G}{1 + t\varepsilon_{AlGaN}/\varepsilon_i d}\right]^{-2}, \quad (9.77)$$

where t, d, ε_i, and ε_{AlGaN} represent the thickness of the dielectric layer under the field plate, thickness of the AlGaN layer, dielectric constant of the dielectric layer, and the dielectric constant of the AlGaN barrier layer, respectively.

The field distribution along the channel with and without the filed plate can be calculated with the aid of commercial software programs or the old fashion, but very elegant and reliable conformal mapping (Mathematica routines are available) techniques. For a modeled HFET, while the maximum field at the gate edge on

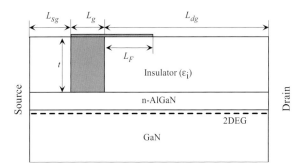

Figure 9.16 Schematic of an AlGaN/GaN structure with a field plate to spread the electric field between the gate and the drain for an increased drain breakdown voltage, where L_{sg}, L_g, L_F, and L_{dg} represent the source–gate distance, the gate length, the plate length extending beyond the gate, and the drain–gate distance, respectively.

the drain side reaches 8.5×10^6 V/cm in the conventional design, the same drops to 6.0×10^6 V/cm with a field plate, representing a 30% reduction in the peak field. The structural and device parameters of the modeled HFET are as follows: AlGaN barrier thickness is 25 nm, gate length is 0.5 μm, channel length is 4.5 μm, and the plate extends to the drain side by 0.6 μm. The HFETs are assumed biased at $V_{DS} = 10$ V and $V_{GS} = -5$ V.

9.8
Anomalies in GaN MESFETs and AlGaN/GaN HFETs

Field effect transistors exhibit anomalies in their output I–V characteristics. Among the causes are surface states and channel carriers being trapped in the wide-bandgap barrier and/or the buffer layers. These phenomena are depicted schematically in Figure 9.17.

The surface states and/or defects play a much more important role in FET performance due in part to the relatively defective nature of the material and polarization field. To avoid confusion, the nomenclature adopted here is as follows: The current collapse is used to describe the case in which the drain current actually goes down to nearly zero and remains there for some time. The slope change in the I–V characteristics will be referred to as kinks. Any variation other than that expected from normal device operation will be referred to as dispersion. A form of this

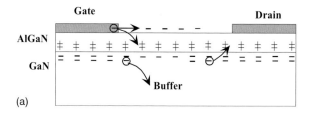

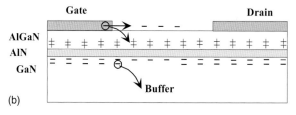

Figure 9.17 (a) Schematic representation of an AlGaN/GaN HFET indicating how the surface charge and charge injection into traps in the buffer layer and defects in AlGaN could serve in depleting conducting channel carriers. For simplicity, the 2DEG is assumed to be due to donors placed in AlGaN. In addition, the regions from which carrier injection takes place are limited to the spacing between the gate and drain electrodes; the exact location is arbitrarily chosen. (b) The same as in (a) except with the insertion of an interfacial AlN layer that would help prevent injection of carriers from the channel to the barrier and/or surface.

dispersion is the current lag in that the drain current does not reach the expected value and degrades with increasing test frequency.

Generally, the trapping effects limit device performance in terms of noise even at relatively low frequencies. In GaN-based FETs, this takes on a different meaning in that dispersion in the output characteristic and transconductance with frequency limits the drain efficiency severely. Some forms of anomalies are thought to come about from recharging of the trapping centers as a result of variation of the gate potential. Charge temporally trapped in the vicinity of the transistor channel can reduce the drain current level. Because the trap release time is very long, the drain current remains affected in that it does not follow the gate voltage at high speeds of modulation, which can be hundreds of kHz. Owing to a strong correlation of the effect with the semiconductor surface treatment, it is assumed that at least some trapping centers are located on the surface.

Filled surface states pinch the channel in the region between the gate and the drain, the effect of which on the output characteristics is schematically shown in Figure 9.18, where the load line and quiescent operating conditions for class A operations are assumed. Also shown are the extremes of DC current as governed by the load line and RF current (shaded) that fails to follow the gate bias, with expected higher drain currents. The RF current can be determined by the use of the so-called

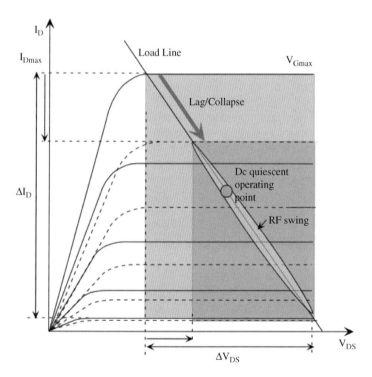

Figure 9.18 Schematic representation of RF current lag superimposed on top of DC drain (solid lines) and RF (dashed lines) I–V characteristics with a DC load line.

load pull tuning. It can also be measured under active loading conditions with the use of a high-speed sampling scope for measuring the output RF voltage, in response to an RF input drive, where the voltage measured can be converted to current knowing the load value.

Current lag due to surface depletion in a GaN MESFET has been treated in some detail. The discussion of MESFET in this context is pertinent here as the channel material in this and HFET device is the same. The carriers for barrier trapping could be provided by tunneling from the gate into the semiconductor and also from the channel into the barrier at high fields. One should not forget that there are traps in the barrier and the charge emission from the barrier traps is also affected by the field. Therefore, the influence of the field must be taken into account in characterizing the barrier traps whose characteristic times of the field-assisted emission may vary from hundreds of nanoseconds to milliseconds. Traps located in the buffer layer generally cause current kinks in the drain current, but one report attempts to link them to current collapse and drain lag. The characteristic time of charge emission determined from the gate lag measurements extends to 10^3 s.

9.8.1
Effect of the Traps in the Buffer Layer

Electrons in the channel can be trapped in the buffer layer by deep centers that produce a negatively charged region in the buffer layer. This is observed in the output characteristics following the application of a high drain voltage, for example, 30 V. The high drain voltage causes the electrons to be captured by the defect states causing depletion in the active channel. It is also possible that the high field could impart sufficient energy to the carriers to overcome the local barriers and be injected into regions outside the conducting channel. In an abrupt junction approximation, one assumes that the traps, which are distributed uniformly with concentration N_T, fill to a depth x_T below the channel–buffer interface. Upon irradiation with light, trapped carriers are photoionized and drift to the channel by the field in the back depletion region, and reverse the process. Reports in the literature indicate that there are two traps at absorption energy thresholds of 1.8 and 2.85 eV, before accounting for the Franck–Condon shift, which represents the difference between thermal and optical activation barriers. Once the Franck–Condon shift is taken into account, the thermal energies for the aforementioned traps become 0.55 and 2.65 eV, respectively. The picture essentially is that carriers, created either by impact ionization by a large drain prebias or by some other means, are captured by defects in the buffer layer near the buffer–channel interface followed by their release due to optical excitation. The carrier capture is governed by thermal capture cross section, whereas the carrier detrapping is governed by optical emission cross section. A schematic representation of this process is shown in Figure 9.19.

At equilibrium, the negative charge trapped in buffer traps must balance the depletion charge in the channel caused by charged traps. A schematic description of an FET channel depleted both by the gate/drain bias and by the charged traps in the buffer layer is shown in Figure 9.20. If the total areal (planar) density of negatively

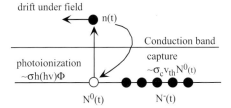

Figure 9.19 Schematic diagram indicating carrier capture, photoionization, and drift. The illumination time-dependent carrier concentration, neutral trap, and charged trap concentrations are represented by $n(t)$, $N^0(t)$, and $N^-(t)$, respectively. Rates for capture and photoionization are also shown.

charged trap is Q_T, depletion width in the channel from the substrate side is w_n, the depth from the channel–buffer interface to the point where the traps are uniformly negatively charged is w_T, the channel doping level is N_d, and the trap concentration is N_T (assumed to be uniform in plane and in depth), the charge balance equation is

$$Q_T = qN_d w_n = qN_T w_T. \tag{9.78}$$

Light illumination causes some of the traps to release their captured electrons returning them to the channel, reducing the depletion depth and thereby increasing the drain current, as depicted in Figure 9.20.

Following the standard derivation of long-channel MESFET current–voltage expressions but with the inclusion of depletion from the charge traps in the buffer layer near the buffer–epilayer interface, one can write for the voltage drop across an incremental segment of the channel:

$$dV = I_D\, dR = \frac{I_D\, dx}{qN_d \mu_n Z [a - w_n(x) - h(x)]}, \tag{9.79}$$

where dR, a, Z, $w_n(x)$, and $h(x)$ represent the resistance of the incremental channel, open-channel thickness, device width, depletion depth caused from the interface side, and depletion depth caused by the gate, respectively. The depletion depths can be found, in depletion approximation, by solving the one-dimensional Poisson equation. Doing so for interface and surface depletion layer thicknesses leads to

$$w_n(x) = \sqrt{\frac{2\varepsilon}{qN_d}\left(\frac{1}{1 + N_d/N_T}\right)[V_c(x) + V_{\text{binter}}]}, \tag{9.80}$$

$$h(x) = \sqrt{\frac{2\varepsilon}{qN_d}[V_c(x) + V_{\text{bi}}]},$$

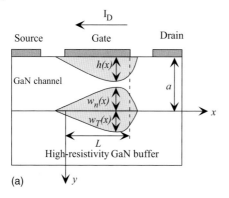

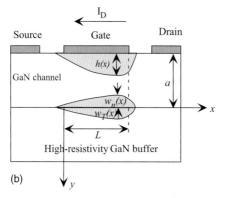

Figure 9.20 Schematic representation of a GaN FET channel depleted both by the gate/drain bias and by the charged traps in the buffer layer. The picture following a large drain bias in dark (a) and in light (b) is depicted. It is assumed that the charged trap density in buffer is lower than doping in the channel. The depletion front particularly near the buffer interface as well the extent of the negatively charge traps that is dependent on distance is arbitrarily drawn.

where V_c represents the channel voltage, which is equal to V_D, the drain voltage, on the drain side for $x = L$ (the gate length) and 0 at the source end for $x = 0$ if the source is grounded. V_{binter} and V_{bi} represent the potentials causing depletion from the interface side and surface side of the device, respectively. Integrating Equation 9.79 with aforementioned boundary conditions:

$$I_D = G_0 \left\{ V_D - \frac{2}{3}\sqrt{\frac{2\varepsilon}{qN_d a^2}} \left[(V_{bi} + V_D)^{3/2} - V_{bi}^{3/2} \right] \right.$$

$$\left. + \frac{1}{\sqrt{1 + N_d/N_T}} \left[(V_{binter} + V_D)^{3/2} - V_{binter}^{3/2} \right] \right\}, \qquad (9.81)$$

where G_0 is the open-channel conductance and is given by $G_0 = qN_d\mu_n Za/L$. The current degradation term due to traps in the buffer layer is included through the V_{binter} term, which is a function of the intensity and duration of light illumination.

For very small drain biases, that is, $V_D \ll V_{bi}, V_{binter}$, the depletion depths become nearly independent of the drain voltage, which leads to linear drain current–voltage relationship in the form of

$$I_D|_{linear} = G_0 V_D \left[1 - \sqrt{\frac{2\varepsilon}{qN_d a^2}} \left(\sqrt{V_{bi}} + \sqrt{V_{binter}}\right)\right]. \tag{9.82}$$

The change in the drain current in the linear region due to light illumination is then given by

$$\begin{aligned}\Delta I_D|_{linear} &= I_D(\Phi, t) - I_{dark}|_{linear} \\ &= G_0 V_D \left[1 - \sqrt{\frac{2\varepsilon}{qN_d a^2}} \left(\sqrt{V_{binter}|_{dark}} - \sqrt{V_{binter}|_{light}}\right)\right], V_{binter}|_{light} \\ &= \frac{1 + N_d/N_T}{2\varepsilon q N_d} [Q(\Phi, t)]^2 \quad \text{and} \quad V_{binter}|_{dark} \\ &= \frac{1 + N_d/N_T}{2\varepsilon q N_d} [Q(0,0)]^2, \end{aligned} \tag{9.83}$$

where Q is the areal density of negatively charged traps.

At $t=0$ just before illumination, the charged trap concentration and the areal trapped charge density at time t following light illumination is given by $Q(t) = qN^-(t)w_T$, where x_T is assumed to be the depth of the region containing traps, which is assumed constant because the charge cannot distribute. Moreover, assuming slow carrier capture (see Figure 9.19), $N^-(t) = N_T \exp[-\sigma(h\nu)t\Phi_{h\nu}]$, where $\sigma(h\nu)$ is the photoionization cross section of the trap. In terms of the relative intensity in light versus dark,

$$\frac{\Delta I(h\nu)}{I_{dark}} = t\Phi_{h\nu} S(h\nu) = \frac{Q_T}{qN_d a - \sqrt{2qN_d V_{bi}} - Q_T}[1 - \exp(\sigma(h\nu)\Phi_{h\nu}t)]. \tag{9.84}$$

By fitting Equation 9.84 and the experimental data, one can get the trap threshold energies. For multiple traps, a new fitting should be done. To obtain concentrations (and photoionization cross sections), the dependence of the increase in the drain current and the total light flux Φt (or in the case of fixed illumination time, the incident light flux Φ) can be fitted.

A configuration diagram representative of the traps under discussion has been suggested as shown in Figure 9.21 for both traps 1 and 2. Pertinent is the coordinate diagram reported in the literature where the classical vibrational barrier energy for nonradiative multiphonon (NMP) capture of electrons from the vicinity of the conduction band edge into trap 1 is very low $(E^T - D)^2/4D = 0.024$ eV. This involves a high efficiency of trap 1 in NMP electron capture processes. V_{T1}, V_c, and V_v represent the trap 1, conduction, and valence bands, respectively. E^T, E^0, and D

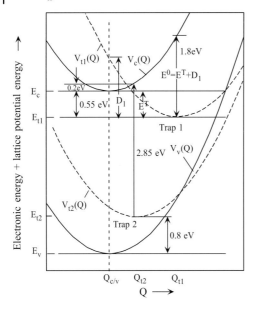

Figure 9.21 Configuration coordinate diagram for traps 1 and 2 in GaN based on the trap energies determined by Klein and Binari. The thermal ionization (electron binding) energy, E^T, is ~0.55 eV (~$E_g/6$) for trap 1 (with a Franck–Condon shift of 1.25 eV) and about 2.65 eV for trap 2 (with a Franck–Condon shift of 0.2 eV). For comparison, for trap 1 (0.55 eV) a trap thermal barrier of $E^T = 0.8$ eV and a Franck–Condon shift of $D = 1.1$ eV have also been reported. In part, courtesy of P.B. Klein.

represent the thermal energy, optical emission barrier, and Franck–Condon shift, respectively, for trap 1.

The above-discussed FET model has been extended by allowing depletion from the gate side as well as the buffer layer side, the latter by negatively charged traps in the buffer layer. The governing equations have been generalized to include the saturated current region, and explicitly included the drain voltage dependence of the occupied trap concentration and of the channel electron concentration:

$$I_D = G_0 \left\{ V_D - \frac{2}{3}\sqrt{\frac{2\varepsilon}{qn_c(V_D)a^2}} \left[(V_{bi} + V_D + V_G)^{3/2} - (V_{bi} + V_G)^{3/2} \right. \right.$$

$$\left. \left. + \frac{1}{\sqrt{1 + n_c(V_D)/N_T(V_D)}} \left[(V_{binter} + V_D)^{3/2} - V_{binter}^{3/2} \right] \right] \right\}, \quad (9.85)$$

where $V_{binter} = (kT/q)\ln[n_c(V_D)N_T(V_D)/n_i^2]$ and $n_c(V_D)$ is the drain voltage-dependent carrier concentration. When $V_D = V_P + V_G = (qn_c a^2/2\varepsilon) - V_{bi} + V_G$ is substituted in Equation 9.85, the drain saturation condition is achieved. The velocity saturation at the drain end of the channel at high drain biases can be accounted for in the form

$$I_D = qN_d v_{sat} Za(1 - u_s), \quad (9.86)$$

where u_s is the normalized depletion layer width with respect to the open-channel thickness, a, at the drain end of the gate. The term u_s can be obtained by equating Equations 9.85 and 9.86 for drain voltage at the onset of drain current saturation assuming a saturation velocity. The saturation drain voltage $V_{DS} = V_p(u_s^2 - u_0^2)$, where $u_0^2 = (V_{bi} - V_G)/V_p$ is the normalized depletion layer width at the source end of the gate, which is simply the depletion layer thickness normalized to the open-channel thickness. Empirical expressions such as $v = \mu_n E/(1 + E/E_c)$ with $v_{sat} = \mu_n E_c$, where E_c is the critical field representing the onset of velocity saturation, can be used to represent the velocity–field characteristic. The effects of carriers generated by impact ionization for $V_D > 30\,V$ that in turn get trapped and subsequently detrapped at progressively higher drain biases have also been observed. The results of such simulations are shown in Figure 9.22, where the experimental and calculated data are shown for -1 and $-3\,V$ gate biases, in dark and under illumination. In addition, the density of occupied traps is shown in the inset. In dark, the I–V curves show significant current anomalies. In the presence of light, the current does not collapse. The solid lines atop the experimental data, which are shown by symbols, represent the calculations.

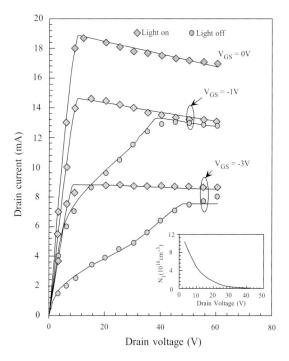

Figure 9.22 Experimental and calculated output I–V characteristics for $V_G = -1$ and $-3\,V$ for a 1.5 μm × 150 μm GaN MESFET considering trapping effects. Calculated and measured data are shown by solid lines and symbols, respectively. The inset shows the variation of occupied trap concentration (N_t) with applied drain–source voltage. The wavelength of illumination used is 470 nm. Courtesy of A.F.M. Anwar.

9.8.2
Effect of Barrier States

In addition to the bulk states, surface states and also the defects in the barrier layer cause anomalies. In this vein, localized trapping centers within the bandgap in the vicinity of the gate where the gate potential defines the energy position of the trap level with respect to the Fermi level have been considered. Consequently, any variation in the gate potential causes a change in the trap occupation, which then causes channel current variation. If the detrapping process is slow compared to the gate AC voltage, the drain current may lag the gate. In FETs, current transients, akin to DLTS in capacitance transients, are used to gain understanding of the traps in terms of ionization energy and capture cross section. The measured ionization energies can be skewed by electric field as shown in Figure 9.23, which also embodies the Poole–Frenkel effect and emission from a trap by phonon-assisted tunneling and direct tunneling processes. It should also be pointed out that the thermal ionization energies are different from the optical ones, particularly if lattice defects are involved, as depicted in Figure 9.21.

In current transients, the FET is typically held at a constant source–drain bias with a steady-state drain current of $I_D^{SS}(V_G)$. The gate voltage is then switched from a high level V_G^{SS} to a low level V_G^{P} for a period of time τ_p during which the channel current drops in response to the gate pulse. The large negative gate bias causes the electrons

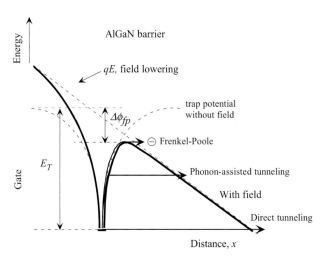

Figure 9.23 Energy diagram of the trapping center, for example, in the AlGaN barrier under the gate, in the presence of the electric field. The potential around a trap in the absence of electric field is shown in dashed lines. Contrastingly, the ones with and without field schematically show the field lowering of the barrier for emission. Possible mechanisms of electron emission are thermal ionization over the field lowered barrier (Poole–Frenkel effect), phonon-assisted tunneling, and direct tunneling in case the barrier width is sufficiently low.

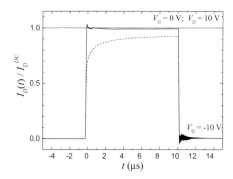

Figure 9.24 Normalized channel current response to the gate pulse $V_G(0 < t < 10\,\mu s) = 0\,V$, after the off state $V_G = -10\,V$. The devices are continuously biased at $V_D = 10\,V$. The drain current is measured with the low insertion impedance current probe. Two traces show devices with and without gate lag. Courtesy of O. Mitrofanov.

to tunnel from gate into the semiconductor and fill the available trap states. After the time τ_p, the gate potential is restored. The transient drain current is measured using a current probe. The results of one such measurement in unpassivated devices with Si doping are presented in Figure 9.24 in the form of the channel current response. In this experiment, the transistor in kept in the pinched off state for $t < 0$ to allow for all available trapping centers to be filled. At $t = 0$, the gate potential is switched to the on state ($V_G = 0\,V$) for $10\,\mu s$. One of the devices in Figure 9.24 shows an instant recovery but the other exhibits obvious gate lag. Following the initial current switching to ~85% of the steady-state level, the drain current in lagging devices slowly completes the full recovery within 50–100 μs.

Naturally, increased temperature increases the rate of current recovery as the trap emission is an activated process. The temperature dependence is illustrated in Figure 9.25 with a series of normalized transients taken at temperatures ranging from 283 to 363 K. As in the case of Figure 9.24 and prior to the measurement, the device is held in the pinch-off state with $V_D = 12\,V$ and $V_G = -11\,V$ for ~10 ms to saturate the traps. As the gate potential is changed to $V_G = 0\,V$, the captured electrons are slowly released from the traps and the resulting drain current transients exhibit long exponential tails. These can then be used to determine the electron emission rate from the traps. The details regarding emission and capture rates, capture cross section, how they are modified by field under the gate, and so on can be found in Chapter 3 of Morkoç (2009).

9.8.3
Correlation between Current Collapse and Surface Charging

Drain current lag (collapse) has been attributed to surface states whose thermal distribution has been studied. The gradual decrease of drain current in Figure 9.26a, which shows the potential drop across a 100 Ω resistor, after applying a $-1\,V$ gate bias and 7.5 V drain bias is caused by a slow accumulation of charge in the surface

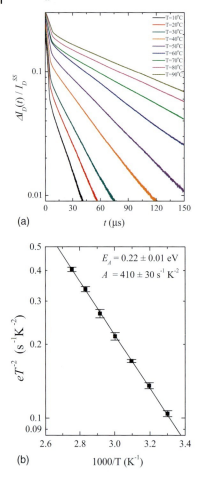

Figure 9.25 Temperature dependence of the trap emission rate. (a) The difference between the steady-state and the actual drain current after switching the gate voltage V_G from $-11\,V$ to $0\,V$ in temperature window ranging from 10 to $90\,°C$ and the source–drain bias of $12\,V$. (b) Experimentally determined values of eT^{-2} versus the inverse temperature. The emission rate is extracted by fitting an exponential decay function to the data. Courtesy of O. Mitrofanov.

states or in the barrier at the gate edge, as shown in Figure 9.27. The drain current in Figure 9.26a collapses by 20% in approximately 30 s to reach the steady-state value of Figure 9.26b. The time it takes for the drain current in Figure 9.26a to reach its steady-state value is comparable to the time required for achieving maximum band bending. The comparable timescales observed for the drain current behavior and surface charging indicate that the excess band bending near the gate is related to the drain current level. Next, $-4\,V$ pulses to the gate (10 ms on time, 150 ms off time) superimposed on the previous bias (gate $-1\,V$, drain $7.5\,V$) were applied (see the inset). The trace in the inset of Figure 9.26 shows that the drain current during

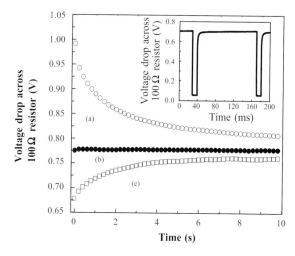

Figure 9.26 Drain current transients due to different biases. (a) Current collapse with bias on (gate: −1 V; drain: 7.5 V; source: 0 V). (b) Steady state after 30 s. (c) Drain current recovery after the superimposed pulsed bias is turned off. *Inset*: Additional current collapse due to superimposed pulsed bias (gate: − 4 V, 10 ms on, 150 ms off).

150 ms when the −4 V bias is off is 12% less than the steady-state value of Figure 9.26b indicating an additional current collapse due to the superimposed pulsed bias. During each negative excursion of the gate drive, the reverse bias between gate and drain becomes higher, which is consistent with this model in that a higher reverse bias causes a larger number of electrons to tunnel to the surface states. All of this excess surface charge does not discharge immediately after the bias is turned off. After turning off the −4 V reverse-bias pulses, the drain current recovers (Figure 9.26c) in about 40 s to the level of Figure 9.26b. If we momentarily set both the gate and drain potentials to 0 V and then reapply the bias, we find that the drain current level is almost the same as that in Figure 9.26b. The

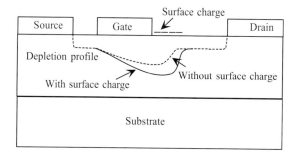

Figure 9.27 A schematic showing the effect of surface charge on the drain current in a field effect transistor.

gate and drain potentials had to be held at 0 V for over 700 s before reapplying the bias to observe a drain current transient that would start at a level similar to the beginning of the transient of Figure 9.26a.

To reiterate, the drain current decreases with increasing drain to source voltage and also with gate bias because this causes a higher reverse bias between the gate and drain that in turn increases tunneling of the electrons from the gate to the surface states near gate edge. Because surface states are involved, the state of surface affects the FET characteristics, which is exacerbated in GaN as surface states change the channel electron density that in turn causes the LO phonon decay time to increase and the carrier velocity to decrease. Two effects combined cause the current to change considerably. Naturally, surface should be passivated with an appropriate dielectric passivant.

9.9
Electronic Noise

The low-frequency and high-frequency noise characteristics of GaN HFETs are very important for microwave applications. Fluctuations or deviations in the form of irregular changes in parameters of interest such as current, voltage, resistance, frequency, and so on from long-term time averages or from periodical time dependence are considered noise. Hot carriers prevalent in FETs cause a hot (nonequilibrium) phonon population to be generated in addition to impact ionization (inherently noisy), which must be considered. Traditionally, before the advent of high-performance nonequilibrium devices, noise to a first degree has been thought of as fluctuations in current flow (*shot noise*), due to recombination–generation process (*generation–recombination noise or simply GR noise*) or due to thermal processes (*thermal or Johnson or Nyquist noise*). In terms of the thermal noise (Nyquist–Johnson noise), any resistor shows spontaneous fluctuations in current caused by universal thermal motion of carriers. The Nyquist theorem treats the thermal noise as the available noise power or a thermal noise source as a universal function of the absolute temperature. The shot noise is caused by the corpuscular nature of electrical charge and is associated with current flow when it is controlled by a barrier such as in p–n junctions, Schottky barriers, across heterojunctions, tunneling structures, barriers induced by nonuniform doping regardless of how small they are, and non-ohmic contacts, and is characteristically "white" in nature (i.e., having equal power across the spectrum).

The noise produced by conductance fluctuations is viewed as $1/f$ noise when it is dominant at low frequencies decaying as $1/f$ with frequency. Generation–recombination noise is generated by random transitions from and to the trap levels from the conduction and/or the valence bands as a fluctuation in the number of mobile carriers. In devices such as FETs, the electrons are not in equilibrium with the lattice (hot electrons) and fast kinetic dissipation takes place in the conduction or valence band. Processes such as energy relaxation, intervalley scattering (if applicable), impact ionization, and so on cause fluctuations resulting in hot electron noise.

The low-frequency noise also manifests itself at high frequencies, mainly as phase noise with deleterious implications in, for example, oscillators. All forms of electronic noise are characterized by a mean squared current fluctuation $\langle(\delta I)^2\rangle$ (or $\langle(\delta V)^2\rangle \equiv (\text{rms } \delta V)^2$ for voltage), measured in series with (or across for voltage noise) the device or sample under test, when a constant voltage (or current) is applied, except for thermal noise, which is present even in thermal equilibrium, with no bias applied. All the other forms of noise present in addition to thermal noise are also known as current noise and are absent in thermal equilibrium. Nevertheless, 1/f noise and GR noise also modulate the root mean square (rms) level of the thermal noise currents (or voltages) in thermal equilibrium, while the available power remains constant at $k_B T$ per Hz, where $k_B = 1.38 \times 10^{-23}$ J/K is Boltzmann's constant and T is the absolute temperature. These two forms of current noise are also called modulation noise, because they modulate the resistance. If a bandpass filter is inserted between the measuring device (usually a quadratic meter) and the noise source, the spectral density of the fluctuations $\langle(\delta I)^2\rangle f \equiv S_I(f)$ (or $S_V(f)$) is obtained by dividing the measured mean square by the bandwidth Δf of the filter. Let us now briefly discuss the nature of the various noise phenomena. More detailed discussion of the topic can be found in Chapter 3 of Morkoç (2009).

In addition to fluctuations that are manifested at low frequencies, many fluctuations and scattering events respond to high frequencies and are viewed as high-frequency noise. Among them of course are thermal noise, shot noise, and avalanche noise. Here we will also use this nomenclature as a prelude to discuss noise producing processes specific to high-frequency FETs brought about by high fields that are present in the channel of an FET and processes that become imperative at high frequencies.

Thermal noise (Johnson noise) comes into play through contact metal resistance and also through the metal–semiconductor contact resistance (metal to semiconductor resistance at the source and drain electrodes) and semiconductor resistance where the electric field is low. Better device designs, most of which are technologically driven, can be employed to lower these resistances and thus the thermal noise. Because the gate provides a barrier to carrier flow and in cases of gate leakage, the fluctuations associated with this barrier-activated current flow induce shot noise. The gate leakage problem is somewhat an issue of technology and can be alleviated to some extent, which in turn would eliminate the shot noise.

The central issue with FETs that have high electric fields in the channel has to do with the random processes involving hot electrons (intervalley scattering, energy gain and loss through interaction with optical phonons, and impact ionization, whichever is applicable) that cause fluctuations in the channel current, which constitute the basis of hot electron noise. This noise source is inevitable.

Hot electrons, which are synonymous with FETs, involved in processes such LO phonon absorption and generation and interband scattering impact ionization introduce fluctuations in the system that manifest themselves as noise, and to be specific this type of noise will be called hot electron noise and the equivalent noise temperature ascribed to it will be called hot electron noise temperature. In MESFETs with three degrees of freedom for carriers, the hot electron noise temperature is different in the directions parallel and normal to the steady-state current flow even in the case of spherical bands and isotropic scattering mechanisms. The noise in the transverse direction is caused by only the velocity fluctuations in an isotropic semiconductor, which is white for frequencies below that corresponding to the mean carrier relaxation time in the context of scattering. This means that the energy relaxation time, which is dependent on electric field, can be estimated from the hot electron transverse noise temperature.

Regardless of the complexity of the problem, it is certain that the shorter the transit time the less random the hot electron processes and thus the lower the noise. The can be accomplished by reduced channel lengths and use of materials with a relatively high carrier velocity. High carrier velocity naturally implies reduced scattering at high fields, and thus lowered fluctuations associated with the relevant processes. On the practical side, empirical expressions have been developed that allow one to bypass the detailed basic physics for a rapid determination of noise with certain material-dependent parameters and knowledge of resistances, among others. More detailed discussion of the topic can be found in Chapter 3 of Morkoç (2009).

9.9.1
FET Equivalent Circuit with Noise

From a practical point of view, the noise associated with FETs is best described by adding noise sources to the equivalent circuit of a noiseless FET of Figure 9.7. The equivalent circuit-based noise treatment differs from the physics-based treatment that utilizes physical processes such as transport, scattering, materials properties, and device dimensions. Electronic noise can be represented by equivalent current or voltage sources as shown in Figure 9.28. The contribution by the admittance, $Y(f)$, and the impedance, $Z(f)$, which are at a certain nonzero temperature, can be modeled by a combination of a noise source and a noiseless passive element. The statement $T=0$ is to underscore the fact that both the admittance, $Y(f)$, and the impedance, $Z(f)$, are noiseless. The noise present is represented by a noise equivalent temperature indicated as T_n.

Let us define the figure of merit universally used to describe the noise performance of a device, the *noise figure* (NF). To express the impact of noise in a meaningful manner, the noise contribution by a noise source is described in terms of the noise to signal ratio at the output port and the noise to signal ratio at the input port. The ratio of the two quantities in the form of a logarithmic function in decibels is the noise figure and can be written as

$$\mathrm{NF}(f) = 10\log_{10}\frac{S_{n2v}}{S_{n2v0}} \quad \text{or} \quad \mathrm{NF}(f) = 10\log_{10}\frac{T'_n(f)}{T_0}\,\mathrm{dB}, \quad (9.87)$$

where S_{n2v} represents the output noise spectrum of the noisy two-port circuit and S_{n2v0} represents the output noise spectrum of the noiseless two-port circuit. For the second definition, $T'_n(f)$ represents the output noise power reduced back to the value at the input of the circuit (meaning it is divided by the amplifier gain in the form $T'_n(f) = T_n/\mathrm{Gain}$) and T_0 represents the input noise power generated by a

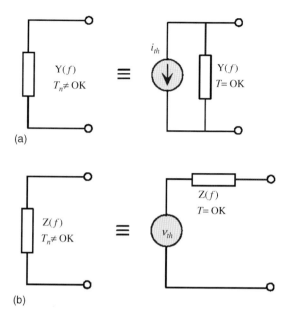

Figure 9.28 Equivalent circuit of (a) a noisy admittance Y represented by a parallel combination of a noise current source i_{th} and noise free admittance $Y(f)$ and (b) a noisy impedance represented by a serial combination of a noise voltage source v_{th} and noise-free impedance.

fictitious input matched resistor that is kept at room temperature (T_0 is assumed to be 290 K). The noise figure can also be defined as, using the above definition, the output noise at equivalent noise temperature divided by that of a fictitious output resistor that is kept at room temperature.

The noise figure in a device can be minimized under optimized load and bias, and the term NF_{min} is used for it. The knowledge of NF_{min} is insufficient to describe the behavior of a two-terminal circuit. Four independent parameters, that is, the minimum noise figure NF_{min}, the equivalent noise resistance R_n, and the real part G_S and the imaginary part B_S of the optimal generator (Z_s) impedance, are necessary to describe the noise behavior of a linear two-port circuit fully.

The noise figure of a noisy linear two-port circuit can be calculated utilizing an applicable equivalent circuit with one current noise source or one voltage noise source at the input circuit as shown in Figure 9.29.

Using the equivalent circuit of Figure 9.29, one can calculate the noise figure as

$$NF = 1 + \frac{S_{nv} + |Z_s|^2 S_{ni} + 2\text{Re}[Z_s S_{vi}]}{4k_B T_0 R_S}, \tag{9.88}$$

where S_{nv} and S_{ni} represent the voltage noise spectrum and current noise spectrum, respectively, and S_{vi} is the cross-correlation spectrum between the noise sources v_n and i_n. As can be noted, the noise figure is independent of both the input impedance,

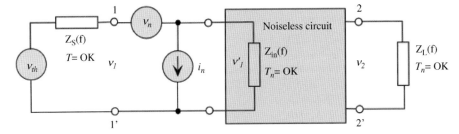

Figure 9.29 A noise equivalent circuit with a noise current and a noise voltage source chained at the input terminal of a noiseless two-port circuit.

Z_{in}, and the load impedance, Z_L. The minimum noise figure NF_{min} can be obtained by optimizing the input source impedance, which is called *noise matching*. It should be mentioned that noise matching conditions are different from power matching conditions. In addition, the noise matching conditions, like the power matching conditions, are dependent on the bias. Generally, lower noise figures are achieved for lower drain currents up to the point beyond which the gain, transconductance, and possible velocity degradation may occur. The latter could be due to carrier injection into traps in buffer layers, among others, which is enhanced as the channel is depleted by gate bias. In short, the maximum available gain may not be available. Rather an associated gain may be in effect for matching conditions leading to the minimum noise figure.

The dependence of NF on matching conditions, that is, deviation from NF_{min}, can be described by

$$NF = NF_{min} + \frac{R_n}{G_S}|Y_S - Y_{S,opt}|^2, \qquad (9.89)$$

with $Y_S = G_S + jB_S$ in terms of the Y parameters, and

$$NF = NF_{min} + \frac{G_n}{R_S}|Z_S - Z_{S,opt}|^2, \qquad (9.90)$$

with $Z_S = R_S + jX_S$ and the equivalent noise conductance $G_n = R_n|Y_{S,opt}|^2$ in terms of the Z parameters.

Therefore, to reiterate just four parameters, the minimum noise figure NF_{min}, the equivalent noise conductance G_n (or the equivalent noise resistance R_n of the transistor), and the real and the imaginary parts of the optimal source admittance ($G_{S,opt}$ and $B_{S,opt}$), are required to fully describe the noise behavior of a linear two-port circuit. The equivalent noise resistance R_n (or the equivalent noise conductance G_n) is a measure of the sensitivity of the noise figure on deviations from the optimum source admittance or impedance leading to the minimum noise figure and can be construed as a noise figure of merit. In practice, one might have to deviate from optimum matching for noise in favor of matching for power or gain.

In the microwave world, the admittance or impedance is normalized with respect to the characteristic impedance Z_0 (typically 50 Ω) and reflection coefficients are determined that in turn are used to determine impedance/admittance parameters on the source or the load side. In this context, the source admittance can be expressed as

$$Y_S = \frac{1}{Z_0} \frac{1-\Gamma_S}{1+\Gamma_S}, \qquad (9.91)$$

where Γ_S is the generator reflection coefficient. Similarly,

$$Y_{S,opt} = \frac{1}{Z_0} \frac{1-\Gamma_{S,opt}}{1+\Gamma_{S,opt}}, \qquad (9.92)$$

where $\Gamma_{S,opt}$ is the optimum generator reflection coefficient at the noise matching conditions for minimum noise figure (both reflection coefficients are complex numbers).

With the help of Equations 9.91 and 9.92, we can construct $|Y_S - Y_{S,opt}|^2$ as

$$|Y_S - Y_{S,opt}|^2 = \frac{4}{Z_0^2} \frac{|\Gamma_S - \Gamma_{S,opt}|^2}{|1+\Gamma_S|^2 |1+\Gamma_{S,opt}|^2}, \qquad (9.93)$$

with which Equation 9.89 can be rewritten as

$$NF = NF_{min} + 4r_n \frac{|\Gamma_S - \Gamma_{S,opt}|^2}{(1-|\Gamma_S|^2)|1+\Gamma_{S,opt}|^2}, \qquad (9.94)$$

where $r_n = R_n/Z_0$ is the normalized equivalent noise resistance.

Expanding further, the contribution to noise at the input side by the FET can be treated as voltage and current sources with an appropriate noiseless resistance added to the circuit as shown in Figure 9.30. Gradually easing into relating the noise to the FET parameters, the thermal noise in an intrinsic FET due to only its channel can be described as

$$\langle i_{th}^2(t) \rangle = 4\pi k_B T_{eff} \frac{1}{R_{ch}} \Delta f, \qquad (9.95)$$

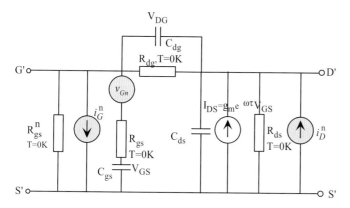

Figure 9.30 High-frequency equivalent circuit of an FET including the noise sources associated with the input and the output circuits.

where T_{eff} represents the effective temperature in the channel of an FET and R_{ch} represents the channel resistance.

The noise equivalent circuit of Figure 9.30 can be used to discern RF noise of an FET, which is based on a *temperature noise model* (TNM) inclusive of gate leakage that is modeled as R_{dg} and R_{gs} in parallel to the intrinsic drain–gate and gate–source elements, respectively. If needed, the impact ionization effects can also be added.

9.9.2
High-Frequency Noise in Conjunction with GaN FETs

The fundamental discussion with respect to the high-frequency noise in FETs in terms of the circuit representation of a noisy FET circuit also involves contributions by the extrinsic elements, mainly the resistances at the gate, source, and drain circuit, in the form of thermal noise. Because the device has considerable gain, the drain resistance noise source may be neglected. Although they have been defined in the thermal (Johnson) noise section, for continuity they are given as

$$\langle v_g^2 \rangle = 4k_B T R_G \Delta f \quad \text{and} \quad \langle v_s^2 \rangle = 4k_B T R_S \Delta f. \tag{9.96}$$

The noise figure, associated gain, and other noise parameters can then be calculated using the method that transforms a fourpole with internal noise sources to a noise-free fourpole cascaded with a preceding noise network in which all noise sources are lumped into one noise current source and one noise voltage source as schematically shown in Figure 9.29. The fourpole contains an equivalent noise conductance G_n, an equivalent nose resistance R_n, and a complex correlation impedance Z_c, which can be expressed as follows:

$$G_n = \frac{R_s + R_{nd}}{|Z_{21}|^2}, \tag{9.97}$$

$$R_n = R_s + R_G + R_{ng} - \left| \frac{Z_{11} - Z_c}{Z_{21}} \right|^2 (R_s + R_{nd}), \tag{9.98}$$

$$Z_c = Z_{11} - Z_{12} \left(\frac{R_s + C^* \sqrt{R_{nd} R_{ng}}}{R_s + R_{nd}} \right), \tag{9.99}$$

where $R_{nd} = \langle v_d^2 \rangle/(4k_B T \Delta f)$, $R_{ng} = \langle v_g^2 \rangle/(4k_B T \Delta f)$, $C^* = \langle v_g v_d^* \rangle/[\langle v_g^2 \rangle \langle v_d^2 \rangle]^{1/2}$, and Z_{ij} are the extrinsic impedance parameters. Finally, the minimum noise figure is calculated using

$$NF_{min} = 1 + 2G_n \left[R_c + \left(R_c^2 + \frac{R_n}{G_n} \right)^{1/2} \right], \tag{9.100}$$

where R_c is the real part of Z_c. The optimum source resistance for minimum noise figure is

$$Z_{opt} = [R_c^2 + R_n/G_n]^{1/2} - jX_c. \tag{9.101}$$

(Again the complex correlation impedance $Z_c = R_c + jX_c$.)

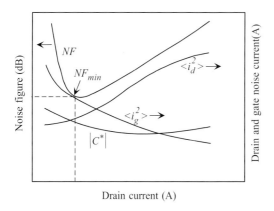

Figure 9.31 The functional dependence of $\langle i_d^2 \rangle \langle i_g^2 \rangle$, C^*, and the noise figure calculated using these parameters as outlined in the text. The drain current leading to minimum noise figure is also indicated.

The functional dependence of $\langle i_d^2 \rangle \langle i_g^2 \rangle$, C^*, and the noise figure calculated using these parameters is shown in Figure 9.31 in order to give the reader a feel as to how the noise figure depends on bias conditions that govern the drain current. The drain current leading to NF_{min} is indicated.

The noise and gain performance of the HFET depend critically upon the device parameters (gate length, carrier mobility, carrier velocity, barrier thickness, etc.). The device optimization for noise is somewhat empirical in nature underscoring the complexity of the problem and many sources of noise. Suffice it to say that reduction of resistances, utilizing layered structures that could minimize fluctuations of any kind, such as 2DEG, which would hinder velocity fluctuations perpendicular to the general direction of current flow, reducing the transit time by scaling dimensions, and also using materials with high velocity, can reduce noise.

It is useful to provide a practical treatment of noise that is somewhat simplified but easier to grasp in the context of device and materials parameters. We began with determination of the open-circuit voltage, $\langle v_{dso}^2 \rangle$, and therefore short-circuit current, $\langle i_d^2 \rangle = \langle v_{dso}^2 \rangle |y_{22}|^2$, sources for the drain circuit and short-circuit noise source for the gate circuit $\langle i_g^2 \rangle$ as

$$\langle i_d^2 \rangle = 4k_B T \Delta f L_g (\alpha Z + \beta I_{ds}) \frac{g_d^2 + \omega^2 C_{dg}^2}{g_d^2} \frac{g_m}{C_{gs}}, \qquad (9.102)$$

$$\langle i_g^2 \rangle = \frac{2k_B T \Delta f \omega^2 C_{gs}^2}{g_m}. \qquad (9.103)$$

The correlation coefficient can be expressed as

$$C^* = C' \frac{y_{22}^*}{|y_{22}|} \quad \text{with purely imaginary component } C' = \frac{\langle i_g v_{dso}^* \rangle}{(\langle i_g^2 \rangle \langle v_{dso}^2 \rangle)^{1/2}}. \qquad (9.104)$$

Z, L_g, g_m, g_d, C_{gs}, C_{dg} are the gate width, gate length, transconductance, drain conductance, gate–source capacitance, and drain–gate capacitance, respectively, and I_{ds} is the DC drain current. Except the geometrical parameters, the rest represent the small-signal parameters. Terms α and β are fitting parameters that are nearly independent of geometrical and materials parameters, and assume the values of $\alpha = 2 \times 10^5$ pF/cm² and $\beta = 1.25 \times 10^2$ pF/mA. The main factor influencing the factor C' is the gate aspect ratio, meaning the gate length divided by the gate to channel separation of HFETs and gate length to epilayer thickness in MESFETs. This parameter has been reported to be close to 0.8 for an aspect ratio of greater than or equal to 5 and 0.7 for a ratio of about 3. By the nature of the device geometry, this factor is larger in HFETs leading to lower noise figures.

Let us refocus on the effect of various parameters on the noise. The small-signal drain–gate capacitance C_{dg} and the drain conductance g_d come into play in determining the drain short-circuit noise current in Equation 9.102 through the term $\delta(\omega) = (g_d^2 + \omega^2 C_{dg}^2)/g_d^2$. For frequencies below $f_0 = g_d/(2\pi C_{dg})$, the $\delta(\omega)$ term is close to unity but increases as ω^2 for frequencies $f \gg f_0$. The goal here is to have small drain short-circuit noise current source and therefore keeping the frequency of operation below f_0 would lead to lower noise. This means that the drain feedback capacitance, C_{dg}, must be minimized by device design. Increasing the output conductance, g_d, would achieve the same result, but is deleterious because it reduces the gain and output impedance of the device making it more difficult to impedance match.

Going nearly fully practical, an empirical expression for the minimum noise figure in FETs at room temperature provided by H. Fukui, which is generally referred to as the Fukui equation, describes the dependence of noise on device and materials parameters (Equation 9.105):

$$\text{NF}_{\min} = 10 \log \left\{ 1 + \frac{k_f f}{f_t} [g_{m0}(R_g + R_s)]^{0.5} \right\} \text{ dB}, \quad (9.105)$$

where g_{m0} is the intrinsic transconductance (mS), R_g is the gate resistance function of device geometry (Ω), R_s is the source resistance function of device geometry (Ω), and k_f is an empirical fitting factor that depends on the semiconductor material used. This fitting factor can be expressed as

$$k_f = 2 \left[\frac{\langle i_d^2 \rangle}{4 k_B T g_m \Delta f} \right]^{0.5}. \quad (9.106)$$

It would be interesting to calculate this factor using the analytical expression of Equation 9.102 with $\delta(\omega) = (g_d^2 + \omega^2 C_{dg}^2)/g_d^2 = 1$, chosen for simplicity. Doing so leads to

$$k_f = 2 \left[\frac{L_g}{C_{gs}} (\alpha Z + \beta I_{ds}) \right]^{0.5}. \quad (9.107)$$

Recognizing the gate capacitance $C_{gs} = \varepsilon Z L_g/t + C_p$, where ε is the dielectric constant of the semiconductor under the gate (AlGaN in the case of AlGaN/GaN

FET), t is the AlGaN thickness, and C_p is the gate fringe capacitance, Equation 9.107 can be rewritten as

$$k_f = 2\left[\frac{t}{\varepsilon Z}\frac{(\alpha Z + \beta I_{ds})L_g}{1 + C_p t/\varepsilon Z L_g}\right]^{0.5}. \tag{9.108}$$

As in the case of a large aspect ratio leading to lower noise, a small gate to channel distance t, which leads to large aspect ratios, is beneficial in reducing the fitting factor, and therefore the noise figure of the device. In this respect, HFETs are better than MESFETs, keeping other parameters the same. Substituting Equation 9.107 into Equation 9.105 leads to

$$\text{NF}_{\min} = 10\log\left\{1 + \sqrt{8\pi}f\left[\frac{L_g}{f_c}(\alpha Z + \beta I_{ds})(R_g + R_s)\right]^{0.5}\right\}\text{dB}. \tag{9.109}$$

Taking advantage of the average velocity in the channel through $f_c(1/2\pi)(\bar{v}/L_g)$, Equation 9.109 can be rewritten as

$$\text{NF}_{\min} = 10\log\left\{1 + 4\pi L_g f\left[\frac{1}{\bar{v}}(\alpha Z + \beta I_{ds})(R_g + R_s)\right]^{0.5}\right\}\text{dB}. \tag{9.110}$$

This expression is a qualitative one linking the minimum noise figure to device parameters where the effects of gate and source resistances are prominently displayed.

In another approach, relevant to the Fukui equation, Delegebeaudeauf et al. developed a relationship of k_f as a function of the optimum bias current I_{opt}:

$$k_f = 2\left[\frac{I_{opt}}{E_c L_g g_m}\right]^{0.5}. \tag{9.111}$$

With the cutoff frequency estimated as $f_t = v_{sat}/2\pi L$, the NF$_{\min}$ can be expressed as

$$\text{NF}_{\min} = 10\log\left\{1 + \frac{4\pi f L_g}{\bar{v}}\left[\frac{I_{opt}(R_g + R_s)}{E_c L_g}\right]^{0.5}\right\}\text{dB}. \tag{9.112}$$

The literature indicates that the above model predictions for minimum noise figure agree well with experimental data up to about 26 GHz in GaN HFETs, but the discrepancy increases with frequency because the Fukui model does not take into account the gate to drain capacitances and the higher order frequency terms. It was estimated that very low noise figures of 0.4 dB at 12 GHz might be feasible from a GaN HFET with an improved F_t, which is dependent on material quality. Suffice it to say that the available data indicate that high-frequency noise performance of GaN HFETs is respectable and is nearly comparable to those of AlGaAs/GaAs HFETs.

9.10
Self-Heating and Phonon Effects

Self-heating is of paramount interest in any device and the HFET is no exception, particularly considering the high-power varieties, either switching or RF. Self-heating is caused by power dissipated within the device, and is modeled via a thermal subcircuit in the form of an added, parallel configuration RC circuit to the source. The circuit consists of a thermal resistance R_{th} that describes the rate of temperature rise with respect to power, in units of K/W, and a thermal capacitance C_{th} that describes the time response of the junction temperature for a given input waveform. Then the static channel temperature can be expressed as

$$T = R_{th}P + T_0. \tag{9.113}$$

Here T_0 is the environment temperature and P is the DC component of $p(t)$.

A plethora of methods have been proposed to deal with the dependence of current on the channel temperature. Large-signal models incorporate the temperature-dependent terms into the drain current analytical expression. Other methods include the transient thermal model and the Laplace heat spread model. An alternative treatment that attempts to separate the self-heating part from the isothermal model follows:

$$I_{DS}(t) = I_{DSO}(t, T_0)[1 - \delta \cdot P(t)^* h_{th}(t)]. \tag{9.114}$$

Here T_0 is the ambient temperature, I_{DSO} is the isothermal current, $h_{th}(t)$ is the impulse response of the electrothermal subcircuit, * denotes convolution, and δ is a fudging parameter and a function of the thermal resistance of the HFET and the temperature dependence of the drain current.

If one assumes a source and drain contact resistivity of 1.2 Ω mm, an ambient temperature $T_0 = 300$ K, assumed to be the case temperature as well, $\delta = 0.2$ W^{-1}, and $R_{th} = 70$ K/W, one can calculate I–V characteristics, the results of which are shown in Figure 9.32a as the dashed lines. Figure 9.32b illustrates the channel temperature under DC bias conditions. The junction temperature can reach 340 K at about 6 W/mm output power level for the particular δ value chosen.

9.10.1
Heat Dissipation and Junction Temperature

A material that can handle very high temperatures and has a high thermal conductivity is well suited for use in power devices. The junction temperature can be set high compared to the case temperature facilitating heat transfer, thanks to the good thermal conductivity of the material between the junction and the case. Owing to the enormous importance of heat, four-dimensional models (*x*, *y*, *z* for spatial dimensions and *T* for temperature) have been developed to cope with power devices, originally for Si devices in general and BJTs in particular as they represent the early power devices. The thermal conductivity concept, particularly in relation to FETs, assumes that heat is dissipated through LA phonons without considering the hot electron to LA phonon conversion. While in materials such as Si and to a lesser extent GaAs, this pathway is applicable to a large extent, it is not so in GaN due to the strong electron–phonon coupling that causes the heat dissipation to take the route of hot electrons to LO phonons, then decay to LA phonons, and then to the case. The slowest process in the aforementioned chain is the decay of LO phonons to LA

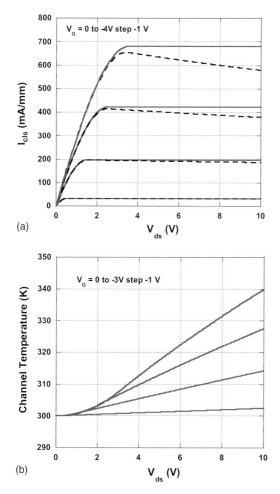

Figure 9.32 AlGaN/GaN HFETs with 1 nm AlN spacer: (a) the calculated I–V characterization considering the influence of access resistances (solid lines) and self-heating (dashed lines); (b) channel temperature under the DC bias conditions employed. Courtesy of Q. Fan.

phonons in the case of GaN. The thermal conductivity as measured and reported throughout most of the literature is a measure of the efficiency of the last stage of the aforementioned process. Heat dissipation without consideration of the processes prior to LA phonon generation is not prudent.

In FETs, the heat can be removed from the junction vertically to the case or through a combination of lateral and vertical heat diffusion, the former requiring only one-dimensional treatment, whereas the latter requiring full 3D spatial treatment. The proximity of gate fingers to each other, the length of each gate finger, and their distance from the case are important device parameters to deal with in terms of heat removal and thus power performance. For example, gates near the center of the chip run higher in temperature as compared to those near the periphery. Similarly, the temperature near the center of the each gate is higher than that at the edge. Luckily, negative temperature coefficients associated with mobility and velocity help prevent thermal runaway. Serendipitously, while the parameters change, the heat is transferred from one material to another, which begins with a region generating the heat followed by its dissipation. The temperature of the heat source is determined by the dissipated DC power minus the RF power extracted. This brings the device efficiency to the forefront as the higher the efficiency the lesser the heat dissipated.

Specific to GaN HFETs, measurements to assess the temperature of a device may be predictors of lifetimes. Infrared (IR) and *micro-Raman thermal mapping* techniques have been used to deduce temperatures of the devices with micrometer spatial resolution. IR measurements on TLM structures have also been used to measure temperature but the micro-Raman thermal mapping technique has shown even higher temperatures. The finite element analysis (FEA) method was then used to model the FET (using a pure thermal as opposed to an electrothermal model), and the linewidth and shift of the micro-Raman signal correlated to the FEA results, but were underestimated (the Stokes–anti-Stokes method also correlates with the linewidth method). 2D electric field corrections have been shown to improve fits. Finally, an Arrhenius plot yielded the mean time to failure (MTTF) and the authors claim that by putting the same power into a device at DC, RF failure can be "measured" using DC biases.

The maximum current density or power density is determined by the ability of the device together with its housing/case to dissipate the heat while keeping the junction temperature somewhat lower than the maximum allowed temperature. In a one-dimensional problem where the temperature is the only variable, the difference between the device and ambient temperature under steady-state conditions is determined by the product of the total device to ambient thermal resistivity and the total power dissipated by the device:

$$\Delta T = T_j - T_{ca} = \left(\frac{W}{\kappa A} + \theta_{ca}\right) P_D, \tag{9.115}$$

where ΔT is the temperature difference between the maximum junction temperature T_j and the case temperature T_{ca}, W is the distance from the junction to the case, κ is the thermal conductivity of the semiconductor material, A is the area of the hot source, assumed to be spatially uniform to allow the use of 1D models, θ_{ca} is the thermal resistance of the case to the ambient, and P_D is the dissipated electrical power that is represented by the DC power in and RF power out. Note that in terms of electrical circuit equivalency, temperature is analogous to voltage and heat flux (dissipated power per unit area) is analogous to current. Therefore, the $W/\kappa A$ (A being the area) term represents the thermal resistance of semiconductor to the case, and θ_{ca} represents the thermal resistance of the case to the ambient. It is implicit in Equation 9.115 that heat dissipation is only from the junction to the case followed by radiation to the ambient unless other cooling

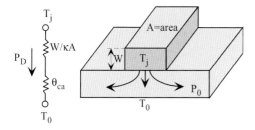

Figure 9.33 A schematic representation along with the electrical circuit equivalent for thermal dissipation described in Equation 9.115. Temperature could be construed as potential (voltage) and the dissipated power as a current in the realm of electrical circuits. A typical value for the case to ambient thermal resistivity is 1 K cm²/W.

mechanisms are employed. A schematic representing a 1D thermal heat dissipation with its electrical equivalent circuits is shown in Figure 9.33.

As mentioned, a convenient technique for measuring not just average temperature but also the spatial distribution of temperature is micro-Raman spectroscopy. This technique utilizes the change in vibrational mode frequency due to temperature. The Raman shift measured on a device is converted to temperature using the following equation:

$$\omega(T) = \omega_0 - \frac{A}{\exp(B\hbar\omega_0/kT) - 1}, \quad (9.116)$$

where $\omega_0 = 571.89 \pm 0.01 \text{ cm}^{-1}$, $A = 20.52 \pm 0.47 \text{ cm}^{-1}$, and $B = 1.04 \pm 0.01$ for the HFET investigated in one particular report to give a sense of the parameters to the reader. The temperature error is less than 10 K at 300 K. As mentioned and as will be discussed in detail next, in GaN FETs, the heat is generated by the electric field and ensuing hot E_1 LO phonons that in turn decay to acoustic phonons help diffuse the heat.

9.10.2
Hot Phonon Effects

For GaN, being a polar semiconductor, the dominant scattering mechanism at high temperatures is polar LO phonon scattering. There are nine optical and three acoustical phonon branches in wurtzitic GaN. The acoustical modes are simple translational modes and optical modes are more complex. The modes with the strongest interaction with electrons are the E_1 and A_1 LO phonons with an energy of about 92 meV in GaN. The E_1 and A_1 modes are very close in energy (about 8 cm^{-1}) and there are three LO phonon modes in wurtzitic GaN. The E_1 mode couples perpendicular to the c-axis, which is to the first extent represented by FETs fabricated in layers on the c-plane, and the A_1 mode couples along the c-direction, which is applicable to vertical p–n structures built on the c-plane. However, if there exist scattering events that cause non-c-plane transport, the phonon wave vector acquires an orthogonal component, causing both phonon modes to be involved. If the angle between the c-axis and phonon wave vector (often designated by q) is θ, the angular dependence of the zone-center LO phonon frequency can be expressed as

$$\omega_{LO}(\theta) = [(\omega_{LO}^{A_1})^2 \cos^2\theta + (\omega_{LO}^{E_1})^2 \sin^2\theta]^{1/2}. \tag{9.117}$$

The LO phonon frequency depends on the scattering geometry and has a frequency lying between that of the A_1(LO) mode (wave vector along the c-axis, or $\theta = 0°$) and that of the E_1(LO) mode ($\theta = 90°$), with the maximum difference being about $8\,cm^{-1}$. Depending on the scattering angle, both modes might be present. In steady state, energy loss by LO phonon emission and electron energy gain from the electric field balance, so the electron transport is at a saturated velocity that precedes the limitation by the intervalley scattering in a multi-conduction valley system such as GaN.

Comparison of the power supplied to electrons in a GaN 2DEG to the amount of power dissipated by an equilibrium distribution of LO phonons (cold phonons) shows that the dissipated power actually exceeds the supplied power by a factor of 30. Obviously, equilibrium distribution of phonons is unreal and therefore the phonon distribution at elevated temperatures meaning "hot" phonons must be considered. Therefore, two sets of phonons exist in the channel: the "hot" modes, representing nonequilibrium "hot phonons," and equilibrium "cold" modes (associated with the lattice temperature). The genesis of the nonequilibrium population of hot phonons is that the time associated with the emission of LO phonons is much shorter than the time associated with the decay of LO phonons into propagating LA modes (Ridley mechanism). Thus, the (relatively stationary–low group velocity) LO phonons tend to build up and an equilibrium distribution cannot describe them.

Investigations have indicated that hot carriers (with a density approximately greater than $10^{17}\,cm^{-3}$) can produce a nonequilibrium distribution of hot phonons, which slow down the rate of energy relaxation. The phonon generation time is about 10 fs for GaN (about 100 fs in GaAs), but decay time is strongly density dependent. The presence of large concentrations of hot phonons reduces mobility through phonon scattering. At high fields where scattering from ionized impurities is unimportant, the dominant effects appear to be the enhancement of electron temperature for a given field, which causes a reduction in carrier velocity.

Conventional treatments of hot phonons involve a forward displaced distribution of the nonequilibrium phonons in momentum space that is caused by the drift of hot electrons. The emission and reabsorption of nonequilibrium phonons have been argued to reduce the overall energy relaxation rates for electrons. The highly polar nature of GaN leads to a strong Fröhlich interaction, as tabulated in Table 9.2, which

Table 9.2 Fröhlich interaction constants for several wurtzite materials and for GaAs.

Parameter	AlN	GaAs	GaN	InN	ZnO
α	0.74	0.075	0.41	0.22	1.04

Courtesy of M. Stroscio. For the wurtzite materials, the values of α are those associated with interactions along the c-axis.

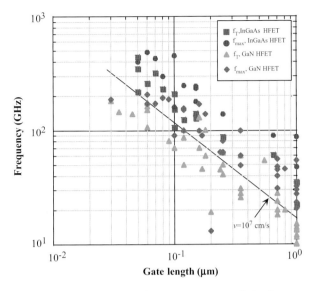

Figure 9.34 Effective carrier velocity, representing GaN, determined from the dependence of current cutoff frequency on gate length (squares). The values of maximum oscillation frequency versus the gate length (circles) are also shown. It should be noted that there is considerable dispersion in the published data of both cutoff frequencies.

likely needs more than the common perturbative treatment used in the typical Monte Carlo calculations used to predict the velocity–field characteristics of nitride binaries and ternaries.

Predictions, albeit without the treatment of hot phonons, for peak velocity are above 3×10^7 cm/s at 140 kV/cm at room temperature and for valley velocity are above 2×10^7 cm/s over a wide range of fields. In contrast, estimates from the current gain cutoff frequency of AlGaN/GaN HFETs place the effective velocity to relatively lower values, as shown in Figure 9.34. To the first extent, the effective electron drift velocity (v_d) can be related to the microwave current gain cutoff frequency through $v_d = 2\pi f_T L_g$, where L_g is the gate length and f_T is the cutoff frequency, the use of which leads to effective velocities of less than about 1×10^7 cm/s. The effective velocity deduced experimentally is not the peak velocity, but is assumed to closely relate to the valley velocity encountered at very high fields.

Velocity deduced from pulsed I–V measurements with short 10–25 ns voltage pulses in a 50 Ω environment is about 3.1×10^7 cm/s at 140 kV/cm at room temperature (see Figure 9.35). However, when the pulse width is increased to 200 ns, the velocity reaches only about 10^7 cm/s at a field of 190 kV/cm, most likely due to hot phonon effects although heating effects should also be considered.

9.10.2.1 Phonon Decay Channels and Decay Time

Optical phonons decay into lower energy optical and/or acoustic modes but energy and momentum conservation must be obeyed. Prior to the advent of phonon lifetime measurements through fluctuations, four possible channels for such a

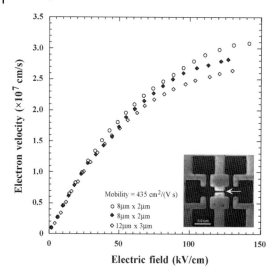

Figure 9.35 Experimentally determined velocity–field characteristics using pulsed I–V measurement using 10 ns input pulses in three different AlGaN/GaN samples. The inset shows the top view of the pattern used for the measurements whose active areas (indicated by the arrow) were 8 μm × 2 μm (open circles), 8 μm × 2 μm (closed diamonds), and 13 μm × 3 μm (open diamonds). When the pulse width is increased to 200 ns, the velocity reaches only 1×10^7 cm/s possibly due to hot phonon effects. Courtesy of J.M. Barker and Dave K. Ferry.

reduction have been proposed. (i) Klemens proposed that the optical phonon decays into two acoustic phonons with opposite momenta (known as the Klemens channel). (ii) Ridley suggested the possibility of a zone-center longitudinal optical mode decaying into a transverse optical mode and a longitudinal acoustic mode. This has been generalized to involve a process whereby the optical mode decays into a lower branch optical mode and an acoustic mode, and this has been termed as the generalized Ridley channel. (iii) Vallée and Bogani proposed that an optical mode may decay into a lower mode of the same branch and an acoustic mode, and this channel will be referred to as the Vallée–Bogani channel. (iv) A zone-center optical mode may also decay into two lower branch optical modes.

The phonon dispersion curves along several high-symmetry directions along with the decay mechanisms for the dominant A_1(LO) phonon (solid and dashed arrows representing the anharmonic interaction via the "Ridley mechanism" LO → TO + LA) in bulk GaN are shown in Figure 9.36. As is expected from the fact that there are four atoms in the unit cell of GaN, there are a total of 12 phonon branches, 3 (low-energy) acoustic and 9 optical. Among these distinctions, the branches can be classified into longitudinal (vibrations in the direction of the phonon wave vector, q; one acoustic and three optical modes) and transverse (vibrations perpendicular to the direction of the wave vector; two acoustic and six optical modes). The acoustic modes are associated with Joule heat and their interaction with electrons dominates at very low fields. At higher fields, electron coupling with acoustic modes becomes

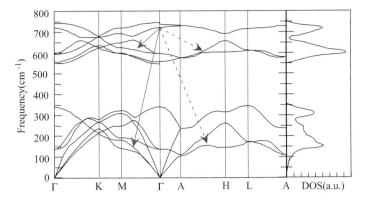

Figure 9.36 Phonon dispersion curves for GaN. The dominant decay routes are indicated by the arrows. Patterned after Barman and Srivastava.

unlikely and the dominant mechanism of electron scattering is scattering (emission or absorption) with the optical modes. It turns out that the electron–TO scattering rate is more than two orders lower than the electron–LO scattering rate, and as such, the LO phonon modes are most important when considering electron scattering mechanisms at moderate to high fields in GaN.

As discussed above, while there are a number of decay channels available to optical phonon modes, the highly ionic nature and large value of the energies of the optical phonon branches (compared to the acoustic branches) of GaN make all but the so-called Ridley mechanism inaccessible. Also, the Ridley process takes place in characteristic timescales much longer than those associated with LO phonon emission or electron–phonon scattering for that matter. Therefore, the density of LO phonons can build up and cause even more scattering between the electrons and these phonons. The TO phonon lifetime is known to be shorter than the LO mode, and additionally electron–TO scattering is known to occur at a rate two orders of magnitude lower than the electron–LO scattering; thus, once the LO phonon has decayed, in the framework of the Ridley mechanism, the effect of the daughter TO mode can be disregarded.

As phonons build up and exceed the equilibrium concentration of phonons, N_0, we refer to them as "hot." The time associated with the disintegration of these hot phonons into the short lifetime TO and more mobile LA phonons can be referred to as the hot phonon lifetime. Clearly, this lifetime is responsible for the density of the phonons that are built up in the channel. Phonon lifetime measurements deduced by time-resolved Raman technique versus the carrier concentration between 10^{16} and 2×10^{19} cm^{-3} indicate decreasing lifetime with increasing carrier concentration, which also include the fluctuation technique, as shown in Figure 9.37. Essentially, the lifetime of $A_1(LO)$ phonon mode decreases from 2.5 ps for 10^{16} cm^{-3} to about 0.35 ps for 10^{16} cm^{-3}, the latter being consistent with that deduced from the 2DEG. Soon, it will be shown that as the bulk electron density is increased further, the LO phonon lifetime goes through a minimum, corresponding to the resonance with plasmons, and then increases again.

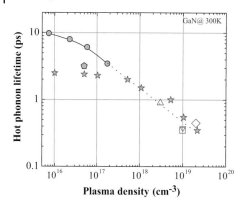

Figure 9.37 Phonon lifetime as a function of the photoexcited electron density, or the plasma density, in the range of slightly under 10^{16} cm^{-3} to slightly over 10^{19} cm^{-3}. The solid stars represent data obtained in bulk GaN by time-resolved Raman scattering, the pentagon and the filled circles also represent Raman data, the upward pointing triangle, diamond, and square represent the data obtained by the fluctuation technique, and the downward pointing triangle is for an AlGaN/GaN heterostructure. Courtesy of A. Matulionis.

The LO phonon decay mechanism is proving to be a rather complex problem now as its dependence on the electron concentration has been established experimentally as illustrated by Matulionis. The empirical $1/\sqrt{n}$ dependence, where n is the electron concentration, of the hot phonon lifetime is beginning to shed some much needed light on the possible (or improbable) LO phonon decay mechanisms. Because the Joule heat can only be removed by acoustic phonons, the acoustic phonons must be involved in the process and in the phonon decay mechanisms. The consideration of energy conservation alone allows the conversion of an LO phonon into four acoustic phonons, but this process is quite improbable as five particles must partake in this event, which is unlikely. Other phonon conversion schemes such as LO $\Rightarrow$ TO + TA/LA do not seem to explain fully the experimentally observed dependence of the hot phonon lifetime on the electron density. Screening of the conversion potential by electrons would lead to an increase in the lifetime with increasing electron density, which is contrary to the experimental observations. The data obtained from both the fluctuation technique and the Raman technique show that the hot phonon lifetime decreases with electron concentration as $1/\sqrt{n}$, which suggests a plasmon-assisted decay of LO phonons into acoustic phonons. Such a model would imply that coupling with plasmons should become more pronounced for carrier concentrations larger than 10^{17}–10^{18} cm^{-3} when the plasmon energy exceeds the acoustic phonon energy as shown in Figure 9.38. Increasing the electron density further to well above 10^{19} cm^{-3} causes the plasmon energy to increase past the LO phonon energy. Therefore, the reduction in the hot phonon lifetime observed with increasing electron density would most likely reverse the course and begin to increase again above 10^{19} cm^{-3}.

The term n above represents the bulk electron density. Estimating the average three-dimensional electron density (3D density) in an HFET channel simply by

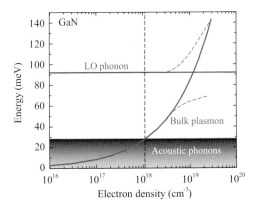

Figure 9.38 Representation of LO phonon, acoustic phonon, and plasmon energies versus the electron concentration in GaN indicating that plasmons would begin to play a role in determining the hot phonon lifetime when their energy exceeds that of the LA phonon energy and would be the dominant mechanism for electron concentrations of around 10^{19} cm^{-3}. Beyond this, the reduction in lifetime with increasing electron concentration would reverse its course and begin to increase with further increase in electron concentration. Courtesy of A. Matulionis.

dividing the sheet density by the effective width of the triangular quantum well at the Fermi energy, we see that densities on the order of 10^{19} cm^{-3} and higher are readily attainable in the GaN channel of an HFET. As such, one might expect the hot phonon lifetimes to be less than 0.35 ps in most 2DEGs. However, this often does not turn out to be the case. Experimental data on the hot phonon lifetime in GaN 2DEGs at low applied fields obtained mainly through the microwave noise technique are illustrated in Figure 9.39 (open symbols). Details of LO phonon lifetime

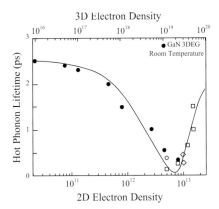

Figure 9.39 Measured low-field hot phonon lifetimes for bulk GaN (closed circles) and various GaN-based 2DEG channels (open symbols) as well as the (Al$_{0.1}$Ga$_{0.9}$N/GaN) camelback channel (closed star) presented in this work.

measurements, particularly those of the fluctuation technique, can be found in Chapter 3 of Morkoç (2009).

In a GaN quantum well, the notion of a bulk-like plasmon mode is not unrealistic if one takes into account that the plasmon wave numbers are of the order of the hot phonon wave numbers ($\sim 1 \times 10^{-9}\,\mathrm{m}^{-1}$) launched by hot electrons, while the well width is several nanometers.

Regardless, for an infinite electron plasma, the plasmon and phonon frequencies are equal at the electron density,

$$n_{res} = \omega_{LO}^2 \frac{m_e^* \varepsilon}{e^2}, \qquad (9.118)$$

where ω_{LO} is the LO phonon frequency and ε is the dielectric constant. The electron density, the resonance density, is estimated to be around $10^{19}\,\mathrm{cm}^{-3}$ in bulk GaN. The coupled mode becomes important near the resonance electron density, $\sim 10^{19}\,\mathrm{cm}^{-3}$, or alternatively near $\sim 5 \times 10^{12}\,\mathrm{cm}^{-2}$ for a typical GaN 2DEG channel if the average 3D density in the channel is estimated as the 2DEG density divided by the width of the triangular well at the Fermi energy. The fastest decay of hot phonons is expected near the plasmon–LO phonon resonance, and thus the lifetime is long at high densities where the plasmon energy exceeds the LO phonon energy. Despite the simplistic nature of the model, it appears to be in reasonable agreement with the nonmonotonic dependence of the hot phonon lifetime on the 2DEG density measured at low fields for various nitride heterostructures (Figure 9.39, open symbols and solid line).

9.10.2.2 Implications for FETs

Emission and absorption of copious quantities of LO phonons constitute the dominant scattering mechanism and thus have inevitable effect on the electron velocity. Note that the group velocity of LO phonons is insignificant, necessitating the need for their decay eventually to LA phonons with high velocity. Consistent with the LO phonon lifetime, the electron velocity deduced from FETs with varying electron concentrations' participation in current transport shows dependence on carrier concentration as shown in Figure 9.40. The cutoff frequency, and thus the carrier velocity, degrades with increasing electron concentration. Note that an electron concentration of $1.4 \times 10^{13}\,\mathrm{cm}^{-2}$ already exceeds the LO phonon–plasmon resonance beyond which the carrier velocity degrades offsetting the benefits of increased carrier concentration.

Balance equations for energy and drift velocity are very instructive in clearly indicating how the electron velocity is limited:

$$\frac{\partial E}{\partial t} = (-eF)v_d - \frac{\hbar \omega_{LO}}{\tau_E(T_E)}, \qquad (9.119)$$

$$\frac{\partial v_d}{\partial t} = \frac{eF}{m_e} - \frac{v_d}{\tau_m(T_E)}, \qquad (9.120)$$

where F is the electric field and $\tau_E(T_E)$ and $\tau_m(T_E)$ are the energy and momentum relaxation times, respectively, as a function of electron temperature. The steady-state

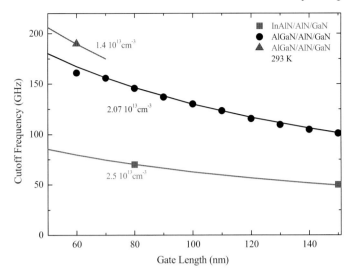

Figure 9.40 The cutoff frequency, a measure of average carrier velocity, versus the gate length in HFET with three different 2DEG concentrations. Note the degradation of the cutoff frequency and thus the carrier velocity with increasing concentration. Even 1.4×10^{13} cm^{-2} exceeds the LO phonon–plasmon resonance. Courtesy of A. Matulionis.

solution of these equations is given by

$$v_d = \sqrt{\frac{\hbar\omega_{LO}}{m_e}}\sqrt{\frac{\tau_m}{\tau_e}}. \tag{9.121}$$

These expressions are not a good match for a 2DEG because of the restriction to a single electron temperature. Degeneracy in a 2DEG precludes the single-temperature model in that relaxation times are not simple functions of temperature. It is clear that in a 2DEG LO phonon scattering plays two roles in the drift velocity: on the one hand, the momentum scattering tends to reduce v_d. This can be envisioned as electrons emitting and reabsorbing phonons at very high rates, which leads to no change in electron energy. This process, however, effectively randomizes the momentum of the electrons, which causes the drift velocity to decrease. On the other hand, the phonons pave the way for the electrons to dissipate their energy. Without phonon scattering, the electrons would ascend in the conduction band where either scattering to satellite valleys or the negative effective mass in the Γ valley would come into the picture. This would result in even more dramatic reduction in drift velocity. Ultimately, though, the feasibility of achieving negative differential resistance is questionable as the optical phonon scattering would preclude the electrons from reaching energies high enough to allow for scattering into upper valleys. The LO phonon scattering turns out to be necessary yet limits the electron drift velocity that can be achieved. The key turns out to lie in the "lifetime" associated with the conversion of the energy contained in the LO modes into modes that can carry the energy away from the channel. Clearly, given the situation, the shorter the LO phonon lifetime, the more beneficial it is to the electron velocity and just as importantly to heat removal from the hot electron region of the channel.

9.10.2.3 Heat Removal in View of Hot Phonons

The classical approach for heat removal is limited to the consideration of LA phonons (Joule heat) through a heat sink. In this approach, a substrate with a high thermal conductivity (such as SiC) is desirable. A careful consideration of phonon effects in GaN, as outlined above, indicates that this is true only for very low supplied powers imparted on the electrons ($>1\,\text{nW}/\text{electron}$). At higher supplied powers present in practically any GaN-based HFET, the electron temperature and thus electrons are capable of emitting LO phonons.

Phonons can be thought of as ordinary boson "gas" of particles, and as such, they have a distribution congruent with a boson gas heated to a temperature T_{ph}:

$$N_{\text{ph}} = \frac{1}{e^{(\hbar\omega/k_B T_{\text{ph}})} - 1}. \tag{9.122}$$

Considering that energy is conserved, meaning supplied energy is equal to dissipated energy, the energy dissipation is given by

$$P_d = \frac{\hbar\omega}{\tau_{\text{sp}}}(1 + N_{\text{ph}})p_- - \frac{\hbar\omega}{\tau_{\text{abs}}} N_{\text{ph}} p_+, \tag{9.123}$$

where τ_{sp} and τ_{abs} are the average times associated with spontaneous emission and absorption of LO phonons by hot electrons, respectively, and p_- and p_+ are the probabilities of finding a hot electron ready to emit ($-$) or absorb ($+$) an LO phonon, respectively:

$$p_{\pm} = \frac{1}{n_{\text{2D}}} \int D(E) f(E) [1 - f(E \pm \hbar\omega_{\text{LO}}) dE], \tag{9.124}$$

where E is electron energy, $f(E)$ is the temperature-dependent Fermi function, $D(E)$ is the density of states, and n_{2D} is the density of the 2DEG. Thus, the power dissipated by a distribution of hot phonons at arbitrary temperatures can be plotted and the phonon temperatures as a function of the (measured) electron temperatures can be graphically obtained (Figure 9.41). The intersection points in Figure 9.41 show that the phonon and electron temperatures are very close to each another (Figure 9.42). Thus, the hot electrons and hot phonons constitute an isolated subsystem (electron interaction with acoustic modes is very weak), and the only avenue to drain energy from the subsystem is through LO mode decay, as schematically illustrated in Figure 9.43.

Reiterating, once an electron emits an LO phonon, the extraordinarily high Fröhlich interaction in GaN means that the likelihood of reabsorption of the LO phonon is very high. Thus, if the emission and reabsorption of LO phonons take place before the LO phonon decay, the energy of the electron would be largely unchanged while the momentum is effectively randomized. It is the slowness of this LO → LA conversion process that gives rise to the so-called phonon bottleneck in GaN, which ultimately limits the performance of the HFET in terms of electron

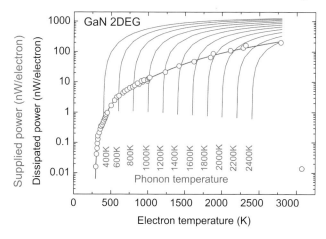

Figure 9.41 Solid lines: power dissipated by hot phonons at a number of phonon temperatures. Symbols and curve: experimental electron temperature as a function of supplied power. The intersection of the curves gives hot phonon temperature for a given hot electron temperature. Courtesy of A. Matulionis.

velocities that can be achieved, and in fact is intimately linked to the reliability of GaN-based HFET devices, which will be discussed.

A long LO phonon lifetime leads to a larger buildup of hot phonons in the channel, and as one would expect, it causes more scattering adversely affecting the overall drift velocity. Figure 9.44 shows the results from a Monte Carlo simulation of

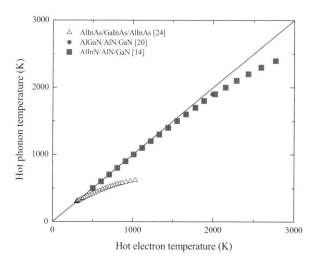

Figure 9.42 Hot electron temperature and hot phonon temperature. The solid line represents that hot phonon temperature is equal to hot electron temperature. It is clear that in GaN channels the hot phonon and hot electron temperatures are nearly equal. Courtesy of A. Matulionis.

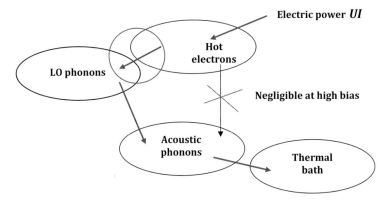

Figure 9.43 Schematic of the dissipation of heat in GaN at high fields. The only means of transferring energy out of the hot electron/hot phonon subsystem is through the hot phonon decay into acoustic modes. Note that if the hot electron and hot phonon temperatures were equal, no energy would be dissipated. Courtesy of A. Matulionis.

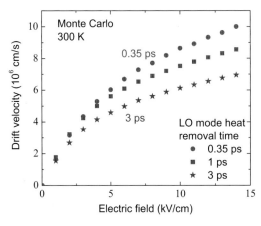

Figure 9.44 Monte Carlo simulation of the electron drift velocity (vertical axis, $\times 10^7$ cm/s) versus applied electric field under the assumption of various LO phonon lifetimes. Courtesy of A. Matulionis.

the drift velocity in an AlGaN/GaN channel versus the hot phonon lifetime. Clearly, a short hot phonon lifetime is desirable for HFETs so that the drift velocity, v_{sat}, can be maximized.

Hot phonons also impact the reliability and stability of HFETs. The decay of hot phonons is the fundamental bottleneck for transferring heat out of the channel of the FET and the fundamental reason for reduced average carrier velocities in GaN channels in HFET devices. This gives rise to a reliability problem as the heat buildup can cause the formation of defects, particularly when some defects are already present. The question then is what is the value of the hot phonon lifetime and how one might attempt to design a device in such a way so as to minimize it.

9.10.2.4 Tuning of the Hot Phonon Lifetime

Ample experimental evidence is provided in this chapter as well as in the literature that the minimum LO phonon lifetime occurs at the resonance of plasmon frequency and LO phonon frequency in GaN. The plasmon frequency is determined by the bulk electron concentration and therefore the vertical distribution of the electrons within a 2DEG channel affects this frequency. This distribution can be changed by the design of the HFET channel or by gate bias or drain bias as the channel field widens the electron distribution. Among the design feature is the dual-channel FET with the so-called Camelback electron distribution. In this approach, a dual channel is formed, for example, by low mole fraction $Al_{0.1}Ga_{0.9}N$ (juxtaposed to the AlN spacer layer) and GaN in place of the GaN channel utilized to spatially spread the electron distribution. This reduces the peak bulk density for a given integrated 2DEG density. Alternatively, the peak bulk density can be kept at the resonance while increasing the integrated 2DEG density. This approach allows one to increase the 2DEG density without the deleterious effect of longer LO phonon lifetimes.

Let us first discuss tuning by the lateral field in the channel for which ungated structures can be used as well as tuning with the gate bias in structures with gates. The hot phonon lifetime controlled by changing the power applied to a gateless channel of an InAlN/AlN/GaN HFET structure is shown in Figure 9.45. This approach allows one to probe the hot phonon lifetime versus sheet density because the electron temperature increases with the increase in power applied to the channel, which in turn causes the wavefunction to spread over a wider spatial range, resulting in a decrease of the value of the bulk carrier density (Figure 9.46). The data associated with the sample having a 2DEG density of $8 \times 10^{12}\,cm^{-2}$ in Figure 9.45 clearly show that with increasing supplied power (the electric field), the

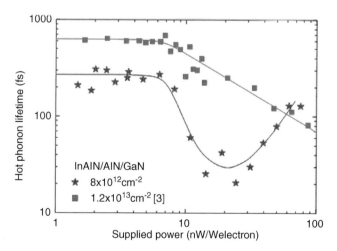

Figure 9.45 Hot phonon lifetime versus applied power for two InAlN/AlN/GaN channels demonstrating the effect of the phonon–plasmon resonance. Courtesy of A. Matulionis.

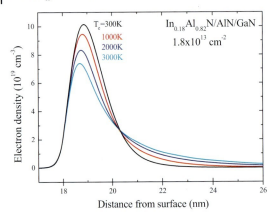

Figure 9.46 Electron density at various electron temperatures. Courtesy of A. Matulionis.

hot phonon lifetime goes through a minimum at the LO phonon–plasmon resonance. While the phonon lifetime continuously decreases with increased supplied power in the sample with a 2DEG density of $1.2 \times 10^{12}\,\mathrm{cm}^{-2}$, the resonance point cannot be reached because of sample degradation. Reduced bulk density explicitly implies reduced plasmon frequency. In other words, with supplied power, owing to spreading of spatial distributions of electrons, the 2DEG density at the LO phonon–plasmon resonance is increased as illustrated in Figure 9.47, in which the solid lines are a fit to the data using an empirical expression:

$$\tau_{LO} = a\left[1 + \frac{b}{\left(\sqrt{n} - \sqrt{n_{res}}\right)^2 + c}\right]^{-1}, \qquad (9.125)$$

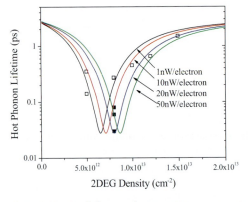

Figure 9.47 Fitted phonon–plasmon resonance curves using Equation 9.125 (solid lines) to the measured phonon lifetimes at low field (open squares), and for selected powers applied to the ungated device used for Figure 9.45. Courtesy of A. Matulionis.

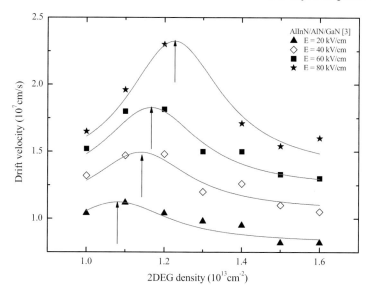

Figure 9.48 Dependence of drift velocity on 2DEG concentration with applied channel field as the parameter. Note the maxima occurring at the LO phonon–plasmon resonance as well as this resonance shifting to higher 2DEG concentration because of carrier heating (spatial carrier spread). Courtesy of A. Matulionis.

where n is the 2DEG density, n_{res} is the 2DEG density at the phonon–plasmon resonance, and a, b, and c are fitting parameters that control the value of the lifetime far from resonance, the "sharpness" of the resonance curve, and the value of the minimum lifetime, respectively. The effect of LO phonon lifetime on the drift velocity as viewed through the 2DEG density as well as the effect on the applied field (as illustrated in Figure 9.47 as well) is exhibited very clearly in Figure 9.48.

Let us now discuss the case with gated structures, actual HFETs, in which the electron density can be varied with applied gate bias. To determine the electron density dependence on the gate bias, capacitance–voltage (C–V) measurements on Schottky diodes, fabricated alongside the HFET devices, can be conducted for estimating the 2DEG density with the aid of

$$n = -2 \left\{ \varepsilon q A^2 \frac{d(1/C^2)}{dV} \right\}. \tag{9.126}$$

Here ε is the dielectric constant ($8.0\varepsilon_0$ for the InAlN barrier used), q is the electron charge, and A is the area of the diode. The electron concentration thus determined as a function of the depletion depth, $\varepsilon A/C$, is plotted in Figure 9.49. An estimate of the 2DEG density as a function of applied voltage can then be obtained by integrating the density versus depth profile with respect to the applied voltage, as shown in the inset of Figure 9.49. The 2DEG density changes linearly with the applied voltage, and the value at which the density approaches zero is estimated to be close to -4 V, which is consistent with the sister transistor pinch-off voltage.

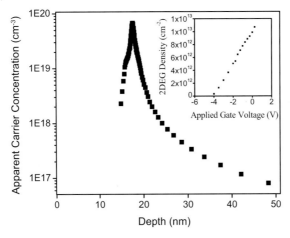

Figure 9.49 Electron density versus depth profile deduced from the capacitance–voltage measurements of a Schottky diode and (inset) integrated electron density for various biases. The values represent the approximate 2DEG density at various biases. Courtesy of J.H. Leach.

To find the effect of hot phonon lifetime on the electron density, the transit intrinsic time deduced from microwave s parameter measurements can be used. After determining $\tau_{int} + \tau_{RC}$ and $\tau_{int} + \tau_D$, from the total transit time (utilizing measured total transit time while the drain current is changed through and also when the channel voltage is changed, details of which are discussed in Chapter 3 of Morkoç (2009)), one can calculate the intrinsic transit times at a given bias. The intrinsic transit time is plotted in Figure 9.50 as a function of the 2DEG density determined from the

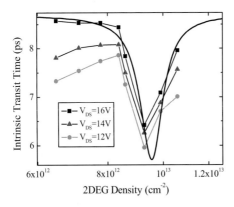

Figure 9.50 Intrinsic transit time as a function of the 2DEG density for three drain biases as well as the best fit using an optimal electron density of 9.5×10^{12} cm^{-2} (thick solid black line). The 2DEG density corresponding to the minimum in the intrinsic transit time is consistent with that corresponding to the minimum in the hot phonon lifetime. Courtesy of J.H. Leach.

C–V measurements (Figure 9.49, inset) for each of the drain biases used in the analysis. A clear minimum in the intrinsic transit time exists for 2DEG densities near 9.3×10^{12} cm^{-2} for each drain bias. The electron velocity corresponding to the minimum transit time is estimated to be $v = L_G/\tau_{\text{int}} \sim 1.75 \pm 0.1 \times 10^7$ cm/s. Also in the figure, we include a fitting after Equation 9.125 (thick solid line); the best fit occurs using an optimal electron density of 9.5×10^{12} cm^{-2}.

The minimum in the intrinsic transit time, or equivalently a maximum in the electron velocity, at a particular 2DEG density is consistent with tuning of the hot phonon lifetime discussed above. As illustrated in Figure 9.39, the hot phonon lifetime exhibits a minimum for a 2DEG density near 6.5×10^{12} cm^{-2} at low fields, which increases with supplied power.

The spatial electron distribution normal to the heterojunction can also be changed by modifying the channel region, for example, employing a combination of AlGaN/GaN as the active region as opposed to GaN only. Self-consistent solutions of the Schrödinger–Poisson equations for HFET structures with and without the Al$_{0.1}$Ga$_{0.9}$N interlayer in the form of the conduction band edge and electron distribution show the spreading of the wavefunction, as displayed in Figure 9.51. In dual-channel one, the conduction band edge changes from the typical quasitriangular well to a pair of quasitriangular wells due to the conduction band offsets at both the AlN/Al$_{0.1}$Ga$_{0.9}$N and Al$_{0.1}$Ga$_{0.9}$N/GaN interfaces. Second, despite the same total charge in the channel (1×10^{13} cm^{-2}), the electrons are effectively spread out in the camelback channel (dual channel) and result in a reduced peak 3D electron density. This is important as the 3D electron density is

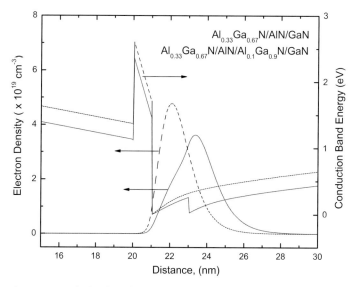

Figure 9.51 Calculated conduction band edge and electron 3D density for a standard (dashed lines) as well as the Al$_{0.1}$Ga$_{0.9}$N camelback channel structure (solid lines). The total numbers of electrons are equal (1×10^{13} cm^{-2}) in each channel.

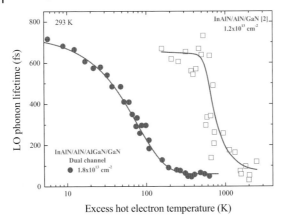

Figure 9.52 The hot phonon lifetime as a function of the excess noise temperature in the reference channel and the camelback channel (circles). Curves guide the eye. Courtesy of A. Matulionis.

believed to be the parameter responsible for hot phonon interaction with plasmons; once the 3D density is reduced, the hot electron–hot phonon system is closer to the plasmon–LO phonon resonance and exhibits shorter hot phonon lifetimes.

In terms of the LO phonon lifetimes, Figure 9.52 illustrates the LO phonon lifetime as a function of the excess noise temperature (electron temperature over the room temperature). The data for the camelback channel (circles) are compared with the control sample (squares). The decrease in LO phonon lifetime from the ∼600 fs down to the ∼60 fs for the reference channel (squares) is due to enhancement of the hot phonon decay when the LO phonon–plasmon resonance is approached upon electron heating. Because the equilibrium plasma frequency exceeds the LO phonon frequency considerably, considerable electron heating is needed to bring the system to resonance in the control channel. The camelback channel contains even higher 2DEG density but not 3D peak density, the profile of which is designed to make the plasma frequency close to the LO phonon frequency. As a result, the decrease in the hot phonon lifetime is observed at moderate hot electron noise temperatures (Figure 9.52, circles). The resonance can be tuned with the electron temperature or the supplied power at the electron densities higher than the resonance value. However, if the electron sheet density is too high, no resonance can be reached for reasonable applied power levels before catastrophic breakdown. On the other hand, below the breakdown, the shorter hot phonon lifetime in the camelback channel would ensure lower LO phonon temperature as compared to the control channel assuming both channels operating at the same power level or, alternatively, at the same LA phonon temperature (the lattice temperature). If the high-field electron transport is limited by the scattering of the LO phonons, the camelback structure is well suited to attain relatively higher electron drift velocity at a given high electric field.

9.11
HFET Degradation

Reliability and stability of any device are of considerable interest and GaN HFETs are no exception. In the realm of devices, the reliability is a measure of the consistency of the method of production of a population of devices. Often, the measure used is the expected lifetime of a population of devices, and in this view, reliability can be defined as the probability that a given device will perform within a specified window for a given period of time. The challenge, therefore, is achieving progressively longer lifetimes as well as process uniformity and repeatability so that an entire population would have the same failure rate. The genesis of the reliability problem in the form of a specific failure mechanism must be illuminated. In GaN, there are several degradation paths with varying roles in device failure, which are accessible under different biasing or environmental conditions. These pathways include high field driven, hot electron/hot phonon driven, and ambient temperature driven (metallurgical) mechanisms. A graphical account of failure mechanisms, excluding wear-out, in GaN-based HFETs is depicted in Figure 9.53.

Degradation experienced early on in the device operation is typically caused by less than stellar materials properties, immature device technology in terms of selection of metals, among others, and also due to effects such as hot electrons. In the case of GaN, hot electrons take a special meaning in that in the on mode hot electrons couple strongly to LO phonons. Due to the phonon bottleneck, discussed in the previous section, the negligible group velocity of LO phonons, and ensuing local heat, which cannot be removed efficiently, devices exhibit premature degradation. Note that the phonon bottleneck causes velocity degradation as well, again discussed

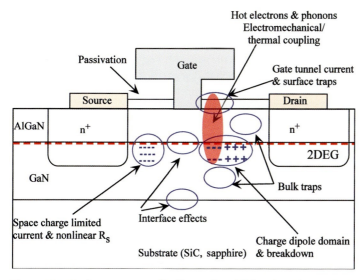

Figure 9.53 Mechanisms potentially affecting the reliability of GaN-based HFETs, excluding the long-term wear-out. In part, courtesy of R. Trew.

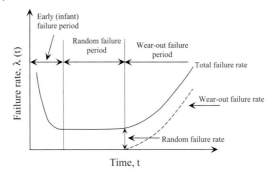

Figure 9.54 A schematic representation of the three failure regions, namely, the infant failure period (experienced very early on), the random failure period, and the wear-out failure period. Courtesy of A. Christou.

in the previous section. Beyond the aforementioned effects, the degradation in the long run is caused by metallurgical issues in general and thus is an activated process to the first extent. In the long-term degradation regime, an accelerated lifetime test can be done by operating the device at high temperatures. In fact, three-temperature (3T) lifetime testing is very popular in that extrapolation to room temperature would determine the lifetime. When performed over many devices, this technique can be used to determine the mean time to failure.

Although great strides have been made, to the extent that some HFET manufacturers boast device lifetimes (expressed as a mean time to failure) of $>10^7$ h (measured under accelerated DC and RF stress tests), a major problem still to be addressed is the vast variation among HFETs on the same wafer. A broader question perhaps entails the viability of the 3T test itself, which is solely used in many instances, since the extracted activation energies vary widely from laboratory to laboratory under both DC and RF stresses.

To gain an appreciation of the standard treatment and attitude toward reliability, the all too common GaAs-based approach is depicted schematically in Figure 9.54. Three regimes of failure termed infant, random, and wear-out can be noted. Some devices suffer the early (infant) failure, caused by failures occurring very early in the operating life of a device and addressed commercially by a "burn-in" procedure, which screens a portion or all of the population of devices, to induce infant failure in the factory, not in the field. This is followed by random failure, which generally occurs at a relatively constant rate. Finally, devices enter the unavoidable realm of wear-out failure. The goal would be to eliminate the infant and random failures altogether and reduce the wear-out failure rate to the extent possible.

As alluded to earlier, a highly popular procedure to probe the reliability is the accelerated life test. This is typically performed at three different but elevated temperatures in which cumulative failure rates are recorded and the mean time to failure for each of the three temperatures is established. The failure is defined as, for example, a reduction in the drain current by 10% for DC-based reliability test. Similar standards for RF performance can be set as well. The MTTF values thus

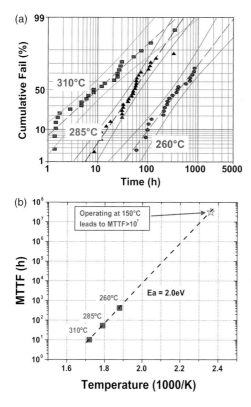

Figure 9.55 (a) 3T lifetime test data showing the cumulative failure rates versus time obtained at three different temperatures, namely, 180, 285, and 310 °C. The σ values for 180 and 285 °C are approximately 1 but that at 310 °C is 1.5, which might imply contribution by infant failure. (b) Arrhenius plot of the MTTF determined at three different temperatures, 180, 285, and 310 °C, which lead to an activation energy of about 2 eV and an extrapolated MTTF value greater than 10^7 h at 150 °C. Courtesy of Dr. A. Hanson (Nitronex).

determined are then converted to a single Arrhenius plot wherein the MTTF values are plotted in log scale as a function of inverse temperature. Extrapolation to a given temperature, namely, the expected operating junction temperature, then determines the expected MTTF for an HFET operating at that temperature, or more applicably that channel temperature, which is always higher than the case temperature (Figure 9.55).

Although widely used, the 3T method is limited in that the peak junction temperature at the heterointerface should be measured as opposed to the surface temperature, which can be difficult to access, particularly when a field plate is used that obstructs the optical access. More fundamentally, the assumption of the 3T test is that the channel temperature (as opposed to the hot electron temperature and/or hot phonon temperature) is dominant and furthermore that the failure mechanism follows an Arrhenius-type dependence, which may not be valid in that in some cases the increased temperature would reduce the rate of degradation as the electron mean free path is reduced. Widely varying activation energies reported in the literature, ranging from about 1 to 2.47 eV, prove this assertion to a great extent. In spite of its shortcomings, accelerated stress tests do provide a means to determine the lifetime of a population without having to operate devices for unreasonably long periods of time, assuming that the increased test temperatures constitute well-controlled degradation wherein failure mechanisms applicable to more typical case temperatures are the same as those that cause degradation during the accelerated life temperature tests.

Meaningful tests are those wherein the devices operate very close to the actual system working conditions (typically under RF bias), so that reliability predictions can be more accurate and representative. Furthermore, it would be prudent not to limit the characterization to classical transistor parameters (drain saturation current I_{DSS}, pinch-off voltage V_p, transconductance g_m, etc.) but to include other parameters such as parasitic resistances, gate diode characteristics, and so on, which can help to accurately identify the actual failure mechanisms. In addition, environmental testing on packaged devices to ascertain the product's robustness to vibration, moisture, shock (i.e., electrostatic discharge), pressure, and so on would be imperative prior to use in real applications.

One should note that meaningful reliability values cannot, in principle, be reduced by a single measure such as an MTTF. To illustrate this very point, a plot of reliability versus time for three sets of mock data with identical MTTF (arbitrarily chosen to be 10^6 h) but different values of σ (the standard deviation of the distribution function) is presented in Figure 9.56. One assumes the failure to follow a log-normal probability distribution function (PDF), which is typically used when fitting measured failure rates of actual devices in the "normal" (random wear-out period) phase. Although the sets of devices presented in Figure 9.56 have identical MTTF, the expected percentages of failed devices range from 99.98 to 70.4% after

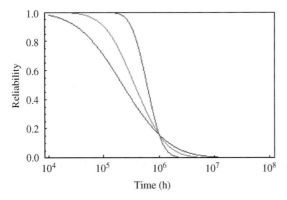

Figure 9.56 Simulated reliability versus time curves for three sets of mock data with failures following log-normal probability distribution functions, each of which has an MTTF of 10^6. Despite having identical MTTF, the three sets of data illustrate that the expected percentages of failed devices after 10^5 h are 99.98, 90.4, and 70.4% for $\sigma = 0.5$, 1, and 1.5, respectively. Courtesy of J.H. Leach.

operating for 10^5 h when σ increases from 0.5 to 1.5. Along this line, other metrics such as T1% or T10% could be more useful (which would be the times when 1 or 10% of the population fails, respectively). Furthermore, the assumption of a lognormal distribution holds only when the failure rate is *constant* in time, meaning the random failure period indicated by the "bottom of the bathtub" in Figure 9.54, which in fact may have nonzero slope. The overall bathtub including infant failure and wear-out phases is generally modeled using Weibull analysis (see Christou (1992)).

While the lack of a standard for GaN-based HFETs makes for difficult comparison of raw MTTF values given by manufacturers, phenomena such as "sudden degradation" or the slow initial degradation further confound the issue. The sudden degradation phenomenon refers to the case where drain current under pinch-off conditions suddenly increases, usually within the first couple of hours of stress, and can be present or absent in devices on the same wafer. Slow initial degradation depicts the cases wherein degradation tends to occur more strongly in, for example, the first 10 h of stress followed by relatively little degradation in the subsequent hundreds of hours of stress. A burn-in treatment could be used to screen devices in either case. The sudden degradation phenomenon has been correlated to higher degrees of initial current collapse as well as higher initial gate leakage current.

In terms of insight into the physical mechanisms causing degradation, two primary classes of degradation mechanisms can be identified: high field driven with the associated hot electron/hot phonon, strain, and thermally driven (metallurgical). At high fields, hot electrons and imbricated hot phonons nucleate degradation at the drain end of the gate where the field is the highest. Field plates mitigate this to some extent but the mitigation disappears if the drain bias is increased further owing to the presence of a field plate. Strain effects coupled with the piezoelectric nature of GaN are of importance as well, a matter that is somewhat muted if the channel is aligned along the nonpolar *a*- and *m*-directions.

In general, regardless of device, strain has a role in the reliability, particularly in terms of RF output power in HFET devices. If the barrier to relaxation is very high, the strain effect can be muted to some extent. However, in GaN HFETs it has been reported that degradation (in the framework of slumping I_{Dmax}) as a function of barrier thickness (and therefore strain) decreased in an $Al_{0.32}Ga N_{0.68}N$ barrier HFET when the thickness of the barrier is decreased from 26 to 13.8 nm. Furthermore, variation in V_{pinch}, which is attributable to trap generation near and under the gate, has been shown to be more pronounced after long-term RF stress in devices with higher Al mole fraction in the barrier. Strain relaxation of AlGaN/(AlN)/GaN through the generation of dislocations can therefore be construed as a source of degradation. Along these lines, the so-called inverse piezoelectric effect has been proposed as a mechanism of degradation of operating devices as well. The effect has to do with the vertical field (due to potential difference between the gate and the channel) and thus would increase the strain in the already tensile-strained AlGaN barrier, an effect confirmed by micro-Raman spectroscopy. The result is increased gate leakage and access resistances as well as decreased I_{Dmax}. TEM investigations following stress have indicated structural degradation at the drain side of the gate where the above-mentioned strain effects increase, but so do the hot electron and hot phonon effects. One question to mull over is the precise role of the gate current in this degradation since electrons that might tunnel out of the gate during application of this high V_{DG} could have kinetic energy sufficient to damage the semiconductor surface.

Thermally driven metallurgical degradation mechanisms, for example, metal–semiconductor diffusion, phase changes in metal stacks, and electromigration within the metal, are sources of potential permanent degradation. These are long-term, thermally activated, permanent degradation mechanisms and can be considered part of the "wear-out" portion of the reliability curve. Electromigration of the gate electrode metal, impurity activation, and contact diffusion effects are fairly well understood, particularly in GaAs-based devices. However, GaN-based devices push the metallization technology to its limit causing some metallurgical changes, particularly under prolonged high-current or high-temperature operation. Ti/Al-based metallization is commonly employed for ohmic contacts. The Al in the contacts is susceptible to oxidation and cracking; therefore, top metal layers of Ti/Au, Ni/Au, Pt/Au, Pd/Au, Mo/Au, Ta/Au, Ir/Au, Nb/Au, TaN, TiN, ZrN, TiB_2, CrB_2, and W_2B_5 have been used to varying degrees of success. In addition, the thicknesses and ratios of various components within the same stack are important. Furthermore, an n^+ top layer atop the AlGaN barrier can be employed to not only reduce the contact resistance but also ensure uniform current distribution, which would make for a more reliable contact.

The reliability investigations of GaN FETs have been relatively less developed in comparison to the GaAs counterparts, not to mention the large current levels/fields involved. As a result, the gate and drain/source metallurgy undergoes changes during sustained operation including phase change. Degradation of this kind is progressive and not reversible, making it imperative for this process to be understood well and incorporated into the physics-based reliability models as well as finding technological improvements. Reports on thermal stability of ohmic contacts are difficult to compare because of parametric dispersion used for thermal stress to ascertain the quality of their ohmic contact metallization, but long-term thermal stability tests (representing aging of the device) have been demonstrated with some success in terms of minimal to no change in the contact resistivity of the ohmic metals in some systems.

As in the case of GaAs-based FETs, the extent of electromigration is dependent on factors such as conductor line properties and inhomogeneities as well as structural features of the conductor layout. Naturally, electromigration must be studied noting that the magnitude of defect transfer depends superlinearly on current density, which poses formidable challenges for GaN-based FETs. High temperatures, particularly applicable to gate and drain metallization, cause mass transport facilitated by short-scale diffusion associated with defects, such as dislocations, grain boundaries, interphase boundaries, and/or external surfaces. It is therefore imperative that one studies and understands fully the mechanisms for electromigration and applies that knowledge to the reliability model.

Schottky gate contact degradation is to large extent a thermally driven process. Metallization schemes such as Ni/Au, Mo/Au, Pt/Au, Ni/Ti/Au, Pt/Ti/Au,

ZrB$_2$/Ti/Au, TiB$_2$/Ti/Au, CrB$_2$/Ti/Au, and W$_2$B/Ti/Au have been used with varying degrees of success. Pt–Au interdiffusion is known to occur at temperatures as low as 200 °C. Although the precise role of the gate leakage is unclear, it has been suggested that a silicide can form in Ni-based Schottky contacts and SiN$_x$ passivation layers at the SiN$_x$–Ni interface. In addition, the nonuniform nature of the silicide results in localized leakage conduction paths, which results in current crowding and gate electrode degradation, an increased instance of the "sudden degradation" phenomenon, and much lower device lifetimes. When subjected to DC stress tests at $T = 200\,°C$ and $V_D = 28\,V$, physical analysis using scanning tunneling electron microscopy (STEM) showed that a thin interfacial layer can form between the gate and the AlGaN barrier with ensuing consumption of/diffusion into the AlGaN barrier. The gate leakage should furthermore be avoided as electrons flowing from the gate could be trapped on the surface or within the underlying layer resulting in "virtual gating" that, among others, increases the effective gate length, and also because leakage would reduce the efficiency and gain, as input power would increase with no coordinate increase in output power. A schematic description of possible electron flow associated with gate leakage is shown in Figure 9.57 with deleterious effects.

As alluded to earlier, the existing defects within the device structure and on its surface play a key role in the reliability of devices. Indeed, the surface states provide a pathway by which electrons tunneling out of the gate under high fields can migrate across the surface and contribute to virtual gating. In addition, gate leakage has been shown to increase in a discrete, irreversible manner when devices are stressed in off-state conditions, which can be correlated to the formation of pits on the device surface, hypothesized to form at the top of conductive threading dislocations. Consequently, importance of minimizing threading dislocations to the extent possible cannot be overstated. In fact, dislocation density can be correlated with accelerated degradation in terms of reduction in drain current and increased gate and drain lags.

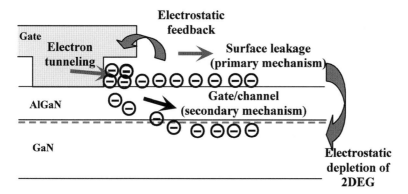

Figure 9.57 Possible gate leakage current paths that contribute to device instability. "Normal" gate leakage produces surface conduction, and at a critical electric field conducting path to channel is formed. In part, courtesy of R. Trew.

Degradation arising from hot electrons and hot phonons plagues GaN in that electrons are imparted with enough kinetic energy to cause physical damage to the crystal in the buffer or barrier layers. The effect of hot electrons can be differentiated from that of the hot phonons by stressing the device under high fields but low electron density. The ensuing degradation is not of the activated process and thus not thermally driven as expected. Degradation studies in the off state as well as at elevated temperatures have been undertaken in HFETs fabricated on bulk GaN substrates. Generation of traps spatially located in both the intrinsic and extrinsic (outside the active gate area) HFET regions was found to be most pronounced for off-state bias stress performed at room/case temperature. Increasing the base plate temperature up to 150 °C decreased trap generation underneath the gate.

Hot electrons cause new traps and manifest themselves as increased gate lag as well as reduced I_{Dmax} and transconductance. Simulations placing traps both in the buffer and on the surface as well as in the buffer, the barrier, and on the surface are sufficient to replicate the experimental degradation. In addition, threshold voltage shifts are typically observed after on-state biasing, which are attributable to some metallurgical change at the gate/semiconductor interface (due to heating), but could also be explained in terms of trapped electrons in the barrier layer under the gate. In any case, it is likely that at least some portion of the degradation that is attributed to hot electrons is in fact dovetailed with hot phonons when the drain current is large (discussed below), which in any case cannot be decoupled, since the ionic nature of GaN yields very strong electron–phonon coupling (perhaps some 30 times stronger compared to that in GaAs). This constitutes a very important source of physical degradation and, as will be shown below, an important impediment to achieving the optimum performance from devices.

9.11.1
Gated Structures: Reliability

To restate, nonequilibrium (hot) optical phonons have very low group velocity and as such tend to remain localized to the region where they are initially emitted. One can imagine that such a localized high density of phonons is likely a place where actual crystalline defects may be formed. Subsequently, the formation of defects would cause observable changes in the performance of the transistor, which has been observed in InAlN-based FETs subjected to high field stress ($V_D = 20$ V) in the dark at room temperature for 20 h. Parameters such as the maximum drain current, peak transconductance, and channel access resistances were monitored to quantify degradation. In addition, the gate leakage current was monitored as well during the stress at a fixed electron density, as a function of the total charge passed through the drain and the gate electrodes.

Figure 9.58 shows the change in the maximum drain current for the case where charge of 1500 mAh/mm is passed through the drain. Note that the applied gate voltage is converted to the 2DEG density at each voltage, obtained from a gated Hall bar measurement. The stars in Figure 9.58 show the same but for reduced drain voltage so that the drain–gate bias ($V_{\text{DG}} = 24$ V) is maintained in order to exclude

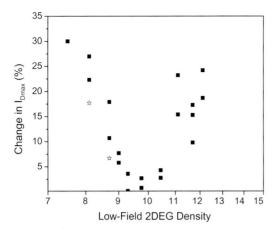

Figure 9.58 Change in maximum drain current after subjecting devices to high-field electrical stress. The change is given for devices that have passed 1500 mAh/mm of charge. The electron density is controlled by the gate bias. The stars represent devices that were stressed at a reduced drain voltage so that the devices were subjected to $V_{DG} = 24$ V, which is the same as that employed for the devices stressed with 2DEG density $\sim 10.5 \times 10^{12}$ cm^{-2}. Courtesy of J. H. Leach.

possible degradation due to high V_{DG} for devices subjected to high negative gate bias, corresponding to electron densities below 9×10^{12} cm^{-3}. The degradation rate exhibits a minimum at electron densities of around 10^{13} cm^{-2}, which is strikingly similar to that of the LO phonon lifetime (Figure 9.39). This is despite the fact that at lower sheet densities, the devices are subjected to lower power densities (and therefore channel temperatures) and additionally to comparable lateral fields still degrade at higher rates. This is consistent with hot phonon-induced degradation.

Additional support for hot phonon-induced degradation comes from low-frequency noise measurements that are particularly useful to address the mobility- and trap-related fluctuations in an HFET channel and its vicinity. This may pave the way to showing how and where the trap generation occurs in the HFET structures. For this particular study, two representative InAlN/AlN/GaN HFETs that have been stressed employing a 2DEG density at the plasmon–LO phonon resonance condition and off the plasmon–LO phonon resonance condition (Figure 9.59b) have been used. Consistent with the above-mentioned investigation, the off-resonance stress causes some gate lag (open circles) in addition to the severe drain current drop compared with the devices stressed at resonant 2DEG density condition (closed diamonds). One can again argue that the hot phonon buildup under the off-resonance bias conditions causes both the drain–gate access region degradation and trap generation in the barrier and buffer due to the thermally isolated subsystem of hot electrons and hot phonons.

The low-frequency phase noise measured for each device relative to a 4 GHz carrier signal using an Agilent 5505 test set is shown in Figure 9.60 in normalized power spectral density (PSD) units. The stress condition corresponding to the

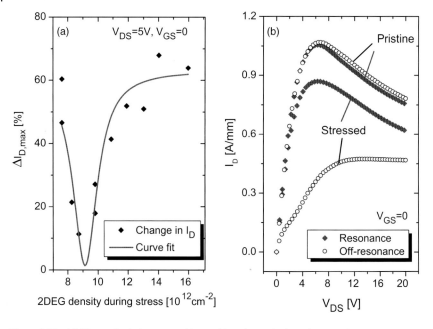

Figure 9.59 (a) Change in drain current ΔI_D due to the stress at $V_{DS} = 20\,V$ versus the 2DEG density during the stress and the Lorentzian fit; the minimum degradation occurs at a sheet density of ~$9.2 \times 10^{12}\,cm^{-2}$ corresponding to a drain current density of 0.55 A/mm. (b) A representative I_D versus V_{DS} graph measured at zero gate bias for two HFETs before and after the stress at $V_{DS} = 20\,V$ when the gate was biased near the hot phonon–plasmon resonance (closed diamonds) and at the off-resonance (open circles). The percentage of drain current degradation can be estimated from the ratio of the current after the stress (lower curve) and the value before the stress (corresponding upper curve). Courtesy of C. Kayis.

resonant 2DEG density ($V_{GS} = -4\,V$) yielded a 12 dB increase in the noise power, whereas the completely off-resonant stress conditions ($V_{GS} = -2\,V$ and $V_{GS} = -5.5\,V$) exhibited an increase of about 30 dB. The change in normalized PSD at 1 kHz as a function of the 2DEG density (under stress) is estimated for different gate biases (Figure 9.61). The trend in the normalized noise data strongly correlates with that of the ΔI_D data (Figure 9.59a).

Consistent with the values of 9.2×10^{12} and $9.3 \times 10^{12}\,cm^{-2}$ obtained from Figure 9.59a and current gain cutoff frequency measurements at high fields (discussed above already), respectively, the reduction in the RF power output (ΔP_{out}) at 4 GHz carrier frequency versus the 2DEG density during stress is shown in the inset of Figure 9.61. The data support the noise data; that is, the devices stressed at a given gate bias demonstrate the minimum power loss at the same optimal, resonant, 2DEG density. High-field values being slightly larger than the low-field resonant 2DEG density (~$6.5 \times 10^{12}\,cm^{-2}$) obtained for 2DEG channels from density dependence of the hot phonon lifetime estimated from the microwave noise technique

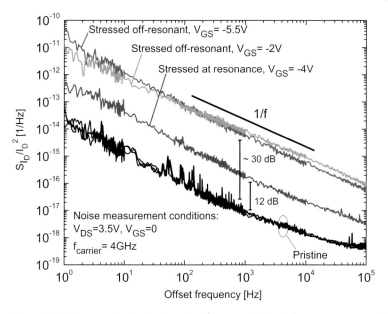

Figure 9.60 The normalized noise data (S_{ID}/I_D^2 versus offset frequency) for devices stressed with 2DEG densities near resonance and off-resonance. The power spectral density shows no peak of a single generation–recombination source for both pristine and stressed devices. Courtesy of C. Kayis.

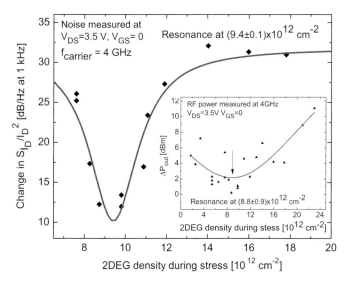

Figure 9.61 Increase in the noise measured at zero gate bias following a 7 h electrical stress at 20 V drain bias versus channel 2DEG density during stress controlled by the gate bias. The resonance observed at a 2DEG density of around 9.4×10^{12} cm^{-2} (fitting line) corresponds to minimum degradation. The inset shows the reduction in the RF output power (ΔP_{out}) at 4 GHz after stress with the fit guiding the eye. Output power was measured by feeding the device output into the spectrum analyzer. The arrow marks the resonant 2DEG density. Courtesy of C. Kayis.

have to do with hot electrons spread over a larger volume in a GaN channel, necessitating a higher 2DEG density to reach the resonance.

9.11.2
Reliability Tests

A typical procedure to ascertain lifetime of various types of devices is the accelerated life test to attain a quick assessment of projected lifetime. The critical issues for accelerated life testing are that they must provide estimates of device reliability in a time much shorter than that required to produce a significant number of failures under normal operating conditions. To be successful, the tests must stress the largest possible number of devices in a controlled manner without introducing artifacts (i.e., unrealistic failure modes). The accelerated tests normally adopted, which must definitely be augmented by RF and pulsed I–V tests, and the various combination thereof as they, among others, stress the gate by pushing to forward operating mode during part of the RF swing, in particular for power HFETs, are, without extensive description of each as they are beyond the scope of this chapter, as follows:

a) *High-temperature storage test* (HTS). Unbiased samples are stored at different temperatures, in order to accelerate thermally activated failure mechanisms, such as interdiffusion processes, occurring at the metal/semiconductor and semiconductor/semiconductor interfaces.

b) *High-temperature operating life test* (HTOL or HTOT). Samples are stored at different temperatures and biased in conditions similar to those experienced by the device during normal operations, aiming at studying the combined effects of thermal and electrical stresses. This approach in the form of the three-temperature plot is used to determine the MTTF Arrhenius plots that are very limited in scope and do not really represent in and by themselves the real operating conditions of the device. Note that they do serve albeit a limited purpose but an important one.

c) *High forward gate current test* (HFGC). The gate Schottky diode is forward biased in order to investigate the effects of high current densities (at high temperatures).

d) *High-temperature reverse-bias test* (HTRB). The gate Schottky diode is reverse biased close to breakdown voltage to observe the cumulative effects of high electric fields and temperatures.

e) *Temperature humidity bias test* (THB) and *highly accelerated stress test* (HAST). Samples are biased in a high-temperature and high relative humidity environment. To increase both temperature and relative humidity, a pressure cooker is employed in the HAST test. Usually, the gate diode is reverse biased to analyze the effects of humidity directly on the chip and the protection capability of the passivation layers and plastic packages.

Sometimes the accelerating factors are not kept constant during the test but are increased at defined times, giving rise to the so-called step stress test, which can be either thermal or electrical. This test can provide very quick information on the limiting conditions for the investigated technology and lead to, with carefully defined time intervals, reliability predictions.

Accelerated DC testing including tests at elevated temperatures does not gauge the device under operating conditions representative of the operating environment. A clear advantage of RF life tests over the above-mentioned tests is that the devices operate very close to the actual system working conditions, so that reliability predictions can be more accurate and representative. In particular, for high-power devices, large RF signals can drive the devices in electrical conditions not experienced during DC life testing (e.g., the gate voltage can be swung in the forward direction). However, it is not easy to control RF working conditions during life testing in that it is possible to introduce spurious failure mechanisms due to overstress, input/output impedance mismatching in particular at high temperature, and so on. Moreover, the failure mechanisms, which are enhanced by RF life testing, such as electromigration caused by forward gate conduction or degradation due to operation close to gate–drain breakdown, can be induced by proper DC tests such as HFGC or HTRB, which can be performed under well-controlled conditions.

For parameter monitoring, it is generally preferable to measure both DC and RF parameters, in spite of the fact that in some cases a clear correlation can be found between DC and RF degradations. In the case of DC parameters, it is, however, recommended not to limit the characterization to classical transistor parameters (drain saturation current I_{dss}, pinch–off voltage V_p, transconductance g_m, etc.) but to include the measurement of other parameters, such as parasitic resistances, gate diode characteristics, and so on, which can help to correctly identify the actual failure mechanisms.

In the thermal characterization methods, accurate evaluation of the channel temperature is needed to correctly accelerate the different failure mechanisms and to evaluate their activation energy. The channel temperature (T_{ch}) of an electronic device is conventionally described as the sum of the case temperature (T_{case}) and the product of the power dissipated (P_D) and the thermal resistance (R_{th}), that is, $T_{ch} = T_{case} + P_D R_{th}$. To evaluate R_{th} of microwave MESFETs, the electrical method based on the current–voltage forward $I(V)$ characteristics of the gate Schottky barrier diode is widely used because it enables $T_{ch}(\Delta V_{gs})$ to be evaluated through a calibration curve, which allows R_{th} to be obtained from the knowledge of $T_{ch}(\Delta V_{gs})$, T_{case}, and P_D. It should be noted that R_{th} is also a function of the temperature.

In operating conditions or during accelerated life tests, however, the power dissipated in the active device areas leads to a nonuniform increase in the device temperature. $T_{ch}(\Delta V_{gs})$ is therefore an unknown weighted average of the temperature distribution on the device and can be very inaccurate, in particular if a small high-temperature spot exists within the structure. The actual temperature distribution on the chip can be measured by liquid crystal techniques or directly observed by means of high lateral resolution IR thermography (IR near-field optical microscopy and micro-Raman spectroscopy mapping can also be used), all of which allow one to detect the thermal gradients caused by local differences in heat dissipation or by structural inhomogeneities.

9.12
HFETs for High-Power Switching

All electric and hybrid vehicles, locomotive trains and associated power distribution systems, and distributed power generation using alternative energy sources in the form of, among others, converters and power conditioning devices have catapulted efficient, compact, and long-lasting switching devices to the forefront of attention in this particular arena. The frequency and power levels vary with applications. In terms of use in industry, thyristors seem to be preferred for high-power and low-frequency applications, insulated gate bipolar transistors for medium-power and medium-frequency applications, and power MOSFETs for high-frequency applications.

At the heart of all the above-mentioned designs is a switching device, or a switch. A switch operates in the "on" or "off" state. The most coveted parameter in the "on" state is the "on" resistance (mΩ cm^2) that must be minimal to avoid losses and heating. In the "off" state, the leakage current must be as small as possible since a high voltage exists across the device, approaching and in some cases exceeding 1000 V. In the "on" state, the amount of current that the switch can carry is an important figure of merit as well. This is replaced with the hold voltage in the "off" state, which is approaching and greater than 1000 V as stated above. As such, the switching losses in the transistor, the HFET in the present case, must be very low, which, among others, is a function of materials properties and switching speed. Many semiconductor-based switches are employed, depending on application, performance, and cost, among which are Schottky rectifiers, pin rectifiers, various varieties of power MOSFETs, bipolar junction transistors, thyristors, and insulated gate bipolar transistors.

While Si-based switching devices have been and are dominant, the GaN-based HFET has shown significantly lower switching losses in conversion devices. Owing to heavy consumer applications, the cost is of prime concern, which provided the impetus for the successful development of GaN-based switch technology on Si(111) in conjunction with C doping of the buffer layer, in addition to using, for example, >6 µm thick buffer for quality reasons, for attaining high resistivity and thus large breakdown voltage. The C source is from the metalorganics used in the growth process, whose concentration in the solid can be manipulated by the chamber pressure in the organometallic vapor-phase epitaxy system used to grow the HFET structure. Higher C concentrations yield relatively larger hold voltages, albeit at the expense of reduced 2DEG concentration and thus larger on resistance, making it imperative to strike a balance. Carbon has also been shown to increase the dynamic on resistance for high-frequency switching (a discrepancy between the on resistance under DC conditions and that at high voltages and high switching frequencies), presumably due to electron trapping and slow detrapping in the buffer layers. A typical C concentration for reaching breakdown voltages of about 1000 V is about mid 10^{19} cm^{-3} for electrode spacings greater than approximately 10 µm. One caveat is that GaN on Si is under tensile strain upon cooling, which is notorious for cracking. Therefore, measures must be taken to avoid cracking through strain

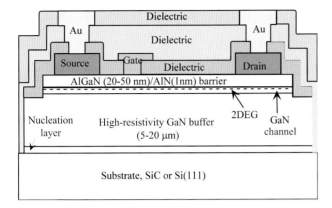

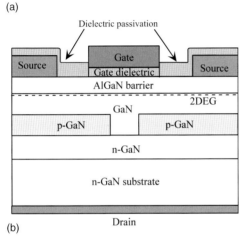

Figure 9.62 Schematic representation of (a) a planar GaN-based HFET (patterned after the Furukawa process) and (b) a vertical variety considered/used for switching applications. Although GaN substrate is shown for the vertical variety, the conducting Si(111) can also be used down below.

balancing approaches such as interspersed AlGaN layers whose thickness also must be optimized as to not cause cracking itself.

With a combination of increased buffer layer thickness (through improved quality) and appropriate selection of electrode spacing (8 μm or greater), breakdown voltages of about 2000 V in planar structures have already been reported. Schematics of a planar (lateral) HFET device and vertical variety for switching applications are shown in Figure 9.62. A top view of a finished switch capable of handling 100 A current with a hold voltage of 1000 V having a buffer layer thickness of 6 μm, a gate–drain spacing of 15 μm, and a total gate width of 340 mm is shown in Figure 9.63.

Critical points in the vertical device are as follows: The n–p junction of the blocking n–p–n structure on the drain side is reverse biased under blocking conditions while that on the source side is forward biased. Therefore, the hold

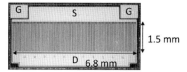

Figure 9.63 Top view of a finished switch capable of handling 100 A current with a hold voltage of 1000 V (15 μm gate–drain spacing and a total gate width of 340 mm). Courtesy of Dr. Ikeda (Furukawa Electric).

voltage will be determined by the properties of the reverse-biased junction that necessitates use of lightly doped n-layer underneath the p-layer in relation to the p-doping. In addition to the p-doping, the p-layer thickness must be large enough to avoid punch-through of this layer under reverse bias. The layer quality of the p-layer declines with increased p-doping, which also critically defines the quality of the n-layer on the top, placing a limit on the hole concentration that can be achieved.

The higher the breakdown voltage of a given material, the easier it is to achieve specific on resistance of a switching device for a given hold voltage as shown in Figure 9.64. Never to be forgotten naturally is the junction temperature, specifically how high a junction temperature can be tolerated. For a given power, this temperature is dependent on the material and specifics of the dimensional properties of the device layout and layer thicknesses as well as heat sinking. The hot phonon bottleneck is mitigated by operating at carrier densities that are at or near resonance in terms of the LO phonon energy and plasmon frequency. Unless the LO phonons decay to propagating LA phonons, the heat is localized and can cause the temperature of the junction where the field is high to be very high. If the phonon bottleneck effect is in the picture, the thermal conductivity of the material, which is

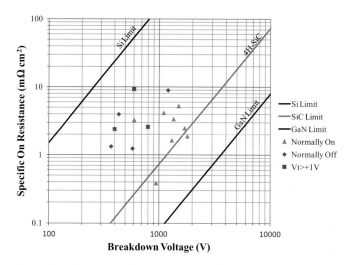

Figure 9.64 Specific on resistance for the candidate semiconductors versus the breakdown voltage, and selected achievements reported in the literature for GaN-based HFETs. Courtesy of J.H. Leach.

representative of LA phonon mode, is basically immaterial. Therefore, the limits presented in Figure 9.64 should be weighed together with the thermal issues. One should keep in mind that the higher the switching speed the larger the heat dissipated.

Normally off lateral devices are advantageous in that no gate voltage would be required for the off state. In the lateral device, the top portion of the AlGaN barrier can be replaced with p-type AlGaN in which case in the on state holes would also be injected into the channel helping to reduce the on resistance. Although holes would reduce the ultimate speed of the device, the practice would still have merit if the on resistance/speed trade-off is advantageous. Alternative schemes for achieving normally off devices include recess etching the region beneath the gate (prior to gate deposition) or implantation of F^- ions beneath the gate. Difficulties in the normally off approaches include crystal quality when p-type layers are employed, difficulties in maintaining low on resistance when using gate recess approaches, diffusion of F^- ions causing V_T shifts, and overall reliability concerns associated with implantation schemes. Despite these concerns, some impressive results have been reported, approaching those achieved by the normally on devices, as shown in Figure 9.64.

In terms of the materials properties, standard figures of merit for candidate materials are commonly used as tabulated in Table 9.3 for competing semiconductor materials. Various figures of merit as well as pertinent parameters are defined in a series of expressions given below.

Drift region conductivity along with its thickness (depletion width) that goes toward the on resistance is given by

$$\sigma_A = \frac{q\mu N_B}{W_m} = \frac{\varepsilon \mu E_M^3}{V_B^2}, \qquad W_m = \frac{2V_B}{E_M}. \qquad (9.127)$$

The numerator of the conductivity equation is commonly referred to as the high-voltage Baliga figure of merit. The on resistance per unit is, therefore, given by

$$R_{on} = \frac{4V_B^2}{\varepsilon \mu E_M^3}. \qquad (9.128)$$

Table 9.3 Figure of merits and pertinent materials properties for candidate semiconductors.

	Bandgap (eV)	Breakdown field, E_B (MV/cm)	Electron mobility, μ_e (cm²/(V s))	Relative dielectric constant, ε_s	Thermal conductivity (W/(cm K))	Baliga figure of merit (high frequency), $\mu_e E_C^2$	Baliga figure of merit (high voltage), $\varepsilon_s \mu_e E_C^3$
Si	1.12	0.25	1500	11.8	1.5	1	1
4H-SiC	3.26	3	1000	10	4.9	96	14
GaN	3.42	3.5	1500	9.6	2.3	163	22

The denominator of the on resistance equation is also, in the same concept, the same figure of merit. This figure of merit is associated with the conductivity losses in a unipolar device for relatively low frequencies.

The high-frequency Baliga figure of merit (BFOM) is given as

$$\text{BFOM} = \varepsilon_s \mu E_B^3 \sqrt{\frac{V_G}{4 V_G^3}}, \qquad (9.129)$$

and as the name implies is related to losses associated with unipolar devices when switching at high frequencies; that is, the losses are primarily associated with the switching itself.

The Johnson figure of merit (JFOM), which gives a measure of the power–frequency product, is expressed as

$$\text{JFOM} = \frac{E_B^2 v_s^2}{4\pi^2}. \qquad (9.130)$$

The Keys figure of merit (KFOM), which is associated with thermal limitations in switching devices, is defined as

$$\text{KFOM} = \kappa \left(\frac{c v_s}{4\pi \varepsilon_s} \right)^{1/2}. \qquad (9.131)$$

The breakdown voltage expression commonly used for Si, Ge, GaAs, and GaP is (that for GaN is given in Equation 9.73)

$$V_B = 60 \left(\frac{E_g}{1.1} \right)^{3/2} \left(\frac{N_B}{10^{16}} \right)^{-3/4}. \qquad (9.132)$$

Appendix 9.A. Sheet Charge Calculation in AlGaN/GaN Structures with AlN Interface Layer (AlGaN/AlN/GaN)

In nitride semiconductors with wurtzite phase, spontaneous and piezoelectric polarization effects are present, which must be taken into account in the balance of boundary condition at the interface ($z = d$). In this respect, the conservation of the normal component of the electrical displacement leads to (see Figure 9.A.1)

$$\varepsilon_1 \cdot F(z = d_+) - \varepsilon_2 \cdot F(z = d_-) = P_2 - P_1 = \Delta P. \qquad (9.A.1)$$

In region I characterized by the AlGaN layer, the Poisson equation can be written as

$$\varepsilon_1 \frac{dF_1}{dz} = q n_{sf}^+ \delta(z + t + w) + q N_D^+, \qquad (9.A.2)$$

Appendix 9.A. Sheet Charge Calculation in AlGaN/GaN Structures | 445

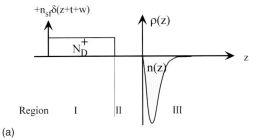

(a)

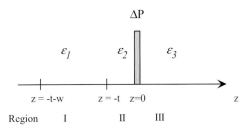

(b)

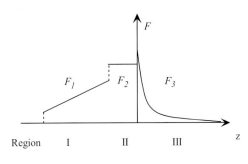

(c)

Figure 9.A.1 (a) Charge distribution and (b) electric field distribution in a c-plane Ga-polarity AlGaN/AlN/GaN heterostructure.

where n_{sf}^+ is the density of ionized surface donor states, an integration of which leads to the electric field in the same region as

$$F_1 = \frac{e}{\varepsilon_1} n_{sf}^+ + \frac{q}{\varepsilon_1} N_D^+ \cdot (z + t + w) + C, \quad (9.A.3)$$

with C being the integration constant. Doing the same for region II leads to

$$\varepsilon_2 \frac{dF_2}{dz} = 0 \quad \text{and} \quad F_2 = C\frac{\varepsilon_1}{\varepsilon_2} + \frac{q}{\varepsilon_2}(n_{sf}^+ + N_D^+ w) = C\frac{\varepsilon_1}{\varepsilon_2} + \frac{q}{\varepsilon_2} n_s,$$

$$\varepsilon_2 \frac{dF_2}{dz} = -qn(z) \quad \text{and} \quad F_2 = -\frac{q}{\varepsilon_2} \int n(z) dz = -\frac{m_C k_B T q}{\pi \varepsilon_2} \sum_l \int_d^z |\psi_1(z)|^2 \, dz \ln$$

$$\times \left[1 + \exp\left(\frac{E_F - E_l}{k_B T}\right)\right] + C'. \quad (9.A.4)$$

According to charge neutrality condition,

$$n_{sf}^+ + N_D^+ d = n_s \quad \text{with} \quad n_s = \int_{z=0}^{\infty} n(z)dz, \tag{9.A.5}$$

which is a constant as the doping level in AlN is taken to be zero. Again doing likewise for region III leads to

$$\varepsilon_3 \frac{dF_3}{dz} = -qn(z) \quad \text{and} \quad F_3 = -\frac{q}{\varepsilon_3}\int^z n(z)dz + C'$$

$$= -\frac{m_C kTq}{\pi \hbar^2 \varepsilon_3} \sum_l \int_0^z |\psi_1(x)|^2 \, dx \ln\left[1 + \exp\left(\frac{E_F - E_l}{kT}\right)\right] + C'. \tag{9.A.6}$$

The constant C' can determined noting that $F_3(z=0) = C'$ as

$$C' = \frac{C\varepsilon_1 + qn_s - \Delta P}{\varepsilon_3}.$$

Noting that $F_2(z=d) = C'$ (from $\varepsilon_1 \cdot F(z=d_+) - \varepsilon_2 \cdot F(z=d_-) = P_2 - P_1 = \Delta P$), we can determine C' by

$$C' = \frac{\Delta P + C\varepsilon_1 + qn_s}{\varepsilon_2}. \tag{9.A.7}$$

Note that for large values of z, $\int_0^z |\psi_1(x)|^2 \, dx = 1$. The constant C can be determined noting that the electric field in the bulk of GaN vanishes, that is, $F_3(z \to \infty) = 0$, as

$$C = -\Delta P/\varepsilon_1. \tag{9.A.8}$$

The electric fields in regions I, II, and III can then be expressed as

$$F_1 = \frac{q}{\varepsilon_1}(n_s + N_D^+ t) + \frac{\Delta P}{\varepsilon_1} + \frac{q}{\varepsilon_1} N_D^+ z, \quad F_2 = \frac{\Delta P}{\varepsilon_2} + \frac{e}{\varepsilon_2} n_s, \quad \text{and} \quad F_3$$

$$= -\frac{q}{\varepsilon_3}\int n(z)dz + \frac{qn_s}{\varepsilon_3}, \tag{9.A.9}$$

with $n_s = \int_{z=0}^{\infty} n(z)dz$.

Further Reading

Ambacher, O., Majewski, J., Miskys, C., Link, A., Hermann, M., Eickhoff, M., Stutzmann, M., Bernardini, F., Fiorentini, V., Tilak, V., Schaff, B., and Eastman, L.F. (2002) *J. Phys.: Condens. Matter*, **14**, 3399–3434.

Ando, T., Fowler, A.B., and Stern, F. (1982) *Rev. Mod. Phys.*, **54**, 437–672.

Bernardini, F. and Fiorentini, V. (2001) *Phys. Rev. B*, **64**, 085207.

Christou, A. (1992) *Reliability of GaAs Monolithic Integrated Circuits*,

John Wiley & Sons, Ltd, Chichester; Christou, A. (1995) *Reliability of GaAs Monolithic Integrated Circuits*, 2nd edn, John Wiley & Sons, Ltd, Chichester.

Fang, F.F. and Howard, W.E. (1966) *Phys. Rev. Lett.*, **16**, 797–799.

Morkoç, H. (2009) *Handbook on Nitride Semiconductors and Devices*, vol. 3, Wiley-VCH Verlag GmbH, Weinheim.

Sze, S.M. and Ng, K. (2007) *Physics of Semiconductor Devices*, 3rd edn, Wiley–Interscience.

Index

a

ab initio calculations 47, 58
absorption 196
– coefficient 194, 197, 246, 267, 273, 281, 293, 306, 307
– – spectral dependence of 198
– LO-phonon absorption 233, 416, 418
– measurement 197
– of nonequilibrium phonons 410
– rate 290–293
air–semiconductor interface 335
AlGaN alloy 13, 14, 326
– $Al_{0.15}Ga_{0.85}N$ 279
– AlGaN/(AlN)/GaN
– – strain relaxation of 431
– $Al_{0.1}Ga_{0.9}N$ camelback channel structure
– – conduction band edge and electron 3D density 425
– AlGaN EBL 238
– AlGaN/GaN channel, I–V characteristics 365
– $Al_{0.46}Ga_{0.54}N$/GaN DBR 327
– AlGaN/GaN heterostructure 414
– – band edge profile 359
– AlGaN/GaN HFETs 384
– – with AlN spacer 407
– – current gain cutoff frequency of 411
– AlGaN/GaN modulation-doped field effect transistor 351
– AlGaN/GaN samples, velocity–field characteristics 412
– AlGaN/GaN structures 383
– – with AlN interface layer 444–446
– – sheet charge calculation 444–446
– AlGaN/InGaN channel 365
– AlGaN layer 364, 383
– – tensile-strained 364
– AlGaN(p)/GaN(p)/AlGaN(n) double-heterojunction device 211

– bandgap *vs.* lattice parameter 16
– barrier thickness 384
– bowing parameter *b* 13
– calculated electron mobility 163
– cladding layers 278, 313
– compositional dependence of lattice parameters 13
– heterojunction 355
– parameters pertinent to mobility calculations 165
– waveguide layers 301
AlInN alloy 14, 15
– bowing parameter value 15
– calculated lattice parameter 15
– compositional dependence of bandgap 15
aluminum nitride (AlN)
– carrier transport 159–161
– experimental bulk modulus and elastic coefficients 11
– mobility, parameters used 134
– optical phonon energies and phonon deformation potentials 11
– spacer layer 421
– structural parameters 6
– thermal expansion 6
– waveguide layers 313
– Wurtzite 28, 29
– – effective masses and band parameters 29, 30
– – parameters related to optical and electrical properties 9, 10
– zincblende AlN, Luttinger parameters 30
amplifier
– classification 377
– efficiency of 376
– HFET amplifier classification and efficiency 374–378
– operating points and load lines 374

Nitride Semiconductor Devices: Fundamentals and Applications, First Edition. Hadis Morkoç.
© 2013 Wiley-VCH Verlag GmbH & Co. KGaA. Published 2013 by Wiley-VCH Verlag GmbH & Co. KGaA.

anisotropy 24, 27, 29, 31, 270, 297, 302, 325
antisite defects 66
Auger recombination 310
– coefficient 220
– rate 219
avalanching 188, 189, 379

b

Baliga figure of merit (BFOM) 443
– high-frequency 444
ballistic transport, probability 233
band alignment 177–179
– band discontinuities
–– in GaN system 178
–– in nitride semiconductors 178
– piezoelectric effect 179
– type I and type II alignment 177, 178
bandgap 30, 35, 51, 65, 78, 86, 159, 177, 183, 198, 209, 220, 222, 287, 349, 379, 392
– direct 28, 150, 194, 222
– energy 7, 9, 12, 150, 272, 301
– optical transitions 287
– wide 63, 64, 78, 210, 220, 267, 326
band-to-band transitions 198
– and efficiency 198–200
– rate of emission of photons 198
BE. *See* bound exciton (BE)
binding energy 70, 82, 85, 86, 204, 390
bipolar junction transistors 440
blocking n–p–n structure 441
Bohr–Sommerfeld quantization condition 167
Boltzmann distribution 346
– for nondegenerate semiconductors 155
Boltzmann equation 121, 126, 128
Boltzmann factors 198
Boltzmann's constant 90, 397
Boltzmann transport equation (BTE) 115, 117
Bose–Einstein distribution 347
bound exciton (BE)
– electronic structure 205
– recombination 204
– transitions 205
breakdown properties 379
breakdown voltage, defined 382
buffer–channel interface 386
BV_{off} values 382

c

Callen's effective ionic charge 130
capacitance–voltage (C–V) measurements 87–94, 423
– electron density *versus* depth profile 424
carrier–carrier scattering 146

carrier concentration 74–78
carrier scattering 117
– acoustic phonon scattering 120, 121
– alloy scattering 134–143
–– energy and mass dependence indices of mobility for 140, 141
–– energy and mass dependence indices of relaxation time 139
– deformation potential scattering 121–123
– for degenerate semiconductors 118
– dislocation scattering 134–143
– impurity scattering 118–120
– mobility as Matthiessen's rule 118
– numericalmethods 118
– optical phonon scattering 126–134
– piezoelectric scattering 124–126
– relaxation time approximation 117
– time evolution of distribution function 117
carrier transport 115
– in alloys 161–164
– in AlN 159–161
– in InN 158, 159
cavity mode dispersion 334
C exciton energy 24
charge neutrality equation 79
chromaticity color coordinates 251
chromaticity diagram 252
circular transmission line method (CTLM) 112
color-rendering index (CRI) 257
– Munsell samples spectra for 259
compensated semiconductor 77
conduction band 30, 78
– conductivity effective mass 70
– deformation potential 168
– density of states effective mass 75
– discontinuity 51, 135, 352, 359
– dispersion 23
– energy 23
– minimum 293, 302
– of p-GaN 240
– s-like and isotropic 297
– structure 236
continuous-wave (CW) semiconductor lasers 270
correlated color temperature (CCT) 253
Coulomb interactions 118, 207, 283
Coulomb screening 230
crystal field splitting affect 24
cubic system 25
current gain cutoff frequency 373
current–voltage characteristics 112, 373
cutoff frequencies 349

CW InGaN multiple quantum well laser diodes 312
– degradation characteristics of 337

d

DBR mirror, maximum reflectivity 330
DC-based reliability test 428
Debye length 183
deep-level transient spectroscopy (DLTS) 65
deformation potentials 23
– scattering 121
2DEG channel 383, 415, 416, 421, 436
2DEG density 423
– intrinsic transit time 424, 425
– low-field 435
2DEG LO phonon scattering 417
density functional theory (DFT) 66
density of states (DOS) 74–78, 267
dielectric reflector-based microcavity structures 327
diffusion constant 117
diffusion current
– under forward bias 190, 191
– under reverse bias 190
digital versatile disks (DVDs) 267
diode current 186
Dirac delta function 294
Dirac distribution function 299
donor–acceptor transitions 206, 207
doping 30, 63–65
– candidates 69, 70
– p-doping 442
– uniform 89
DOS. *See* density of states (DOS)
drain breakdown voltage 381
– temperature dependence 381
drain current 350, 361, 363, 372, 376, 382, 386, 393, 403
– degradation 436
– due to the stress 436
– lag 393
– maximum 434, 435
– recovery 395
– transients 393, 395
– – in field effect transistor 395
drain delay 372
drain–gate voltage 382
drain saturation voltage 362
drain–source current 367
drain–source voltage 362, 391
drift–diffusion current for electrons 116
drift–diffusion model 115
drift mobility 74, 143
– electron 160, 161
– as a function of temperature 159
– low-field 151
– polar optical phonon limited 160
– theoretical electron 147
drift region conductivity 443

e

EBL barrier
– overflow electron current 234
edge-emitting injection laser
– far-field characteristics 280
effective density, of states in conduction band 76
effective mass (EM) approximation 350
eigenvalues problem 49
Einstein expansion 286
Einstein's coefficients 194–196, 289
electroluminescence (EL) 193, 227
electron density
– doping-induced 32
– at electron temperatures 422
– by gate potential 169
– photoexcited 414
– total sheet 354
– in triangular well region 166
electron–electron scattering 147
electron–hole pair generation
– threshold energy for 381
electron–hole plasma 307
electronic band structure 18
electronic noise 396–405
– FET equivalent circuit 398–402
– GaN FETs, high-frequency noise 402–405
– hot electrons 397, 398
– noise production 397
  Nyquist theorem 396
– thermal noise 397
electron mobility 349, 443
– in AlGaN 163
– GaN 2D system 168
electron paramagnetic resonance (EPR) 65
electron–phonon interaction 27
electron–TO scattering 413
electron transport properties 349
ELO process. *See* epitaxial lateral overgrowth (ELO) process
emission 196
– spontaneous 193
empirical/analytical device modeling 367
energy balance 151, 153, 195
energy distribution, of electrons in energy band 78
energy transfer rate 152
epitaxial lateral overgrowth (ELO) process 312, 313

f

Fabry–Pérot resonator 328, 329
Fang–Howard variational wavefunction 168
feedback capacitance 371
feedback time constant 373
Fermi-Dirac statistics 288, 291
– distribution function 75, 294, 300, 354, 355
– for electrons and holes 294
Fermi energy 65–67, 75, 79, 415, 416
Fermi integral 94, 95
Fermi level 65, 101, 359
– in conduction band 76
– for zero interface 360
FET. See field effect transistor (FET)
f factor 75
field effect transistor (FET) 349, 384
– electronic noise
– – equivalent circuit 398–402
– – GaN, high-frequency noise 402–405
– equivalent circuit gleaned 369
– high-frequency equivalent circuit 368, 401
– hot electrons 397
– modeling of 366
– planar devices 379
field emission Auger electron spectroscopy (FEAES) 340
field emission SEM (FESEM) 340
finite element analysis (FEA) method 408
formation energies
– as a function of Fermi level for native point defects in GaN 66
– of point defects vs. Fermi level 65
Fourier expansion 352
Franck–Condon shift 386
free carriers 70
– losses 281
free-to-bound transitions 205, 206
Fresnel loss, efficiency 216
Fröhlich interaction 410, 418
Fukui noise equation 405
full-potential linearized augmented plane wave (FPLAPW) 303

g

GaAs/AlGaAs system 358
GaAs-based FETs 432
gallium nitride 5

EQE. See external quantum efficiency (EQE)
excitation power density 230
exciton–polariton dispersion 334
external quantum efficiency (EQE) 211, 232, 236, 311, 312

– AlGaN/GaN channel, $I–V$ characteristics 365
– based devices 432
– calculated lattice parameter 16
– calculated mobility of 143–147
– Callen's effective ionized charge for 132
– capacitance vs. voltage for a GaN Schottky device in 93
– donor–acceptor transitions 206, 207
– electrically pumped VCSELs 332
– electron mobility for 163
– experimental and calculated elastic coefficients, bulk modulus and 8, 12
– formation energies as a function of Fermi level for 66
– free and bound excitons 201
– free-to-bound transitions 205, 206
– GaN/AlGaN heterostructure
– – schematic band structure 353
– GaN/AlGaN short-period superlattices (SLS) 313
– GaN-based 2DEG channels 415
– GaN-based LED introduction 253
– GaN-based microcavities 327
– GaN-based modulation-doped FET 360
– GaN-based switch technology 440
– GaN 2DEG channel 416
– GaN/InGaN channel 364–366
– GaN LEDs functions
– – lightemitting diodes (LEDs) 247
– GaN–substrate interface 59
– Heckman diagrams applied to 17
– heterojunction 355
– intrinsic and extrinsic optical transitions 201
– limited mobility, parameters used 134
– longitudinal piezoelectric field 324
– LO phonon angular zone-center frequency 131
– optical transitions 200
– – bound excitons 204, 205
– – excitonic transitions 200–203
– – strain effects 203, 204
– phonon dispersion curves 413
– plasmon frequency and LO phonon frequency 421
– polar optical phonon limited mobility in 128, 129
– p-type 99
– quantum well 416
– semipolar orientations 323
– structural parameters 6
– template
– – representative edge-emitting laser, schematic 268

– theoretical electron drift mobility 147
– two-dimensional transport in n-type GaN 164–166
– – de Broglie wavelength 165
– – parameters pertinent to mobility calculations 165
– velocity–field characteristics 412
– wafers 337
– waveguide layers 313
– wurtzite (WZ) GaN 270
– Wz GaN quantum well structure
– – conduction/heavy hole/light hole band 298, 300
GaN. *See* gallium nitride
GaN-based HFETs 349
– candidate semiconductors
– – figure of merits and pertinent materials properties 443
– – *vs*. breakdown voltage 442
– low-frequency and high-frequency noise characteristics 396
– mechanisms potentially affecting 427
– modified equivalent circuit 370
– reliability and stability 427
GaN FETs channel
– reliability investigations 432
– schematic representation 388
– trap energies
– – configuration coordinate diagram 390
GaN MESFETs/AlGaN 384–386
– barrier states, effect of 392, 393
– buffer layer, traps effect 386–391
– current collapse/surface charging, correlation 393–396
– current lag 386
– low-frequency and high-frequency noise characteristics 396
Ga-polarity 58
– single AlGaN–GaN interface 51–56
gate depletion region 369
gate–drain bias 380
gate–drain diode 369
gate fingers 407
gate leakage current paths 380, 397
– device instability 433
gate–source capacitance 371
gate voltage 359–361, 375, 382, 394, 434, 443
– Schottky barrier 359
generalized gradient approximation (GGA) 65, 66
generator reflection coefficient 401
Green's function method 115
group III nitrides. *See* nitrides

h

Hall coefficient 73, 154, 155
Hall factor 74, 155, 156
– for various scattering mechanisms 157
Hall measurements 71
– differential 158
– Hall bar geometry used for 73
Hamiltonian equation 357
heat dissipation 420
heat sinking 378
heterojunction field effect transistors (HFETs) 48, 349
– $Al_{0.3}Ga_{0.7}N$/GaN, output characteristic for 366
– AlGaN/InGaN, *I-V* characteristics for 365
– amplifier classification and efficiency 373–378
– analytical description of 358–364
– breakdown voltage
– – field plate for 383, 384
– cutoff frequency 417
– degradation 427–439
– – DC-based reliability test 428
– – failure regions, schematic representation of 428
– – gated structures 434–438
– – gate leakage current paths 433
– – hot electrons 434
– – lack of standard for GaN 431
– – reliability tests 438, 439
– – stability and reliability 427
– – thermally driven metallurgical degradation mechanisms 432
– – 3T lifetime test data 429
– drain voltage/drain breakdown mechanisms 378–382
– electronic noise 396–405
– – FET equivalent circuit 398–402
– – GaN FETs, high-frequency noise 402–405
– – hot electrons 397, 398
– – noise production 397
– – Nyquist theorem 396
– – thermal noise 397
– equivalent circuit models 366
– – cutoff frequency 370–373
– – small-signal equivalent circuit modeling 367–370
– GaN/InGaN channel 364–366
– GaN MESFETs/AlGaN 384–386
– – barrier states, effect of 392, 393
– – buffer layer, traps effect 386–391
– – current collapse/surface charging, correlation 393–396

－－ low-frequency and high-frequency noise characteristics 396
– gate of a submicron 371
– gate to channel separation 404
– heterointerface charge 350–358
– high-power switching 440–444
– nitride semiconductor 376
– noise and gain performance 403
– operation principles of 350
– phonon effects 405–426
– Schrödinger–Poisson equations 425
– self-heating 405–426
– small-signal modeling 367
– transconductance 363
heterojunctions. *See* n-p heterojunctions; p-n heterojunctions
HFETs. *See* heterojunction field effect transistors (HFETs)
highly accelerated stress test (HAST) 438
high-temperature operating life test (HTOL) 438
high-temperature reverse-bias test (HTRB) 438
high-temperature storage test (HTS) 438
Hooke's law 22
hot electrons 233, 434
– temperature 419
hot phonons 420
– lifetime 424, 436
– power dissipation 419
– temperature 419
hot probe 71
hydride vapor phase epitaxy (HVPE) 267
hydrogen
– and impurity trapping at extended defects 67, 68
– ionization energy 70

i

InAlGaN quaternary alloy 15–17
– bandgap 15
– calculated lattice parameter 16
– lattice bandgap *vs.* lattice parameter 16
– relationships between composition and bandgap 15
– ternaries constructed for quaternary 15
InAlN/AlN/GaN channels
– HFETs, 2DEG density 435
– hot phonon lifetime *versus* applied power 421
indium nitride 10, 29, 30
– bandgap 15
– carrier concentration vs Hall mobility 159
– carrier transport 158, 159

– effective masses and band parameters 31
– electron drift mobility 160, 161
– Hall mobility 160
– mobility, parameters used 134
– thermal conductivity 10
– wurtzitic, parameters related to electrical and optical properties 12
– zincblende (cubic) form 10
InGaN alloy 14
– bandgap energy 13
– b parameter 14
– channel 364
– compositional dependence 14
– electron mobility 163
– energy bandgap 14
– $In_{0.20}Ga_{0.80}N$ LEDs
－－ relative EQE of m-plane 232
– $In_{0.10}Ga_{0.90}N$ SEI layer 236
– high-resolution X-ray diffraction 14
– InGaN LEDs spanning 210
– InGaN MQW active layer 314
– InGaN QW diode laser heterostructures 312
– lattice parameter 14
– parameters derived and used in mobility calculations 164
– quantum wells 58, 59, 303, 320
InGaN laser 317
– current, evolution of 322
– mode hopping 321
– net modal gain *vs.* the injection current 320
InN. *See* indium nitride
integrated electroluminescence (EL) intensity 242
internal quantum efficiency (IQE) 209, 218, 311
internal strain 22
inversion domains (IDs) 338
ionization energy 70
IQE. *See* internal quantum efficiency (IQE)
IR near-field optical microscopy 439
isoelectronic defects 205
isotropic parabolic conduction band 23

j

Jahn–Teller displacement 66
Johnson figure of merit (JFOM) 444
Johnson noise. *See* Thermal noise
joint density of states (JDOS) 270, 301
– definition 292
– direct transition selection rule 294
– energy, gain 298
Joule heating 109, 338, 341, 379, 412. *See also* LA phonons

k

Kelvin probe microscopy measurements 356
Keys figure of merit (KFOM) 444
k.p theory 23

l

LA phonons 406
– temperature 426
Laplace heat spread model 406
laser beam divergence angle 333
laser diodes
– aging test 339
– characteristic temperature of 343
– lifetimes 336
laser emission spectra 332
laser lift-off (LLO) technology 255, 328
laser oscillation, delay of 311
lattice vibrations 121
leakage current 342
LEDs. See light-emitting diodes (LEDs)
light–current–voltage 316
light emission intensity vs. generation rate 226
light-emitting diodes (LEDs) 193, 209
– AlInGaP system performance 211
– atomic resolution image of DH 245
– BG_GR_R approach 263
– blue and red 259
– chromaticity coordinates 251–253
– chromaticity diagram 260
– color matching functions 251
– color-mixing efficiency 215
– color perception 250, 251
– color temperature 251–253
– conduction band, schematic 240
– current crowding 247–249
– current distribution 249
– degradation 253–255
– electrical efficiency 215
– equivalent circuit 248
– external quantum efficiency 211, 215
– extraction efficiency 215
– flip-chip design 255, 256
– GaN-based 212
– – efficiency reduction 246
– $In_{0.15}Ga_{0.85}N$ region and $Al_{0.15}Ga_{0.85}N$ EBL 238
– high-brightness 213
– high-power packages for lighting applications 257
– InGaN-based 244
– injection current density 238
– integrated EL efficiency 243
– internal quantum efficiency 215, 218
– low to high injection regime transition 225
– luminance 215
– luminescence conversion/white light generation 257
– – color-rendering index 258, 259
– – leds/phosphor(s) combining 262–265
– – white light from multichip 259–261
– luminous efficacy 250
– luminous flux 215
– luminous intensity 215
– monochromatic radiation, luminous efficacy 251
– multiple-chip 262
– nitride-based 210
– optical output power 214
– – Auger recombination 219, 220
– – carrier overflow 231–235
– – continuity/rate equations 223–230
– – efficiency 215–217
– – external efficiency 217, 218
– – internal quantum efficiency 218, 219
– – polarization effects on electron overflow 238, 239
– – radiative recombination 222, 223
– – SEI, optimization of 239–241
– – SRH generation/recombination process 220, 221
– – staircase electron injector (SEI) 235–238
– optical power generation 250
– overflow current percentile 239
– packaging 255–257
– phosphor solution 265
– photodiodes/photodetectors 214
– photoluminescence (PL) experiment 226
– photopic vision 250
– range of wavelength 210
– R_BG_BB approach 262
– rod sensitivity eye 250
– scattering efficiency 215
– spatial carrier and light distribution 212
– surface recombination effect 244–247
– surface-recombination velocity 246
– threading dislocation, effect of 247
– uses 209
– UV InGaN and GaN LEDs functions 247
– UV wavelengths 211
– voltage drop 249
– wall plug efficiency 215
– white light generation 209
– white light output emission spectrum 261
– white light single-phosphor conversion, emission spectrum of 264
– YAG:Ce phosphor 254
– yellow die, conformal coating of 265

light illumination 387, 389
Liouville equation 115
LLO. *See* laser lift-off (LLO) technology
local density approximation (LDA) 65
LO phonons 233, 415, 427
– absorption 233
– decay mechanism 414
– frequency 410, 416
– – emission and absorption 416
– lifetime
– – hot phonon lifetime 426
– plasmon resonance 422
Lorentzian broadening lineshape 296
Lorentzian fit 436
luminous energy 250
Luttinger-like parameters 24
Luttinger parameters 25
Lyddane–Sachs–Teller relation 130

m

maximum oscillation frequency 373
Maxwell's equations 274, 343, 344
MBE-grown AlGaN samples 272
– extraordinary index dispersion 272
mean time to failure (MTTF) 322, 408, 428–431, 438
– Arrhenius plot 408
MESFETs. *See* metal semiconductor field effect transistors (MESFETs)
metallization 109, 110
metal n-type semiconductor system 99
metal p-type semiconductor
– fictitious matching of 100
– pair 99
– system, metal work function 100
metal–semiconductor
– junctions, current flow in 101–107
– resistance 108
metal–semiconductor band alignment 97–100
metal semiconductor field effect transistors (MESFETs)
– anomalies in GaN MESFETs 384–386
– current–voltage expressions 387
– gate length to epilayer thickness 404
– microwave 439
– wurtzite polytype GaN MESFET, output characteristics for 380
metal Vertical Photon LED (MVP-LED) structure 257
Mg acceptor 66
microcavities (MC) 333
– structure 328
micro-Raman spectroscopy 409, 431
– mapping 439

mobility measurement 71–74
– van der Pauw pattern 74
Monte Carlo simulation 121, 366, 379, 411
– drift velocity in an AlGaN/GaN channel *versus* hot phonon lifetime 420
– of electron drift velocity 420
– full-band ensemble 379
Mott density 336
MTTF. *See* mean time to failure (MTTF)
multiple conduction layer mobilities, delineation of 156, 158
multiple quantum well (MQW) 278
– MQW-active region 318
– MQW InGaN active layer 279

n

NF. *See* noise figure (NF)
Nichia chemical laser 338
Nichia laser structure 314
nitrides 1
– axial lattice parameter 4
– ball-and-stick stacking model of crystals 2
– basal plane lattice parameter 4
– bond length 4
– crystal structure 1–5
– Hermann–Mauguin notation 1
– labeling of planes in hexagonal symmetry 3
– Schoenflies notation 1
– wurtzite structure 1
– – parameters related with electrical and optical properties 7
– wurtzitic metal nitride structure 4
nitride semiconductors 5, 33
– elastic constants and spontaneous polarization charge 33
nitrogen antisite defect 66
noise
– electronic 396–405
– – FET equivalent circuit 398–402
– – GaN FETs, high-frequency noise 402–405
– – hot electrons 397, 398
– – noise production 397
– – Nyquist theorem 396
– – thermal noise 397
– equivalent circuit 400
– measurement conditions 437
– voltage source 399
noise figure (NF) 398
– dependence of 400
– functional dependence of 403
noninjected facet (NIF) 339
nonlinearity
– nonlinear polarization in quantum wells 57, 58

– polarization 35–42
– reduced by linearization 375
– in spontaneous polarization 40
nonpolar
– optical phonon scattering 127
– orientations 59, 323
nonradiative multiphonon (NMP) capture, of electrons 389
n–p heterojunctions
– abrupt heterojunction under equilibrium 181
– band diagrams 179, 183
n-polarity 32
n-type semiconductor 77, 79–84, 109
Nyquist–Johnson noise 396

o

ohmic contact resistance 107
– determination 109–113
– specific contact resistivity 107, 108
open-circuit voltage 403
optical emission rate *vs* injection current density 229
optical excitation density 227, 229
optically detected magnetic resonance (ODMR) 65
optical phonon scattering 126
– nonpolar 126, 127
– polar 127–134
optical transitions 200–205
– energy 296, 301
– matrix (OTM) element 297

p

Packaging, LEDs 255–257
– cross-sectional schematic flipchip-mounted high-luminance 256
Pauli's exclusion principle 196
phonon lifetimes 414, 422
phonon–plasmon resonance 422
phosphor conversion LED (pcLED) 253
photoluminescence (PL) 65, 193, 197, 312
photometry 250
photon
– absorbtion 269
– density 217, 284–286, 290, 343, 346
– – determination 285, 343–348
– – emission rate depends on 198
– flux density 217
– induced carbon build up 341
– mode density 285
– momentum 287
– quantization 333
piezoelectric

– constants 33, 43, 58, 124, 144, 162, 164, 165, 170, 364
– field strength *vs*. orientation 60
– polarization 32–35
– – in hexagonal symmetry 34
– – nonlinearities in 42–46
– – quaternary alloys 34, 35
– scattering 120
planar GaN-based HFET 441
Planckian blackbody radiation 252
Planckian locus 251, 253, 259
Planck's blackbody radiation distribution law 286
Planck's formula 217, 286
p–n heterojunctions
– abrupt heterojunction under equilibrium 181
– band diagrams 179
– current–voltage characteristics 185, 186
– – avalanching 188, 189
– – diffusion current 189–191
– – diode current under reverse bias 186
– – Poole–Frenkel and Schottky effects 187, 188
– electrostatic characteristics 179–185
– ideal in equilibrium, energy band diagram 183
Poisson ratio 23
Poisson's equation 115, 180, 181, 350, 354
polariton lasing 326, 336
polariton–polariton interaction 336
polarization 18, 23, 31, 32
– in heterostructures 46–51
– nonlinearity of 35–42
– piezoelectric 32–35
– in quantum wells 56, 57
– spontaneous 35, 65
polar optical phonon (POP) scattering 120, 127–134
Poole–Frenkel effect 392
positron annihilation (PA) 65
power added efficiency (PAE) 376
power amplifier (PA) 376, 377
power dissipation 378
power gain compression point 377
power spectral density (PSD) units 435
Pt–Au interdiffusion 433
p-type conductivity 66, 67
p-type semiconductor 77, 78, 84–87

q

quantitative mobility spectrum analysis (QMSA) 159, 173, 174

quantum confined Stark effect (QCSE) 46, 59, 60, 230
quantum well (QW) heterostructures 267
quantum wells 304
– InGaN/GaN 58, 59
– nonlinear polarization 57, 58
– polarization in 56, 57
quasicubic approximation 25, 204, 297
quasicubic model 23–25
quasi-Fermi levels 307
QW InGaN lasers, indium compositions 320

r

Rabi splitting 333, 334
radiative recombination 193
radiometry 250
Rayleigh–Ritz method 352
reactive ion etching (RIE) 158
recombination 193. *See also* Auger recombination
– radiative 193
refractive index dispersion 271
RE^{3+}:YAG emission 263
RF breakdown 379
RF current lag 385
RF signal 379
ridge laser, light intensity 319
Rydberg (Ry) energy 297

s

saturating velocity–field characteristics 151
scanning transmission electron microscopy (STEM) 244, 433
scattering, at high fields 147–152
– energy-dependent relaxation time and large radiative recombination coefficient B 153–155
– Hall factor 155, 156
– transport at high fields 152, 153
scattering mechanisms 116
Schottky barriers 183, 350, 359, 379
– built-in voltage 365
– height 379
– potential energy diagram/current flow mechanisms 101
Schottky contact–n junction 379
Schottky diodes 380, 423
Schottky effect 187
Schottky gate contact degradation 432
Schrödinger equation 167, 350, 352, 357
Schrödinger–Poisson equations 53, 54
screening effect 130
Seebeck effect 71
SEI. *See* staircase electron injector (SEI)

Sellmeier dispersion formula 271
– *B* and *C* coefficients 272
semiconductor lasers 267, 307
– absorption coefficient
– – absorption rate to 290
– – relating spontaneous emission rate 290, 291
– – stimulated emission rate 290
– bulk layers, optical gain 289, 290
– continuous-wave (CW) 270
– cross-sectional SEM images of 328
– degradation 337–343
– digital versatile disks (DVDs) 267
– far-field pattern 280, 281
– Fermi–Dirac statistics 288
– gain in quantum wells 299–302
– gain measurement
– – nitride lasers 304
– – optical pumping 304–306
– GaN-based LD design and performance 312–317
– InGaN injection lasers, gain spectra 317–321
– injection lasers, analysis of
– – with simplifying assumptions 307, 308
– laser diodes (LDs) 267
– light-emitting diodes (LEDs) 267
– loss/threshold/cavity modes 281–283
– microcavity fundamentals 328–333
– mode hopping 321, 322
– nonpolar/semipolar orientations 323–325
– optical gain 283–286, 302–304
– optical microscope image 327
– optical transitions in direct bandgap 287
– photon density, determination 343–348
– polariton lasers 333–336
– principles of 268–270
– – TE-mode 274–276
– – TM-mode 276–279
– – waveguide problem, analytical solution 273, 274
– – waveguiding 270–272
– quantum efficiency 311, 312
– quantum well (QW) 267
– recombination lifetime 309–311
– semiconductor realm 291–298
– thermal resistance 322, 323
– threshold current 306, 307
– vertical cavity surface-emitting lasers (VCSELs) 325–336
semiconductor nitride family 359
semiconductor resistance 108, 109
– built-in potential 109
– defined 108

– p-type contact resistance 109
– Si-doped n-typeGaN cracks thickness and 109
semiconductor sheet resistance 112
semiconductor statistics 74–78
SemiLEDs metal Vertical Photon LED (MVP-LED) 257
semipolar orientations 59, 60
Shockley–Read–Hall (SRH) generation–recombination current. *See also* p–n Junctions
– generation/recombination process 218, 220
Shubnikov–de Haas (SdH) measurements 357
Si-based switching devices 440
Si doping 393. *See also* doping
Single particle model 283
SiO_2/Si_3N_4 quarter-wave reflector 330, 331
Sony InGaN/GaN/AlGaN DH laser structure
– cross-sectional view of 315
Sony ridge laser, light *vs.* current characteristics 315
spectral emission 198
sphalerite (zincblende) GaN 26, 27. *See also* gallium nitride
– effective masses for 27
– predictions of Luttinger parameters 26, 27
– valence band 26
spin-dependent scattering 154
spin–orbit interaction 19, 24, 203
spin splitting 24
spontaneous
– emission 193
– polarization 35
staircase electron injector (SEI) 235
– electron overflow percentiles 241
– one-layer/two-layer, schematics 241
Stern–Howard wavefunctions 166
stiffness constants 23
strain effects 203, 204
strain–stress relationship 22
strain tensor 22

t

temperature
– dependence of Wurtzite GaN bandgap 26
– humidity bias test 438
– noise model 402
thermal characterization methods 439
thermal dissipation, electrical circuit equivalent 409
thermally driven metallurgical degradation mechanisms 432

thermal noise 397, 402
thermal resistance 323
thermodynamic equilibrium 284
Ti/Al-based ohmic contacts 113
transient thermal model 406
transistor, DC characteristic 374
transmission line model (TLM) 109, 112
transport properties of carriers, in semiconductors 115. *See also* two-dimensional transport
transverse electric (TE) polarizations 271, 325
– external reflection/transmission coefficients 277
– internal reflection for 278
– material gain for 305
– optical field profile of 276
– wave guidance for 274
transverse magnetic (TM) polarizations 271, 276, 325
– external reflection/transmission coefficients 277
– graphical approach 275
– internal reflection for 278
– reflection coefficient 277
– wave guidance for 274
transverse mode-stabilization
– far-field emission pattern 318
trap emission rate, temperature dependence of 394
trapping center, energy diagram 392
two-dimensional transport, in n-type GaN 164–166
– electron mobility in AlGaN/GaN 2D system 168–170
– inverse 2DEG electron mobility *vs.* temperature 169
– magnetotransport, and mobility spectrum 173
– numerical two-dimensional electron gas mobility calculations 170–172
– α parameter *vs.* electron density 169
– QMSA spectrum 174
– scattering in 2D systems 166–168
two-valley conduction band system 152

u

UV InGaN functions
– internal quantum efficiency 248
– lightemitting diodes (LEDs) 247

v

valence band 24
– densities 77
– structures 24

Vallée–Bogani channel 412
van der Pauw geometry 73, 74
VCSELs. *See* vertical cavity surface-emitting lasers (VCSELs)
Vegard's law 13, 35, 58
vertical cavity surface-emitting lasers (VCSELs) 325, 326, 331, 332, 333
voltage distribution 110

w
wall-plug efficiency 317
Wentzel–Kramers–Brillouin approximation 232
wide-bandgap semiconductors 78
Wigner distribution function 115
wurtzite crystal
– first Brillouin zone 19
– splitting of valence band 20
– τ point valence and conduction bands 21
wurtzite (WZ) GaN. *See* gallium nitride
wurtzite polytype GaN MESFET, output characteristics 380
wurtzitic metal nitride structure 4

y
YAG:Ce phosphor 254
YAG-pcLEDs 254
Yellow die, conformal coating of 265

z
Zinc blende (ZB) structures 297